An Introduction to Spatial Data Science with GeoDa

Volume 1 – Exploring Spatial Data

This book is the first in a two-volume series that introduces the field of spatial data science. It offers an accessible overview of the methodology of exploratory spatial data analysis. It also constitutes the definitive user's guide for the widely adopted GeoDa open-source software for spatial analysis. Leveraging a large number of real-world empirical illustrations, readers will gain an understanding of the main concepts and techniques, using dynamic graphics for thematic mapping, statistical graphing, and, most centrally, the analysis of spatial autocorrelation. Key to this analysis is the concept of local indicators of spatial association, pioneered by the author and recently extended to the analysis of multivariate data.

The focus of the book is on intuitive methods to discover interesting patterns in spatial data. It offers a progression from basic data manipulation through description and exploration to the identification of clusters and outliers by means of local spatial autocorrelation analysis. A distinctive approach is to spatialize intrinsically non-spatial methods by means of linking and brushing with a range of map representations, including several that are unique to the GeoDa software. The book also represents the most in-depth treatment of local spatial autocorrelation and its visualization and interpretation by means of GeoDa.

The book is intended for readers interested in going beyond simple mapping of geographical data to gain insight into interesting patterns. Some basic familiarity with statistical concepts is assumed, but no previous knowledge of GIS or mapping is required.

Key Features:

- Includes spatial perspectives on cluster analysis
- Focuses on exploring spatial data
- Supplemented by extensive support with sample data sets and examples on the GeoDaCenter website

This book is both useful as a reference for the software and as a text for students and researchers of spatial data science.

Luc Anselin is the Founding Director of the Center for Spatial Data Science at the University of Chicago, where he is also the Stein-Freiler Distinguished Service Professor of Sociology and the College, as well as a member of the Committee on Data Science. He is the creator of the GeoDa software and an active contributor to the PySAL Python open-source software library for spatial analysis. He has written widely on topics dealing with the methodology of spatial data analysis, including his classic 1988 text on Spatial Econometrics. His work has been recognized by many awards, such as his election to the U.S. National Academy of Science and the American Academy of Arts and Science.

An Introduction to Spatial Data Science with GeoDa

Volume 1 – Exploring Spatial Data

Luc Anselin

CRC Press
Taylor & Francis Group
Boca Raton London New York

CRC Press is an imprint of the
Taylor & Francis Group, an **informa** business
A CHAPMAN & HALL BOOK

Designed cover image: Luc Anselin

First edition published 2024
by CRC Press
2385 NW Executive Center Drive, Suite 320, Boca Raton FL 33431

and by CRC Press
4 Park Square, Milton Park, Abingdon, Oxon, OX14 4RN

CRC Press is an imprint of Taylor & Francis Group, LLC

Library of Congress Cataloging-in-Publication Data

Names: Anselin, Luc, 1953- author.
Title: An introduction to spatial data science with GeoDa / Luc Anselin.
Description: First edition. | Boca Raton, FL : CRC Press, 2024. | Includes bibliographical references and index. | Contents: Volume 1. Exploring spatial data -- Volume 2. Clustering spatial data. | Summary: "This book is the first in a two-volume series that introduces the field of spatial data science. It offers an accessible overview of the methodology of exploratory spatial data analysis. It also constitutes the definitive user's guide for the widely adopted GeoDa open source software for spatial analysis. Leveraging a large number of real-world empirical illustrations, readers will gain an understanding of the main concepts and techniques, using dynamic graphics for thematic mapping, statistical graphing, and, most centrally, the analysis of spatial autocorrelation. Key to this analysis is the concept of local indicators of spatial association, pioneered by the author and recently extended to the analysis of multivariate data."--Provided by publisher.
Identifiers: LCCN 2023048617 | ISBN 9781032229188 (volume 1 ; hardback) | ISBN 9781032229621 (volume 1 ; paperback) | ISBN 9781003274919 (volume 1 ; ebook) | ISBN 9781032713021 (volume 2 ; hardback) | ISBN 9781032713168 (volume 2 ; paperback) | ISBN 9781032713175 (volume 2 ; ebook)
Subjects: LCSH: Spatial analysis (Statistics) | Spatial analysis (Statistics)--Data processing. | GeoDa (Computer file)
Classification: LCC QA278.2 .A565 2024 | DDC 519.5/3--dc23/eng/20240201
LC record available at https://lccn.loc.gov/2023048617

ISBN: 978-1-032-22918-8 (hbk)
ISBN: 978-1-032-22962-1 (pbk)
ISBN: 978-1-003-27491-9 (ebk)

DOI: 10.1201/9781003274919

Typeset in Latin Modern font
by KnowledgeWorks Global Ltd.

Publisher's note: This book has been prepared from camera-ready copy provided by the authors.

To Emily

Contents

List of Figures

Preface

This two-volume set is the long overdue successor to the *GeoDa Workbook* that I wrote almost twenty years ago (Anselin, 2005a). It was intended to facilitate instruction in spatial analysis and spatial regression by means of the `GeoDa` software (Anselin et al., 2006b). In spite of its age, the workbook is still widely used and much cited, but it is due for a major update.

The update is two-fold. On the one hand, many new methods have been developed or original measures refined. This pertains not only to the spatial autocorrelation indices covered in the original Workbook but also to a collection of newer methods that have become to define *spatial data science*. Secondly, the `GeoDa` software has seen substantial changes to become an open-source and cross-platform ecosystem that encompasses a much wider range of methods than its *legacy* predecessor.

The two volumes outline my vision for an *Introduction to Spatial Data Science*. They include a collection of methods that I view as the core of what is *special* about *spatial* data science, as distinct from applying data science to spatial data. They are not intended to be a comprehensive overview but constitute my personal selection of materials that I see as central to promoting *spatial thinking* through teaching spatial data science.

The level in the current volume is introductory, aimed at my typical audience, which is largely composed of researchers and students (both undergraduate and graduate) who have *not* been exposed to any geographic or spatial concepts or have only limited familiarity with the subject. So, by design, some of the treatment is rudimentary, covering basic concepts in GIS and spatial data manipulation, as well as elementary statistical graphs. I have included this material to keep the books accessible to a larger audience. Readers already familiar with these topics can easily skip to the core techniques.

I believe the two volumes offer a unique perspective, in that they approach the identification of spatial patterns from a number of different standpoints. The first volume includes an in-depth treatment of *local indicators of spatial association*, whereas Volume 2 focuses on *spatial clustering* techniques. The main objective is to indicate where a *spatial* perspective contributes to the broader field of data science and what is unique about it. In addition, the aim is to create an intuition for the type of method that should be applied in different empirical situations. In that sense, the volumes serve both as the complete user guide to the `GeoDa` software and as a *primer* on spatial data science. However, in contrast with the original Workbook, spatial regression methods are not included. Those are covered in Anselin and Rey (2014) and not discussed here.

Most methods contained in the two volumes are treated in more technical detail in the various references provided. With respect to my own work, these include Anselin(1994; 1995; 1996; 1998; 1999; 2005b), Anselin et al. (2002), Anselin et al. (2004), Anselin et al. (2006b), and, more recently, Anselin (2019a; 2019b; 2020), Anselin and Li (2019; 2020) and Anselin et al. (2022). However, a few methods are new and have not been reported elsewhere or are discussed here in greater depth than previously appeared. In this volume, these include the co-location map and the local neighbor match test.

The methods are illustrated with a completely new collection of seven sample data sets that deal with topics ranging from crime, socio-economic determinants of health, and disease spread, to poverty, food insecurity and bank performance. The data pertain not only to the U.S. (Chicago) but also include municipalities in Brazil (the State of Ceará) and in Mexico (the State of Oaxaca), and community banks in Italy. Many of these data sets were used in previous empirical analyses. They are included as built-in *Sample Data* in the latest version of the `GeoDa` software.

The empirical illustrations are based on Version 1.22 of the software, available in Summer 2023. Later versions may include slight changes as well as additional features, but the treatment provided here should remain valid. The software is free, cross-platform and open-source and can be downloaded from https://geodacenter.github.io/download.html.

Acknowledgments

This work would not have existed without the tremendous efforts by the people behind the development of the `GeoDa` software over the past twenty-some years. This started in the early 2000s in the Spatial Analysis Laboratory at the University of Illinois, with major contributions by Ibnu Syabri, supported by the NSF-funded Center for Spatially Integrated Social Science (CSISS). Later, the software development was continued at the GeoDa Center of Arizona State University with Marc McCann as the main software engineer. The last ten years, Xun Li has served as the lead developer of the software. In addition, he has also been a close collaborator on several methodological refinements of the LISA approach. Julia Koschinsky has been on the team for some twenty years, as a constant inspiration and collaborator, starting at UIUC and most recently at the Center for Spatial Data Science of the University of Chicago. Xun and Julia have been instrumental in the migration of `GeoDa` from a closed-source Windows-based desktop software to an open-source and cross-platform ecosystem for exploring spatial data. Julia in particular has been at the forefront of refining the role of ESDA within a scientific reasoning framework, which I have tried to represent in the book.

In addition, I would like to thank Lara Spieker from Taylor & Francis Group for her expert guidance in this project.

Over the years, the research behind the methods covered in this book and the accompanying software development has been funded by grants from the U.S. National Science Foundation, the National Institutes for Health and the Centers for Disease Control, as well as by institutional support by the University of Chicago to the Center for Spatial Data Science.

Finally, Emily has been patiently living with my GeoDa obsession for many years. This book is dedicated to her.

Shelby, MI, Summer 2023

About the Author

Luc Anselin is the Founding Director of the Center for Spatial Data Science at the University of Chicago, where he is also the Stein-Freiler Distinguished Service Professor of Sociology and the College. He previously held faculty appointments at Arizona State University, the University of Illinois at Urbana-Champaign, the University of Texas at Dallas, the Regional Research Institute at West Virginia University, the University of California, Santa Barbara, and The Ohio State University. He also was a visiting professor at Brown University and MIT. He holds a PhD in Regional Science from Cornell University.

Over the past four decades, he has developed new methods for exploratory spatial data analysis and spatial econometrics, including the widely used local indicators of spatial autocorrelation. His 1988 *Spatial Econometrics* text has been cited some 17,000 times. He has implemented these methods into software, including the original SpaceStat software, as well as GeoDa, and as part of the Python PySAL library for spatial analysis.

His work has been recognized by several awards, including election to the U.S. National Academy of Sciences and the American Academy of Arts and Sciences.

1

Introduction

Spatial data are special in that the location of the observations, the *where*, plays a critical role in the methodology required for their analysis. Two aspects in particular distinguish spatial data from the standard independent and identically distributed paradigm (i.i.d.) in statistics and data analysis, i.e., *spatial dependence* and *spatial heterogeneity* (Anselin, 1988; 1990). Spatial dependence refers to the similarity of values observed at neighboring locations, or "everything is related to everything else, but closer places more so," known as Tobler's first law of geography (Tobler, 1970). Spatial heterogeneity is a particular form of structural change associated with *spatial subregions* of the data, i.e., showing a clear break in the spatial distribution of a phenomenon. Both spatial dependence and spatial heterogeneity require a specialized methodology for data analysis, generically referred to as *spatial analysis*.

Spatial data science is an emerging paradigm that extends spatial analysis situated at the interface between spatial statistics and geocomputation. What the term actually encompasses is not settled, and the collection of methods and software tools it represents is also sometimes referred to as geographic data science or geospatial data science (Anselin, 2020; Comber and Brunsdon, 2021; Singleton and Arribas-Bel, 2021; Rey et al., 2023). The concept is closely related to, overlaps somewhat with and has many methods and approaches in common with fields such as geocomputation (Brunsdon and Comber, 2015; Lovelace et al., 2019), cyberGIScience (Wang, 2010; Wang et al., 2013), and, more recently, GeoAI (Janowicz et al., 2020; Gao, 2021).

This two-volume collection is intended as an introduction to the field of spatial data science, emphasizing data exploration and visualization and focusing on the importance of a *spatial* perspective. It represents an attempt to promote spatial thinking in the practice of data science. It is admittedly a selection of methods that reflects my own biases, but it has proven to be an effective collection over many years of teaching and research. The first volume deals with the *exploration* of spatial data, whereas the second volume focuses on *spatial clustering* methods.

The methods covered in both volumes work well for so-called small to medium data settings, but not all of them scale well to *big* data settings. However, some important principles do scale well, like local indicators of spatial association. Even though data sets of very large size have become commonplace and arguably have been the drivers behind a lot of methodological development in modern data science, this is not always relevant for spatial data analysis. The point of departure is often big data (e.g., geo-located social media messages), but eventually, the analysis is carried out at a more spatially aggregate level, where the techniques covered here remain totally relevant.

The methodological approach outlined in this first volume supports an abductive process of exploration, a dynamic interaction between the analyst and the data with the goal of obtaining new insights. The focus is on insights that pertain to *spatial* patterns in the data, such as the *location* of interesting observations (hot spots and cold spots), the presence of

DOI: 10.1201/9781003274919-1

structural breaks in the spatial distribution of the data, and the comparison of such patterns between different variables and over time.

The identification of the patterns is intended to provide cues about the types of processes that may have generated them. It is important to appreciate that exploration is not the same as explanation. In my opinion, exploration nevertheless constitutes an important and necessary step to obtain effective and falsifiable hypotheses to be used in the next stages of the analysis. However, in practice, the line between pure exploration and confirmation (hypothesis testing) is not always that clear, and the process of scientific discovery may move back and forth between the two. I return to this question in more detail in the closing chapter.

The two volumes are both an introduction to the methodology of spatial data science and the definitive guide to the `GeoDa` software. This software represents the implementation of my vision for a gradual progression in the exploration of spatial data, from simple description and mapping to more structured identification of patterns and clusters, culminating with the estimation of spatial regression models. It came at the end of a series of software developments that started in the late 1980s (for a historical overview, see Anselin, 2012).

`GeoDa` is designed to be user-friendly and intuitive, working through a graphical user interface, and therefore it does not require any programming expertise. Similarly, the emphasis in the two volumes is on spatial concepts and how they can be implemented through the software, but it does not deal with geocomputation as such.

A distinctive characteristic of `GeoDa` is the efficient implementation of dynamically linked graphs, in the sense that one or more selected observations in a "view" of the data (a graph or map) are immediately also selected in all the other views, allowing interactive linking and brushing of the data (Anselin et al., 2006b). Since its initial release in 2003 (through the NSF-funded Center for Spatially Integrated Science), the software has been adopted widely for both teaching and research, with close to 600,000 unique downloads at the time of this writing.

In the remainder of this introduction, I first provide a broad overview of the organization of this first volume. This is followed by a quick tour of the `GeoDa` software and a listing of the sample data sets used to illustrate the methods.

1.1 Overview of Volume 1

The first volume is organized into five main parts and an Epilogue as a sixth, offering a progression from basic data manipulation, through description and exploration, to the identification of clusters and outliers by means of spatial autocorrelation analysis. It closes with some reflections on the limits of exploration and its role in scientific discovery. As mentioned, spatial clustering methods are covered in Volume 2.

The six parts are:

- Spatial data wrangling
- EDA and ESDA
- Spatial weights
- Global spatial autocorrelation

- Local spatial autocorrelation
- Epilogue

Part I deals with basic data operations for both tabular and spatial data, covered in two chapters. The material includes a review of the distinctive characteristics of spatial data, how to create spatial layers inside `GeoDa`, as well as essential transformations and data queries. There is also a rudimentary discussion of a range of basic GIS operations, such as projections, converting between points and polygons, and spatial joins. Even though `GeoDa` is not (and not intended to be) a GIS, this functionality has been included over the years in response to user demand.

Part II covers the principles behind exploratory data analysis (EDA) and its spatial counterpart, exploratory spatial data analysis (ESDA). This includes six chapters. Three of these are devoted to map use in various degrees of complexity, starting with basic mapping concepts and moving to statistical maps and maps for rates. The other three chapters deal with conventional (non-spatial) EDA, in the form of univariate and bivariate data exploration, multivariate data exploration and space-time exploration. The core idea here is to leverage linking and brushing between various graphical representations (*views* of the data), which is central to the architecture of `GeoDa`.

The remaining three main parts deal with the topic of spatial autocorrelation. First, in Part III, three chapters are devoted to spatial weights, both contiguity-based and distance-based spatial weights, and various spatial weights operations. These are essential pre-requisites for the computation of the global and local spatial autocorrelation indices covered in Parts IV and V.

Part IV contains three chapters on global spatial autocorrelation, centered around the Moran scatter plot as a visualization device. The basic concepts are covered, as well as more advanced applications and extensions to a bivariate setting. The third chapter provides an overview of some non-parametric techniques, such as a spatial correlogram.

Part V includes an in-depth treatment of local spatial autocorrelation, spread over five chapters. It starts with the introduction of the concept of a LISA and the Local Moran statistic. The second chapter deals with other local spatial autocorrelation statistics, such as the Local Geary and the Getis-Ord statistics. The next two chapters outline extensions to the multivariate domain and to discrete variables. These chapters contain material that was only fairly recently developed. The last chapter of Part V reviews density-based clustering methods applied to point locations, such as DBScan and HDBScan.

The Epilogue offers some thoughts on the limits of the exploratory perspective. This includes an assessment of the role of data exploration in aiding with scientific discovery and scientific reasoning, the limits of spatial analysis, and reproducibility in the exploratory framework as implemented in the `GeoDa` software.

An Appendix includes detailed preference settings for the software and an outline of the complete menu structure. To close, a brief discussion is offered of the new scripting possibilities through the `geodalib` library.

The division of the material in two volumes follows my own teaching practice. The first volume corresponds to what I cover in an *Introduction to Spatial Data Science* course, whereas the second volume matches the content of a *Spatial Cluster Analysis* course. The volumes are also designed to constitute a self-study guide. In fact, a previous version was used as such for remote teaching during the Covid pandemic (in the form of laboratory workbooks, available at https://geodacenter.github.io/documentation.html).

In addition to the material covered in the two volumes, the GeoDaCenter Github site (https://geodacenter.github.io) contains an extensive support infrastructure. This includes detailed documentation and illustrations, as well as a large collection of sample data sets, cookbook examples, and links to a YouTube channel containing lectures and tutorials. Specific software support is provided by means of a list of *frequently asked questions* and *answers to common technical questions*, as well as by the community through the *Google Groups Openspace* list.

1.2 A Quick Tour of GeoDa

Before delving into the specifics of particular methods, I provide a broad overview of the functionality and overall organization of the `GeoDa` software. The complete toolbar with icons corresponding to a collection of related operations is shown in Figure 1.1. Each icon is matched by a menu item, detailed in Appendix B. The menu and user interface can be customized to several languages (details are in Appendix A). The default is English, but options are available for Simplified Chinese, Russian, Spanish, Portuguese and French, with more to come in the future.

With each toolbar icon typically corresponds a drop-down list of specific functions. The structure of the drop-down list matches the menu sub-items (Appendix B).

Figure 1.1: GeoDa toolbar icons

The organization of the toolbar (and menu) follows the same logic as the layout of the parts and chapters in the two books. It represents a progression in the exploration, from left to right, from support functions to queries, description and visualization, and more and more formal methods, ending up with the estimation of actual spatial models in the regression module (not covered here).

A brief overview of each of the major parts is given next. This also includes the spatial clustering functionality, which is discussed more specifically in Volume 2.

1.2.1 Data entry

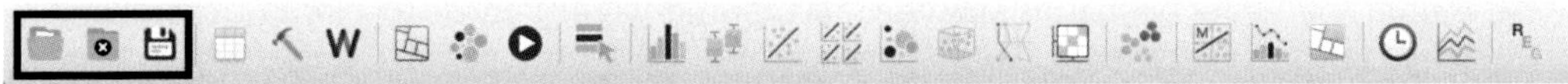

Figure 1.2: Data entry

The three left-most icons, highlighted in Figure 1.2, deal with data entry and general input-output. This includes the loading of spatial and non-spatial (e.g., tabular) data layers from a range of GIS and other file formats (supported through the open-source GDAL library). In addition, it offers connections to spatial databases, such as PostGIS and Oracle Spatial. It also supports a **Save As** function, which allows the software to work as a GIS file format converter. Further details are provided in Chapter 2.

1.2.2 Data manipulation

Figure 1.3: Data manipulation/table

Functionality for data manipulation and transformation is provided by the **Table** icon, highlighted in Figure 1.3. This allows new variables to be created, observations selected, queries formulated and includes other data table operations, such as merger and aggregation, detailed in Chapter 2.

1.2.3 GIS operations

Figure 1.4: GIS operations/tools

Spatial data operations are invoked through the **Tools** icon, highlighted in Figure 1.4. These include many GIS-like operations that were added over the years to provide access to spatial data for users who are not familiar with GIS. For example, point layers can be easily created from tabular data with X,Y coordinates, point in polygon operations support a spatial join, an indicator variable can be used to implement a dissolve application, and reprojection can be readily implemented by means of a **Save As** operation. Specific illustrations are included in Chapter 3.

1.2.4 Weights manager

Figure 1.5: Weights manager

The **Weights Manager** icon, Figure 1.5, contains a final set of functions that are in support of the analytical capabilities. It gives access to a wide range of weight creation and manipulation operations, discussed at length in the chapters of Part III. This includes constructing spatial weights from spatial layers, as well as loading them from external files, summarizing and visualizing their properties, and operations like union and intersection.

1.2.5 Mapping and geovisualization

Figure 1.6: Mapping and geovisualization

The mapping and geovisualization functionality is represented by four icons, highlighted in Figure 1.6: the **Map** icon, **Cartogram**, **Map Movie** and **Category Editor**. The mapping function supports all the customary types of choropleth maps, as well as some specialized features, such as extreme value maps, co-location maps and smoothed maps for rates. The cartogram is a specialized type of map that replaces the actual outline of spatial units by a circle, whose area is proportional to a given variable of interest. Animation, in the sense of moving through the locations of observations in increasing or decreasing order of the value for a given variable is implemented by means of the map movie icon. Finally, the category editor provides a way to design custom classifications for use in maps as well as in statistical graphs, such as a histogram. Details are provided in Chapters 4 through 6.

1.2.6 Exploratory data analysis

Figure 1.7: Exploratory data analysis

The next eight icons, grouped in Figure 1.7, contain the functionality for exploratory data analysis and statistical graphs. This includes a **Histogram**, **Box Plot**, **Scatter Plot**, **Scatter Plot Matrix**, **Bubble Chart**, **3D Scatter Plot**, **Parallel Coordinate Plot** and **Conditional Plots**. These provide an array of methods for univariate, bivariate and multivariate exploration. All the graphs are connected to any other open window (graph or map) for instantaneous linking and brushing. This is covered in more detail in Chapters 7 and 8.

1.2.7 Space-time analysis

Figure 1.8: Space-time analysis

The exploration of space-time data, treated in Chapter 9, is invoked by means of the icons on the right, highlighted in Figure 1.8. This includes a **Time Editor**, which is required to transform the cross-sectional observations into a proper (time) sequence. In addition, the **Averages Chart** implements a simple form of treatment analysis, with treatment and controls defined over time and/or across space.

1.2.8 Spatial autocorrelation analysis

Figure 1.9: Spatial autocorrelation analysis

Spatial autocorrelation analysis is invoked through the three icons highlighted in Figure 1.9. The first two pertain to global spatial autocorrelation. The left-most icon corresponds to various implementations of the Moran scatter plot (Chapters 13 and 14). The middle icon invokes nonparametric approaches to visualize global spatial autocorrelation as a spatial correlogram and distance scatter plot (Chapter 15).

The third icon contains a long list of various implementations of local spatial autocorrelation statistics, including various forms of the Local Moran's I, the Local Geary c, the Getis-Ord statistics and extensions to multivariate settings and discrete variables. The local neighbor match test is a new method based on an explicit assessment of the overlap between locational and attribute similarity. Details are provided in the chapters of Part V.

1.2.9 Cluster analysis

Figure 1.10: Cluster analysis

Finally, cluster analysis is invoked through the icon highlighted in Figure 1.10. An extensive drop-down list also includes the density-based cluster methods DBScan and HDBScan, which are treated in this volume under local spatial autocorrelation (Chapter 20).

The other methods are covered in Volume 2. They include dimension reduction, classic clustering methods and spatially constrained clustering methods. The last items in the drop-down list associated with the cluster icon pertain to the quantitative and visual assessment of cluster validity, including a new cluster match map (see Volume 2).

1.3 Sample Data Sets

As mentioned in the Preface, the methods and software are illustrated by means of empirical examples that use seven new sample data sets. They are available directly from inside the `GeoDa` software through the **Sample Data** tab of the input/output interface (see Figure 2.2).

The specific data sets are:

- *Chicago Carjackings* (n = 1,412)
  - point locations of carjackings in 2020 (Chicago Open Data Portal)
  - see Chapters 2 and 3
- *Ceará Zika*, municipalities in the State of Ceará, Brazil (n = 184)
  - Zika and Microcephaly infections and socio-economic profiles for 2013–2016 (adapted from Amaral et al., 2019)
  - see Chapters 4–6, 10 and Part III of Volume 2 (Spatial Clustering)
- *Oaxaca Development*, municipalities in the State of Oaxaca, Mexico (n = 570)
  - poverty and food insecurity indicators and census variables for 2010 and 2020 (CONEVAL and INEGI) (based on the same original sources as Farah Rivadeneyra, 2017)
  - see Chapters 7–9, 12, 14 and 16–17
- *Italy Community Banks* (n = 261)
  - bank performance indicators for 2011–17 (used by Algeri et al., 2022)
  - see Chapters 11–12, 15 and 20, as well as in Part I of Volume 2 (Dimension Reduction)
- *Chicago Community Areas*, CCA Profiles (n = 77)
  - socio-economic snapshot for Chicago Community Areas in 2020 (American Community Survey from the Chicago Metropolitan Agency for Planning – CMAP – data portal)
  - see Chapters 13 and Chapter 5 of Volume 2 (Hierarchical Clustering Methods)

- *Chicago SDOH*, census Tracts (n = 791)
  - socio-economic determinants of health in 2014 (a subset of the data used in Kolak et al., 2020)
  - see Chapters 18–19 and Chapters 6 and 7 of Volume 2 (Partitioning Clustering Methods and Advanced Clustering Methods)
- *Spirals* (n = 300)
  - canonical data set to test spectral clustering
  - only used in Volume 2 (Chapter 8, Spectral Clustering)

In addition, a few auxiliary files are employed to illustrate basic data handling operations in Chapters 2 and 3, such as a boundary layer for Chicago community areas and input data files in comma-separated text format. These files are available from the GeoDaCenter sample data site at https://geodacenter.github.io/data-and-lab/.

Further details are provided in the context of specific methods.

Part I

Spatial Data Wrangling

2

Basic Data Operations

In this and the following chapter, I introduce the topic of *data wrangling*, i.e., the process of getting data from its raw input into a form that is amenable for analysis. This is often considered to be the most time consuming part of a data science project, taking as much as 80% of the effort (Dasu and Johnson, 2003). Even though the focus in this book is on *analysis* and not on data manipulation per se, I provide a quick overview of the functionality contained in `GeoDa` to assist with these operations. Increasingly, data wrangling has evolved into a field of its own, with a growing number of operations turning into automatic procedures embedded into software (Rattenbury et al., 2017). A detailed discussion of this topic is beyond the scope of the book.

The coverage in this chapter is aimed at novices who are not very familiar with spatial data manipulations. Most of the features illustrated can be readily accomplished by means of dedicated GIS software or by exploiting the spatial data functionality available in the `R` and `Python` worlds. Readers knowledgeable in such operations may want to just skim the materials in order to become familiar with the way they are implemented in `GeoDa`. Alternatively, these operations can be performed outside `GeoDa`, with the end result loaded as a spatial data layer.

In the current chapter, I focus on essential input operations and data manipulations contained in the **Table** functionality. In the next chapter, I consider a range of basic GIS operations pertaining to *spatial* data wrangling.

To illustrate these features, I will use a data set with point locations of car jackings in Chicago in 2020. The *Chicago Carjackings* data layer is available from the **Sample Data** tab in the `GeoDa` file dialog (Figure 2.2).

In addition, in order to replicate the detailed steps used in the illustrations, three original input files are needed as well. These are available from the GeoDa-Center sample data site. They include a simple outline of the community areas, *Chicago_community_areas.shp*, as well as comma delimited (csv) text files with the socio-economic characteristics (*Chicago_CCA_Profiles.csv*), and the coordinates of the car jackings (*Chicago_2020_carjackings.csv*). The sample data site also contains the detailed listing of the variable names.

2.1 Topics Covered

- Load a spatial layer from a range of formats
- Convert between spatial formats
- Create a point layer from coordinates in a table
- Create a grid layer

DOI: 10.1201/9781003274919-2

- Become familiar with the table options
- Use the Calculator Tool to create new variables
- Variable standardization
- Merging tables
- Use the Selection Tool to select observations in a table
- Use a selection shape to select observations in a map

GeoDa Functions

- File > Open
- File > Save
- File > Save As
- Tools > Shape > Points from Table
- Tools > Shape > Create Grid
- Table > Edit Variable Properties
- Table > Add Variable
- Table > Delete Variable(s)
- Table > Rename Variable
- Table > Encode
- Table > Setup Number Formatting
- Table > Move Selected to Top
- Table > Calculator
- Table > Selection Tool
- File > Save Selected As
- Map > Unique Values Map
- Map > Selection Shape
- Map > Save Selection

Toolbar Icons

Figure 2.1: Open | Close | Save | Table | Tools

2.2 Spatial Data

Spatial data are characterized by the combination of two important aspects. First, there is information on variables, just as in any other statistical analysis. In the spatial world, this is referred to as *attribute* information. Typically, it is contained in a *flat* (rectangular) table with observations as rows and variables as columns.

The second aspect of spatial data is special and is referred to as *locational* information. It consists of the precise definition of spatial objects, classified as points, lines or areas (polygons). In essence, the formal characterization of any spatial object boils down to the description of X-Y coordinates of points in space, as well as of a mechanism that spells out how these points are combined into spatial entities.

For a single point, the description simply consists of its coordinates. For areal units, such as census tracts, counties, or states, the associated polygon boundary is defined as a series of line segments, each characterized by the coordinates of their starting and ending points. In other words, what may seem like a continuous boundary, is turned into *discrete* segments.

Traditional data tables have no problem including X and Y coordinates as columns, but as such cannot deal with the boundary definition of irregular spatial units. Since the number of line segments defining an areal boundary can easily vary from observation to observation, there is no efficient way to include this in a fixed number of columns of a flat table. Consequently, a specialized *data structure* is required, typically contained in a geographic information system or GIS.

Several specialized formats have been developed to efficiently combine both the attribute information and the locational information. Such spatial data can be contained in files with a special structure, or in spatially enabled relational data base systems.

I first consider common GIS file formats that can serve as input to `GeoDa`. This is followed by an illustration of simple tabular input of non-spatial files. Finally, a brief overview is given of connections to other input formats.

2.2.1 GIS files

Historically, a wide range of different formats have been developed for GIS data, both proprietary as well as open-source. In addition, there has been considerable effort at standardization, led by the Open Geospatial Consortium (OGC).[1] `GeoDa` leverages the open-source `GDAL` library[2] to support input and output of many of the most popular formats in use today.

While it is impossible to cover all of these specifications in detail, I will illustrate three specific formats here. First is the use of the proprietary *shape file* format of the leading GIS vendor ESRI.[3] In addition, the open-source *GeoJSON* format[4] will be covered, as well as the *Geography Markup Language* of the OGC, a standard XML grammar for defining geographical features.[5]

In `GeoDa`, one can load both polygon and point GIS data, but in the current implementation, line files are *not* supported (e.g., to represent road networks).

2.2.1.1 Spatial file formats

Arguably, the most familiar proprietary spatial data format is the *shape file* format, developed by ESRI. The terminology is a bit confusing, since there is no such thing as *one* shape file, but there is instead a collection of three (or four) files. One file has the extension *.shp*, one *.shx*, one *.dbf* and one *.prj* (with the projection information). The first three are required, the fourth one is optional, but highly recommended. The files should all be in the same directory and have the same file name, except for the file extension.

In the open-source world, an increasingly common format is *GeoJSON*, the geographic augmentation of the JSON standard, which stands from *JavaScript Object Notation*. This format is contained in a text file and is easy for machines to read, due to its highly structured nature.

Finally, the *GML* standard, or Geographic Markup Language, is a XML implementation that prescribes the formal description of geographic features.

[1]https://www.ogc.org
[2]https://gdal.org
[3]https://www.esri.com/library/whitepapers/pdfs/shapefile.pdf
[4]https://geojson.org
[5]https://www.ogc.org/standards/gml

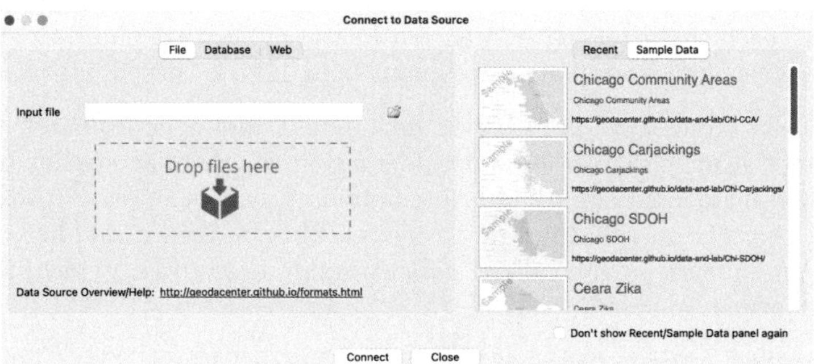

Figure 2.2: Connect to Data Source dialog

Figure 2.3: Supported spatial file formats

A detailed discussion of the individual formats is beyond the current scope. All are well-documented, with many additional resources available online. Although it is always helpful, there is no need to know the underlying formats in detail in order to use `GeoDa`, since the interaction with the data structures is handled under the hood.

The main file manipulations are invoked from the **File** item in the menu, or by the three left-most icons on the toolbar in Figure 2.1.

2.2.1.2 Polygon layers

Since `GeoDa` is particularly geared to the exploration of areal unit data, the input of a so-called *polygon layer* is illustrated first. Any spatial layer present as a file can be loaded by invoking **File > Open File** from the menu, or by clicking on the left-most **Open** icon on the toolbar in Figure 2.1.

This brings up the **Connect to Data Source** dialog, shown in Figure 2.2. The left panel has **File** as the active input format. Other formats are **Database** and **Web**, which are briefly covered in Section 2.2.3. The right panel shows a series of **Sample Data** data that are included with `GeoDa`. In addition, after some files have been loaded in the current application, the **Recent** panel will contain their file names as well. Files listed in either panel can be loaded by simply clicking on the corresponding icon.

The small folder icon to the right of the **Input file** box brings up a list of supported file formats, as in Figure 2.3. In this first example, the top item in the list is selected, **ESRI Shapefile (*.shp)**.

To illustrate this feature, the four files associated with the *Chicago_community_areas* shape file must be available in a working directory (they must be downloaded from the GeoDaCenter sample data site).

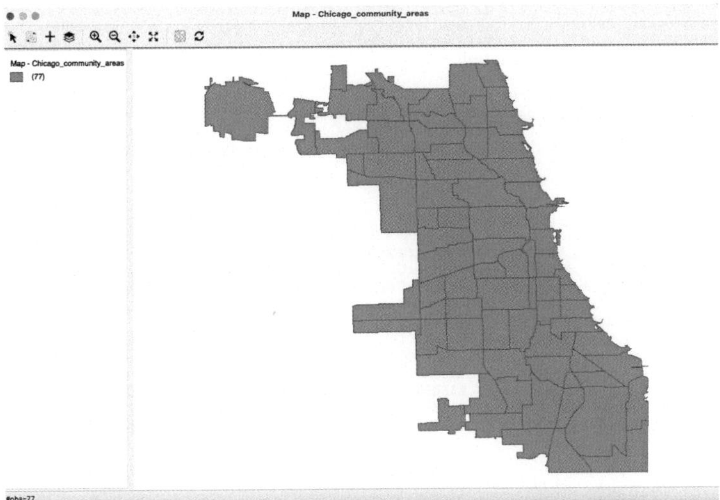

Figure 2.4: Themeless polygon map


```
{
"type": "FeatureCollection",
"name": "Chicago_community_areas",
"crs": { "type": "name", "properties": { "name": "urn:ogc:def:crs:OGC:1.3:CRS84" } },
"features": [
{ "type": "Feature", "properties": { "area_num_1": "35", "area_numbe": "35", "community":
"DOUGLAS", "shape_area": 46004621.158100002, "shape_len": 31027.0545098, "districtno": 7,
"district": "South Side" }, "geometry": { "type": "MultiPolygon", "coordinates":
[ [ [ [ -87.609140876178941, 41.844692502653977 ], [ -87.609148747578075, 41.844661598424032 ],
[ -87.609161120412594, 41.84458961193954 ], [ -87.609167662158384, 41.844517177323162 ],
[ -87.60916860600166, 41.844456260738305 ], [ -87.609150121993977, 41.844238716598113 ],
[ -87.609072412492893, 41.844194738881015 ], [ -87.609006271478208, 41.844106469286963 ],
[ -87.608965021721602, 41.844043457551152 ], [ -87.608915663906146, 41.84395529375054 ],
[ -87.608899801189878, 41.843873616495323 ], [ -87.608867013718623, 41.843804382800478 ],
[ -87.608851434244897, 41.843697606960866 ], [ -87.608810892810936, 41.843571847766412 ],
```


Figure 2.5: Example GeoJSON file contents

Using the navigation dialog and conventions appropriate for each operating system, the shape file can be selected from this directory. This opens a new map window with the spatial layer represented as a themeless choropleth map, as in Figure 2.4. The number of observations is shown in parentheses next to the small green rectangle in the upper-left panel, as well as in the status bar at the bottom (**#obs = 77**).

The current layer is cleared by clicking on the **Close** toolbar icon, the second item on the left in Figure 2.1, or by selecting **File > Close** from the menu. This removes the base map. At this point, the **Close** icon on the toolbar becomes inactive.

A more efficient way to open files is to select the file name in the directory window and to drag it onto the **Drop files here** box in the dialog. Even easier is to load a one of the sample data sets or a recently used one, where a simple click on the associated icon in the **Sample Data** or **Recent** tab suffices.

In contrast to the shape file format, which is binary, a GeoJSON file is simple text and can easily be read by humans. As shown for the *Chicago_community_areas.geojson* file from the sample data site in Figure 2.5 (this file must be downloaded to a working directory), the *locational* information is combined with the attributes. After some header information follows a list of `features`. Each of these contains `properties`, of which the first set consists

```
41.8452665489062 -87.6111225641177 41.8452664389385 -87.6109165492239 41.845266448213 -87.6094061454063
41.8452665039859 -87.6094094918227 41.8452177332683 -87.6093765809228 41.8451533826366 -87.6091408761789
41.844692502654</gml:posList></gml:LinearRing></gml:exterior></gml:Polygon></gml:surfaceMember></
gml:MultiSurface></ogr:geometryProperty>
      <ogr:area_num_1>35</ogr:area_num_1>
      <ogr:area_numbe>35</ogr:area_numbe>
      <ogr:community>DOUGLAS</ogr:community>
      <ogr:shape_area>46004621.1581000015139579772949218750000</ogr:shape_area>
      <ogr:shape_len>31027.0545098000002326443791389465332</ogr:shape_len>
      <ogr:districtno>7</ogr:districtno>
      <ogr:district>South Side</ogr:district>
    </ogr:Chicago_community_areas>
  </ogr:featureMember>
  <ogr:featureMember>
    <ogr:Chicago_community_areas gml:id="Chicago_community_areas.1">
      <gml:boundedBy><gml:Envelope><gml:lowerCorner>-87.6126242724032 41.8168137705722</
gml:lowerCorner><gml:upperCorner>-87.5921528387939 41.8313662468685</gml:upperCorner></gml:Envelope></
gml:boundedBy>
      <ogr:geometryProperty><gml:MultiSurface
gml:id="Chicago_community_areas.geom.1"><gml:surfaceMember><gml:Polygon
gml:id="Chicago_community_areas.geom.1.0"><gml:exterior><gml:LinearRing><gml:posList>-87.5921528387939
41.8169293462668 -87.5923080508337 41.8169321089497 -87.5948918343729 41.8169406679124 -87.5952614717272
41.8169427647923 -87.5959594527106 41.8168331429737 -87.5960713448987 41.8168328321002 -87.5961924032806
41.8168329229749 -87.5962488053834 41.8168329418813 -87.5963998526963 41.8168171023798 -87.5964634709017
41.8169077284035 -87.5968043283174 41.8169268189399 -87.5968828203083 41.8169279822056 -87.5975100001849
```

Figure 2.6: Example GML file contents

of the different variable names with the associated values, just as would be the case in any standard data table. The final item refers to the `geometry`. This includes the `type`, here a `MultiPolygon`, followed by a list of X-Y `coordinates`. In this fashion, the spatial information is integrated with the attribute information.

To view the corresponding map, the *Chicago_community_areas.geojson* file name can be selected in its directory and dragged onto the **Drop files here** box. This brings up the same base map as in Figure 2.4.

2.2.1.3 File format conversion

The file just loaded was originally specified in the GeoJSON format. It can be easily converted to a different format by means of the **File > Save As** functionality. For example, to change it into a GML format file (e.g., for use in a different program), **Geographic Markup Language (*.gml)** can be selected from the drop-down list of available formats, shown in Figure 2.3.

This will yield a text file in the GML XML format, illustrated in Figure 2.6. It shows the characteristic `< >` and `</ >` delimiters of the markup elements, typical of XML files. In the file snippet in the figure, the top lines pertain to the geography of the first polygon, ended by `</ogr:geometryProperty>`. Next follow the actual observations, with variable names and associated values, finally closed off with `</ogr:featureMember>`. After this, a new observation is listed, delineated by the `<ogr:featureMember>` tag, followed by the geographic characteristics. Again, this illustrates how spatial information is combined with attribute information in an efficient file format.

In sum, the **File > Save As** feature in `GeoDa` turns the program into an effective GIS format converter.

2.2.1.4 Point layers

In the same fashion as for polygon layers, spatial data layers containing point locations can be loaded for the file formats listed in Figure 2.3. As before, the file is either selected explicitly from the proper directory, or the file name is dragged directly into the **Drop files here** box in the dialog.

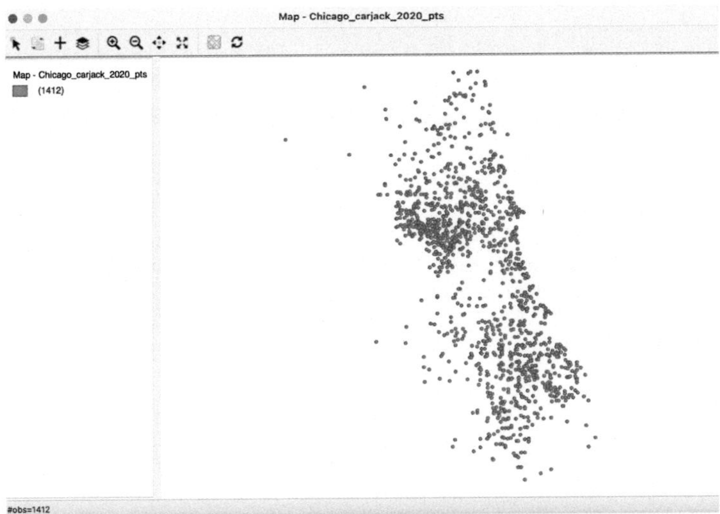

Figure 2.7: Themeless point map

The point map in Figure 2.7 shows the locations of the 1,412 carjackings that occurred in the City of Chicago during the year 2020. It is generated by clicking on the **Chicago Car Jackings** icon in the **Sample Data** tab, or by dragging the file name *Chicago_carjack_2020_pts.shp* from a working directory that contains the shape file.

The shape of the city portrayed by the outline of the points is slightly different from that in the polygon map in Figure 2.4. This is due to a difference in projections: the point map is in the State Plane Illinois East NAD 1983 projection (EPSG 3435), whereas the polygon map uses decimal degrees latitude and longitude (EPSG 4326). This important aspect of spatial data is often a source of confusion for non-GIS specialists. `GeoDa` provides an intuitive interface to deal with projection issues. I return to this topic in Section 2.3.1.1 below and in Section 3.2 in the next chapter.

2.2.2 Tabular files

In addition to GIS files, `GeoDa` can also read regular non-spatial tabular data. While this does not allow for *spatial* analysis (unless coordinates are contained in the table, see Section 2.3.1), all non-spatial operations and graphs are supported. Specifically, all standard techniques of exploratory data analysis (EDA) can be applied, as covered in Chapters 7 and 8 in this Volume. This does not necessitate a map layer.

The data for the Community area socio-economic profiles are contained in a comma-separated file (csv format) on the sample data site (the Chicago CCA Profiles are also available as a spatial layer in the **Sample Data** tab). Selecting this file (after downloading it to a working directory) generates the dialog shown in Figure 2.8.

Since a csv file is pure text, there is no information on the type of the variables included in the file. `GeoDa` tries to guess the type and lists the **Data Type** for each field, as well as a brief preview of the table. At this point, the type can be changed before the data are moved into the actual data table (see also Section 2.4.1.1).

Instead of a base map, which is the default opening window for spatial data, the data table is brought up in a spreadsheet-like format (see also Figure 2.9 for an illustration). In

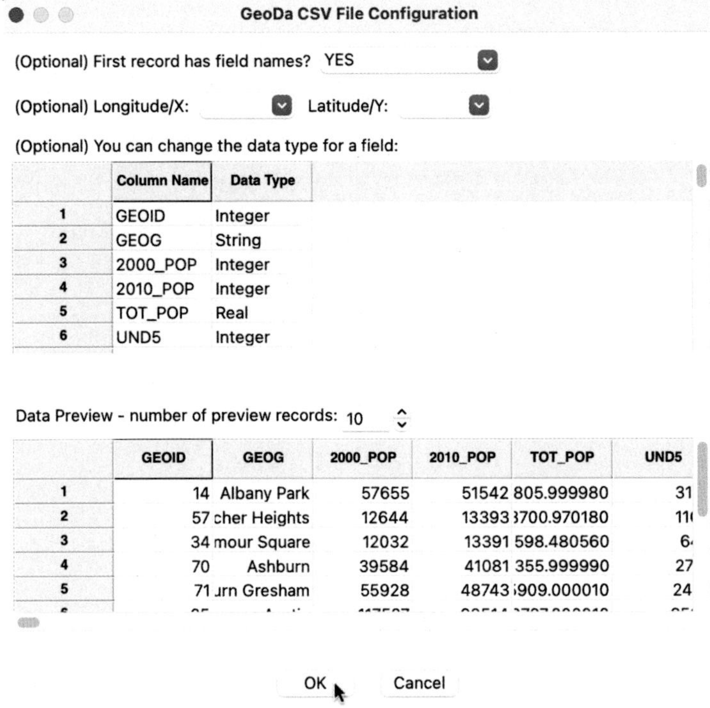

Figure 2.8: CSV format file input dialog

addition to comma-separated files, GeoDa also supports tabular input from dBase database files, Microsoft Excel and Open Document Spreadsheet (*.ods) formatted files. As is the case for the various GIS formats, the **File > Save As** command allows for the ready conversion from one tabular format to another (e.g., from csv to dBase).

An additional feature of the input dialog in Figure 2.8 is inclusion of optional **Longitude/X** and **Latitude/Y** drop-down lists. With coordinate variables specified, this will create a point layer. As such, it provides an alternative to the approach outlined in Section 2.3.1.

2.2.3 Other spatial data input

In addition to file input, GeoDa can connect to a number of spatially enabled relational data bases, such as PostgesQL/PostGIS, Oracle Spatial and MySQL Spatial. This is available through the **Database** button in the **Connect to Data Source** interface, the middle tab in the dialog shown in Figure 2.2. Each data base system has its own requirements in terms of specifying the host, port, user name, password and other settings. Once the connection is established, the data can also be saved to a file format. In the current version of GeoDa, the data base connection is limited to loading a single table. So far, no other SQL commands are supported.

A third data source option is referred to as **Web** in the interface, the right-most tab in Figure 2.2. It allows data to be loaded directly from a **GeoJSON URL**, or, alternatively, from the older web feature server **WFS URL**.

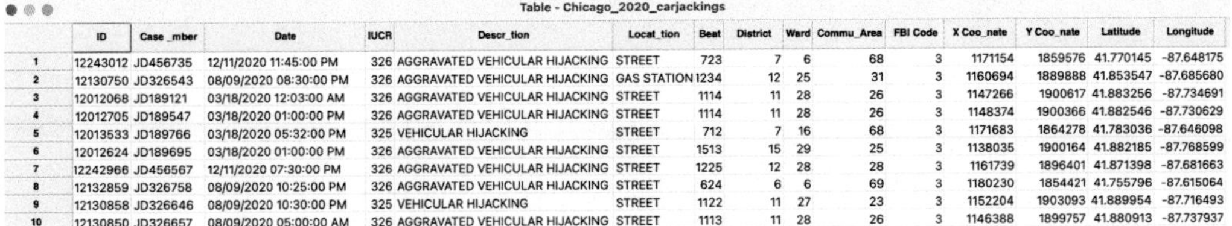

Table - Chicago_2020_carjackings

	ID	Case _mber	Date	IUCR	Descr_tion	Locat_tion	Beat	District	Ward	Commu_Area	FBI Code	X Coo_nate	Y Coo_nate	Latitude	Longitude
1	12243012	JD456735	12/11/2020 11:45:00 PM	326	AGGRAVATED VEHICULAR HIJACKING	STREET	723	7	6	68	3	1171154	1859576	41.770145	-87.648175
2	12130750	JD326543	08/09/2020 08:30:00 PM	326	AGGRAVATED VEHICULAR HIJACKING	GAS STATION	1234	12	25	31	3	1160694	1889888	41.853547	-87.685680
3	12012068	JD189121	03/18/2020 12:03:00 AM	326	AGGRAVATED VEHICULAR HIJACKING	STREET	1114	11	28	26	3	1147266	1900617	41.883256	-87.734691
4	12012705	JD189547	03/18/2020 01:00:00 PM	326	AGGRAVATED VEHICULAR HIJACKING	STREET	1114	11	28	26	3	1148374	1900366	41.882546	-87.730629
5	12013533	JD189766	03/18/2020 05:32:00 PM	325	VEHICULAR HIJACKING	STREET	712	7	16	68	3	1171683	1864278	41.783036	-87.646098
6	12012624	JD189695	03/18/2020 01:00:00 PM	326	AGGRAVATED VEHICULAR HIJACKING	STREET	1513	15	29	25	3	1138035	1900164	41.882185	-87.768599
7	12242966	JD456567	12/11/2020 07:30:00 PM	326	AGGRAVATED VEHICULAR HIJACKING	STREET	1225	12	28	28	3	1161739	1896401	41.871398	-87.681663
8	12132859	JD326758	08/09/2020 10:25:00 PM	326	AGGRAVATED VEHICULAR HIJACKING	STREET	624	6	6	69	3	1180230	1854421	41.755796	-87.615064
9	12130858	JD326646	08/09/2020 10:30:00 PM	325	VEHICULAR HIJACKING	STREET	1122	11	27	23	3	1152204	1903093	41.889954	-87.716493
10	12130850	JD326657	08/09/2020 05:00:00 AM	326	AGGRAVATED VEHICULAR HIJACKING	STREET	1113	11	28	26	3	1146388	1899757	41.880913	-87.737937

Figure 2.9: Coordinate variables in data table

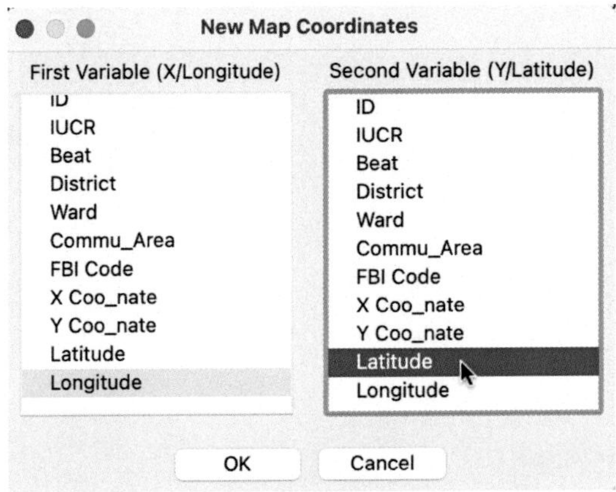

Figure 2.10: Specifying the point coordinates

2.3 Creating Spatial Layers

In addition to loading spatial layers from existing files or data base sources, new spatial layers can be created as well. This is implemented through the **Tools** functionality.

It is invoked from the menu as **Tools > Shape**, or, alternatively, from the **Tools** icon on the toolbar, the right-most icon in Figure 2.1. Two different operations are supported: one is to turn X-Y coordinates from a table into a point layer, **Tools > Shape > Points from Table**, the other creates a rectangular shape of grid cells, **Tools > Shape > Create Grid**.

2.3.1 Point layers from coordinates

Point layers are created from two variables contained in a data table that can serve as X and Y coordinates. This is illustrated with the file *Chicago_2020_carjackings.csv*, downloaded from the GeoDaCenter sample data collection. This file contains the same information as *Chicago Carjacking* in the **Sample Data** tab, but is a simple comma-separated text file.

Once loaded, the contents of the data table are shown in a spreadsheet-like format, as illustrated in Figure 2.9. In the example, the latitude and longitude of the point locations (in decimal degrees) are included as the two last columns in the table. Even though these variables are commonly referred to as *lat-lon*, latitude is actually the vertical dimension (Y), while longitude is the horizontal dimension (X).

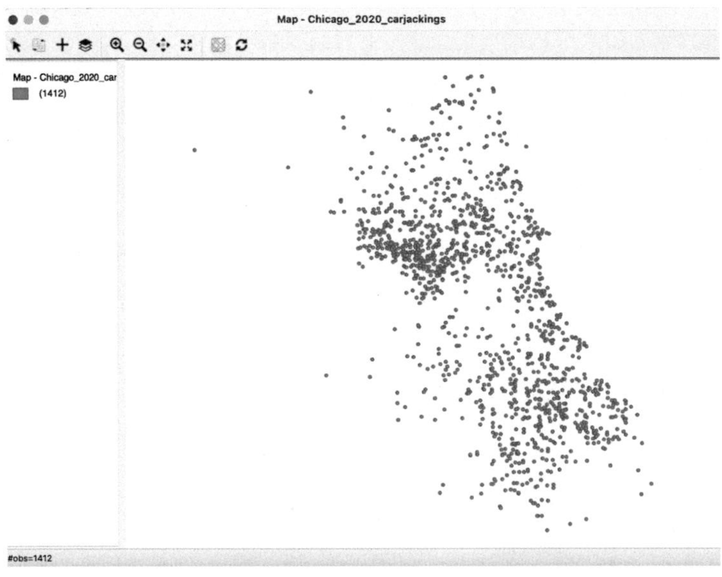

Figure 2.11: Point layer from coordinates

Figure 2.12: Blank CRS

The **Tools > Shape > Points from Table** command opens up a dialog that lists all numerical variables contained in the data table. In the example, shown in Figure 2.10, the variables to be selected are **Longitude** and **Latitude**.

After specifying the coordinate variables and clicking on **OK**, a point map appears, as in Figure 2.11.

The outline of Chicago suggested by the points is different from the one in Figure 2.7. Again, this is due to the difference in projections, which is discussed next.

At this stage, the point layer can be saved as a GIS file, such as a shape file, using the familiar **File > Save As** command (the file name should differ from any existing carjackings point shape file, or the latter will be overwritten).

2.3.1.1 A first note about projections

At the bottom of the dialog to save the just created layer is an option to enter **CRS (proj4 format)**. This refers to the projection information, or, more precisely, to the *coordinate reference system*. Projections will be covered in more detail in the next chapter, but at this point it is useful to be aware of the implications.

As mentioned, a careful comparison of the point pattern in Figure 2.11 and in Figure 2.7 reveals that they are not identical. While the relative positions of the points are the same, they appear more spread out in Figure 2.11. The reason for the discrepancy is the use of different projections. The point shape file in Figure 2.7 is based on the *projected* X-Y coordinates, whereas the just created point layer uses the latitude-longitude decimal degrees.

Figure 2.13: CRS with projection information

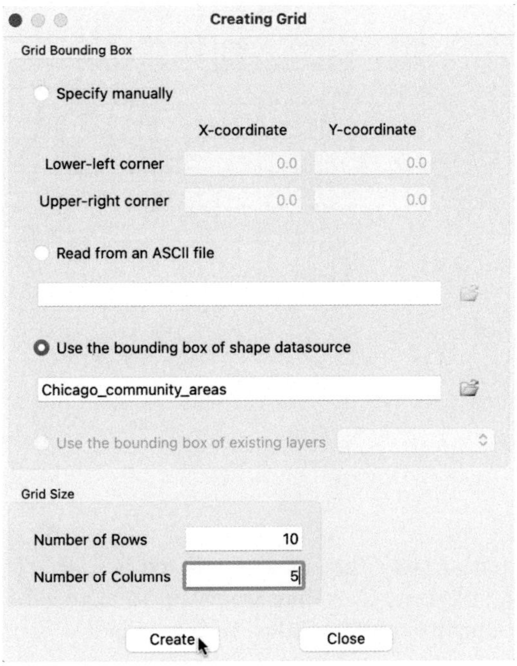

Figure 2.14: Grid creation dialog

As a default, the *Coordinate Reference System* (CRS) entry in the file save dialog will be empty, as in Figure 2.12. In other words, since nothing was specified other than the coordinates, there is no projection information. As a result, if the layer was saved as a shape file, it would *not* include a file with a **prj** file extension.

The lack of projection information severely hampers a range of data manipulations. Specifically, for computations such as Euclidean (straight line) distance, the decimal degrees are inappropriate and either special calculations must be used (such as great circle distance), or the coordinates need to be reprojected.

Even though latitude-longitude decimal degrees are not an actual projection, there is a corresponding CRS. While in and of itself this information is not that meaningful, including it when saving the file makes future reprojections easy. The CRS in **proj4** format for decimal degrees is (this is discussed in more detail in the next chapter):

+proj=longlat +datum=WGS84 + no_defs.

Including this entry in the CRS box (see Figure 2.13) when saving the file will ensure that future reprojections will be possible.

In other words, in the absence of specific projection information, the best practice is to create a spatial layer using decimal latitude-longitude degrees and recording the CRS from Figure 2.13 when saving the information as a GIS format file.

This is pursued in greater technical detail in the next chapter.

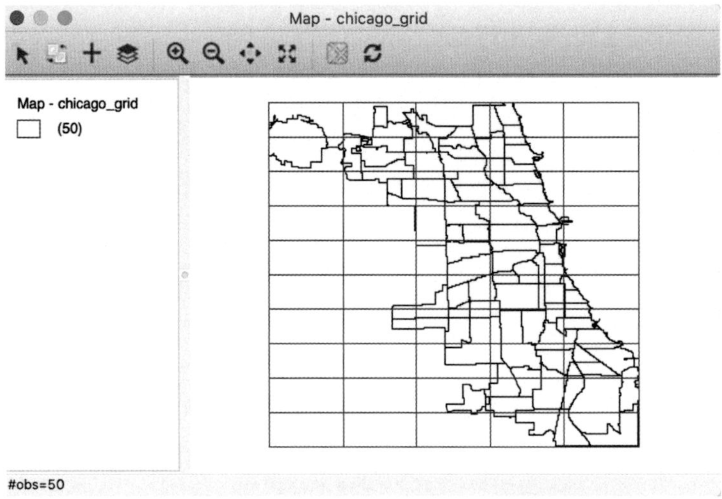

Figure 2.15: Grid layer over Chicago community areas

2.3.2 Grid

The second type of spatial layer that can be created is a set of grid cells. Such cells are rarely of interest in and of themselves, but are useful to aggregate counts of point locations in order to calculate point pattern statistics, such as quadrat counts.

The grid creation is one of the few instances where `GeoDa` functionality can be invoked from the menu or toolbar without any table or layer open. From the menu, **Tools > Shape > Create Grid** brings up a dialog that contains a range of options, shown in Figure 2.14. The most important aspect to enter is the **Number of Rows** and the **Number of Columns**, listed under **Grid Size** at the bottom of the dialog. This determines the number of observations. In the example, the entries are **10** rows and **5** columns, for 50 observations.

There are four options to determine the **Grid Bounding Box**, i.e., the rectangular extent of the grids. The easiest approach is to use the bounding box of a currently open layer or of an available shape data source. The **Lower-left** and **Upper-right** corners of the bounding box can also be set manually, or read from an ASCII file.

In the example in Figure 2.14, the Chicago community area boundary layer, contained in the *Chicago_community_areas.shp* file is shown as the source for the bounding box.

After the grid layer is saved as a file, it can be loaded into `GeoDa`. In Figure 2.15, the result is shown, superimposed onto the community areas layer (how to implement this will be covered in the next chapter). Clearly, the extent of the grid cells matches the bounding box for the area layer.

2.4 Table Manipulations

A range of variable transformations is supported through the **Table** functionality. This requires that the data table associated with a spatial layer is open. In contrast to the other

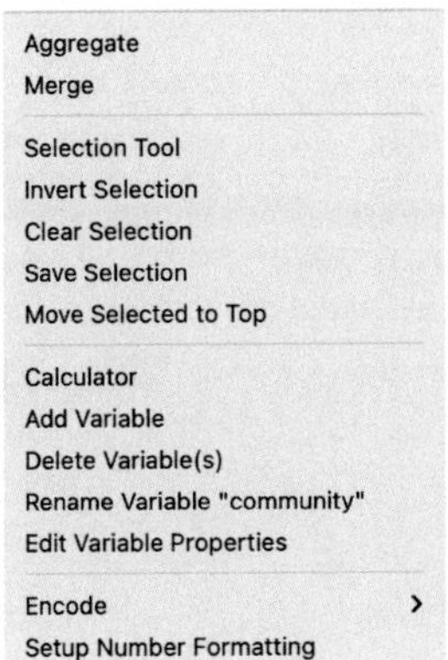

Figure 2.16: Table options

operations in `GeoDa`, opening a data table can only be accomplished by selecting a toolbar icon, i.e., the second icon from right on the toolbar shown in Figure 2.1.

With the table open, the options can be accessed from the menu, by selecting the **Table** menu item, or by right clicking on the table itself. This brings up the list shown in Figure 2.16.

In addition to a range of variable transformations, the options menu also includes the **Selection Tool**, which is the way in which data queries can be carried out. I discuss this in more detail in Section 2.5. The more commonly used variable manipulations are covered in what follows.

2.4.1 Variable properties

2.4.1.1 Edit variable properties

One of the most used initial transformations pertains to getting the data into the right format. Often, observation identifiers, such as the FIPS code used by the U.S. census, are recorded as character or string variables. In order to use these variables effectively as observation identifiers, they need to be converted to a numeric format. This is accomplished through the **Edit Variable Properties** option.

After this option is invoked by clicking on it, a window appears with a list of all variables, listing their **type**, as well as a number of other properties, like precision.

For example, consider the data table associated with the *Chicago_community_areas* community area boundaries, shown in Figure 2.17, after selecting the **Edit Variable Properties** option.

In the table, the area identifiers – both **area_num_1** and **area_numbe** (which are identical) – are left-aligned in their columns. This indicates they are character variables and not in a numeric format.

By selecting the proper **integer** format from the drop-down list, the change is immediate. As mentioned, the most common use of this functionality is to change **string** variables to a **numeric** format.

In order to make the change permanent, the table needs to be saved, either by selecting the middle icon in the toolbar in Figure 2.1, or by means of **File > Save**. Note that this will

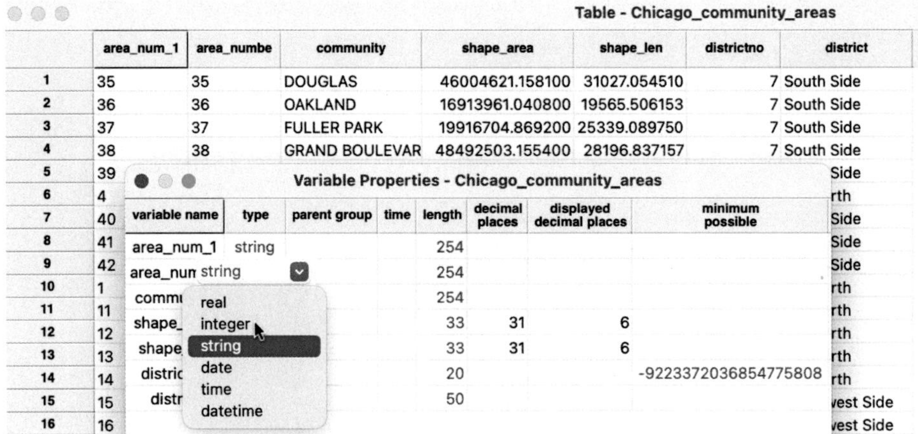

Figure 2.17: Edit variable properties

overwrite the current files. To avoid this behavior, select **File > Save As** and specify a new file name.

To facilitate data table operations like **Merge**, the community area identifier must be changed to **integer**.

2.4.1.2 Formatting date/time variables

One special case where the **Edit Variable Properties** option is extremely useful is when information on the date and time of an observation is included in the data table. Typically, this is contained in a string variable, which does not allow for any operations such as searching or sorting. Instead, a special date/time format must be used, which requires converting from string to this format.

For example, this is the case for the observations in the *Chicago Carjackings* points data set. The third column of the data table in Figure 2.18 shows the **Date** as a **string** variable. Specifically, the first entry is **12/11/2020 11:45:00 PM**, which lists the month, day and year, separated by a forward slash (/), followed by the time in hours, minutes and seconds (separated by a colon) and AM or PM.

Selecting **datetime** from the drop-down list will change the format into a standard time expression, as in Figure 2.19. Note how the order of the information has changed to year-month-day and the time is in 24 hours, minutes and seconds (no more AM or PM). At this point, the date information can be manipulated by means of the **Date/Time** functions in the calculator (see Section 2.4.2.5).

However, there is a catch. `GeoDa` recognizes many date-time formats, but not all. The default ones are listed under the **Data** tab of the **GeoDa Preference Setup** (see also Appendix A). Sometimes, the particular format used is not part of the default. It therefore may have to be added to the list, as shown in Figure 2.20. For example, the proper entry for the carjacking format is **%m/%d/%Y %I:%M:%S %p**, the last item in the **Date/Time formats** list. This follows the standard format for date and time information.[6]

[6]Specifically, %m stands for two digit month (including a leading zero), %d for two digit day (again, including a leading zero), %Y for the year in full four digits, %I for the hour in 12 hour segments (this requires AM or PM in addition), %M for minutes in two digits, %S for seconds in two digits and %p for AM or PM.

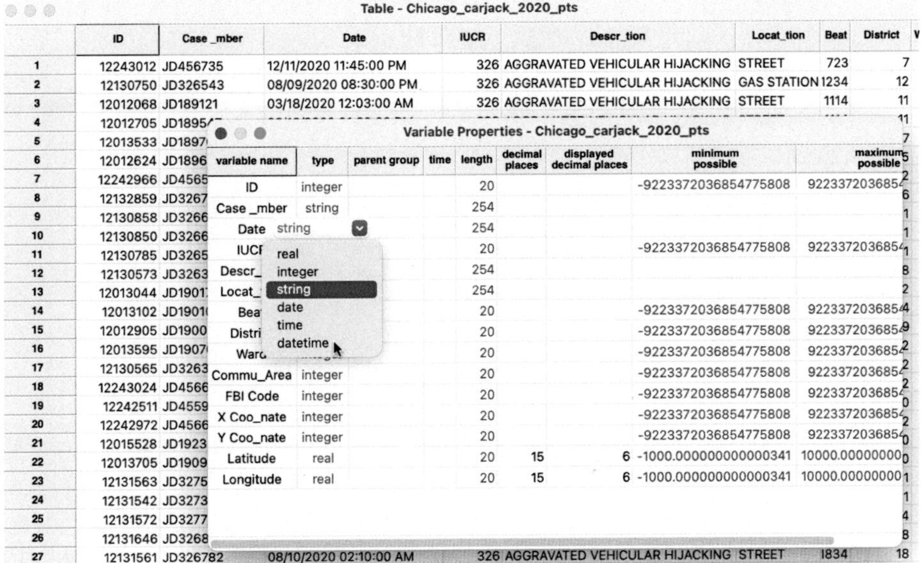

Figure 2.18: Create a date/time variable

	ID	Case _mber	Date
1	12243012	JD456735	2020-12-11 23:45:0
2	12130750	JD326543	2020-8-9 20:30:0
3	12012068	JD189121	2020-3-18 12:3:0
4	12012705	JD189547	2020-3-18 13:0:0
5	12013533	JD189766	2020-3-18 17:32:0

Figure 2.19: Date/time format

Once a customized format is included into the setup interface, it will be recognized in all future instances of **datetime** format conversion.

2.4.1.3 Other variable operations

The other variable operations listed in the drop-down list in Figure 2.16 are mostly self-explanatory. For example, a new variable can be included through **Add Variable** (its value is obtained through the **Calculator**, see Section 2.4.2), or variables can be removed by means of **Delete Variable(s)**. **Rename Variable** operates on a specific column, which needs to be selected first.

Two more obscure options are **Encode** and **Setup Number Formatting**. The default encoding in `GeoDa` is **Unicode (UTF-8)**, but a range of other encodings that support non-western characters are available as well. The default numeric formatting is to use a period for decimals and to separate thousands by a comma, but the reverse can be specified as well, which is common in Europe.

2.4.1.4 Operations on the table

The columns, rows and cells of the table can be manipulated directly, similar to spreadsheet operations. For example, the values in a column can be sorted in increasing or decreasing order by clicking on the variable name. A > or < sign appears next to the variable name to

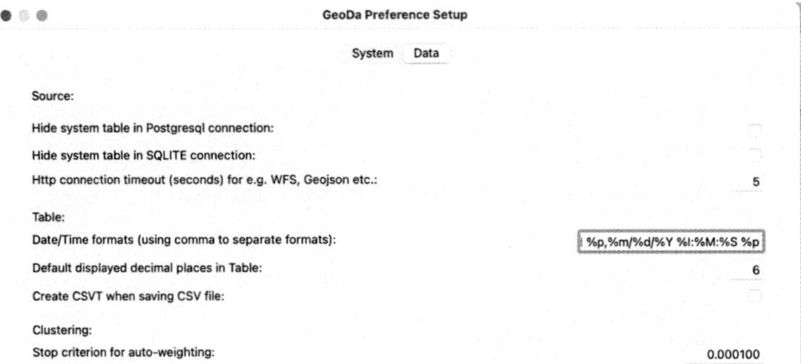

Figure 2.20: Customizing the date/time format

indicate the sorting order. The original order can be restored by *sorting* on the row numbers (the left-most column).

Individual cell values can be edited by selecting the cell and entering a new value. As is always the case in GeoDa, such changes are only permanent after the table is saved.

Specific observations can be selected manually by clicking on their row number (selection through queries is covered in Section 2.5). The corresponding entries will be highlighted in yellow. Additional observations can be added to the selection in the usual way, for example, by means of shift-click (or another key combination, depending on the operating system). The selected observations can be located at the top of the table for easy reference by means of **Move Selected to Top**.

Finally, the arrangement of the columns in the table can be altered by dragging the variable name and moving it to a different position. This is often handy to locate related variables next to each other in the table for easy comparison when the original layout is not convenient.

2.4.2 Calculator

The **Calculator** functionality provides a rudimentary interface to create new variables, transform existing variables and carry out a number of algebraic operations. It is limited to one operation at a time, so it is not suitable for bulk programming.

The **Calculator** interface contains six tabs, dealing, respectively, with **Special** functions, **Univariate** and **Bivariate** operations, **Spatial Lag**, **Rates** and **Date/Time** operations. **Spatial Lag** and **Rates** are advanced functions that are discussed separately in Chapters 6 and 12.

The calculator interface is shown in Figure 2.21, illustrated for the **Bivariate** operations. Each tab at the top includes a series of functions, either creating a new variable (**Special**), operating on a single variable (**Univariate**), two variables (**Bivariate**), or values contained in a special **Date/Time** format. The specific functions are selected from the **Operator** drop-down list.

Each operation begins by identifying a target variable in which the result of the operation will be stored. Typically, this will be a new variable, which is created by means of the **Add Variable** button located to the right of **Result**.[7] This brings up an interface in which the **Name** for the new variable is specified, as well as its **Type**, its precision, and where to

[7]A new variable can also be created directly from the table options menu as **Table > Add Variable**.

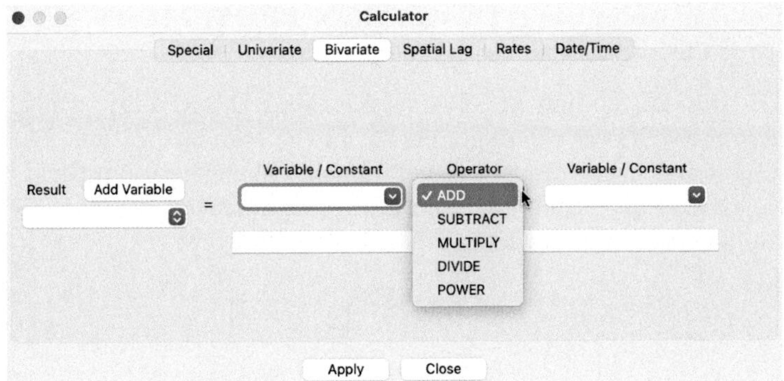

Figure 2.21: Calculator interface

insert it into the table. The default is to place a new variable to the left of the first column in the table, but any other variable can be taken as a reference point (i.e., the new column is placed to the left of the specified variable). In addition, the new variable can also be placed at the right most end of the table (the last option in the list).

With the target variable specified, the actual calculations can be carried out by selecting the operation from the drop-down list and choosing the variables involved. Again, to make the new variable permanent, the data table needs to be saved.

The various operations are briefly reviewed next.

2.4.2.1 Special functions

Special functions create a new variable, either a random variable, **NORMAL** or **UNIFORM RANDOM**, or the rank order of the current observations. The latter is generated by the **ENUMERATE** function. This yields a new variable with the *current order* of the observations. It is especially useful to retain the order of observations after sorting on a given variable, such as the rank order corresponding to a given statistic.

2.4.2.2 Univariate functions

Operations that pertain to a single variable are included in the **Univariate** drop-down list. **ASSIGN** allows a variable to be set equal to any other variable, or to a constant (the typical use). The next five operations are straightforward transformations, including **NEGATIVE** (change the sign), **INVERT**, **SQUARE ROOT**, and base 10 and natural **LOG**.

SHUFFLE randomly permutes the values for a given variable to different observations. This is an efficient way to implement *spatial randomness*, i.e., an allocation of values to locations, but where the location itself does not matter (any location is equally likely to receive a given observation value).

2.4.2.3 Variable standardization

The univariate operations also include five types of variable standardization. The most commonly used is undoubtedly **STANDARDIZED (Z)**. This converts the specified variable such that its mean is zero and variance one, i.e., it creates a z-value as

$$z = \frac{(x - \bar{x})}{\sigma(x)},$$

with $\bar{x}$ as the mean of the original variable x, and $\sigma(x)$ as its standard deviation.

A subset of this standardization is **DEVIATION FROM MEAN**, which only computes the numerator of the z-transformation.

An alternative standardization is **STANDARDIZED (MAD)**, which uses the *mean absolute deviation* (MAD) as the denominator in the standardization. This is preferred in some of the clustering literature, since it diminishes the effect of outliers on the standard deviation (see, for example, the illustration in Kaufman and Rousseeuw, 2005, pp. 8–9). The mean absolute deviation for a variable x is computed as:

$$\text{mad} = (1/n) \sum_i |x_i - \bar{x}|,$$

i.e., the average of the absolute deviations between an observation and the mean for that variable. The estimate for mad takes the place of $\sigma(x)$ in the denominator of the standardization expression.

Two additional transformations that are based on the range of the observations, i.e., the difference between the maximum and minimum. These the **RANGE ADJUST** and the **RANGE STANDARDIZE** options.

RANGE ADJUST divides each value by the range of observations:

$$r_a = \frac{x_i}{x_{max} - x_{min}}.$$

While **RANGE ADJUST** simply re-scales observations in function of their range, **RANGE STANDARDIZE** turns them into a value between zero (for the minimum) and one (for the maximum):

$$r_s = \frac{x_i - x_{min}}{x_{max} - x_{min}}.$$

Note that any standardization is limited to one variable at a time, which admittedly is not very efficient. However, most analyses where variable standardization is recommended, such as in the multivariate clustering techniques covered in Volume 2, include a transformation option that can be applied to all the variables in the analysis simultaneously. The same five options as discussed here are available to each analysis.

2.4.2.4 Bivariate operations

The bivariate functionality includes all classic algebraic operations. The target variable is selected from the drop-down list on the left in Figure 2.21. Then the respective variables (or a constant) are entered in the dialog with the appropriate operation selected from the list. This includes **ADD**, **SUBTRACT**, **MULTIPLY**, **DIVIDE** and **POWER**.

2.4.2.5 Date and time functions

The calculator also contains limited functionality to operate on date and time fields. In practice, this can be quite challenging, due to the proliferation of formats used to define variables that pertain to dates and time, as encountered in Section 2.4.1.2.

The **Date/Time** operators implement functions to extract specific parts from the data format and store them as integer variables. These include **Get Year**, **Get Month**, **Get Day**, **Get Hour**, **Get Minute** and **Get Second**. The resulting integer variable can then be used to carry out queries (see Section 2.5.1).

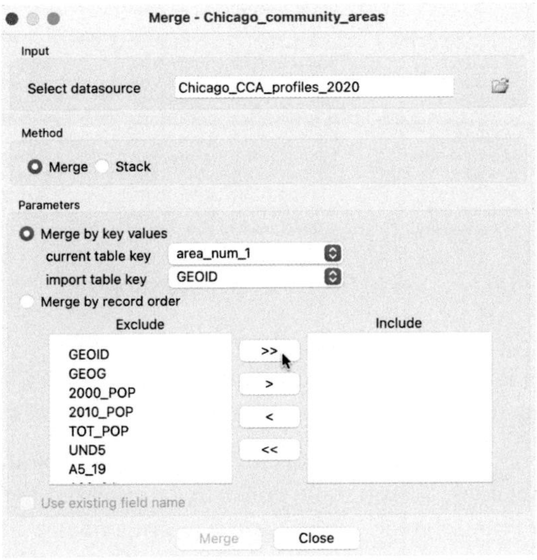

Figure 2.22: Merge dialog

2.4.3 Merging tables

An important operation on tables is the ability to **Merge** new variables into an existing data set, or, in other words, to *join* an external table to the currently active one.

For example, the *Chicago_community_areas.shp* file only contains information on the community area boundaries, but no actual data. On the other hand, the *Chicago_CCA_profiles_2020.csv* file contains a wealth of socio-economic data, but no *spatial* information other than the ID number for the community area. In order to combine the spatial information with the non-spatial data, the two tables need to be joined.

With the community area boundaries loaded, the **Table > Merge** option brings up the dialog shown in Figure 2.22.

The first step is to load the file to be merged (**Select datasource**). In Figure 2.22 this is *Chicago_CCA_profiles_2020.csv*. The default method is **Merge**, but **Stack** is supported as well. The latter operation is used to add *observations* to an existing data set.

Rather than relying on the order of the observations in both data sets, it is best practice to carry out a merging operation by selecting a **key**. This is a variable that contains (numeric) values that match the observations in both data sets. Using a key is superior to merging **by record order**, since there is often no guarantee that the actual ordering of the data in the two sets is the same, even though it may seem so in the interface (such as the table view). In the example, **area_num_1** is the key in the first data set (assuming it has been reformatted to integer) and **GEOID** the key in the second.

The variables to be merged are selected by moving their names from the **Exclude** box to the **Include** box. Using the » button selects all of them.

Upon clicking the **Merge** button at the bottom of the dialog, the new variables will be added to the current data table, unless there is a potential conflict of variable names. This conflict can be two-fold. First, when variables have identical names in both data sets, they need to be differentiated. If that is the case, alternatives are suggested in a dialog. These alternatives are not always the most intuitive and can be readily edited in the entry box.

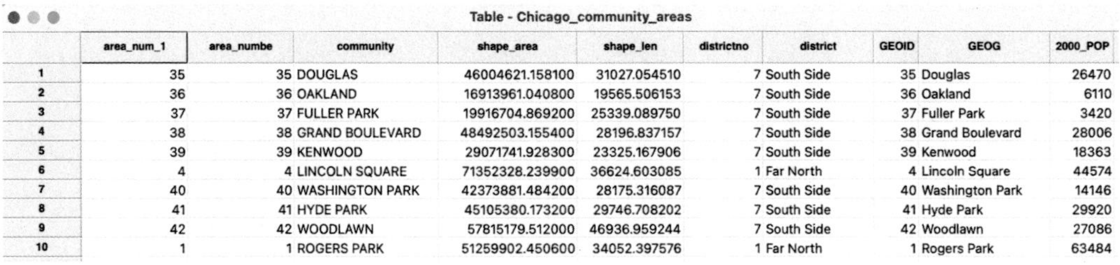

Table - Chicago_community_areas

	area_num_1	area_numbe	community	shape_area	shape_len	districtno	district	GEOID	GEOG	2000_POP
1	35	35	DOUGLAS	46004621.158100	31027.054510	7	South Side	35	Douglas	26470
2	36	36	OAKLAND	16913961.040800	19565.506153	7	South Side	36	Oakland	6110
3	37	37	FULLER PARK	19916704.869200	25339.089750	7	South Side	37	Fuller Park	3420
4	38	38	GRAND BOULEVARD	48492503.155400	28196.837157	7	South Side	38	Grand Boulevard	28006
5	39	39	KENWOOD	29071741.928300	23325.167906	7	South Side	39	Kenwood	18363
6	4	4	LINCOLN SQUARE	71352328.239900	36624.603085	1	Far North	4	Lincoln Square	44574
7	40	40	WASHINGTON PARK	42373881.484200	28175.316087	7	South Side	40	Washington Park	14146
8	41	41	HYDE PARK	45105380.173200	29746.708202	7	South Side	41	Hyde Park	29920
9	42	42	WOODLAWN	57815179.512000	46936.959244	7	South Side	42	Woodlawn	27086
10	1	1	ROGERS PARK	51259902.450600	34052.397576	1	Far North	1	Rogers Park	63484

Figure 2.23: Merged tables

One exception to this is when you do not want to merge in the duplicate fields. Checking the box **Use existing field name** makes sure that only the original columns are kept.

A second potential conflict is when the variable name to be included contains more than 10 characters. In csv input files, there are no constraints on variable name length, but the **dbf** format used for the attribute values in a shape file format has a limit of 10 characters for a variable name. Again, an alternative will be suggested in the dialog, but it can be easily edited (as long as it stays within the 10 character limit).[8]

The merged tables are shown in Figure 2.23. By keeping the identifiers for both data sets in the merged table (respectively **area_num_1** – in the first column – and **GEOID** – in the 8th column), one can easily verify that the observations are lined up correctly.

As before, the merger only becomes permanent after a **File > Save** operation.

Note that the merge operation in `GeoDa` implements a *left join*, in the sense that only observations that match the ID in the current table are included in the merged table. Specifically, if the table to be merged does not contain certain observations, the corresponding entries in the merged table will be missing values. Also, if the table to be merged contains observations that are not part of the current table, they will be ignored. Other forms of joins are not currently implemented.

2.5 Queries

While queries of the data are somewhat distinct from data wrangling per se, *drilling down* the data is often an important part of selecting the right subset of variables and observations. In order to select particular observations or rows in the data table, the **Selection Tool** is used. This is arguably one of the most important features of the table options.

The *Chicago Carjackings* data set is used to illustrate the selection functionality. First, a few adjustments are needed: the **Date** column must be reformatted to **datetime**, and new variables for the **Month** and the **Day** must be included (using **Table > Calculator** as in Section 2.4.2.5).

[8]The same problem can occur when using **File > Save As** to convert a csv file to the dBase format. Here too, a dialog will suggest alternative variable names, which can be edited.

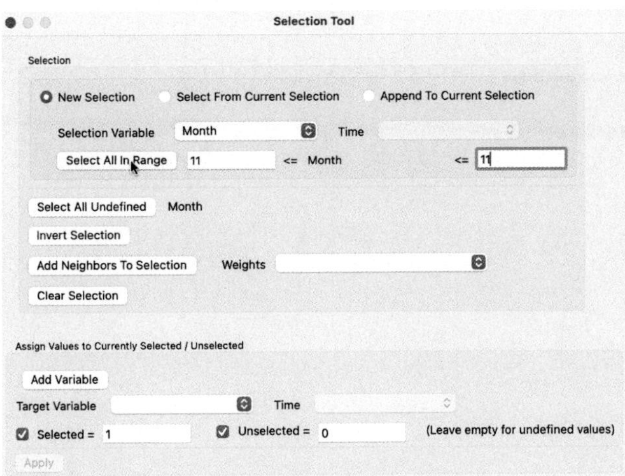

Figure 2.24: Selection Tool

2.5.1 Selection Tool

Queries can be carried out by invoking **Table > Selection Tool** or right clicking on the table to select this option. The interface, shown in Figure 2.24, supports quite complex searches and operations, even though it may seem somewhat rudimentary.

2.5.1.1 New selection

The main panel of the interface deals with the selection criteria. In Figure 2.24, a **New Selection** has been started, as indicated by the radio button in the top line. To select the carjackings that occurred in the month of November, the **Selection Variable** must be set to the newly created variable **Month**. Also, the beginning and end value of the **Select All in Range** option must be set to **11** (for November). Clicking on the button will select the observations that meet the selection criterion.

The selected observations are immediately highlighted in the table. For clarity, they can also be moved to the top of the table by means of **Table > Move Selected to Top**. Simultaneously, the selected observations are also highlighted in the themeless point map. More precisely, they retain their original shading, while unselected observations become transparent.[9]

This is illustrated in Figure 2.25. The status bar lists that 207 observations have been selected. In addition, the selected observations are also immediately highlighted in any other open graph or map. This is the implementation of *linking*, which will be discussed in more detail in Chapter 4.

2.5.1.2 Other selection options

The **Selection Tool** contains several more options to construct a more refined query. For example, a second criterion can be chosen to **Select From Current Selection**. This could be used to select a particular day of the month of November (provided the corresponding integer variable was created). Alternatively, to combine observations from both November *and* December, **Append To Current Selection** is appropriate.

[9]This behavior can be changed in the **GeoDa Preferences** settings, see Appendix A.

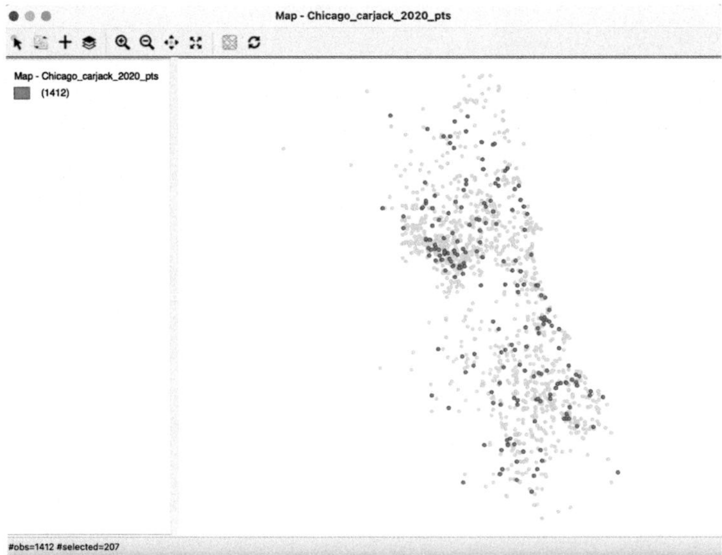

Figure 2.25: Selected observations in themeless map

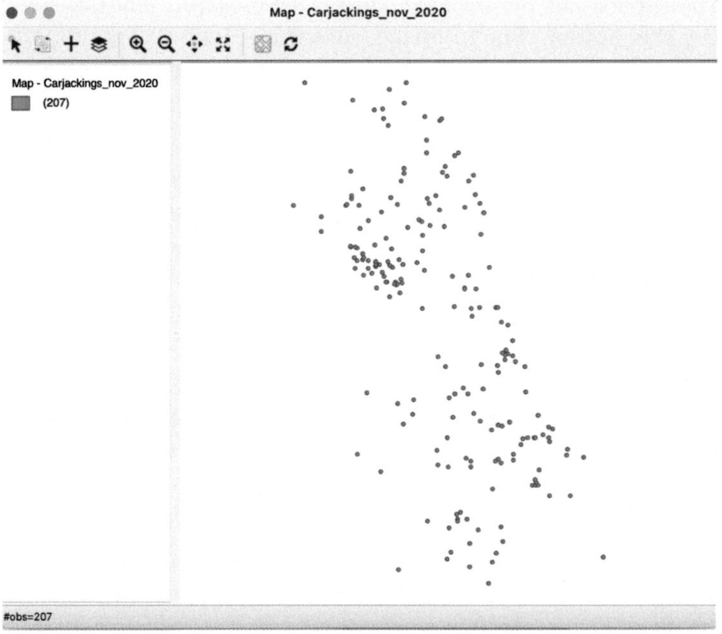

Figure 2.26: Selected observations in new map

A useful feature is to use **Invert Selection** to choose all observations *except* the selected ones. For example, this could be used to choose all months but November.

Often, this approach is the most practical way to *remove* unwanted observations, since there is *no* **Save Unselected** function. First, the observations to be removed are selected, followed by inverting the selection. At this point, **File > Save Selected As** can be employed to create a new data set (see Section 2.5.3).

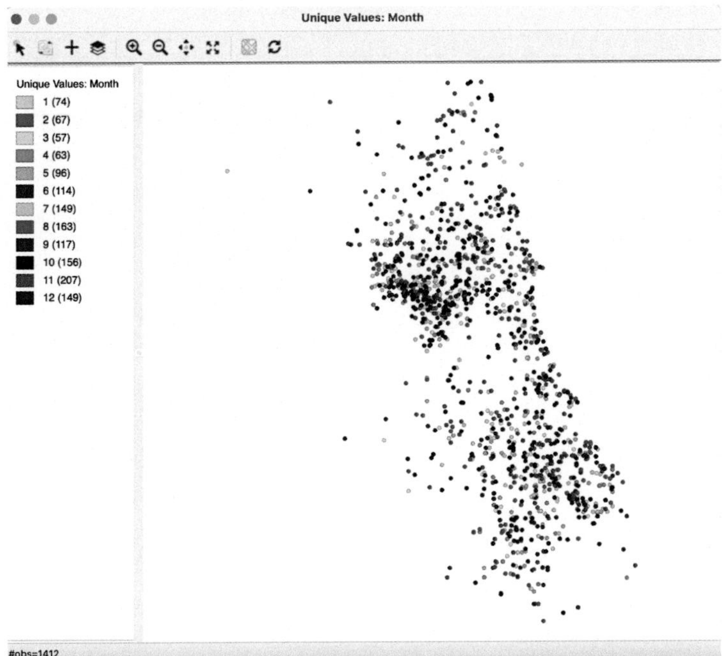

Figure 2.27: Car thefts by month

The same approach can also be used in combination with **Select All Undefined**, which identifies the observations with missing values for a given variable. The inverted selection can then be saved as a data set without missing values.

The **Add Neighbors To Selection** option will be discussed in the chapters dealing with spatial weights, in Part III.

2.5.2 Indicator variable

Once observations are selected in the table, a new *indicator variable* can be created that typically holds the value of 1 for the selected observations, and 0 for the others, as shown in the bottom panel of Figure 2.24. More generally, any combination of values for presence/absence can be specified in the dialog.

With a **Target Variable** specified, clicking on **Apply** will add the 0–1 values to the table. This then makes the variable available for use as an indicator or conditioning variable in a range of statistical analyses, including conditional plots (covered in Part II) and analysis of variance.

2.5.3 Save selected observations

Arguably one of the most useful features of the selection tool in terms of data wrangling is the ability to save the selected observations as a new data set. This is accomplished by means of **File > Save Selected As**. For example, a new file can be specified to save just the observations for November, shown in Figure 2.26. The point pattern has the same shape as the selected observations in Figure 2.25, but the number of observations is now listed as 207.

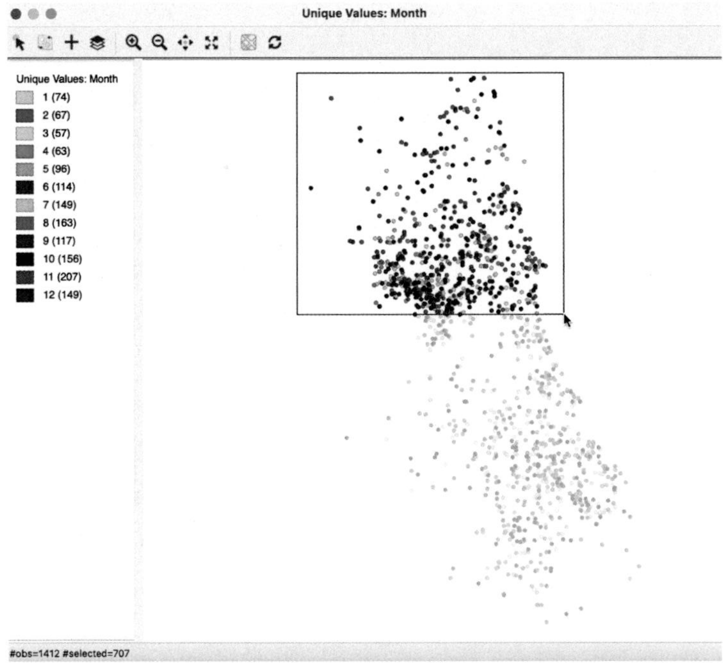

Figure 2.28: Selection on the map

2.5.4 Spatial selection

In addition to selecting observations by means of **Table > Selection Tool**, queries can also be constructed visually from any open map.

To illustrate the selection feature, a simple map of the points classified by month is shown in Figure 2.27.[10] The number of observations in each category is listed next to the corresponding legend item, with **(207)** as the value for November.

2.5.4.1 Selection by shape

The map selection tool – the pointer icon at the left of the map window toolbar – is the default interaction with a map view. Observations are selected by clicking on them or by drawing a selection shape around the target area. The default selection shape is a **Rectangle**, as in Figure 2.28, but **Circle** and **Line** are available as well. The particular shape is chosen by selecting **Selection Shape** from the map options menu (right click on the map).

Identical to table selection, all selected observations are highlighted in the table in yellow and in any other open map or graph (through linking). Also, they can be saved to a new data set using **File > Save Selected As**, in the same way as for a table selection.

The selection can be inverted by clicking on the second left-most icon in the map toolbar. This works in the same way as for table selection.

[10]This is implemented by means of the **Map > Unique Values Map** command, with **Month** as the variable for the unique values. A detailed discussion of mapping functionality is deferred until Chapter 4.

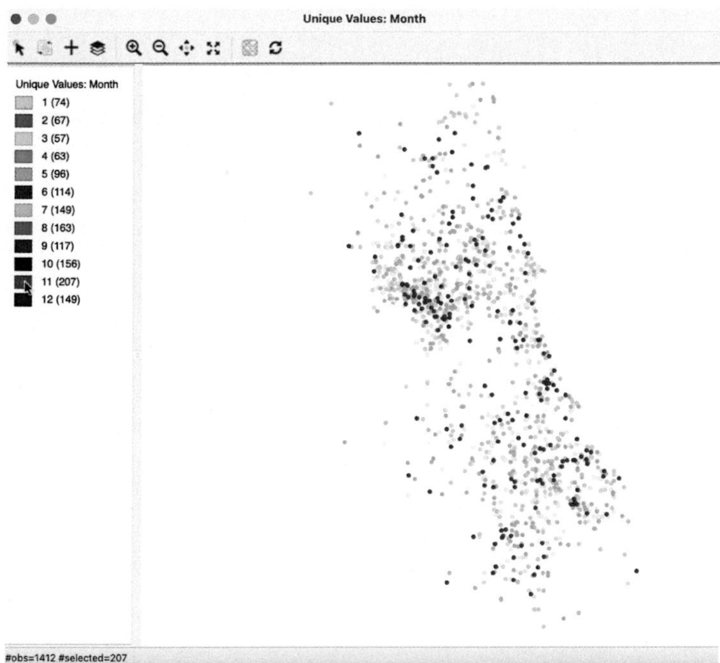

Figure 2.29: Selection on a map category

2.5.4.2 Selection on map classification

In addition to using a selection shape on a map, observations that fall into a particular map classification category can be selected by clicking on the corresponding legend icon. Map classifications are discussed in more detail in Chapter 4, but Figure 2.29 illustrates how the observations for the month November are selected by clicking on the small rectangle next to **11**. The selected points match the pattern in Figure 2.25.

2.5.4.3 Save selection indicator variable

Finally, as is the case for a table selection, a new indicator variables can be saved to the table, with, by default, a value of 1 for the selected observations, and 0 for the others. This is invoked from the map options (right click on the map) by selecting **Save Selection**.

The default variable name is **SELECTED**, but this can be easily changed, as can the values assigned to selected and unselected. After the indicator variable is added to the table, it can be made permanent through a **File > Save** command, in the usual fashion.

3

GIS Operations

By design, GeoDa is primarily focused on *analysis* and is not intended to be a geographic information system (GIS). However, over the years and in response to user requests, a range of spatial data manipulation functions have been incorporated. These include efficient treatment of projections, conversion between points and polygons, the computation of a minimum spanning tree, spatial aggregation and spatial join through multi-layer support. These functions are not intended to replace the range of operations that can be carried out in a *proper* GIS, but rather serve as an alternative for those without access to or familiarity with a GIS.

In this chapter, I present a quick overview of this functionality, again primarily aimed at novices in the field. Readers familiar with GIS operations can skim over the material to check on how these elementary spatial data manipulations are handled in GeoDa. The discussion is limited to the essentials. More extensive treatments are offered in standard GIS texts, such as Law and Collins (2018), Brunsdon and Comber (2019), Shellito (2020) and Madry (2021), among many others.

The various operations will be again be illustrated with the *Chicago Carjackings* data set from the **Sample Data**, as well as with the Community Area boundary layer from the GeoDaCenter sample data site, *Chicago_community_areas.shp*.

3.1 Topics Covered

- Understand how projections are expressed in a coordinate reference system (CRS)
- Be able to efficiently change from one projection to another
- Create shape centers (mean center and centroid) for polygons
- Construct Thiessen polygons for a point layer
- Compute a minimum spanning tree based on the max-min distance between points
- Operate on more than one layer
- Dissolve areal units and compute aggregate values
- Compute aggregate values based on a common indicator in a table
- Implement a spatial join to carry out point in polygon operations
- Visualize the common link between two layers

GeoDa Functions

- File > Save As
- Map > Shape Centers
- Map > Thiessen Polygons
- Map > Minimum Spanning Tree
- Map > Add Map Layer

DOI: 10.1201/9781003274919-3

- Map > Map Layer Settings
- Tools > Dissolve
- Table > Aggregate
- Tools > Spatial Join
- Map Layer Settings > Set Highlight Association
- Map Layer Settings > Change Associate Line Color

3.2 Projections

Spatial observations need to be *georeferenced*, i.e., associated with specific geometric objects for which the location is represented in a two-dimensional Cartesian coordinate system (i.e., on a flat map). Since all observations originate on the surface of the three-dimensional near-spherical earth, this requires a transformation from three to two dimensions.

The transformation involves two concepts that are often confused by non-geographers, i.e., the geodetic datum and the projection. The topic is complex and forms the basis for the discipline of *geodesy*. A detailed treatment is beyond the current scope, but a good understanding of the fundamental concepts is important. The classic reference is Snyder (1993), and a recent overview of a range of technical issues is offered in Kessler and Battersby (2019).

The basic building blocks are degrees (and minutes and seconds) latitude and longitude that situate each location with respect to the equator and the Greenwich Meridian (near London, England). Longitude is the horizontal dimension (x) and is measured in degrees East (positive) and West (negative) of Greenwich, ranging from 0 to 180 degrees. Since the U.S. is west of Greenwich, the longitude for U.S. locations is negative. Latitude is the vertical dimension (y) and is measured in degrees North (positive) and South (negative) of the equator, ranging from 0 to 90 degrees. Since the U.S. is in the northern hemisphere, its latitude values will be positive. Latitude and longitude are typically given as decimal degrees, but, if not, a conversion from degrees, minutes and seconds is straightforward.

In order to turn a geographic description, such as an address, into latitude-longitude degrees, it is necessary to adhere to a so-called *geodetic datum*, a three-dimensional coordinate system or model that represents the shape of the earth. Currently, the most commonly used datum is the World Geodetic System of 1984, WGS 84, which represents the earth as an ellipsoid (and not as a perfect sphere). In North America, an alternative is NAD 83, the North American Datum of 1983. In practice, for U.S. locations, there is not much difference between the two. Both these standards are about to be replaced by reference frames that take advantage of Global Navigation Satellite Systems (GNSS).[1]

The second step in the process of georeferencing consists of converting the latitude-longitude coordinates to Cartesian x-y coordinates in a planar system, using a cartographic projection. Hundreds of projections have been developed, each addressing different aspects of the mathematical problem of converting a three-dimensional object (on a sphere or ellipsoid) to two dimensions (a flat map). In this conversion, every projection involves distortion of one or more fundamental properties of geographic objects: angles (sometimes confused with shape), area, distance and direction. It is important to be aware of this limitation, since many investigations rely on the computation of variables such as distance, or density

[1] For details, see https://geodesy.noaa.gov/datums/newdatums/index.shtml.

(which involves area). The use of an inappropriate projection or distance metric may yield misleading results, especially for analyses that cover large areas (e.g., continent-wide).

For our purposes, three aspects are most important. One is to recognize whether spatial coordinates are projected (typically in units of feet or meters), or unprojected (i.e., in decimal degrees). To confuse matters more, the latter are sometimes referred to as a geographic projection, even though there is no projection involved. It is important to keep in mind that latitude and longitude are not expressed in distance units, but are degrees.

For example, a graph showing locations with longitude as the x-axis and latitude as the y-axis, treated as if they were regular distance units, can be misleading, even though it is seen quite commonly in publications. It ignores the fundamental property that latitude and longitude are degrees (angles). Similarly, the calculation of Euclidean distance is only supported for projected coordinates and should not be performed on longitude-latitude pairs. In many instances, `GeoDa` will generate a warning when an attempt to compute distances with decimal degrees is made, but this cannot be detected in all situations.

A second aspect is to be aware of the characteristics of a particular projection that is being used. Specifically, it is important to know whether the projection respects properties such as area (equal area) or distance (equidistant), although such properties typically only pertain to a subset of the projected map.[2]

A final important aspect is to make sure that layers are in the same projection when combined. `GeoDa` has some functionality to reproject layers on the fly to accomplish this, but it is not fail-safe.

3.2.1 Coordinate reference system

A *coordinate reference system* or CRS is a formal representation of location. It typically contains both a datum and a particular projection, as well as some information on the location of the coordinate center, units of measurement and related items. A CRS is identified by a code, such as an *EPSG code*, referred to as such because it was originally developed by the European Petroleum Survey Group (EPSG).

Interestingly, the *lack* of a projection, i.e., coordinates as simple latitude-longitude decimal degrees, has an EPSG code of **4326**. As mentioned, this is sometimes referred to as a geographic projection, even though strictly speaking no projection is involved. Nevertheless, identifying latitude and longitude coordinates with a CRS of EPSG 4326 will provide a valid point of departure, Without it, there would be no information on what the coordinates represent. This is critical for use in later transformations (see Section 3.2.3).

`GeoDa` uses the **proj4** convention to specify the CRS. This is compatible with the open-source PROJ library upon which the projection functionality is built.[3] Specifically, in order to save a spatial layer that is expressed as latitude-longitude decimal degrees, the proper entry in the CRS box is:

+proj=longlat +ellps=WGS84 +datum=WGS84 +no_defs

3.2.2 Selecting a projection

Non-geographers are often at a loss when faced with specifying an appropriate projection. An excellent resource in this respect is the web site *spatialreference.org*, which contains

[2] For a detailed technical discussion, see, for example Kessler and Battersby (2019).

[3] https://proj.org

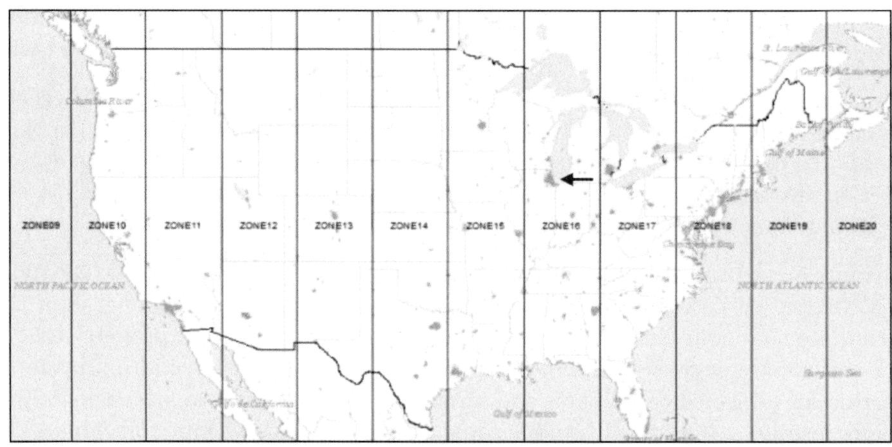

Figure 3.1: UTM zones for North America (source: GISGeography)

literally hundreds of projection definitions that can be easily searched.[4] For each projection, there is a summary of its properties and a list of its CRS specification in a range of different formats. For example, this includes the format used by ESRI in the *.prj files (part of the shape file specification), as well as the proj4 specification used by `GeoDa`.

While it is perfectly fine to use locations expressed as latitude-longitude decimal degrees, it is imperative to use the proper mathematical operations when computing properties like area and distance. For the latter, it is necessary to use great circle distance (arc distance), which is expressed in terms of the angles represented by latitude and longitude (see Chapter 11). In addition, the proper conversion of the great circle distance from angular units to distance unit (e.g., miles or kilometers) needs to differentiate by degree latitude. The implementation in `GeoDa` is only approximate and uses the distances at the equator.

In practice, it is preferred to use projected coordinates so that Euclidean distance operations and area calculations are straightforward to perform. For North America, a useful projection is *universal transverse Mercator*, or UTM. The country is divided into parallel zones, as shown in Figure 3.1. With each zone corresponds a specific projection that is represented by a CRS (e.g., an ESPG code).[5]

Let's say we need a projection for locations in Chicago (indicated by the arrow on the figure). From the map, we can see that this city is located in UTM zone 16 (north – there is a southern hemisphere equivalent). Searching the *spatialreference.org* site for this projection yields a list of specifications (i.e., combinations of different datums). For example, for a WGS84 datum we can find an associated EPSG code of **32616**. In the proj4 format, the corresponding CRS is:

+proj=utm +zone=16 +ellps=WGS84 +datum=WGS84 +units=m +no_defs

3.2.2.1 Specifying the CRS

As illustrated in the previous chapter, the **File > Save As** dialog contains a entry box at the bottom to specify the CRS for a spatial layer (see Figures 2.12 and 2.13). This will yield

[4]https://spatialreference.org

[5]For more details on the fundamentals behind the UTM projection, see, for example, the GISGeography site at https://gisgeography.com/utm-universal-transverse-mercator-projection/.

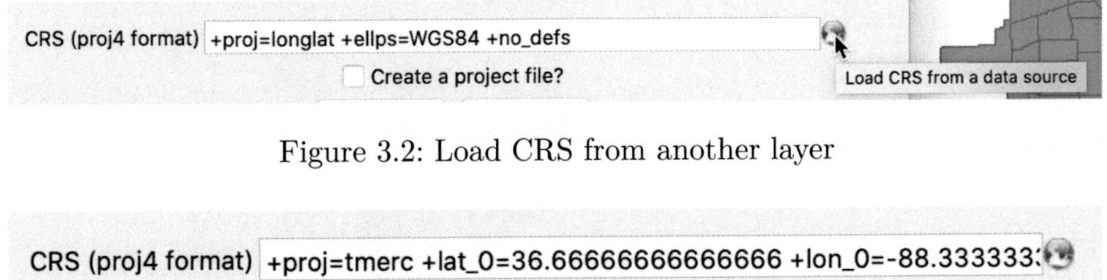

Figure 3.2: Load CRS from another layer

Figure 3.3: New CRS from another layer

a new layer with the proper projection. An alternative approach that avoids the need to type in the actual proj4 specification is considered next.

3.2.3 Reprojection

To avoid the manual entry of the correct CRS code, the **Save As** process in `GeoDa` provides an alternative that does not need an explicit specification, but requires the presence of another layer with the desired projection. The CRS information is then copied from that other layer and used in the *reprojection* of a current layer.

For example, consider the 77 Chicago community areas expressed in latitude-longitude coordinates in Figure 2.4. We saw how the locations of the carjackings in Figure 2.7 suggested a somewhat different shape for the city, because it was expressed in the State Plane Illinois East NAD 1983 projection (EPSG 3435). To convert the polygon community areas to the same projection, we could enter the proper proj4 specification in the CRS box.

An alternative is to copy the CRS information from a different layer. For example, after opening the community area file (*Chicago_community_areas.shp*) and invoking **File > Save As**, the CRS box shows the proj4 specification for latitude-longitude degrees, as in Figure 3.2. Instead of typing in the new specification, the small globe icon to the right can be selected to load a CRS specification from another file.

This brings up a file load interface into which the file name for the point shape file with the carjacking locations can be specified (*Chicago_carjack_2020_pts.shp*). The easiest way to accomplish this is by dragging the file name into the **Drop file here** area. Once the file name is loaded, the contents of the CRS box change to the new proj4 specification, as in Figure 3.3.

After saving the new file (e.g., as *community_areas_proj.shp*), the current project should be closed. A new project is started by loading the just created projected layer. The corresponding themeless base map is as in Figure 3.4. The more compressed shape matches the layout of the car jacking locations in Figure 2.7.

3.3 Converting Between Points and Polygons

So far, the geography of the community areas has been represented by their actual boundaries. However, this is not the only possibility. Equivalently, a representative point for each area can be specified, such as a *mean center* or a *centroid.* In addition, new polygons can be

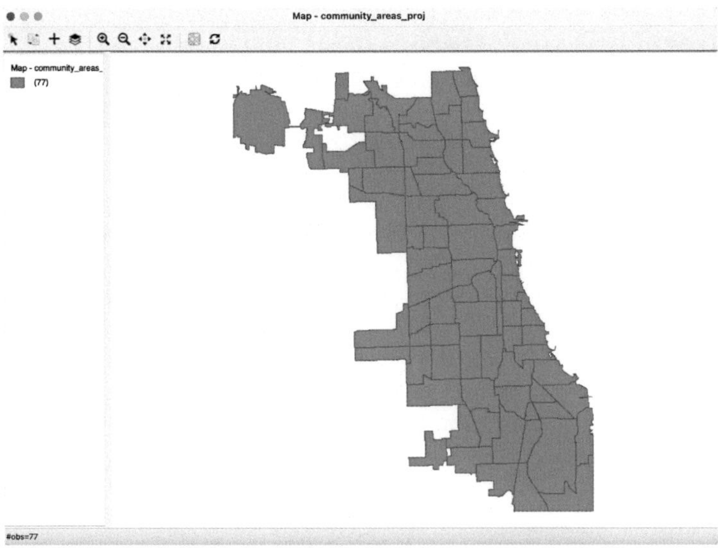

Figure 3.4: Reprojected Chicago community areas layer

constructed to represent the community areas as tessellations around those central points, such as *Thiessen polygons*. The key factor is that all three representations are connected to the same cross-sectional data set. As discussed in more detail in later chapters, for some types of analyses it is advantageous to treat the areal units as points, whereas in other situations Thiessen polygons form a useful alternative to dealing with an actual point layer.

The center point and Thiessen polygon functionality is invoked through the options menu associated with the map view (right click on the map to bring up the menu). This is illustrated with the just created projected community area map (Figure 3.4).

3.3.1 Mean centers and centroids

Two concepts are in common use to represent the *center* of a polygon. One is the *mean center*, which is simply the (unweighted) average of the coordinates of the vertices, so $(1/k)\sum_i x_i$ and $(1/k)\sum_i y_i$ (where k is the number of vertices). The notion of a *centroid* is more complex, as it represents the center of gravity of the polygon. A classic exercise is to create a cutout of the polygon and to find the centroid as the location where a pin would hold it up in a stable equilibrium.

The main difference between the mean center and the centroid is that the latter takes into account the area of the polygon. It is computed as an area weighted average of the centers of the set of triangles that represent the polygon. This so-called triangulation of a polygon is a standard operation in computational geometry (see, for example, Worboys and Duckham, 2004; de Berg et al., 2008). Once the triangulation is obtained, the centers of the triangles (the average of their coordinates) and their areas can be readily computed to obtain the centroid.[6]

The calculation of the mean center and centroid can yield strange results when the polygons are highly irregular (not convex), have holes, or represent a so-called multi-polygon situation

[6] `GeoDa` uses the centroid implementation from the GEOS library, see https://libgeos.org/doxygen/classgeos_1_1algorithm_1_1Centroid.html.

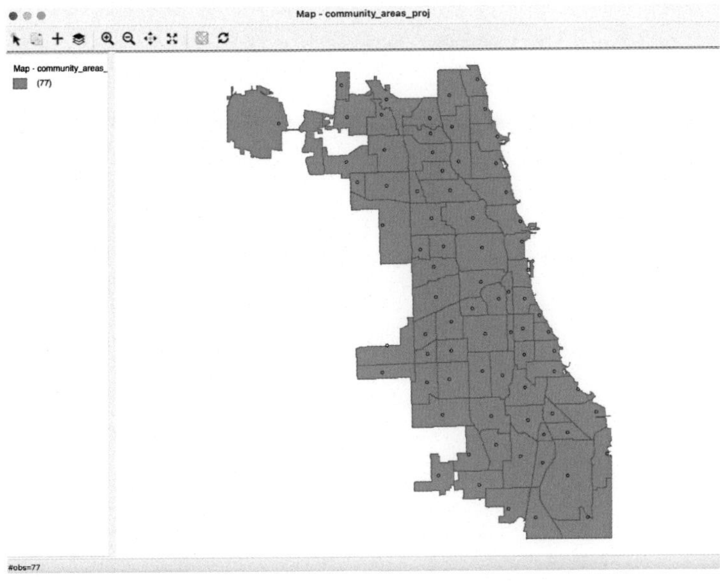

Figure 3.5: Display mean centers on map

Table - community_areas_proj

	area_num_1	community	COORD_XM	COORD_YM	COORD_XC	COORD_YC
1	35	DOUGLAS	1179689.829421	1883721.006180	1179003.714605	1883317.347826
2	36	OAKLAND	1184062.064716	1878677.154315	1183252.465747	1879210.393552
3	37	FULLER PARK	1175380.844929	1873967.239201	1175334.896726	1873799.967375
4	38	GRAND BOULEVARD	1178953.957640	1875020.216596	1179293.930099	1875240.895750
5	39	KENWOOD	1186861.343895	1873901.133388	1185215.715250	1873821.358957
6	4	LINCOLN SQUARE	1158035.853898	1933874.752722	1159860.137090	1934204.133018
7	40	WASHINGTON PARK	1179392.651258	1867316.145479	1179336.836488	1867737.225511
8	41	HYDE PARK	1188438.446739	1868441.915586	1186318.036497	1868427.912559
9	42	WOODLAWN	1188529.102906	1862568.050516	1185652.676323	1862878.113577
10	1	ROGERS PARK	1165694.188338	1947647.842447	1164480.367097	1946794.665848

Figure 3.6: Add mean centers to table

(different polygons associated with the same ID, such as a mainland area and an island belonging to the same county). In those instances, the centers can end up being located outside the polygon. Nevertheless, the shape centers are a handy way to convert a polygon layer to a corresponding point layer with the same underlying geography. For example, as covered in Chapter 11, they are used under the hood to calculate distance-based spatial weights for polygons.

3.3.1.1 Shape Center options

In `GeoDa`, the **Shape Center** option from the map view has three main functions, applied to either the **Mean Centers** or the **Centroids**. These are: **Add to Table**, **Display** and **Save**.

The simplest is to **Display** the center points on the map. This is purely illustrative and the points are not available for any further manipulation, in contrast with the multi-layer option discussed in Section 3.6. For example, in Figure 3.5, the mean centers of the Chicago community areas are displayed as small circles on top of the default themeless map from Figure 3.4.

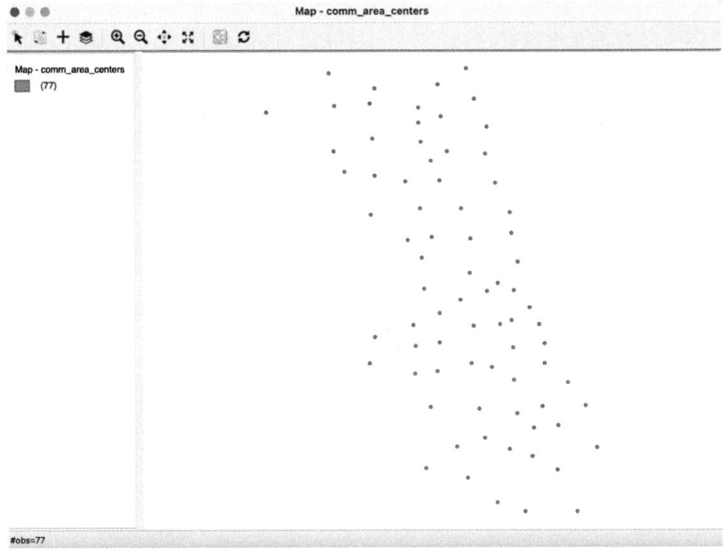

Figure 3.7: Chicago community area mean center layer

A second option is to **Add** the coordinates of the central points **to Table**. This brings up a dialog that prompts for the variable names for the coordinates, with **COORD_X** and **COORD_Y** as defaults. For example, the mean center coordinates could be set to **COORD_XM** and **COORD_YM** and the centroids to **COORD_XC** and **COORD_YC**. In Figure 3.6, the associated coordinate values are shown in the table (after some reshuffling of columns). Close examination of the values reveals only minor differences between the two types of mean centers in this particular case, in part because the community areas have mostly fairly convex shapes (the coordinates are in meters).

Finally, the **Save** option results in the usual prompt for a file name (e.g., *comm_area_centers.shp*) and creates the new layer. In the example, the point layer associated with the community area mean centers is as in Figure 3.7.

3.3.2 Tessellations

A tessellation is the exhaustive covering of an area with polygons or tiles. Of particular interest in spatial analysis are so-called Thiessen polygons or Voronoi diagrams (for an extensive discussion, see Okabe et al., 2000).

Thiessen polygons are a regular tiling of polygons centered around points, such that they define an area that is closer to the central point than to any other point. In other words, the polygon contains all possible locations that are nearest neighbors to the central point, rather than to any other point in the data set.[7] The polygons are constructed by considering a line perpendicular at the midpoint to a line connecting the central point to its neighbors. The latter is referred to as Delaunay triangulation and is in fact the *dual* problem to the tessellation.

In sum, any point layer can be converted to a polygon layer with the same geography by constructing Thiessen polygons around the points.

[7] In economic geography, this corresponds to the notion of a market area, assuming a uniform distribution of demand.

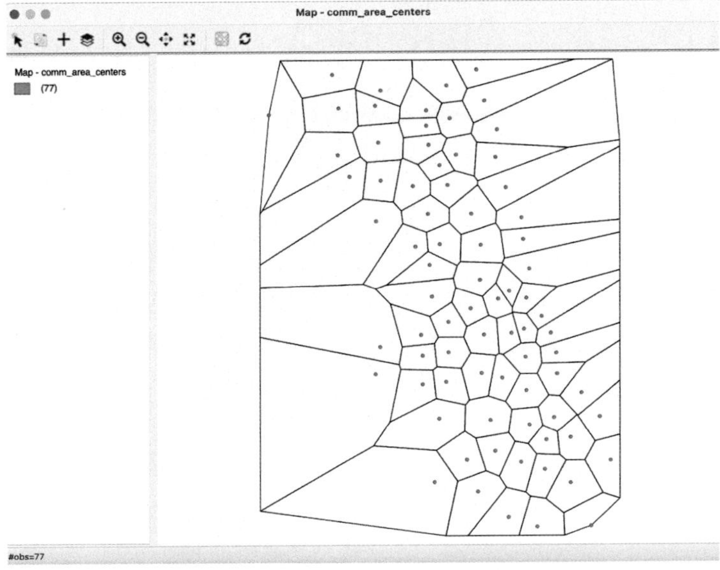

Figure 3.8: Display Thiessen polygons

3.3.2.1 Thiessen Polygons options

The **Thiessen Polygons** functionality is available as an option in the map view for point layers. As for the mean centers, there are three options, although only two of those are used regularly.

The **Display** option simply draws the polygons around the points, as shown in Figure 3.8 for the mean center points from Figure 3.7. As in the case of the display of mean centers, the resulting polygons cannot be used for any other operations.

The **Save** option will create a new layer with the polygons.[8] This layer can then be used in the same way as any other polygon layer, e.g., to create maps or carry out spatial analyses.

3.4 Minimum Spanning Tree

3.4.1 Concept

A central concept in the treatment of spatial autocorrelation is the so-called *spatial weights matrix*, discussed in detail in Part III of this Volume. For now, it is sufficient to know that this matrix shows the neighbor relations between observations. An element of the matrix in row i and column j, w_{ij}, is non-zero if i and j are neighbors (using one of many possible definitions). Technically, this is the same as an *adjacency matrix* for a graph or network, where the observations are *nodes* and the non-zero weights correspond to the *edges* that

[8]A third option is to **Save Duplicate Thiessen Polygons to Table**. This is only available when there is a problem in the calculation of the polygons due to multiple observations with the same coordinates. This may happen when considering different housing units in a high rise, or when the precision of the coordinates is insufficient to distinguish between all points. The option adds an additional column to the table, labeled **DUP_IDS**, which contains a value of 1 for those points that have multiple observations with the same coordinates.

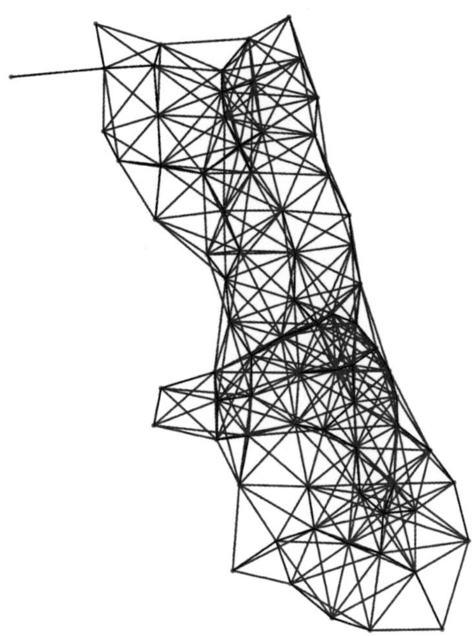

Figure 3.9: Connectivity graph of mean centers

connect the nodes. In the treatment of spatial weights, the graph structure will be called the *connectivity graph*.

For example, in Figure 3.9, the connectivity graph is shown for the community area mean centers from Figure 3.7. This graph represents a distance-band weights matrix with a cut-off distance that corresponds to the largest nearest neighbor distance between points, the so-called max-min distance criterion. This ensures that each observation (node) has at least one neighbor.

A Minimum Spanning Tree (MST) is a subset of this graph that constitutes a *tree*, i.e., "a connected, undirected network that contains no loops" (Newman, 2018). Importantly, this means that all nodes are connected (no isolates or islands) and there is exactly one path between any pair of nodes (no loops). As a result, a tree associated with n nodes contains exactly $n - 1$ edges. The MST is a special tree in that the total sum of the edge lengths (typically a weight) is minimized. It constitutes a simplification of the full network structure that fulfills a minimum cost objective.

For example, the MST computed from the network structure in Figure 3.9 is shown in Figure 3.10. The complexity of the graph is greatly reduced and now consists of only 76 edges (for 77 observations or nodes).

The MST is similar to, but distinct from the solution to the so-called traveling salesperson problem, which looks for a route through the network that achieves a minimum cost objective, but visits each node just once. This implies that the nodes in a solution to the traveling salesperson problem cannot have a degree (i.e., the number of edges connected to it) larger than two. In Figure 3.10, the maximum degree is four.

A classic solution to the MST problem is achieved by *Prim's algorithm* (Prim, 1957), illustrated in the following worked example. The MST plays a major role in some of the spatial clustering procedures discussed in Volume 2.

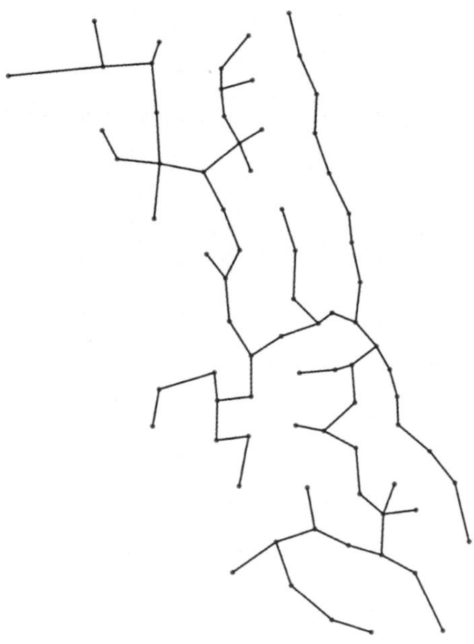

Figure 3.10: Mininum spanning tree for connectivity graph of mean centers

ID	X	Y
1	10	10
2	25	20
3	25	55
4	40	15
5	40	30
6	55	15
7	55	40
8	70	30
9	85	30

Figure 3.11: Toy example point coordinates

3.4.1.1 Worked example – Prim's algorithm for MST

A small toy example consisting of 9 points is used to illustrate the mechanics of Prim's MST algorithm. The coordinates of the points are listed in Figure 3.11. The corresponding inter-point distance matrix is given in Figure 3.12. The inter-point distances form the basis to construct a *connectivity graph* for a distance band that ensures that each point is connected to at least one other point. In the example, this distance is 29.15.[9] The corresponding graph representation is given in Figure 3.13, where each point is a node and the edges show the connections and associated distances.

Prim's algorithm proceeds by starting at a random node and selecting the shortest edge to the next node. For example, starting with node 1, there is only a single edge, such that the first path is from node 1 to 2. At node 2, there are two edges, 2–4 with length 15.81, and 2–5 with length 18.03. Therefore, the next link is between 2 and 4. At this point, the shortest path follows the red lines as in Figure 3.14.

[9]Connectivity graphs are discussed in more detail in the chapters dealing with spatial weights in Part III.

	1	2	3	4	5	6	7	8	9
1	0.00	18.03	47.43	30.41	36.06	45.28	54.08	63.25	77.62
2		0.00	35.00	15.81	18.03	30.41	36.06	46.10	60.83
3			0.00	42.72	29.15	50.00	33.54	51.48	65.00
4				0.00	15.00	15.00	29.15	33.54	47.43
5					0.00	21.21	18.03	30.00	45.00
6						0.00	25.00	21.21	33.54
7							0.00	18.03	31.62
8								0.00	15.00
9									0.00

Figure 3.12: Inter-point distance matrix

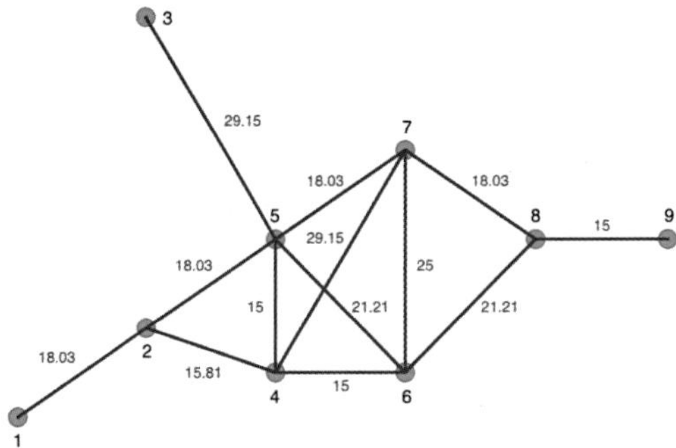

Figure 3.13: Connectivity graph

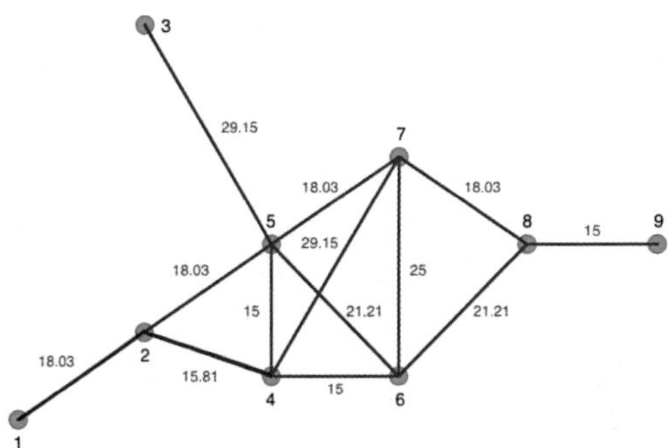

Figure 3.14: Initial steps in Prim's algorithm

At node 4, the two edges have the same length, so 4 could be connected to either 6 or 5. The order is immaterial, since the length of 15 is the smallest in the network and will have priority over any other edge. At this point, the path consists of 1–2, 2–4, 4–6 and 4–5. The next step connects 5 to 7 (18.03), then 7 to 8 (18.03) and 8 to 9 (15). The only remaining unconnected node is 3, which is linked to 5 (29.15). The resulting MST follows the red lines in Figure 3.15. Every node is visited and the total distance of the path is minimized. The solution consists of eight edges, which is $n - 1$. The MST solution is not unique since there can be several trees that achieve the same minimum cost solution.

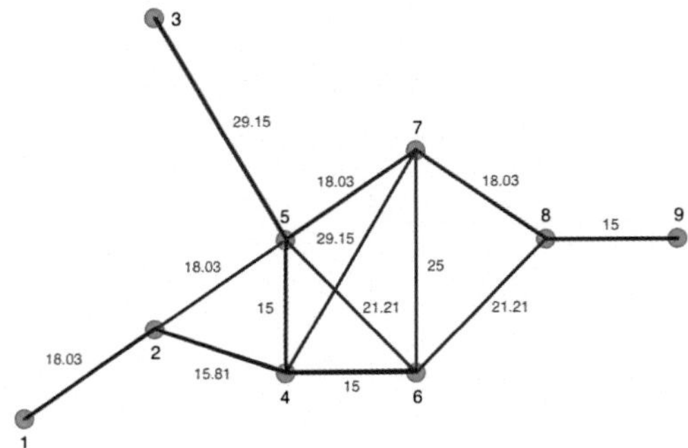

Figure 3.15: Minimum spanning tree

3.4.2 Minimum Spanning Tree options

Minimum Spanning Tree is one of the entries in the options menu associated with a map view for a point layer (right click on the layer to invoke the menu). There are two main options: **Display Minimum Spanning Tree**, which simply shows the edges on the point map, and **Save Minimum Spanning Tree**. The latter saves the MST information in the **gwt** spatial weights format (see Chapter 11), which makes it available for use in other analyses.

In addition, there are two options to (slightly) change the format of the MST display: **Change Edge Thickness** (with three options: **Light**, **Normal** – the default – and **Strong**), and **Change Edge Color**. The display in Figure 3.10 has the **Edge Thickness** as **Strong** and the color changed to blue (from the default black).

3.5 Aggregation

Spatial data sets often contain identifiers of larger encompassing units, such as the state that a county belongs to, or the census tract that contains individual household data. Such *nested* spatial scales readily lend themselves to *aggregation* to the larger unit. The aggregation computes observations for the new spatial scale as the sum, average or other operation applied to the values at the lower scale. Note that this only makes sense if the underlying variables are *spatially extensive*, such as counts (total population, total households). Without proper re-weighting, the aggregation gives misleading results for variables expressed as medians or percentages (e.g., median household value, median income).

`GeoDa` supports two different ways to compute aggregates from the smaller units. One approach also creates a *dissolved* layer, i.e., a spatial layer that includes both a map for the larger spatial units and a data table with the corresponding aggregated observations. The second approach is limited to computing a new table with the aggregate information. It does not generate a spatial component.

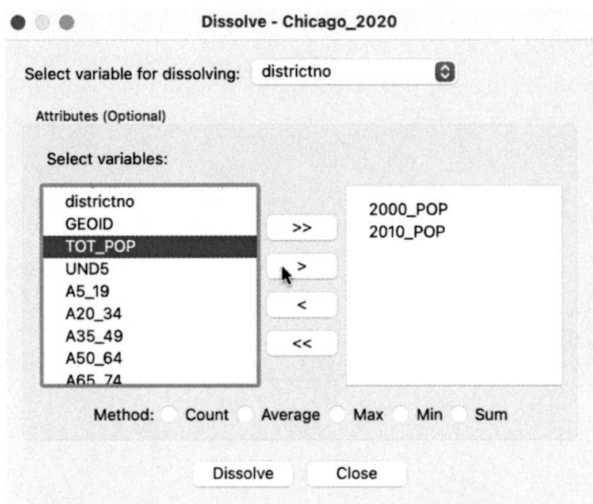

Figure 3.16: Dissolve dialog

3.5.1 Dissolve

The **Dissolve** functionality is part of the **Tools** menu (**Tools > Dissolve**), or can be accessed by selecting the **Dissolve** option from the **Tools** icon on the toolbar, the right-most icon in Figure 2.1.

To illustrate this feature, the point of departure is the merged data set created in Section 2.4.3 (e.g., saved as *Chicago_2020.shp*). This data set contains socio-economic data for the 77 Chicago community areas. In addition, it also has an identifier that associates each community area with a larger encompassing *district*, i.e., the variable **districtno**. The corresponding layout is given by the themeless base map from Figure 2.4

The dialog, shown in 3.16, requests three important pieces of information. The most critical is the **variable for dissolving**, i.e., the key that indicates the larger scale to which the observations will be aggregated (here, **districtno**). Next follows the selection of the variables for which the aggregate values will be calculated. In the example in Figure 3.16, these are **2000_POP**, **2010_POP** and **TOT_POP** (the population in 2020 from ACS). The variables are selected by double clicking on them, or by selecting them and then using the > key to move them to the right-hand side column.

Finally, we need the proper aggregation **Method** must be selected. The available options are **Count**, **Average**, **Max**, **Min** and **Sum**. The **Count** function doesn't actually compute any aggregate values, but provides a count of the number of smaller units included in the larger unit (e.g., the number of community areas in each district). For example, **Sum** will yield the population totals for each district.

After activating the **Dissolve** button, the usual file creation dialog is produced to specify a file type and file name, e.g., *Chicago_district_2020.shp*.

Figure 3.17 shows the outline of the nine districts (in red), together with the original community area boundaries (in blue). This is an illustration of a multiple layer operation, since both the community area layer and the district layer are combined on the same map. The multilayer functionality is further detailed in Section 3.6.1. The associated table (Figure 3.18) lists the population totals in the three census years for the district aggregates. The table also includes a new variable, **AGG_COUNT**, which indicates the number of community areas aggregated into the respective district.

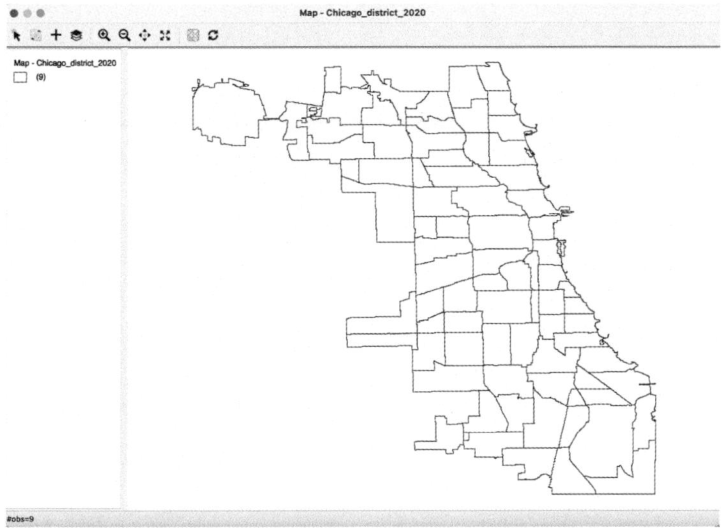

Figure 3.17: Dissolved districts map

Table - Chicago_district_2020

	districtno	AGG_COUNT	2000_POP	2010_POP	TOT_POP
1	1	12	487735	453704	470570
2	2	5	316830	302404	316777
3	3	6	283845	276594	277922
4	4	3	98708	131157	156702
5	5	9	537987	481491	469746
6	6	12	414936	389547	363731
7	7	12	299422	257920	262896
8	8	6	191393	177988	176055
9	9	12	264812	224793	214123

Figure 3.18: Dissolved districts table

3.5.2 Aggregation in table

The same aggregation function is also available as an option for the table, invoked as **Table > Aggregate** from the menu, or, alternatively, from the list of table options, shown in Figure 2.16. The main difference with the **Dissolve** function is that the results are only available as a table, without an associated dissolved map.

The dialog requests the same three features and operates in the same way as shown in Figure 3.16.[10] The result is saved as a table, which contains the same information as shown in Figure 3.18.

[10]The order of the items is slightly different, and the aggregation variable is referred to as **Select key**, but otherwise the operation is identical to that in Figure 3.16.

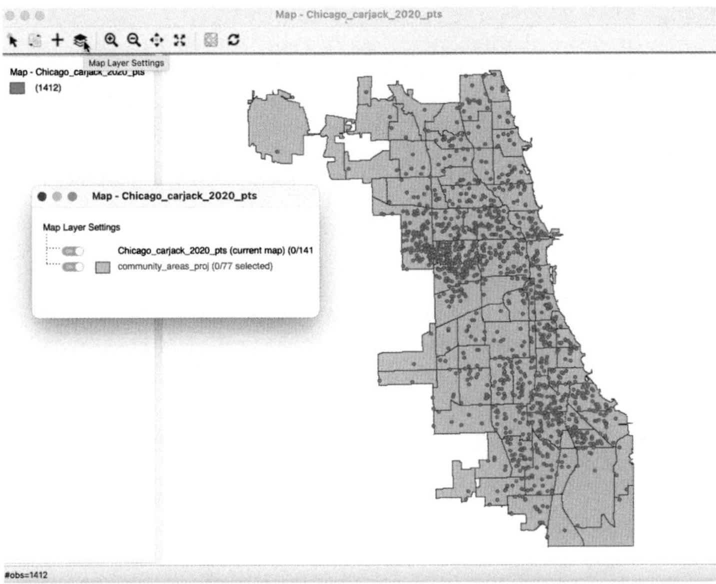

Figure 3.19: Map layer settings

3.6 Multi-Layer Support

So far, the focus has been on a single spatial layer, associated with one cross-sectional data set. Over time, GeoDa has added some limited functionality to handle multiple layers, to facilitate operations such as a *spatial join* between two layers (see Section 3.7).

While multiple layers can be displayed in the same map window, interaction is limited to a single layer, the so-called *current layer*. This pertains to table manipulations, such as selection and variable transformation but also to the full range of mapping and analytical functionality. The current layer is always the layer that was loaded first, irrespective of the actual order of the layers displayed in the map window.

Before delving deeper into the interaction between layers, the basics of the multiple layer logic in GeoDa are outlined.

3.6.1 Loading multiple layers

The starting point in any multi-layer operation is the *current layer*, i.e., the data set first loaded. For example, to display the carjacking locations together with the boundaries of the Chicago community areas, one of these layers must be the starting point, such as the points layer (shown Figure 2.7). An additional layer is loaded by means of the **Add Map Layer** option. This is represented by the large + icon, third from left in the map window toolbar.

This brings up the usual data input interface, to which the projected community area boundaries are loaded (e.g., *community_areas_proj.shp*, shown in Figure 3.4). The result is displayed in Figure 3.19, with the point locations portrayed on top of the community area outlines. The **Map Layer Settings** option (fourth icon from the left on the map window toolbar) brings up a small dialog that lists the order of the layers and indicates the **current map**.

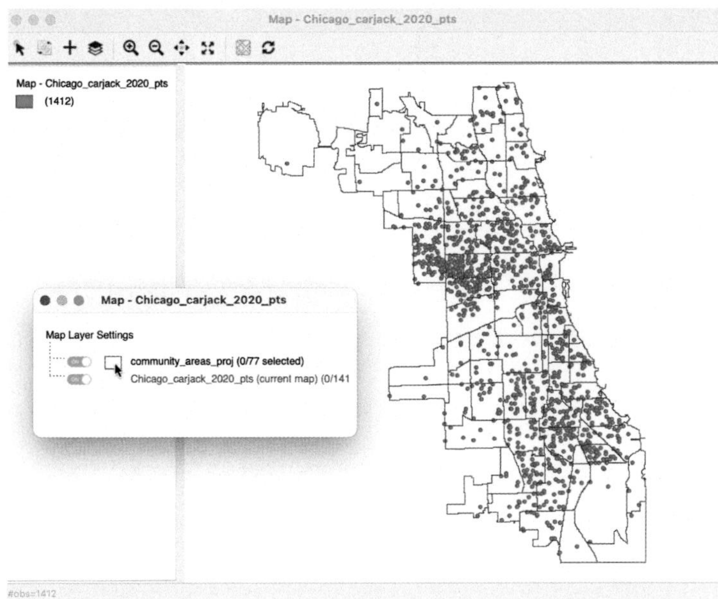

Figure 3.20: Polygon layer on top with opacity=0

It is important to keep in mind that the visual properties (such as color and opacity) of the current layer are managed by the map legend (see Chapter 4). The properties of the other layers are controlled through the **Map Layer Settings**.

3.6.1.1 Map layer setting options

A right click on the small legend rectangle associated with each layer (other than the current layer) brings up a list of options that pertain to the visual representation of that layer. The two most often used functions are **Change Fill Color** and **Change Outline Color**. In addition to setting the color, it may be necessary to adjust the **opacity** in order to make sure that the relevant information is visible. Specifically, setting opacity to 0 makes the areas invisible and only the boundary outlines are shown. The latter can be turned off by unchecking the **Outline Visible** option. The **Only Map Boundary** option results in a dissolve, in that only the outer outline of the polygon layer is shown. This operation is purely visual and does not involve any aggregation, in contrast to what was the case in Section 3.5.1.

The interface also includes options to set the association highlight, which is covered in Section 3.8. The final item in the dialog is to **Remove** the layer.[11]

3.6.1.2 Layer position

The position of the layers in the map view can be altered by means of the **Map Layer Settings** interface. One grabs the entry for a layer and drags it to a different position. For example, in Figure 3.20, the community area layer is moved to the top (this requires setting the opacity to 0 in order for the points to show). However, even though the polygon layer is now on top, the point layer remains the **current map** layer, since it was loaded first.

[11]In addition, it is possible to change the map window scope by using **Zoom to Selected** and **Layer Full Extent**.

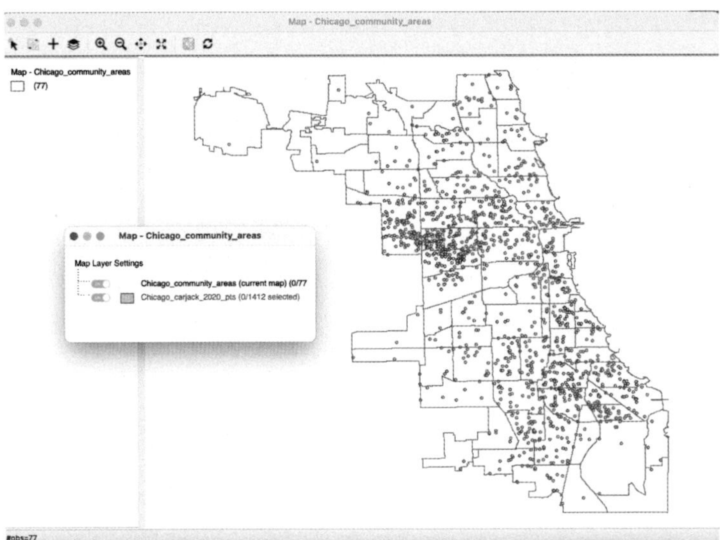

Figure 3.21: Points reprojected to lat-lon

3.6.2 Automatic reprojection

In the previous example in Figure 3.19, the two layers that combined were expressed in the same projection. More precisely, the projected community area boundaries were superimposed on the point layer.

However, as long as layers have projection information associated with them, any added layer will be *reprojected* to the projection of the *current* layer. This is for visualization only, and it does not change the projection information in the files of the reprojected layers.

For example, in Figure 3.21, the order in which the layers are loaded is reversed from Figure 3.19. The community area layer (*Chicago_community_areas.shp*) is loaded first (i.e., it is the current layer). This layer is not projected, but expressed in lat-lon coordinates. To this the point layer *Chicago_carjack_2020_pts.shp* is added, which is projected.

Initially, the point layer is invisible, since it is positioned *under* the polygon layer. The default green coloring of the community areas masks the points underneath. After setting the **opacity** of the top layer to zero, the points become visible.

In Figure 3.21, a slightly more spread out pattern can be noticed relative to Figure 3.19. Under the hood, the point layer has been reprojected to lat-lon decimal degrees. This is only for visualization in the map view and does not change the CRS associated with the added layers.

3.6.3 Selection in multiple layers

It is important to keep in mind that any *selection* only pertains to the **current map** layer, which may or may not also be the top layer. For example, continuing with the previous illustration (Figure 3.21), any selection will pertain to the community area layer, which is the **current map** layer. This is irrespective of which layer is shown as the top layer.

In Figure 3.22, the selection status bar indicates that 14 observations were selected. If the point layer was moved to the top, this would give the same result.

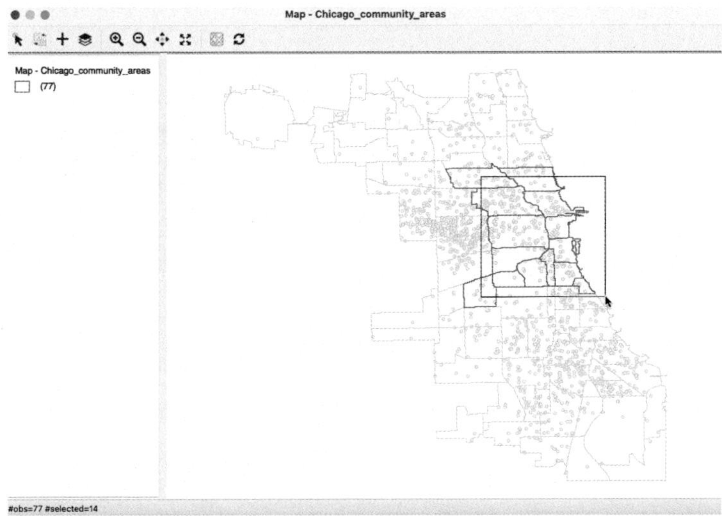

Figure 3.22: Select on current map layer (community areas)

This logic may seem a bit counter-intuitive at first, but it is based on the linking and brushing architecture implemented in `GeoDa` (see Chapters 4 and 7 for a more extensive discussion). Currently, the linking operation works only for a single layer, which is always the first loaded layer.

3.7 Spatial Join

The multi-layer infrastructure allows for the calculation of variables for one layer, based on the observations in a different layer. This is an example of a *point in polygon* GIS operation. There are two applications of this process. In one, referred to as **Spatial Assign**, the ID variable (or any other uniquely identifying variable) of a spatial area is *assigned* to each point that is within the area's boundary. The reverse of this process is **Spatial Count**, i.e., a count or other form of aggregation over the number of points that are within a given area, similar to the operation of **Aggregation** and **Dissolve** (see Section 3.5).

Even though the default application is simply assigning an ID or counting the number of points, more complex assignments and aggregations are possible as well. For example, rather than just counting the points, an aggregate over the points can be computed for any variable, such as the mean, median, standard deviation, or sum across the point observations for that variable.

It is important to keep in mind the *order* in which the layers are loaded into the multi-layer setup. For spatial assign, the point layer is first, since the areal unit identifier is assigned to each point. As mentioned before, only the table of the first loaded layer is available for manipulation or analysis. So, in order to add the ID code to the point layer table, it must be loaded first.

For spatial count, the reverse is the case. The polygon layer needs to be loaded first, since the number of points (or any other aggregation) is added to the table with observations for the areal units.

Figure 3.23: Spatial join dialog – areal unit identifier for point

Table - Chicago_carjack_2020_pts

	ID	Case _mber	Date	Month	Day	IUCR	Descr_tion	Locat_tion	Beat	District	Ward	Commu_Area	SA
1	12243012	JD456735	2020-12-11	12	11	326	AGGRAVATED VEHICULAR HIJACKING	STREET	723	7	6	68	68
2	12130750	JD326543	2020-8-9	8	9	326	AGGRAVATED VEHICULAR HIJACKING	GAS STATION	1234	12	25	31	31
3	12012068	JD189121	2020-3-18	3	18	326	AGGRAVATED VEHICULAR HIJACKING	STREET	1114	11	28	26	26
4	12012705	JD189547	2020-3-18	3	18	326	AGGRAVATED VEHICULAR HIJACKING	STREET	1114	11	28	26	26
5	12013533	JD189766	2020-3-18	3	18	325	VEHICULAR HIJACKING	STREET	712	7	16	68	68
6	12012624	JD189695	2020-3-18	3	18	326	AGGRAVATED VEHICULAR HIJACKING	STREET	1513	15	29	25	25
7	12242966	JD456567	2020-12-11	12	11	326	AGGRAVATED VEHICULAR HIJACKING	STREET	1225	12	28	28	28
8	12132859	JD326758	2020-8-9	8	9	326	AGGRAVATED VEHICULAR HIJACKING	STREET	624	6	6	69	69
9	12130858	JD326646	2020-8-9	8	9	325	VEHICULAR HIJACKING	STREET	1122	11	27	23	23
10	12130850	JD326657	2020-8-9	8	9	326	AGGRAVATED VEHICULAR HIJACKING	STREET	1113	11	28	26	26
11	12130785	JD326555	2020-8-9	8	9	326	AGGRAVATED VEHICULAR HIJACKING	STREET	1113	11	28	26	26

Figure 3.24: Community area identifier (SA) in point table

This functionality is invoked from the menu as **Tools > Spatial Join**, or by selecting the **Spatial Join** option from the **Tools** icon in Figure 2.1.

3.7.1 Spatial assign

To illustrate the spatial assign application, the layer arrangement from Figure 3.19 is considered, with the (projected) car jacking locations loaded first, followed by the (projected) community area boundaries.

Invoking **Tools > Spatial Join** generates the dialog shown in Figure 3.23. This differs slightly between the spatial assign and spatial count functionalities (see Figure 3.26 for the latter).

The dialog shows how the **current map** layer, **Chicago_carjack_2020_pts** is joined to another layer. In the example, there is only one other layer, i.e., **community_areas_proj**, but in general, there could be several layers to select from. The **ID Variable** list shows all the variables that could serve as unique identifiers for the areal units. In the example, there are three variables listed, but only the fist two pertain to the community areas in the map layer (all integer variables are listed by default). In Figure 3.23, **area_num_1** is selected as the ID variable.

The next dialog requests the variable name of the spatial assign variable, which is **SA** by default. After specifying the variable name, it is added to the data table for the point layer. In Figure 3.24, the new **SA** column is listed next to the original **Commu_Area** variable from the point location data base. A quick check confirms that they are the same. The difference is that the values for the **SA** variable were derived from a spatial point in polygon operation, joining the two layers, whereas the original variable was included during data entry.

In cases where the point in polygon operation results in a mismatch (e.g., due to differences in precision of the location coordinates in the two layers), a value of -1 is used for the

Figure 3.25: Point in polygon mismatch

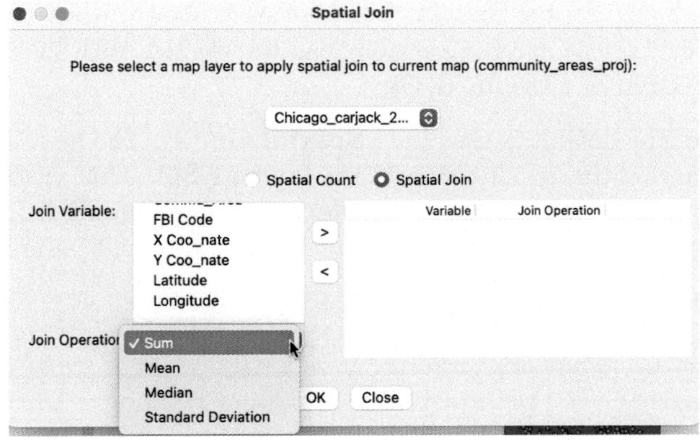

Figure 3.26: Spatial join dialog – aggregation options

spatial assign. In the example, there are three such points, all situated at the very edge of a polygon. This is illustrated in Figure 3.25 for the two mismatches at the southern and south-western edge of the city (one other mismatch is in the very north).

Depending on the goals of the analysis, one could either eliminate the points from the data set, or manually edit the values in the table after zooming in on the actual location.

The new spatial assign variable can now be used to aggregate observations, as in Section 3.5. However, the dissolve operation does not work, since there is no boundary information associated with the points. Their spatial information consists of the coordinates only, which do not lend themselves to a dissolve operation.

As always, a **File > Save** or **File > Save As** operation must be used to make the addition of the spatial assign variable permanent.

3.7.2 Spatial count

To illustrate the spatial count feature, the order of the two layers from Figure 3.19 is reversed. The *community_areas_proj* layer is loaded first (since the counts will be added to the

	area_num_1	area_numbe	community	shape_area	shape_len	districtno	district	SC
1	35	35	DOUGLAS	46004621.158100	31027.054510	7	South Side	19
2	36	36	OAKLAND	16913961.040800	19565.506153	7	South Side	4
3	37	37	FULLER PARK	19916704.869200	25339.089750	7	South Side	10
4	38	38	GRAND BOULEVAR	48492503.155400	28196.837157	7	South Side	31
5	39	39	KENWOOD	29071741.928300	23325.167906	7	South Side	11
6	4	4	LINCOLN SQUARE	71352328.239900	36624.603085	1	Far North	3
7	40	40	WASHINGTON PAR	42373881.484200	28175.316087	7	South Side	9
8	41	41	HYDE PARK	45105380.173200	29746.708202	7	South Side	11
9	42	42	WOODLAWN	57815179.512000	46936.959244	7	South Side	20
10	1	1	ROGERS PARK	51259902.450600	34052.397576	1	Far North	7

Figure 3.27: Spatial count in polygon area table

polygon table observations), followed by *Chicago_carjack_2020_pts*. Note that in order to make the point layer visible, it needs to be moved to the top.[12]

The **Tools > Spatial Join** command generates the dialog shown in Figure 3.26. In the default setup, the **Spatial Count** radio button is checked, which does not require any further input. In the figure, the **Spatial Join** option is shown, which is essentially the same as aggregation (Section 3.5). Several variables can be selected, with the aggregation following **Sum**, **Mean**, **Median** or **Standard Deviation**.

In the illustration, the default is used, i.e., **Spatial Count**. The next dialog asks to specify the variable name for the result, with the default as **SC**. This yields a new variable in the table, which shows the number of car jackings occurring in each community area, as illustrated in Figure 3.27. As before, the addition only becomes permanent after a **File > Save** or **File > Save As** operation.

3.8 Linked Multi-Layers

In the previous example, the connection between two layers is used to compute a new variable (e.g., the number of points in a polygon). An additional feature of the multi-layer architecture is that a *connection* can be established between observations in two layers. A typical application of this feature is to show the connection between two point layers, e.g., where one contains the location of headquarters and the second of branches, connected by a *shared* corporate ID. The latter is critical, since the observations in the two layers must share a common *key*.

The linkage is established through the **Highlight Association** options in the **Map Layer Settings** dialog. There are three aspects of this: **Set Highlight Association**, **Clear Highlight Association** and **Change Associate Line Color**.

Once the linkage is established, it can be highlighted for any *selection* in the **current map** layer. Each observation in the current map is connected to the matching observation in the other layer, either a point location or the centroid of a polygon. It is important to remember that any selection pertains to the current map layer, irrespective of which layer is shown on top in the map view.

[12] Alternatively, the opacity of the polygon layer can be reduced to zero to make the points visible.

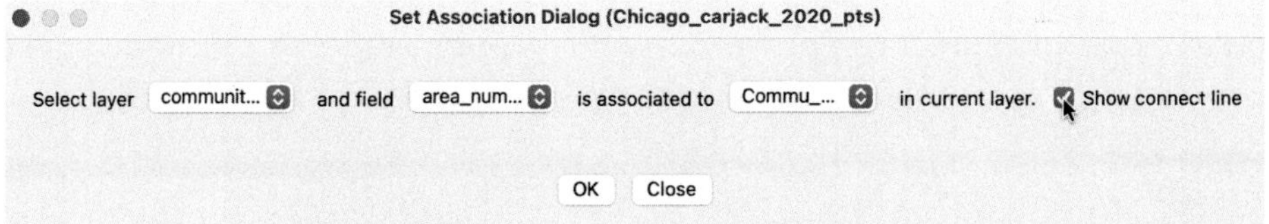

Figure 3.28: Set association dialog

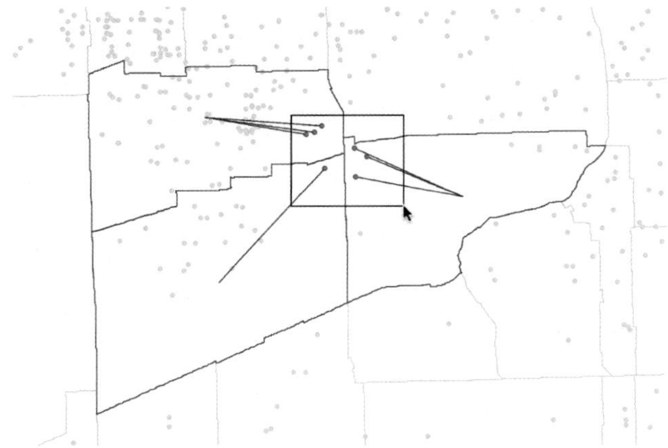

Figure 3.29: Linked selection

3.8.1 Specifying an inter-layer linkage

The layer arrangement from Figure 3.19 has the (projected) car jacking locations loaded first (current map), followed by the (projected) community area boundaries. Observations in both layers share the community area identifier, **Commu_Area** in the point layer, and **area_num_1** in the polygon layer.

Selecting **Set Highlight Association** from the options generates the dialog shown in Figure 3.28. This interface brings all the elements together to connect observations in the two layers through a series of drop-down lists.

The point of departure is the current layer, shown at the top of the dialog (**Chicago_carjack_2020_pts**). The first drop-down list, **Select layer** is to select the layer to link with. In the example, there is only one option, **community_areas_proj**. Next is the key for that layer (**field**), **area_num_1**, and the corresponding key in the current layer (**associated**), **Commu_Area**. Finally there is a check box to activate connection lines between the two layers.

3.8.2 Visualizing linked selections

The choices in the dialog of Figure 3.28 establish the connection between the two layers, but this is only activated through an actual selection in the current map (see Section 2.5.4). For example, in Figure 3.29, point locations have been selected that belong to three different community areas. A line connects each selected point with the centroid of the linked area. The default color is red, but this can be readily changed through the **Change Associate Line Color** option.

The operation can also be reversed, i.e., by making the polygon layer the current map from which observations can be selected, showing the linked points.

Part II

EDA and ESDA

4

Geovisualization

In this first of three chapters dealing with mapping, I begin to explore the concept of geovisualization. Before getting into specific methods, I start with a brief discussion of the larger context of exploratory data analysis (EDA) and especially its spatial counterpart, exploratory *spatial* data analysis (ESDA). Central to this and to its implementation in `GeoDa` are the concepts of *linking* and *brushing*.

The technical discussion begins with an overview of key aspects of thematic map construction, followed by a review of traditional map classifications, i.e., quantile maps, equal interval maps and natural breaks maps. More statistically inspired maps are discussed in Chapter 5. Conditional maps are considered in the treatment of conditional plots in Chapter 8. The particular problem of mapping rates or proportions is covered in Chapter 6.

The common map types are followed by a discussion of various mapping options in `GeoDa` that allow interaction with the map window and the creation of output for use in other media. The chapter closes with an introduction of the implementation of custom classifications and the use of the *project file*.

Even though there is substantial mapping functionality in `GeoDa`, it is worth noting that it is *not* a cartographic software. The main objective is to use mapping as part of an overall framework that interacts with the data in the process of exploration, through so-called *dynamic graphics* (elaborated upon in Section 4.2). By design, maps in `GeoDa` do not have some standard cartographic features, such as a directional arrow, or a scale bar, since they are intended to be part of an interactive framework. However, any map can be saved as an image file for further manipulation in specialized graphics software.

As in the two previous chapters, the discussion here is aimed at novices, who are less familiar with basic cartographic principles. Others may just want to skim the material to see how the functionality is implemented in `GeoDa`. The treatment focuses on gaining familiarity with essential concepts, sufficient to be able to carry out the various operations in `GeoDa`. More technical details can be found in classic cartography texts, such as Brewer (2016), Kraak and Ormeling (2020) and Slocum et al. (2023), among others. Also highly recommended for those not familiar with mapping is Monmonier's *How to lie with maps* (Monmonier, 2018), which provides an easy to read overview of critical aspects of the visualization of spatial data.

To illustrate the various techniques, a new data set is used. It contains information on Zika infection and Microcephaly in municipios of the state of Ceará in northeastern Brazil. This constitutes a subset of the data for the whole of Brazil reported on in Amaral et al. (2019). In addition to the incidence of the two diseases in 2016, the data set also contains several socio-economic indicators from the Brazilian Index of Urban Structure (IBEU) for 2013 (see Amaral et al., 2019, for a detailed definition of each variable). *Ceará Zika* is included as a `GeoDa` sample data set.

DOI: 10.1201/9781003274919-4

4.1 Topics Covered

- Situate the role of geovisualization within a larger exploratory context
- Appreciate the importance of linking and brushing in exploratory spatial data analysis
- Create commonly used thematic maps
- Change legend colors
- Manipulate the map by zooming, panning and selection
- Add a background layer (base map) to a thematic map
- Save map classifications as a categorical variable
- Save the map as an image
- Create a custom classification
- Use the project file in `GeoDa`

GeoDa Functions

- Map > Quantile Map
  - select number of categories
- Map > Natural Breaks Map
  - select number of categories
- Map > Equal Intervals Map
  - select number of categories
- Map toolbar options
- Map selection
  - select all observations in a legend category
- Map options
  - Change current map type
  - Save Categories
  - Save the map as an image
  - Change the look of a map
  - Make map outlines invisible
- Map > Create New Custom
  - Category Editor
- File > Save Project
  - saving a custom classification in the project file

Toolbar Icons

Figure 4.1: Maps and Rates | Cartogram | Map Movie | Category Editor

4.2 From EDA to ESDA

The mapping functionality outlined in this and the next chapters is part of an overall framework to guide and facilitate learning from spatial data, which also includes a range of traditional statistical graphs (see Chapters 7 and 8). `GeoDa` implements general ideas from exploratory data analysis (EDA) and its spatial extension, exploratory spatial data analysis (ESDA). To put these various concepts into context, next, I briefly discuss the evolution of EDA and visual analytics as well as how geovisualization and ESDA fit into this framework. I close with a brief introduction of the important concepts of linking and brushing.

4.2.1 Exploratory data analysis

EDA is generally viewed to have originated in the 1960s as a reaction by computationally oriented statisticians to the primary focus on mathematics and modeling in their discipline. It was first comprehensively outlined in the classic book by John Tukey (Tukey, 1977). His presentation stressed the value of investigating the raw data by means of a range of graphic devices, several of which were developed by Tukey himself. Well-known examples include the box plot and stem and leaf plot.

The objective of EDA is to create effective tools to guide the analyst in the discovery of information in the data, specifically, "indications of unexpected phenomena" or to "display the unanticipated" (Tukey, 1962; Tukey and Wilk, 1966), or even to "discover potentially explicable patterns" (Good, 1983). In this context, EDA is often contrasted with confirmatory data analysis, or CDA. This reflects the traditional distinction in knowledge discovery between an inductive approach (data first, hypothesis later) and a deductive approach (hypothesis/model first, data later). However, with the emphasis on visual exploration (see also Tufte, 1997), EDA is really about an *abductive* approach, where the interaction between data exploration and human perception leads to the detection of patterns jointly with the formulation of hypotheses (e.g., Gahegan, 2009).

Early approaches to visualizing data go back to Greek times, although major innovations did not occur until the work of William Playfair during the late 18th and early 19th century (Friendly, 2008). However, practical visual exploration of large data sets had to wait for the development of powerful computer graphics hardware. This allowed direct interaction with the data shown on a computer screen through so-called dynamic graphics (Becker et al., 1987; Cleveland and McGill, 1988; Cleveland, 1993). Dynamic graphics allowed the data to be represented simultaneously by means of different *views* (Buja et al., 1996), i.e., graphs, tables, charts and even maps that focus on various aspects of the data distribution. Especially when dealing with high-dimensional data, insight is gained through careful manipulation of the views, such as linking, focusing and arranging (for a recent review of the range of techniques, see Chen et al., 2008).

In the computer science literature, a similar and largely parallel development occurred in the form of *visual analytics*, an approach to knowledge discovery based on the use of statistical graphics and other visual tools to facilitate pattern recognition and data mining (Thomas and Cook, 2005; Kielman et al., 2009). As in EDA, in visual analytics much attention is paid to human-computer interaction so as to facilitate analytical reasoning in order to "detect the expected and discover the unexpected" (Kielman et al., 2009, p.245).

Whereas early on EDA and CDA were mostly seen as opposing strategies to gain knowledge, more recently attempts are being made to bring the two closer together. Specifically, a concern that unstructured *exploration* may lead to spurious results has yielded methods to augment EDA in order to provide some measure as to how unusual the findings may be. For example, Buja et al. (2009) and Wickham et al. (2010) suggest a *Rorschach test* and a *line up* to compare what outcomes may be generated under a specific null hypothesis to assess how unusual the graphs and charts obtained in exploration actually are.[1] A Bayesian

[1]The permutation approach toward inference in spatial autocorrelation analysis can be viewed as an implementation of the line up approach. See Chapters 13 and 16.

perspective is taken by Hullman and Gelman (2021), who suggest the use of a graph as a model check. This topic is the subject of ongoing discussion and debate. I revisit some of these ideas in the postscript (Chapter 21).

4.2.2 Mapping as exploration

Historically, cartography, the science (and art) of map making, has focused on the map as a *presentation* or an expository device (Monmonier, 1993). In this context, the map is an end product, created to represent findings, but it is not part of the analytical process itself. In recent years, while this aspect is still important, attention has shifted to making the map an integral part of knowledge discovery from data. This is variously referred to as *geovisualization*, *geovisual analytics*, or even *geospatial visual analytics*. This effort consists of a combination of new analytical tools, visual representations and their software implementation. It represents a shift away from the map as an end product to the integration of the map and a spatial focus in an interactive process of data exploration. In other words, the map becomes one of the views manipulated in a dynamic graphics environment (for a historical perspective, see Anselin, 2005b; 2012).[2]

4.2.3 Exploratory spatial data analysis

Exploratory spatial data analysis (ESDA) similarly incorporates the map and spatial information as an integral part of the data exploration process, but the focus is on *spatial* patterns. As such, ESDA can be viewed as "a collection of techniques to describe and visualize spatial distributions, identify atypical locations or spatial outliers, discover patterns of spatial association, clusters or hot spots and suggest spatial regimes or other forms of spatial heterogeneity" (Anselin, 1999).

ESDA techniques, as implemented in GeoDa thus augment the map as a view of the data with targeted searches for spatial patterns, while leveraging global and local spatial autocorrelation statistics (spatial dependence). In addition, the discovery of the location of structural breaks in the spatial distribution (spatial heterogeneity) is facilitated (early overviews of this literature are contained in Anselin, 1994; 1998; 1999). In Volume 2, spatial constraints are introduced into multivariate clustering techniques.

4.2.4 Linking and brushing

Two concepts central to the way ESDA is implemented in the architecture of GeoDa are so-called *linking* and *brushing* (Anselin et al., 2006b). In Chapter 2, it was shown how observations can be selected, either by creating a query in the *selection tool* (Section 2.5.1), or by means of a *spatial selection* in a map view (Section 2.5.4).

When a selection is made in a map, such as in Figure 2.28, the corresponding observations are also highlighted in the table. The same happens in the other direction as well. When observations are selected in the table, the matching locations (points or polygons) are highlighted in any open map view. The connection between the selections in all open windows is referred to as *linking*. This works not just for a map and associated table (as is the case in many GIS) but also simultaneously for any map view and all the statistical graphs (see Chapters 7 and 8).

[2]Early discussions of these concepts are included in, among others, Haslett et al. (1991), Dykes (1997), MacEachren and Kraak (1997) and MacEachren et al. (1999). Overviews of various techniques can be found in Dykes et al. (2005), Kraak and MacEachren (2005), Rhyne et al. (2006), Andrienko et al. (2011) and Andrienko et al. (2018), among others.

A dynamic version of this process is *brushing*, first proposed for scatter plots in the statistical literature by Stuetzle (1987) and Becker and Cleveland (1987). It was further extended to choropleth maps by Monmonier (1989).

The idea behind brushing is that the selection tool (e.g., the selection rectangle on a map) becomes a moving object. As the rectangle moves over the map, the collection of selected objects is immediately updated. In this way, one can move the selection *brush* over the map (or any statistical graph) and assess the effect of the changing selection.

In `GeoDa`, the concept of brushing is combined with linking in the sense that the updated selection is instantaneously transmitted to all the open windows through the linking process. This provides a very powerful visual tool to assess the effect of the changing selection on various aspects of the spatial and statistical distributions, in both univariate and multivariate settings. The linking-brushing combination is critical to support ESDA. The map plays a central role in this process as an interactive visualization tool, discussed in more detail in the remainder of the chapter.

4.3 Thematic Maps – Overview

4.3.1 Choropleth map

A thematic map is commonly known as a *choropleth* map, from the Greek word *choros*, which stands for area or region. Even though sometimes referred to as chloropleth map, the latter term is incorrect. A choropleth map is a means to visualize the spatial distribution of a variable over discrete areal units.

The proper design of a map involves the manipulation of several parameters, including scale, symbols, legend, intervals and colors (Monmonier, 2018). As mentioned in the chapter introduction, this topic is much too broad to be covered in-depth here. Therefore, only some essential elements are touched upon.

The role of datum and projection was introduced in Section 3.2. The scale of a map, i.e., the correspondence between a distance unit on the map and a matching distance in real-life is an aspect that is ignored in `GeoDa`, since any map window can be resized and/or subject to zooming in and out. This operation changes the scale of the displayed map on the fly. As a result, it becomes impractical to portray these continuous changes in scale.

The three map design parameters that play an important role in `GeoDa` are classification, legend and color. These are briefly reviewed next.

4.3.2 Map classification

A map classification is the process of *binning* the observations from a continuous distribution into discrete categories. Each of these then corresponds to a different color (or shading) on the map. There have been proposals to visually represent the full continuity of the distribution, e.g., by special cross-hatching (Tobler, 1973) or the use of a full color spectrum (Brewer, 1997). However, this quickly becomes impractical for larger data sets. Alternatively, symbols with different sizes and/or colors can be used as well. `GeoDa` only supports discrete categories.

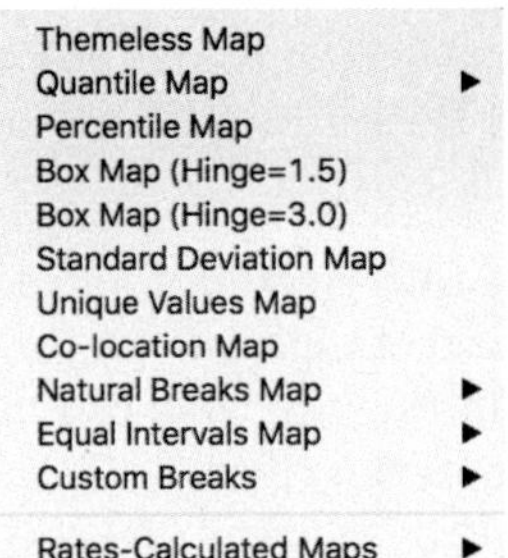

Figure 4.2: Mapping options

In terms of designing a classification, important decisions pertain to the number of categories and how the cut-off or break points are determined. These can be based on external (or exogenous) considerations, such as thresholds determined by policy (e.g., income categories), or, more commonly, endogenously. Endogenous classification exploits the properties of the underlying distribution to come up with meaningful break points. Common methods use the quantiles of the distribution, in a *quantile map*, or divide the range of the variable into a number of *equal intervals*, similar to what is customary for a histogram. An approach based on a clustering logic results in so-called *natural breaks*, i.e., optimal groupings of observations into categories that are most similar internally and maximize the distinction between the categories. These traditional methods are discussed in Section 4.4.

Other classifications that emphasize extreme observations are covered in Chapter 5.

4.3.3 Legend and color

The legend is the way in which the map classification is symbolized, typically positioned next to the map itself. There are two important aspects to the legend. One is the choice of color, which can have major impacts on the perception of value and pattern in the map. For example, red colors are typically associated with *hot* locations, whereas blues are associated with *cold*. More importantly, red is often used to represent or imply danger, especially in political maps. In `GeoDa`, the selection of default color hues and gradations follows the *ColorBrewer* recommendations, which can be found at *colorbrewer2.org* (Brewer et al., 2003; Harrower and Brewer, 2003).

The second aspect is the gradation of values represented in the legend bar. Three broad categories can be distinguished. For maps where the values represented are ordered and follow a single direction, from low to high, a *sequential* legend is appropriate. Such a legend typically uses a single hue and associates higher categories with increasingly darker values. For example, this is the case for the three traditional map classifications reviewed in Section 4.4.

In contrast, for the extreme value maps in Chapter 5, the focus is on the central tendency (mean or median) and how observations sort themselves *away* from the center, either in downward or upward direction. An appropriate legend for this situation is a *diverging* legend, which emphasizes the extremes in either direction. It uses two different hues, one for the downward direction (typically blue) and one for the upward direction (typically red or brown).

Finally, for categorical data, no order should be implied (no high or low values) and the legend should suggest the equivalence of categories. For such discrete categories, a *qualitative* legend is appropriate, as in Section 5.3.1.

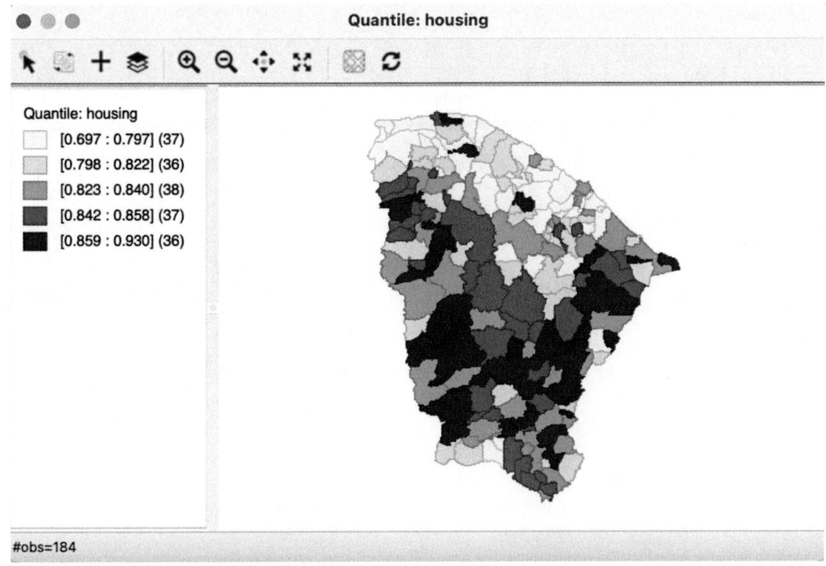

Figure 4.3: Quintile map for housing index, Ceará

4.3.4 Implementation

A thematic map is created in `GeoDa` from the **Map** menu item, or by selecting the left-most toolbar icon in Figure 4.1, **Maps and Rates**. This yields a list of options, shown in Figure 4.2.

In the current chapter, the focus is on traditional map classifications, i.e., the **Quantile Map**, **Equal Intervals Map** and **Natural Breaks Map**. The other options are treated in the next two chapters.

4.4 Common Map Classifications

4.4.1 Quantile map

A quantile map is based on sorted values for a variable that are then grouped into bins. Each bin has the same number of observations, the so-called quantile. The number of bins corresponds to the particular quantile, e.g., four bins for a *quartile* map, or five bins for a *quintile* map, two of the most commonly used categories.

This is illustrated in Figure 4.3 with a quintile map for the housing index in the Ceará data set. The index is scaled between zero and one, with higher values corresponding to better housing conditions.[3] This map is obtained in two steps. First, the classification **Quantile Map > 5** is selected from the options in Figure 4.2. This brings up a dialog to choose the variable to be mapped, in this example **housing**.

[3]The housing condition index is based on the proportion of people living in shanty towns, number of bedrooms with a maximum of two people, number of households with a maximum ratio of four people per restroom, proportion of households whose walls are made of brick or appropriate wood, and proportion of inadequate households. See Amaral et al. (2019) for further details.

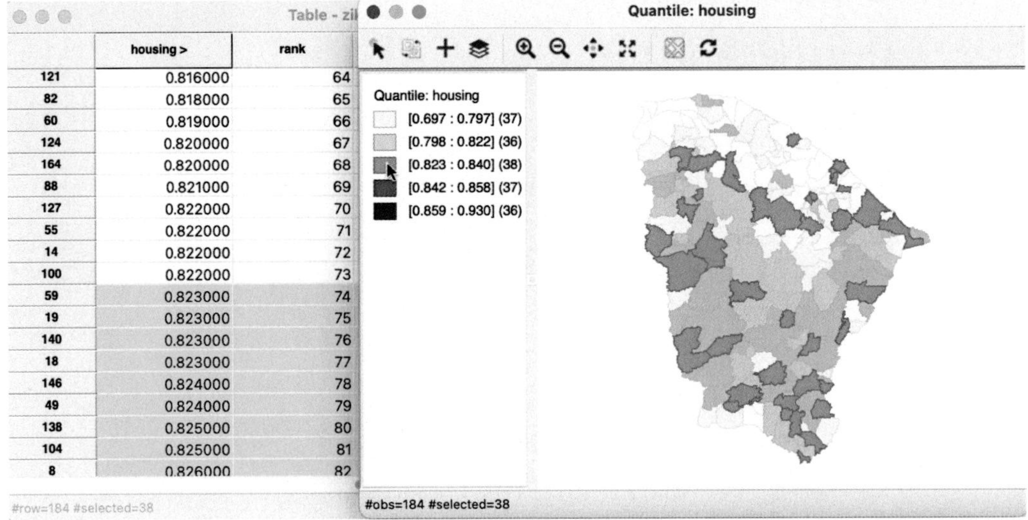

Figure 4.4: Tied values in a quantile map

The legend, to the left of the map, shows the type of classification (**quantile**) and lists the name of the variable (**housing**). The five legend categories correspond with increasingly darker shades of brown. The range of values included in each category is listed in square parentheses and the number of observations in round parentheses. The map suggests that lower values for the housing index tend to be concentrated in the north of the state, with the highest (darkest) values occupying a band in the southern part.

In a quintile map, each category should contain one fifth of the total number of observations, or, $184/5 = 36.8$, so roughly 37 observations. However, in the legend in Figure 4.3, the number of observations varies from 36 to 37 and 38. This is examined more closely in Section 4.4.1.1.

Another important characteristic of the quantile map is that the *range* of values in each category is not constant. In the example, this varies from 0.1 in the lowest quintile to 0.016 for the fourth. The larger the range, the more *heterogeneous* the observations are that were grouped into the same category. In other words, very distinct values may be associated with the same color on the map, which could easily create misleading impressions about the characteristics of the spatial distribution.

4.4.1.1 The problem with quantile maps

In a non-spatial analysis, the computation of quantiles is straightforward. One sorts the data from low to high and looks up the value for the given quantile. In the example, those would be at observations ranked 37, 74, 111, etc. A quick check in the table yields the values 0.797 for observation ranked 37 (actual observation 58), and 0.823 for observation ranked 74 (actual observation 59). However, the second category in the legend goes from 0.798 (observation ranked 38) to 0.822 (observation ranked 73). The true cut-off value of 0.823 is moved to the third quintile.

A closer inspection of Figure 4.4 illustrates the problem. The third category is selected from the legend in the map, with the corresponding observations highlighted in yellow in the table. The observations for **housing** have been sorted (as indicated by the > next to the variable name), with the corresponding **rank** shown in the next column.

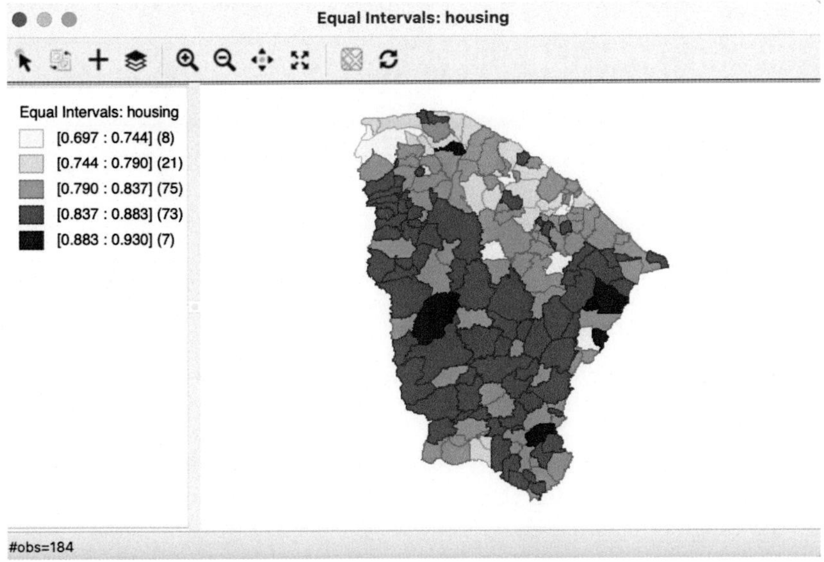

Figure 4.5: Equal intervals map for housing index, Ceará

In a non-spatial analysis, the correct value for the quintile is 0.823, without having to necessarily specify the actual observation that matches this value. However, in a *spatial* analysis, such as a thematic map, each location/observations must be allocated to a map category. In the example, observations ranked 74 (actual 59), 75 (actual 19), 76 (actual 140) and 77 (actual 18) all have the same value. In a thematic map, one cannot arbitrarily decide which of these should be in category 2 and which in category 3. This is the problem of *ties* in the ranking that forms the basis for the computation of the quantiles.

`GeoDa` uses a heuristic that assigns tied observations to the next higher map category. As a result, the second category in Figure 4.3 contains only 36 observations, whereas the third category contains 38.

Even though it is often used as the default setting in thematic mapping software, a quantile map should be interpreted with caution. Widely different value ranges for the quantile categories could mask underlying heterogeneity. In addition, the existence of ties can create problems in practice. For example, when a large number of observations have the same value, some quantile categories may turn out to be empty (`GeoDa` moves tied observations to the next higher category). This is the case for the Zika and Microcephaly incidence variables in the Ceará example data set, where many municipios have an incidence of zero (see also Section 5.3.1).

4.4.2 Equal intervals map

The observations for the housing index go from a minimum of 0.697 to a maximum value of 0.93, resulting in a range of 0.233. In an equal intervals map, this range is divided into a number of bins of equal size. For example, for a five category map, that would yield intervals of $0.233/5 = 0.0466$.

Figure 4.5 illustrates the result for the housing variable with five categories. This is produced as **Equal Intervals Map > 5**, followed by selecting **housing** (the previously selected variable will be the default in the variable selection dialog). The overall design of the map window is the same as for the quantile map.

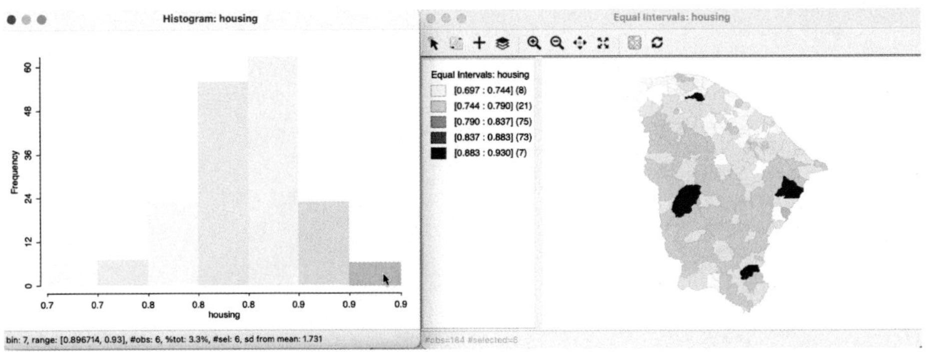

Figure 4.6: Equal intervals map and histogram

In contrast to the quantile map, each category now contains greatly varying numbers of observations, ranging from 7 for the highest category to 75 for the middle category. As intended, the range of each category is exactly 0.0466. Both the lowest and the highest categories contain much fewer observations than in the quantile map.

The patterns suggested by the equal intervals map are quite distinct from those in the quantile map. The overall impression of a band of the lowest index value municipios in the north has been replaced by a scattering of 8 observations, not showing any apparent systematic pattern. Also, the band of 36 observations with the highest category in the quantile map has been replaced by a seven locations, not grouped in any particular way.

The equal intervals map follows the same classification logic as the histogram, illustrated in Figure 4.6. In the Figure, the two graphs are located side by side, with the highest bar selected in the histogram. Through the process of linking, this results in the seven observations from the fifth map category being selected in the map. In practice, inspecting the histogram (or box map) for the variable under consideration is often instructive in suggesting whether a quantile map or equal intervals map is appropriate for the distribution in question.

4.4.3 Natural breaks map

A natural breaks map uses a nonlinear algorithm to group observations such that the within-group homogeneity is maximized, following the path-breaking work of Fisher (1958) and Jenks (1977). In essence, this is a clustering algorithm in one dimension to determine the break points that yield groups with the largest internal similarity, i.e., the smallest internal variance (see also Volume 2). The algorithm to obtain the optimal break points is quite complex, and for large data sets heuristics based on sampling strategies may be necessary (Rey et al., 2013, 2017).

An example is shown in Figure 4.7 for the **housing** variable, using **Natural Breaks Map > 5** to obtain five intervals.

The format of the legend in the natural breaks map differs slightly from the one used in the previous two map types. Each interval is depicted as half open, with the lower value included – shown by the left square bracket [– and the upper value excluded – shown by the right parenthesis). Similarly, the bounds of the lowest and highest category are shown, not the lowest and highest value as before.

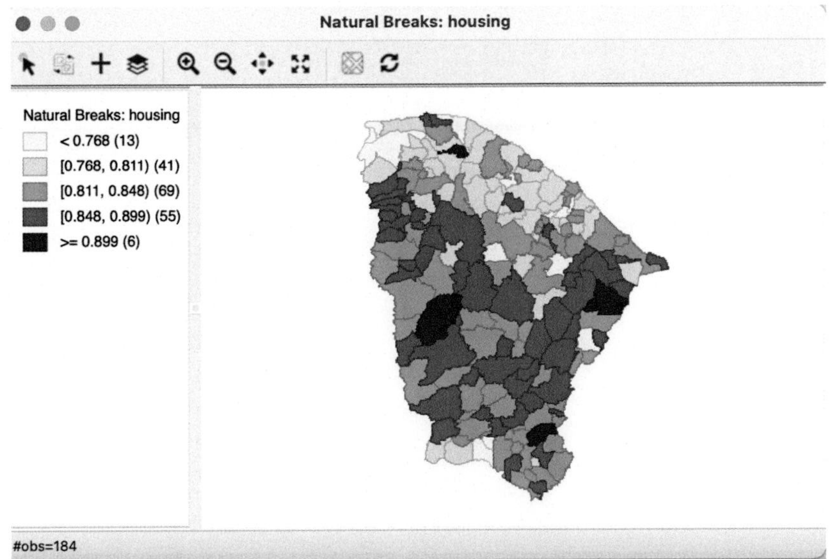

Figure 4.7: Natural breaks map for housing index, Ceará

The patterns suggested by the map have several similar features to the equal intervals map, in that the middle categories contain most observations, and the lowest and highest categories are small. However, the intervals are far from equal, ranging from 0.031 for the fifth category to 0.071 for the first.

The natural breaks bring out a northern pattern for the second category, similar to what was suggested by the first quintile in the quantile map, and again there is a seeming band of higher values. However, in contrast to what the quintile map suggested, the associated observations are not the extremes.

In practice, it is important to go beyond using a single map type, and to compare the similarities and differences between the patterns suggested by the various map classifications.

4.5 Map options

Once a map is created, there are three types of options available. One set pertains directly to the manipulation within the current map window, such as zooming, panning and selecting. This is implemented by means of the icons on the map toolbar (see Figure 4.10 and Section 4.5.2). The second set of options is triggered by right clicking on the map window itself. This invokes the options dialog shown in Figure 4.8, some aspects of which were already encountered in Chapter 3.

The top item (**Change Current Map Type**) brings up the same list with map types as invoked by clicking on the **Map** icon on the main toolbar (see the list in Figure 4.2). Selecting a different map type from a current map window precludes the need to choose a variable, but it also overwrites the current map.

The various features are considered in further detail below, except for the rates functions (**Rates** and **Save Rates**), and operations related to the spatial weights (**Connectivity**),

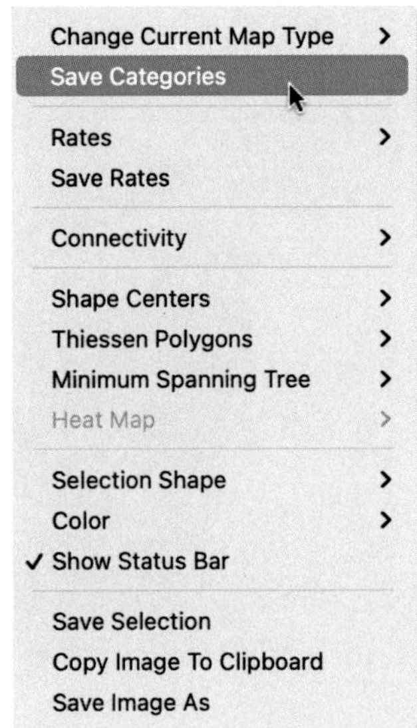

Figure 4.8: Map options

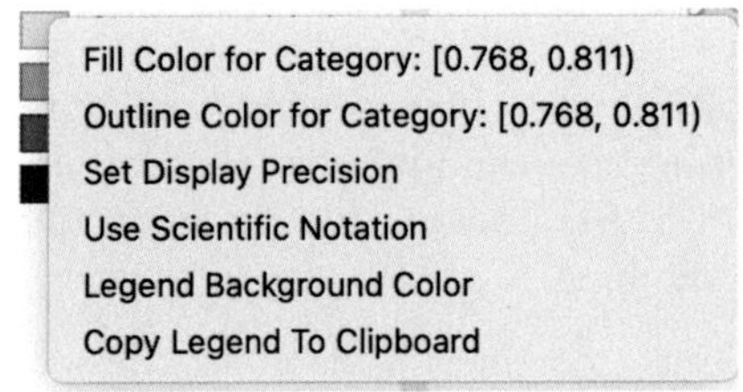

Figure 4.9: Legend options

which are covered respectively in Chapters 6 and 10. Also, the options in the fourth group (**Shape Centers**, etc.), were introduced earlier in Section 3.3.[4]

A final type of customization pertains to the map legend.

4.5.1 The map legend

The map legend color shades are generated automatically from a Color Brewer palette. However, they are also fully customizable, although this is generally not recommended.

A right click on one of the legend rectangles brings up the list of options shown in Figure 4.9.[5] Each of the items in the list allows for the customization of an aspect, i.e., **Fill Color**, **Outline Color**, **Display Precision**, **Scientific Notation** and **Background Color** using the standard interfaces specific to each operating system (color wheels, as well as ways to set an explicit RGB, or CMYK value). In addition, the **Copy Legend to Clipboard** option makes it possible to then paste an image of the legend into any other document.

However, it should be pointed out that for most applications, the shades suggested by Color Brewer should not be changed.

[4]The **Heat Map** option is covered in Chapter 20.

[5]A single click selects the observations that belong to that category.

4.5.2 The map toolbar

The map toolbar contains ten icons that facilitate selection and viewing of the map, shown in Figure 4.10.

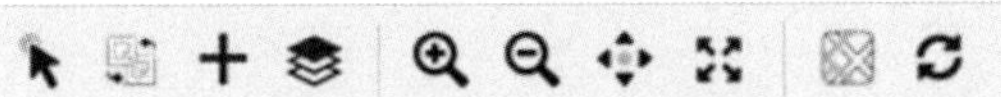

Figure 4.10: Map toolbar

From left to right, the icons allow the following actions:

- **Select**, to select one or more (using shift click) observations, or an observation region by means of a selection shape (the default is a rectangle, see Section 2.5.4); this is the default operation and works in the same fashion as selection on any graph in `GeoDa`
- **Invert Select**, switches to the complement of the current selection (same functionality as **Table > Invert Selection**, see Section 2.5.1.2)
- **Add Map Layer**, add another layer to the current map window, discussed in Section 3.6.1
- **Map Layer Settings**, multiple map layer interface (see Section 3.6.1)
- **Zoom In**, zooms in on the map by drawing a rectangle for the new map extent
- **Zoom Out**, zooms out of the map by repeatedly clicking on the map
- **Pan**, implements panning by dragging the map in any given direction with the pointer
- **Full Extent**, returns the map to its default full extent
- **Base Map**, allows the map to be superimposed onto a base layer that contains roads and other realistic geographic features (see Section 4.5.3)
- **Refresh** the image (in case something went wrong)

These functions are mostly standard and self-explanatory, except for the base map, which is introduced next.

4.5.3 Map base layer (base map)

The choropleth maps created by `GeoDa` are abstractions and lack context. To provide a way to connect the spatial distribution of a variable to the underlying reality on the ground, a *base layer* can be added. This shows geographical features such as roads, rivers and lakes, as well as names of places. The base layer uses the *map tile* concept, essentially a collection of scalable background images (it is referred to as a base map, although it is actually a picture). There are now several sources of such base layers available on the internet. `GeoDa` supports a limited number of these tile layer sources.

In order to add a base layer to the current map window, an active internet connection must be available (without internet, it does not work). The **Base Map** icon is the second from the right in the map toolbar (Figure 4.10). This brings up a list of options, shown in Figure 4.11. The default setting is **No Basemap**.

Currently, `GeoDa` supports six main sources for base layers: **Carto**, **ESRI**, **HERE**, **OpenStreetMap**, **Stamen** and **Other (China)**.[6] Each of these has several sub-options, which can also be customized (see Section 4.5.3.1).

[6]**Other (China)** contains tile data primarily aimed at users in China.

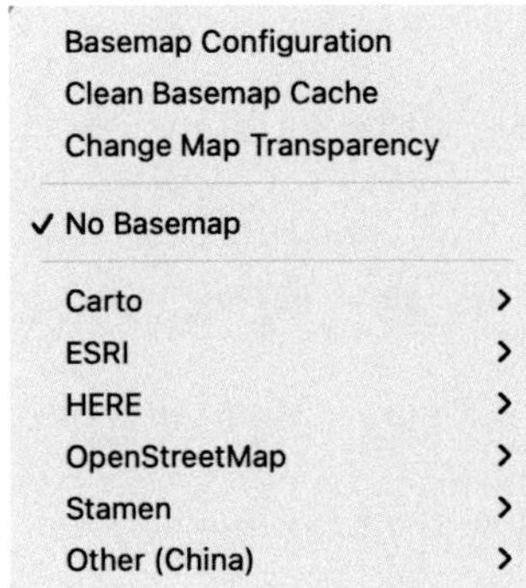

Figure 4.11: Map base layer options

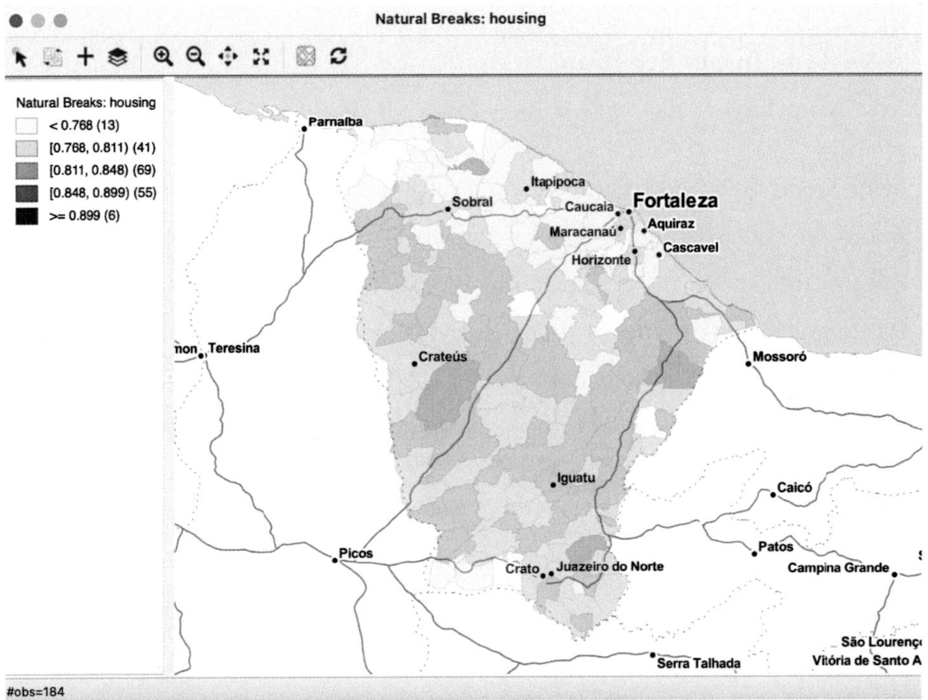

Figure 4.12: Stamen TonerLight base map, default map transparency

For example, continuing with the Natural Breaks map for the housing variable (Figure 4.7), a base map that uses the **Stamen > TonerLite** option (second from bottom in Figure 4.11) with the default map transparency would provide the background illustrated in Figure 4.12. This adds the names of several places as well as some major roads (and the ocean) as the geographical context for the thematic map.

However, the default map transparency of 0.69 does not do justice to the features of the thematic map. The **Change Map Transparency** option (third item in the list in Figure 4.11) provides a slider bar to change the transparency between 1.0 (only the base layer visible) and 0.0 (only the thematic map visible).[7]

[7]Note that the transparency pertains to the base layer, not to the thematic map, hence for a larger value, less of the thematic map will be visible.

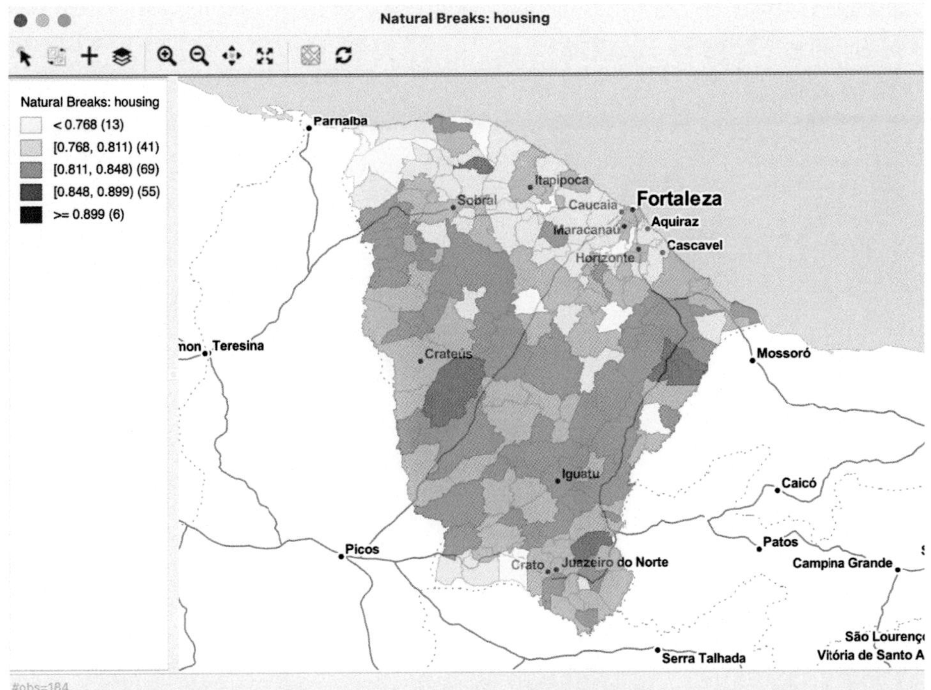

Figure 4.13: Stamen TonerLight base map, 0.40 map transparency

To illustrate this effect, Figure 4.13 shows the same base layer with the transparency set to 0.40. The features of the thematic map are clearly more visible. In practice, some trial and error may be needed to find a configuration that finds the proper compromise between the thematic map and the background layer.

The various base layer sources have quite distinct design features and focus on different aspects of the geography: some have more detailed street information, others include natural features. For example, Figure 4.14 shows a base layer from the **ESRI > WorldStreetMap** option (with transparency set to 0.40). This provides a visual impression that is quite distinct from that in the previous maps.

4.5.3.1 Detailed base layer options

Each of the six base layer sources has several sub-options. These are:

- **Carto**: **Light**, **Dark**, **Light (No Label)**, **Dark (No Label)**
- **ESRI**: **WorldStreetMap**, **WorldTopoMap**, **WorldTerrain**, **Ocean**
- **HERE**: **Day**, **Night**, **Hybrid**, **Satellite**[8]
- **OpenStreetMap**: **MapNik**
- **Stamen**: **Toner**, **TonerLite**, **WaterColor**
- **Other (China)**: **GaoDe**, **GaoDe (Satellite)**

[8]The **HERE** base map support is provided under a generic account for all `GeoDa` users. This account has general usage restrictions, so users may want to set up their own accounts for the **HERE** platform. This can be accomplished by means of the **Basemap Configuration** option. A free **HERE** account can be obtained from https://developer.here.com. The basemap can then be configured with the **App ID** and **App Key**.

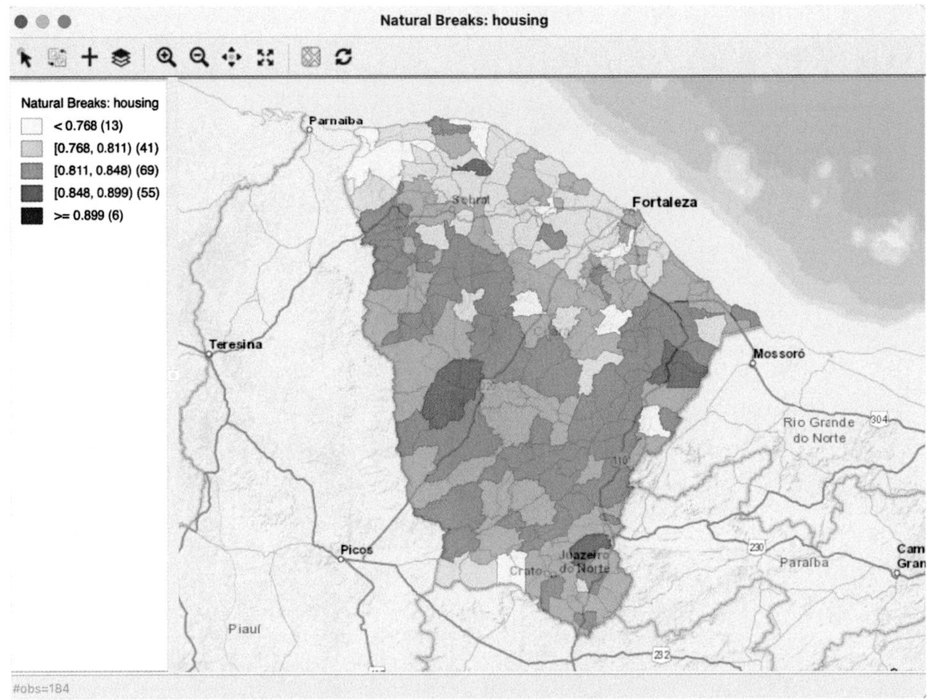

Figure 4.14: ESRI WorldStreetMap base map, 0.40 map transparency

The **Basemap Configuration** option (at the top of Figure 4.11) provides a way to further customize the sources for the base layer map tiles. Each entry corresponds with an entry in the layer options following a format **group_name.basemap_name,basemap_url.**[9] In a typical application, these entries should not be touched. On the other hand, experts are free to customize.

Finally, in a pinch, it may be necessary to use the **Clean Basemap Cache** option when things go wrong (the second option in 4.11).

4.5.4 Saving the classification as a categorical variable

The classification associated with a given map can be added to the data table as a new categorical variable by means of the **Save Categories** option (the second option in Figure 4.8). This brings up a variable selection dialog in which the name for the new variable can be specified (the default is **CATEGORIES**).

Upon selecting OK, a new field is added to the data table. The categories are labeled from low to high, starting with the value 1. In the example that uses natural breaks for the housing variable (Figure 4.7), the range would go from 1 to 5. The resulting categorical variable can be used as input into a **Unique Values Map** or a **Co-location Map** (see Sections 5.3.1 and 5.3.2).

Note that there is no metadata associated with this new variable. As a result, all information is lost regarding important properties, such as the variable on which the classification is based, the type of classification, and the ranges associated with each category. To some

[9]For example, the entry for **Stamen > TonerLite** is: **Stamen.TonerLite,** https://maps.stamen.com/#toner/12/37.7706/-122.3782.

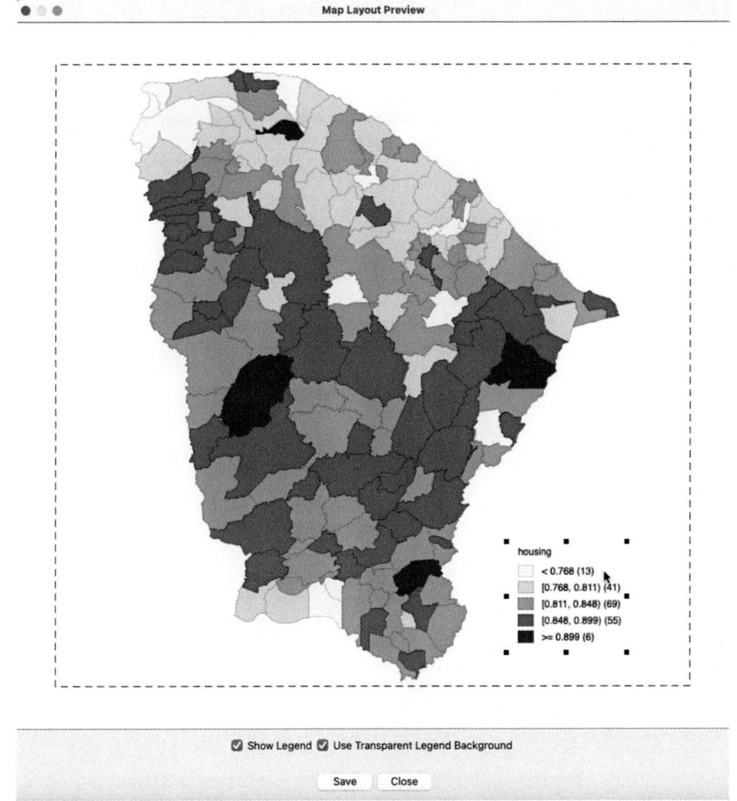

Figure 4.15: Saved Image As dialog

extent, in practice, one can compensate for this lack of metadata by creative naming of variables, although this is limited by the 10 character constraint.

4.5.5 Saving the map as an image

Some other options to save features of the map are provided by the bottom three items in the menu shown in Figure 4.8. **Save Selection** creates an indicator variable for the selected *observations*, as covered in Section 2.5.4.3. For example, this allows one to create a separate indicator variables for each category in the map classification, which is often useful for later statistical analyses.[10]

Copy Image to Clipboard is straightforward and allows for the image to be pasted directly into other documents.

The **Save Image As** option is the closest `GeoDa` comes to publication quality output. It allows for limited manipulation of the map view to then be saved in a standard image output format: **png** (the default), **bmp**, or **SVG**. For example, in Figure 4.15, the legend in the natural breaks map of Figure 4.7 was moved from the top left to the lower right corner. Limited customization is possible, such as moving and resizing legend box, changing its transparency, or even removing it altogether. Also, the image dimensions can be set precisely and the resolution specified (in dpi – the default is 300).

[10]Select the category by clicking on its legend rectangle.

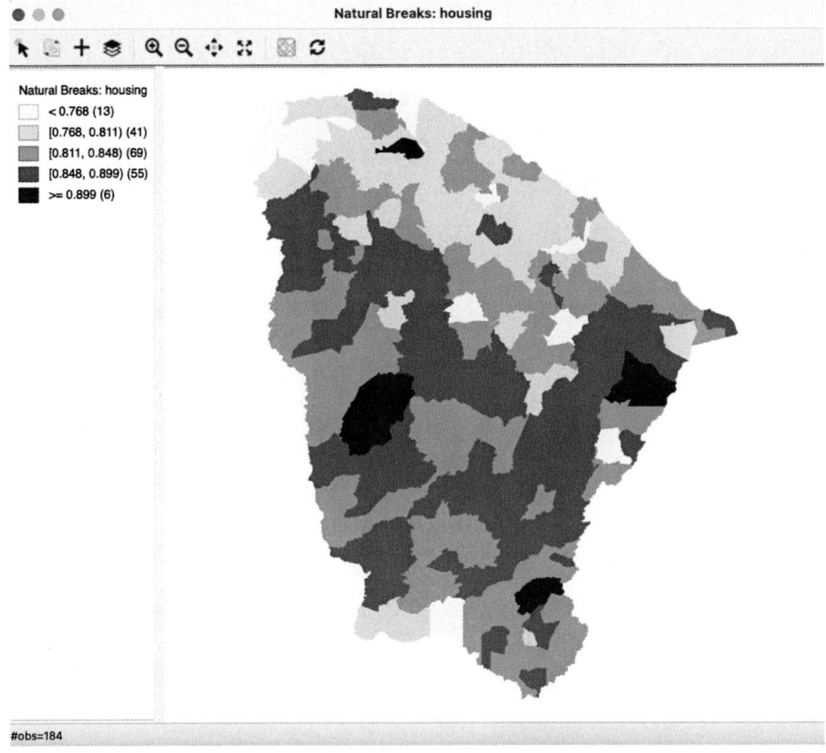

Figure 4.16: Map without outlines visible

The saved image file can be further manipulated in specialized graphics software.

4.5.6 Other map options

The remaining three options in the Figure 4.8 menu pertain to the look of the map view. **Show Status Bar** is on by default, which means that the number of observations and the size of any selection will be displayed on the status bar. The **Selection Shape** is set by default to **Rectangle**, with the other two options as **Circle** and **Line** (see Section 2.5.4.1).

Finally, the **Color** options determine how the map is displayed. By default, the polygon outlines (i.e., in our example, the boundaries of the municipios in Ceará) are shown, with **Outlines Visible** checked,

By unchecking this option, the outlines disappear. This is particularly useful when the map contains many small areas that will tend to be dominated by their boundary lines. In the example in Figure 4.16, this results in an impression of larger regions compared to Figure 4.7, because many adjoining municipios end up in the same category according to the natural breaks classification.

One further option is to **Show Map Boundary**, only available with **Outlines Visible** turned **off**. This option is particularly useful when only a few polygons are colored in a map with many observations, for example, in the local cluster maps (Chapters 16 to 19). Turning off the outlines in such an instance makes the polygons floating in white space. With the **Show Map Boundary** option turned on, the initial outline of the overall map is retained.

The **Background Color** determines the background against which the map is drawn. In most instances, the default of white is the best choice.

4.6 Custom Classifications

So far, the classifications considered for the map legend have been *endogenous*, i.e., they were derived from the distribution of the variable under consideration (e.g., quantile, equal interval, natural breaks). However, such endogenous classification may not be that insightful when comparing the spatial pattern of different variables (on comparable scales, e.g., indices between 0 and 100), or when assessing the change in spatial pattern for a single variable over time. In addition, sometimes substantive concerns dictate the cut points, rather than data driven criteria. For example, this may be appropriate when specific income categories are attached to a policy. In those instances, the policy-based values for the classification are important, rather than what the internal data distribution would suggest.

When assessing the spatial distribution of a variable over time, the endogenous classifications are relative and re-computed for each time period, based on the variable distribution at that time. This can yield misleading impressions when the interest is in the evolution of the actual values. For example, when mapping crime rates over time, in an era of declining rates, the observations in the upper quartile in a later period may have crime rates that correspond to a much lower category in an earlier period. This would not be apparent using the endogenous classifications, but could be visualized by setting the same break points for each time period, i.e., an *exogenous* classification.

In GeoDa, this is accomplished through the **Category Editor**.

4.6.1 Category editor

The **Category Editor** is a complex tool that allows for the creation of a fully customized classification scheme. It can be invoked in a number of different ways. One option is to use the main menu, as **Map > Custom Breaks > Create New Custom**. A second way is as the bottom item from map options menu in Figure 4.2. Alternatively, the category editor can also be invoked directly by selecting the right-most icon on the toolbar in Figure 4.1.

4.6.1.1 Design

The custom categories are designed through the interface in Figure 4.17, shown after all editing has been completed. The initial settings and editing operations will be discussed below.

There are slight differences in the behavior of the dialog, depending on whether it is invoked from the toolbar, or as an option from an existing map. In the latter case, the associated map will be updated as a new classification is being developed. When invoked from the toolbar, there is nothing to update, but the initial variable may not be meaningful. It is the first variable in the data table, which is often just an identifier for each observations.

When the dialog opens (or at any point when the **New** button is selected), the first query is for the **New Categories Title**, where a name for the new classification must be specified (the default is **Custom Breaks**). In the example in Figure 4.17, **Custom1** is the name of the new classification.

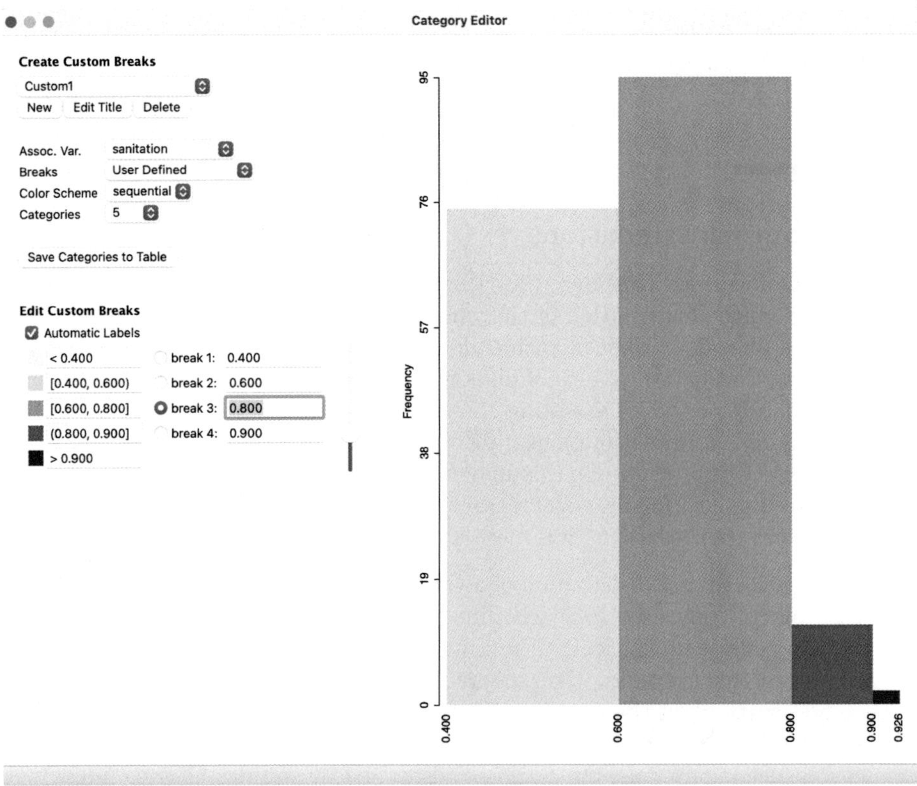

Figure 4.17: Category editor interface

There are three major functions in the interface. First is the general definition of the classification, carried out through the items in the **Create Custom Breaks** panel at the top. This includes the naming of the categories, here **Custom1**.

The computations for the classification and their visualization in a histogram are based on the values for a specific variable, the **Assoc Var.**, here **sanitation**. When the custom breaks are invoked from an existing map, the variable is entered automatically, but it is possible to change it from the drop-down list.

As the breaks are edited, the impact on the associated *histogram* for the classification variable is shown instantly on the right hand side of the dialog.

Other important categories in the interface are whether the **Breaks** are **User Defined** (the default), or follow a traditional classification (**Quantile**, the default, **Unique Values**, **Natural Breaks**, or **Equal Intervals**). These choices may seem counter intuitive for a custom break editing operation, but it allows for the creation of custom labels for the categories (Section 4.6.1.3).

The **Color Scheme** for the legend is either **sequential**, **diverging** (the default), **thematic** (i.e., categorical) or **custom**. This provides an automatic selection of legend colors based on the ColorBrewer palettes. However, the color for each category can also be specified by clicking on its box below **Edit Custom Breaks** by means of the standard color editor (see Section 4.5.1). Finally, the number of **Categories** must be specified (the default is 4).[11]

[11]The classification can also be saved as a new variable in the table, using the **Save Categories to Table** button.

In the example, in Figure 4.17, the color scheme has been set to **sequential** and the number of **Categories** to **5**. The main effort consists of determining appropriate break points, which is considered next.

4.6.1.2 Editing break points

The variable **sanitation** is a constructed index with values between 0 and 1. The observed range for the municipios in the State of Ceará goes from 0.417 to 0.926. These values are shown at, respectively, the left-most point and right-most point at the bottom of the histogram in Figure 4.17. An initial set of break points is suggested, which is almost never the final result. Instead of the suggested values, some absolute cut-offs must be entered, such as 0.4, 0.6, 0.8 and 0.9. This will allow for the comparison of the absolute position of each municipio across different variables, not just their relative position (which is produced by all endogenous classifications).

As new values for the break points are entered, the histogram is immediately updated. The values can be typed in, or obtained by moving the slider bar to the right.

The new break points are immediately associated with the specific custom break definition, without a saving operation. However, they can only be preserved through the use of a project file (see Section 4.6.3). The corresponding distribution is shown in the histogram in Figure 4.17.

4.6.1.3 Custom category labels

When new break points are defined, the category labels are immediately updated to the corresponding intervals, as long as **Automatic Labels** is checked. With that box unchecked, any new label can be specified. This new label will then be used in any map legend.

4.6.2 Applications of custom categories

To illustrate the application of custom categories, consider the variables **sanitation** and **infra** (infrastructure). Both are obtained from the Brazilian Index of Urban Structure (IBEU) for 2013. They take on values from 0 to 1. The relative position of the municipios in Ceará on those two variables can be visualized by means of a quintile map, as in Figure 4.18. The spatial patterns are different, but some high performing municipios overlap.

Contrast this pattern with the maps in Figure 4.19, based on the absolute cut-off points of 0.4, 0.6. 0.8 and 0.9. Clearly, the municipios do not perform as well on the infrastructure index as they do on the sanitation index. For example, the latter has 0 observations in the lowest category (< 0.4), whereas for infrastructure there are 16 such observations. Also, infrastructure has 0 municipios in the top category (> 0.9), whereas sanitation has 2. This allows for the comparison of not only the relative performance of the municipios but also how they meet some absolute standards (admittedly arbitrary in this example).

4.6.3 Saving the custom categories – the Project File

When a project is closed, the information on the custom classification is lost, unless it is first saved in a so-called **Project File**. This file, associated with a geographic input layer, contains information on various variable transformations and other operations, most importantly dealing with spatial weights (see Chapter 10). It also contains the definition of any custom categories that were created. If this definition is not saved in a project file,

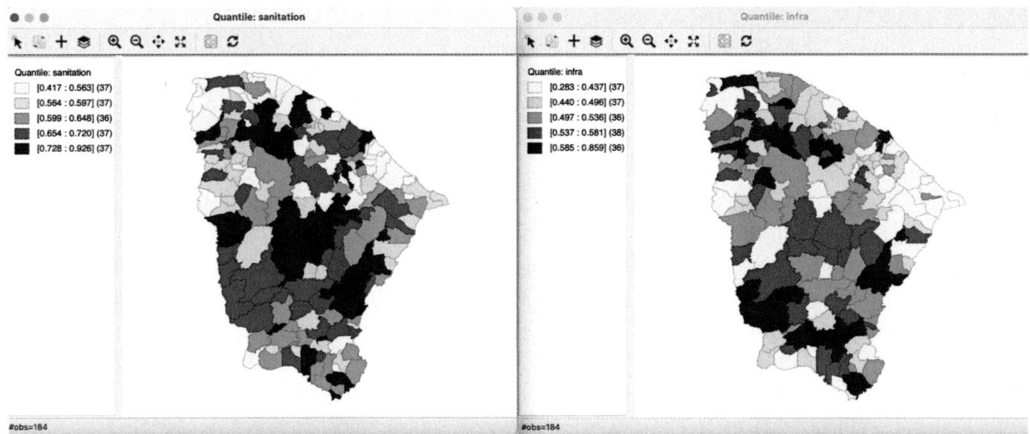

Figure 4.18: Quintile maps for sanitation and infrastructure

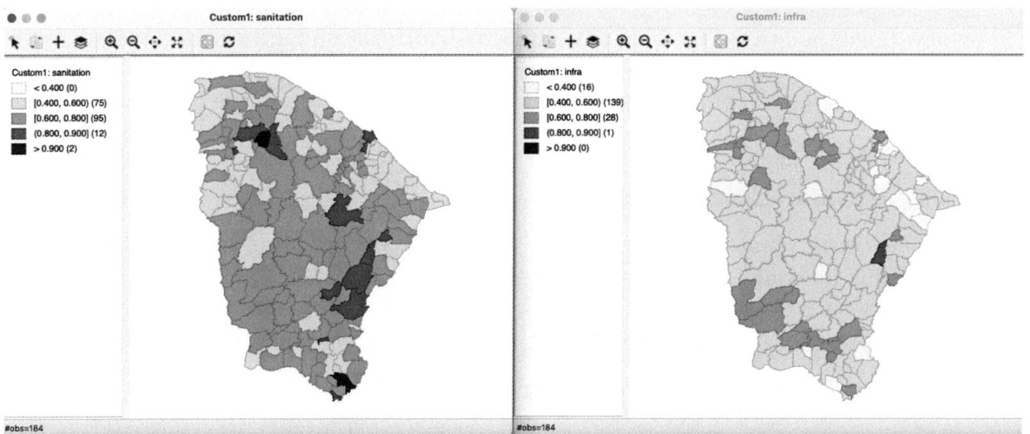

Figure 4.19: Thematic maps for sanitation and infrastructure using custom classification

then it will be lost and will need to be recreated from scratch the next time the data set is analyzed.

The project file is created from the menu as **File > Save Project**. The file is saved with a file extension of **gda**. It is a text file that includes XML encoding. For example, a project file associated with the Ceará data could be **zika_ceara.gda**. Closer inspection reveals the section pertaining to **custom_classifications** in Figure 4.20.

The custom classification section contains all the aspects needed for the definition of the custom category.

When an analysis is started with the project file as input (e.g., instead of a shape file) then all transformations and custom categories become immediately available as an option for any map or histogram.

```
<custom_classifications>
    <classification_definition>
        <title>Custom1</title>
        <type>custom</type>
        <num_cats>5</num_cats>
        <automatic_labels>true</automatic_labels>
        <assoc_db_fld_name>sanitation</assoc_db_fld_name>
        <uniform_dist_min>0.41699999999999998</uniform_dist_min>
        <uniform_dist_max>0.92600000000000005</uniform_dist_max>
        <breaks>
            <break>0.40000000000000002</break>
            <break>0.59999999999999998</break>
            <break>0.80000000000000004</break>
            <break>0.90000000000000002</break>
        </breaks>
        <names>
            <name>&lt; 0.400</name>
            <name>[0.400, 0.600)</name>
            <name>[0.600, 0.800]</name>
            <name>(0.800, 0.900]</name>
            <name>&gt; 0.900</name>
        </names>
        <colors/>
        <color_scheme>sequential</color_scheme>
    </classification_definition>
</custom_classifications>
```

Figure 4.20: Custom category definition in project file

5

Statistical Maps

In this Chapter, I continue the exploration of mapping options with a focus on statistical maps, in particular maps that are designed to highlight extreme values or outliers. Some of these classifications were originally introduced in `GeoDa` and are gradually being adopted by other exploratory software (e.g., the Python PySAL library). Their layout illustrates the emphasis on statistical exploration rather than cartographic design. Specifically, this includes the **Percentile Map**, **Box Map** (with two options for the hinge) and the **Standard Deviation Map**.

Other topics covered in this Chapter include maps for categorical variables in the form of a **Unique Values Map**. Construction of this type of map does not involve a classification algorithm, since it uses the integer values of a categorical variable itself as the map categories. The **Co-location Map** is an extension of this principle to multiple categorical variables.

The Chapter closes with a brief discussion of the **Cartogram** and map animation (**Map Movie**), which move beyond the traditional choropleth framework to visualize a spatial distribution.

I continue to use the *Ceará Zika* sample data set to illustrate the various features.

5.1 Topics Covered

- Create extreme value statistical maps
- Identify outliers using the box map and standard deviation map
- Create a map for a categorical variable
- Examine multivariate co-location patterns with a co-location map
- Construct and interpret a cartogram
- Visually explore patterns using map animation

GeoDa Functions

- Map > Percentile Map
- Map > Box Map
  - Set hinges as 1.5 or 3.0
- Map > Standard Deviation Map
- Map > Unique Values Map
- Map > Co-location Map
- Map > Cartogram
  - improving the cartogram fit
- Map > Map Movie
  - setting animation controls

DOI: 10.1201/9781003274919-5

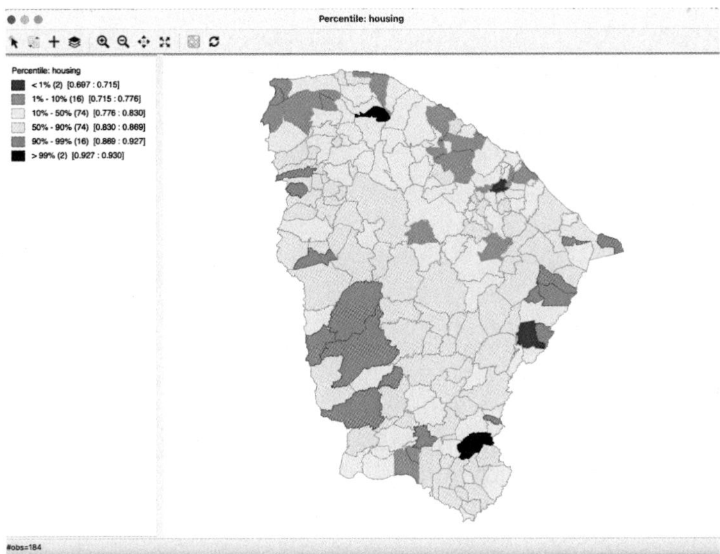

Figure 5.1: Percentile map

5.2 Extreme Value Maps

Extreme value maps are variations of common choropleth maps where the classification is designed to highlight observations at the extreme lower and upper end of the scale. The objective is to identify and highlight *outliers*. These maps were developed in the spirit of *spatializing EDA*, i.e., adding spatial features to commonly used approaches in non-spatial EDA (Anselin, 1994).

`GeoDa` currently supports three such map types in the **Map** menu: a **Percentile Map**, a **Box Map** and a **Standard Deviation Map**. These are briefly described below. Only their distinctive features are highlighted, since they share all the same options with the traditional thematic map types (Section 4.5). It should be noted that the extreme value maps are examples of a *diverging* legend, whereas the conventional maps have a *sequential* legend.

The extreme value maps are the third to sixth items in the map menu of Figure 4.2.

5.2.1 Percentile map

The percentile map is a variant of a quantile map that would start off with 100 categories. However, rather than having these 100 categories, the map classification is reduced to six ranges, the lowest 1%, 1–10%, 10–50%, 50–90%, 90–99% and the top 1%. This is shown in Figure 5.1 for the **housing** variable from the Ceará sample data set.

This map is created by selecting **Map > Percentile Map** from the map menu or the map toolbar icon and specifying the variable.

Compared to the traditional map classifications discussed in Section 4.4, the extreme values are much better highlighted. Both the bottom and top 1% contain two observations, with a very small range, respectively 0.697 to 0.715 and 0.927 to 0.930. Interestingly, these extreme

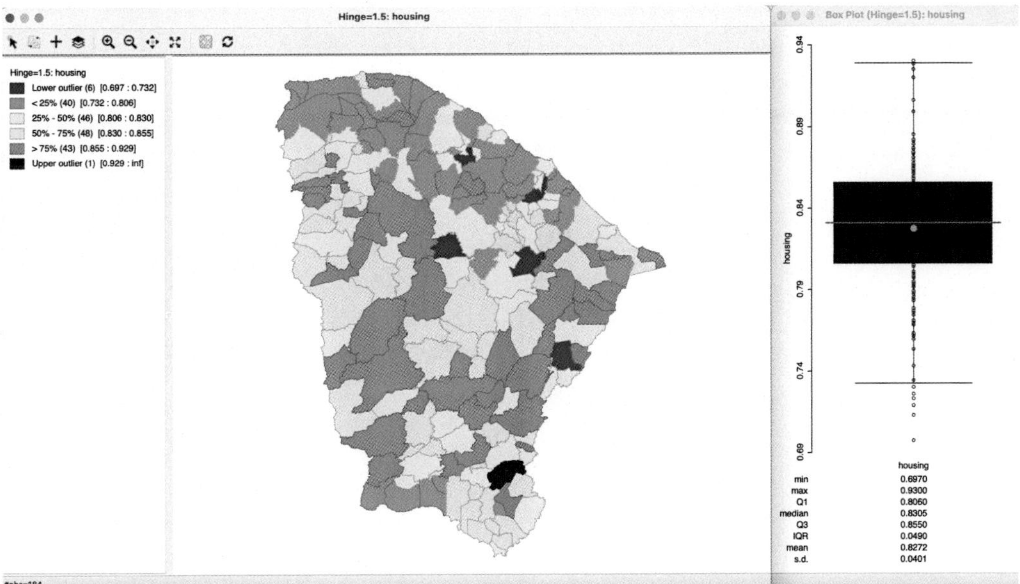

Figure 5.2: Box map, hinge = 1.5, with box plot

values are not closely together in space, countering any suggestion of clustering. However, at least in one case (the northernmost upper percentile observation), a very high value is surrounded by much lower ones, suggesting the potential of a *spatial outlier*.

With the focus on the large range of middle values (i.e., 10–50 percentile and 50–90), two large bands of adjoining municipios in the same category seem to manifest themselves, with below median values in the northeast and above median values forming a U-shape in the center of the state.

As for any quantile map, there are some drawbacks to this approach related to ties and potential heterogeneity of values within the same category. In addition, a percentile map only makes sense when there are more than 100 observations, which is the case here. It is particularly useful to identify the location in space of the truly extreme observations.

5.2.2 Box map

Whereas the category bounds in the percentile map are to some extent arbitrary, in the box map these are connected to the visualization of the distribution of a variable in a box plot (see Chapter 7).

The box map (Anselin, 1994) is thus the mapping counterpart of the idea behind a box plot. The point of departure is again a quantile map, or, more specifically, a *quartile* map. The four categories are extended to six bins, to separately identify the lower and upper outliers. The definition of outliers is a function of a multiple of the inter-quartile range (IQR), the difference between the values for the 75 and 25 percentile. As is customary, there are two options for these cut-off values, or *hinges* in a box plot: 1.5 and 3.0. The box map uses the same convention.

The box map is created by selecting **Map > Box Map (Hinge=1.5)** from the map options and specifying the variable. For the Ceará housing index variable, this yields the map shown

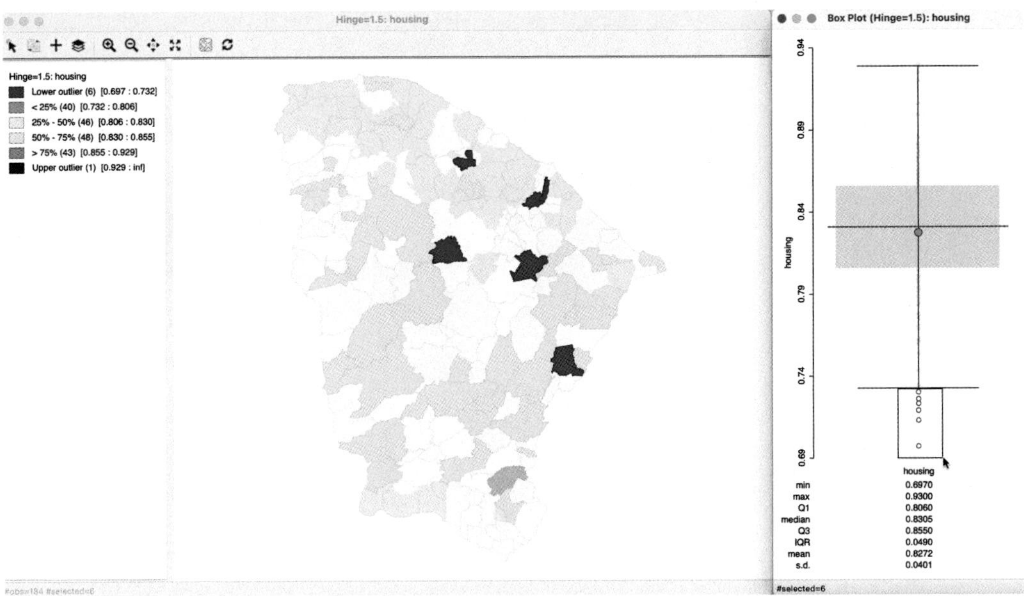

Figure 5.3: Outliers in box plot and box map

in the left panel of Figure 5.2. For easy reference, the corresponding box plot is contained in the right panel.

Compared to a standard quartile map, the box map in Figure 5.2 separates out six lower outliers from the other 40 observations in the first quartile. They are depicted in dark blue. Similarly, it separates a single upper outlier from the 43 other observations in the upper quartile. The upper outlier is colored dark red. The main focus of interest in a box map is to identify the extent to which the outliers show any kind of spatial pattern. In this example, there does not seem to be a suggestion of clustering, but possibly of the presence of spatial outliers (this is explored more formally in Chapters 16 to 19). This constitutes the *spatial* perspective that the *box map* adds to the data exploration.

To further illustrate the correspondence between the box plot and the box map, in Figure 5.3, the six lower outliers are selected in the box plot in the right hand panel. Through linking, these are also highlighted in the map and correspond exactly to the lower outlier category. Again, there is a fundamental difference between a traditional box plot where outlying observations are identified, and the box map, where their *location* is taken into account as well.

From the current map, the box map can be switched between the hinge criterion of 1.5 and 3.0 by opening the options menu (right click on the map) and selecting **Change Current Map Type > Box Map (Hinge = 3.0)**. Alternatively, a new map window can be opened from the main menu or map toolbar icon by selecting **Box Map (Hinge=3.0)** as the option.

The box map is arguably the *preferred method* to quickly and efficiently identify outliers and broad spatial patterns in a data set, although it does suffer form the same drawbacks as any other quantile map.

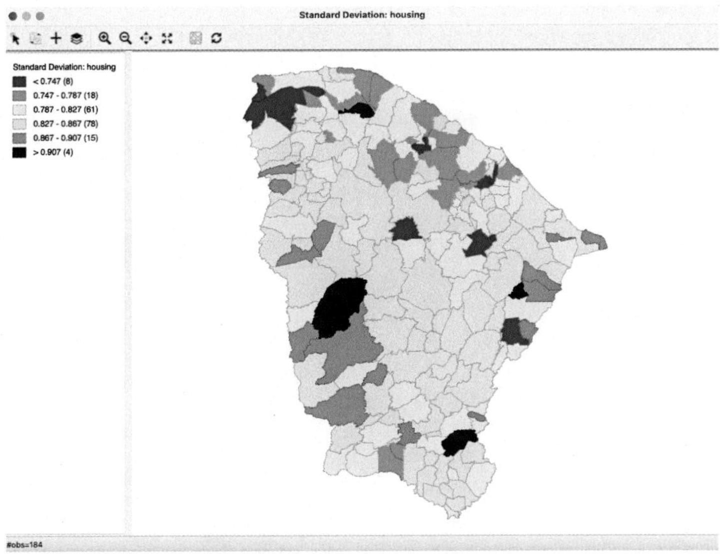

Figure 5.4: Standard deviation map

5.2.3 Standard deviation map

The third type of extreme values map is a standard deviation map. In some way, this is a *parametric* counterpart to the box map, in that the standard deviation is used as the criterion to identify outliers, instead of the inter-quartile range.

In a standard deviation map, the variable under consideration is represented as standard deviational units (with mean 0 and standard deviation 1). This is equivalent to the z-standardization (see Section 2.4.2.3).

The number of categories in the classification depends on the range of values, i.e., how many standard deviational units cover the range from lowest to highest. It is possible that some categories do not contain any observations, since there may be gaps in the distribution for a given standard deviational range.[1]

The relevant option is **Map > Standard Deviation Map**, which yields the map in Figure 5.4 for the Ceará housing index example.

In the example, there are six categories, each with a range of 0.040. There are eight observations that are more than two standard deviational units away from the mean in the lower direction, colored dark blue. As in the other extreme value maps, they do not seem to show any particular spatial pattern.

In the upward direction, there are four observations more than two standard deviational units from the mean, which is quite a bit more than identified with the box map. The one observations in the north seems to be both an outlier in the value distribution as well as a *spatial* outlier. Otherwise, there is no indication of any clustering of the extremes.

On the other hand, when focusing on the central values in the distribution, again there is the suggestion of a northern band of lower values and a U-shape pattern of observations within one standard deviation above the mean.

[1]For example, this would be the case when outlying observations are more than one standard deviational unit away from the rest of the distribution.

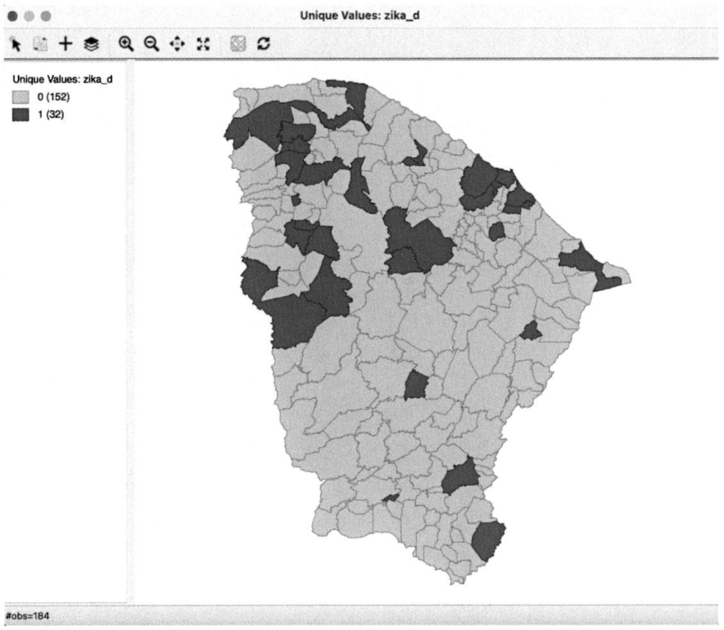

Figure 5.5: Unique values map, Zika indicator

In practice, it is often useful to compare the outliers identified by both the non-parametric (box map) and parametric (standard deviation map) approaches. Of particular importance is whether extreme values show an interesting spatial pattern, which can then be more formally investigated by means of the local spatial autocorrelation indicators in Chapters 16 to 19.

5.3 Mapping Categorical Variables

In the discussion so far, the variable of interest (**housing**) was continuous, with a clear ordering from low to high. However, one often deals with indicator variables (0–1), or, even more generally, with variables that represent multiple categories. For the latter, each category is distinguished by a different integer value, but those values are typically not meaningful in and of themselves. Most importantly, the numerical values do not imply any ordering of the categories.

For a single variable under consideration, a map of its spatial distribution is created through the **Unique Values Map** option, the seventh item in the list in Figure 4.2. The functionality is illustrated with an indicator variable constructed from the cumulative incidence of Zika during the first three quarters of 2016 in the state of Ceará (**zika_d**).

A **Co-location Map** (eighth item in Figure 4.2) extends this to the comparison of multiple categorical variables, using the logic of *map algebra* (Tomlin, 1990). This is illustrated with a comparison of the cumulative incidence of Zika (**zika_d**) with the presence of Microcephaly in the fourth quarter of 2016 (**mic_d**).

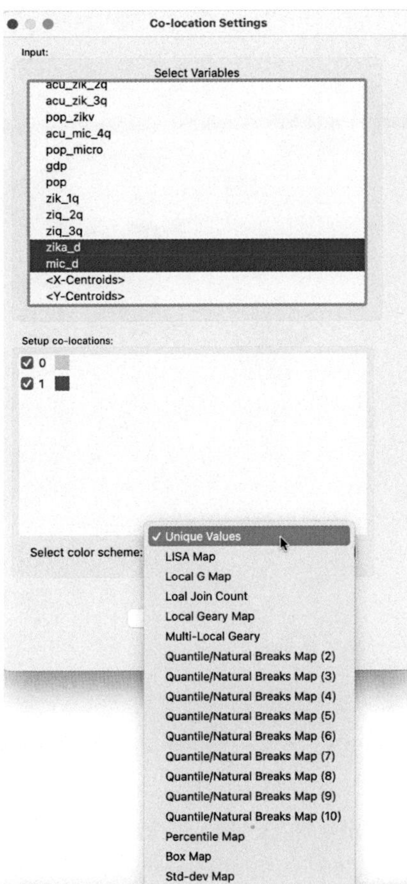

Figure 5.6: Co-location map variable selection, Zika and Microcephaly indicators

5.3.1 Unique values map

A map for a categorical variable is created by invoking **Map > Unique Values Map** and specifying the variable. Note that only variables that take on integer values are included in the drop-down list.[2]

In the example in Figure 5.5, the categorical variable is **zika_d**. It only takes on the values of 1 (presence) and 0 (absence), reflected in the two colors in the map legend. The legend colors are generated from the *ColorBrewer categorical map palette*.[3] As mentioned in Section 4.5.1, they can also be customized.

The map indicates that 32 of the 184 municipios had an incidence of Zika during the period under consideration. There seems to be a suggestion of a greater presence in the northern part of the state, with some groupings consisting of adjoining locations. A more formal assessment of these patterns is investigated in Chapter 19.

All the standard map options apply to the unique values map. One additional option is to change the order (and color) of the categories. Since the categories are just that, and the

[2] If a variable that should be categorical appears to be missing from the list, it may have been formatted as real. This can be readily changed using **Edit Variable Properties** in the table, see Section 2.4.1.1.

[3] http://colorbrewer2.org/#type=qualitative&scheme=Accent&n=3

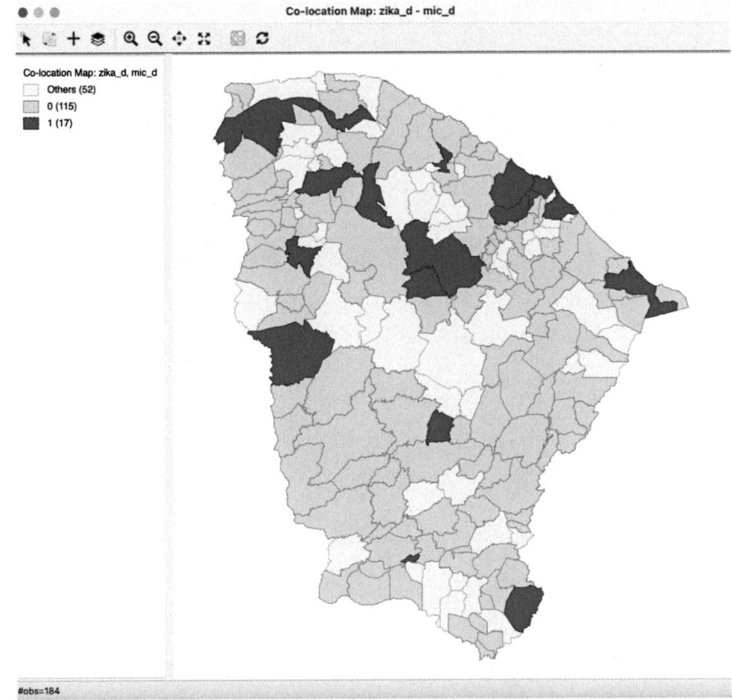

Figure 5.7: Co-location map, Zika and Microcephaly incidence

numerical values associated with them do not imply any ordering, they can be changed. This is accomplished by *grabbing* the associated legend rectangle and moving it up or down in the list.

Such reordering can be handy in a so-called *cluster map* where the categories correspond to different cluster classifications. The comparison of two cluster maps can be facilitated by moving the categories around so that the same colors more or less correspond to the same locations.[4]

5.3.2 Co-location map

A co-location map combines the information from multiple categorical variables into a unique values map that shows those locations where the categories match. This is an example of map algebra, but applied to irregular spatial units rather than the more customary raster data.[5]

In essence, the process boils down to finding those locations where the *codes* for different categorical variables match. This is handled slightly differently for the simple case of binary variables and the more complex situation with multiple categories.

5.3.2.1 Binary categories

When the variables under consideration take on binary values, a co-location map can be constructed *by hand* as a unique values map of the product of the respective indicator

[4]This is covered in Volume 2.

[5]An exhaustive treatment can be found in Tomlin (1990).

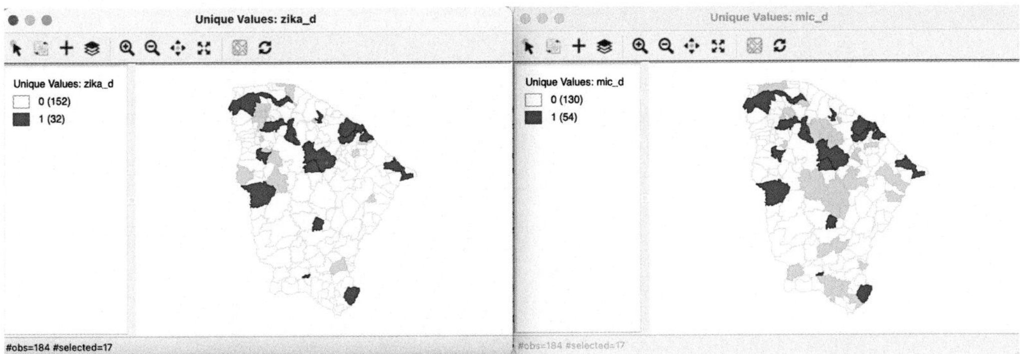

Figure 5.8: Selection of co-located observations

variables. Only those locations that take on a value of 1 for all variables will be coded as one in the resulting unique values map.

However, a co-location map is different in that it also provides information on matches of locations with 0 for all variables, as well as on the locations of mismatches. In sum, rather than just two categories (match and no match), there are three: match of 1, match of 0 and mismatch.

The map is invoked in the usual way as **Map > Co-location Map**, which brings up a dialog to specify the different variables to be considered, as in Figure 5.6. The interface is used to select the variables but also to choose the associated legend structure. Several options are provided, ranging from **Unique Values**, the suggested default in this case, to several customized legends associate with different types of maps (such as extreme values maps) and visualizations of spatial analyses (e.g., LISA Map, see Chapter 16).

In our example, the **Unique Values** default will do fine.

The two variables selected are indicator variables for the presence of respectively Zika (**zika_d**) and Microcephaly (**mic_d**). The corresponding co-location map is shown in Figure 5.7.

The legend contains three categories. One pertains to those locations with a common occurrence of 1 (17 observations), and another to those observations that share a value of 0 (115 observations). The third category (in grey) highlights the locations where there is a mismatch between the values (52 observations).

The logic of the co-location map is further illustrated in Figure 5.8. It shows the selected observations in each of the respective unique values maps that correspond to locations with a value of 1 in the co-location map. There are 17 observations selected in each of the maps. Clearly, these are the only locations where both maps have a dark blue color.

5.3.2.2 Multiple categories

When the variables pertain to multiple categories, the logic of the co-location map is the same, but it must be applied with caution. It is based on the equality of the categorical codes for each variable.

It is up to the user to ensure that the categories across variables are meaningful, since the *co-location* is based on the variables having the same code. For example, this is useful when comparing the extent to which the quartiles across different variables occur at the same

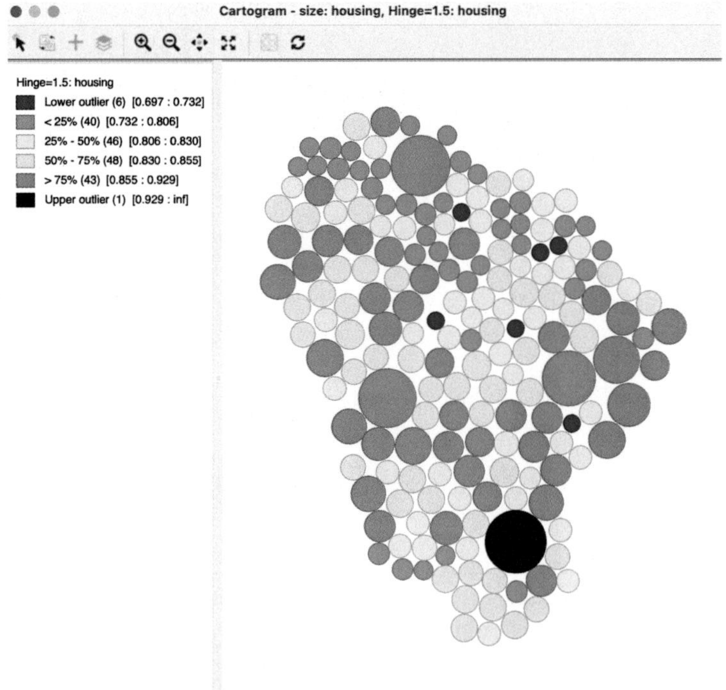

Figure 5.9: Cartogram

locations. Or, to assess whether significant patterns of local spatial autocorrelation match across multiple variables. But it is also very easy to generate nonsensical results, for example, when the labels are not comparable.

One can change the label and color of a category by moving the legend rectangle up or down in the legend. This again highlights that the category values do not have any intrinsic numerical value.

5.4 Cartogram

5.4.1 Principle

A cartogram is a map type where the original layout of the areal unit is replaced by a regular geometric shape (usually a circle, rectangle, or hexagon) that is proportional to the value of the *variable* for the location. This is in contrast to a standard choropleth map, where the size of the polygon corresponds to the *area* of the location in question. The cartogram has a long history and many variants have been suggested, some quite creative (see Dorling, 1996; Tobler, 2004, for an extensive discussion of various aspects of the cartogram). In essence, the construction of a cartogram is an example of a nonlinear optimization problem, where the geometric forms have to be located such that they reflect the topology (spatial arrangement) of the locations as closely as possible.

`GeoDa` implements a circular cartogram, in which the areal units are represented as circles, whose size (and color) is proportional to the value observed at that location. The changed

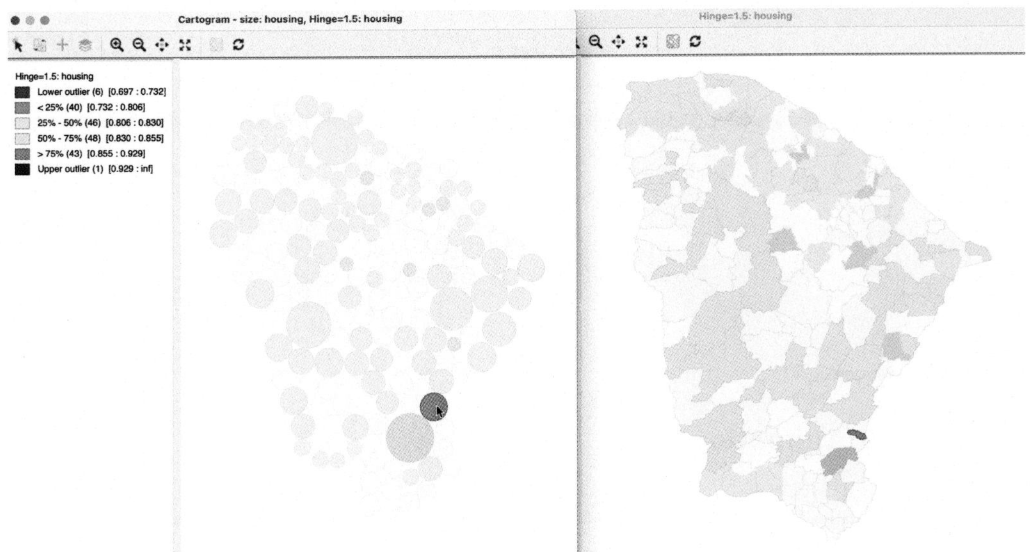

Figure 5.10: Linked Cartogram and map

shapes remove the misleading effect that the area of the unit might have on perception of magnitude.

5.4.2 Creating a cartogram

The cartogram is created from the main menu, as **Map > Cartogram**, or by selecting the second item in the toolbar shown in Figure 4.1.

This brings up a dialog to specify the **Cartogram Variables**, both for the **Circle Size** and the **Circle Color**. In most applications, those should be the same variable. In the example shown in Figure 5.9, the variable used is again the **housing** index.

The default is a **Box Map** classification for the circle colors (with hinge at 1.5 IQR).

The cartogram is most insightful when used in conjunction with a regular choropleth map. Selecting an observation in the cartogram then immediately links it with the corresponding area in the choropleth map. In Figure 5.10, this identifies an observation in the upper quartile as a tiny municipio in the south-east of the state. Note how the location of the observation in the cartogram is not identical to that in the thematic map.

Except for multi-layer and base map, the cartogram has all the same options as a regular choropleth map, with one addition. The positioning of the circles in the cartogram is the result of a non-linear optimization approach. The algorithm attempts to locate the center of the circle as close as possible to the centroid of the areal unit with which it corresponds, while respecting the contiguity structure as much as possible. As illustrated in Figure 5.10, this is not perfect. There is no unique solution to this problem, and it may be useful to experiment with further iterations that will slightly reposition the circles. This is implemented in the **Improve Cartogram** option. A number of different iteration options are listed, together with the estimated time. The latter is particularly useful for larger data sets. Each additional set of iterations will result in a slight reshuffling of the circles. In most applications, the initial layout is sufficient for linking and brushing.

Figure 5.11: Animation control panel

5.5 Map Animation

Animation is a method to discover patterns in data by moving through observations in a systematic fashion. A typical example is to consider maps or graphs for the same variable over time (see Chapter 9). Alternatively, a given map or graph can be explored by moving through the observations from low to high, or from high to low, either individually (highlighting one at a time), or cumulatively (taking up increasingly larger parts of the map or graph).

In `GeoDa`, animation is implemented through the **Map Movie** functionality (referred to as such for historical reasons). It is invoked from the menu as **Map > Map Movie** or by selecting the third icon in the toolbar shown in Figure 4.1.

This brings up the **Animation** dialog, shown in Figure 5.11, the control center through which the various aspects of the animation are manipulated. The first item to specify is the **Variable**, selected from a drop-down list, here **sanitation**.

At the bottom of the dialog are the main controls: the start > button, step-by-step forward » or backward «, whether the animation loops or stops at the end, an option to **Reverse** the progress, the speed of the animation, and whether the order followed is ascending or descending. The defaults are usually good, with **Cumulative** checked (i.e., the selection grows as the animation progresses) and **Ascending order**.

Once the forward button is activated, each observation is selected in turn, starting with the lowest value. This selection is not only for the map, but, through linking, for all currently active windows. The slider in the **Animation** dialog moves from left to right. Under the variable name, the currently selected observation and its value are listed.

The animation tool can be paused at any point, reversed, changed from continuous change to step-by-step, etc., using the controls provided. The main point of the animation is to visually check for any patterns, such as all the lowest or highest values occurring in one location, or an increase in value that follows a given spatial trend (e.g., core-periphery, or East-West). Of course, this visual impression is only that and will need to be confirmed with the more formal pattern detection methods covered in later chapters.

A particular powerful feature of the animation tool is to assess the spatial relationship between two (or more) variables. This is carried out by selecting one as the variable in the animation control, but following the selected observations for a different variable.

In the example in Figure 5.12, the progression is based on the variable **sanitation**. Two quintile maps are shown side by side, one for **sanitation** and one for **infra** (see also Figure 4.18).

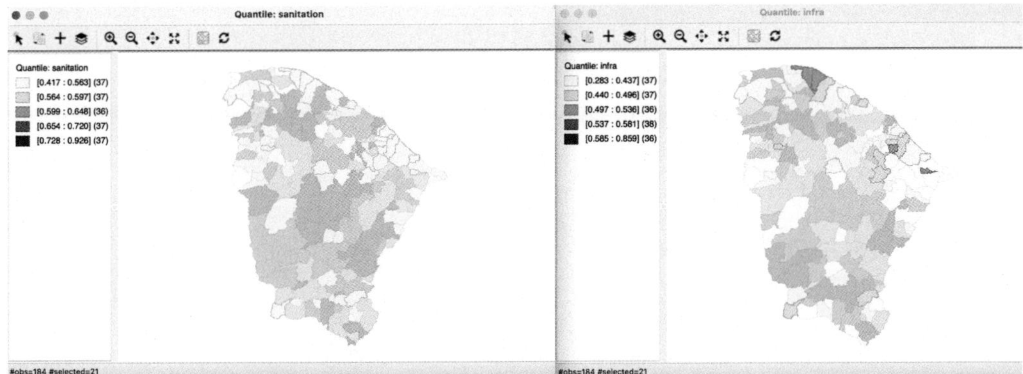

Figure 5.12: Linked animation sanitation-infrastructure

As shown in Figure 5.11, the current cut-off value for **sanitation** is 0.536. This results in 21 of the lowest observations to be highlighted in the map on the left. Through linking, the corresponding observations in the map on the right, for **infra**, are highlighted as well. This allows for the comparison of the cumulative locations of the values from low to high between the two variables. More precisely, one assesses the extent to which the *colors* for the matching locations in the two maps follow the same pattern. In the example, several of the lowest observations for **sanitation** belong to a higher quintile in the map for **infra**.

An extension to multiple maps and graphs is straightforward.

6

Maps for Rates

In Chapters 4 and 5, the maps pertained to a single variable. In the current chapter, I deal with some special aspects related to the mapping of *rates* or *proportions*. In GeoDa, such variables can be constructed on the fly, by specifying a numerator and a denominator. The numerator is typically a type of event, such as the incidence of a disease, and the denominator is the *population at risk*. Such data have broad applications in public health and criminology. However, several of the principles covered can be equally applied to any ratio, such as an unemployment rate (number of unemployed people in a given region as a share of the labor force), or any other per capita measure (e.g., gross regional product per capita).

The chapter contains three main sections. First, the creation of a *raw rate* or *crude rate* map is considered, which is the same as the types of maps considered so far, except that the rate is calculated on the fly. Next, the focus shifts to so-called *excess risk* maps, which compute a measure that compares the value at each location to an overall average, highlighting *extreme* observations. Such excess risk is known under different terms in various literatures, such as a *standardized mortality rate* (SMR) in demography (Preston et al., 2001), or a *location quotient* (LQ) in regional economics (McCann, 2001).

Finally, attention shifts to the important topic of *variance instability* that pertains to any rate measure, and the associated concept of *rate smoothing*. In essence, the precision of the rate as an estimate for the underlying *risk* depends on the size of the denominator. In practice, this means that rates estimated for small populations (e.g., rural areas) may have large standard errors and provide imprecise estimates for the actual risk. This may lead to erroneous suggestions of extreme values, such as the presence of outliers. Rate smoothing techniques use a Bayesian logic to *borrow strength* and adjust the small area estimates. There is a large literature in statistics dealing with such techniques (e.g., Lawson et al., 2003). Here, the discussion will be limited to the basic principle, illustrated by the most common form of *Empirical Bayes* smoothing.

I continue to use the *Ceará Zika* sample data set to illustrate the various features.

6.1 Topics Covered

- Create thematic maps for rates
- Assess extreme rate values by means of an excess risk map
- Understand the principle behind shrinkage estimation or smoothing rates
- Apply the Empirical Bayes smoothing principle to maps for rates
- Compute crude rates and smoothed rates in the table

DOI: 10.1201/9781003274919-6

GeoDa Functions

- Map > Rates-Calculated Map
  - Raw Rate
  - Excess Risk
  - Empirical Bayes
  - saving calculated rates to the table
- Table > Calculator > Rates
  - Raw Rate
  - Excess Risk
  - Empirical Bayes

6.2 Choropleth Maps for Rates

The distinguishing characteristic of the implementation of maps for rates or proportions in GeoDa is the explicit computation of the ratio by specifying variables for the numerator and the denominator. The map types are the same as those reviewed in Chapters 4 and 5. In addition, there are some options to save the rates in the table.

Before delving into the specifics, I first discuss the important difference between *spatially extensive* and *spatially intensive* variables. Only the latter should be used in a choropleth map. Next, I outline the implementation of the rate maps and some important options.

6.2.1 Spatially extensive and spatially intensive variables

Many variables in empirical studies are directly correlated with the size of the observational unit. Examples include total population, number of housing units, count of crimes, etc. In essence, everything else being the same, larger observational units should have larger values for such variables. They are therefore referred to as *spatially extensive*, being directly related to the *size* of the observational unit. Unless all spatial units have the same area, this can lead to spurious suggestions of importance or outlier values. Consequently, choropleth maps, which use the *area* of the spatial unit to represent observations, are not an appropriate application for spatially extensive variables.

Instead, the variables should be standardized in some fashion, so that they reflect intrinsic variation, rather than variation in the size of the observational unit. This is readily accomplished by dividing by some measure of size. The classic example is to use the *population at risk*, such as the population exposed to the risk of a disease. This implies that the rate is interpreted as an estimate for the underlying *risk* (see Section 6.4.1). Alternatives are total area (resulting in *density* measures), or some other total, such as total population (resulting in *per capita* measures), without the implication of measuring an underlying risk. The resulting *spatially intensive* variables are appropriate for thematic mapping.

More formally, if O_i is the value for the numerator in area i, and P_i is the corresponding denominator, then the raw or crude rate or proportion follows as:

$$r_i = \frac{O_i}{P_i}.$$

To illustrate this concept, consider the box maps in Figures 6.1 and 6.2, depicting, respectively, population (**pop**) and population density (**popdens**), and GDP (**gdp**) and GDP per capita

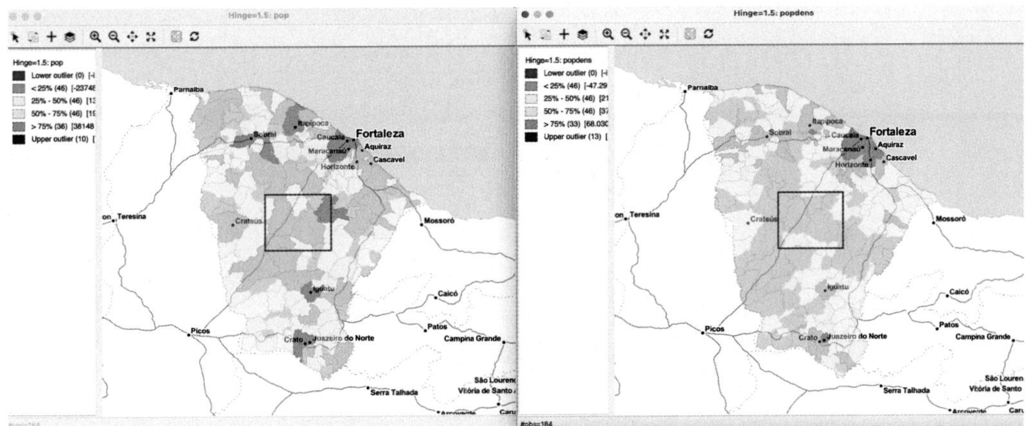

Figure 6.1: Population and population density

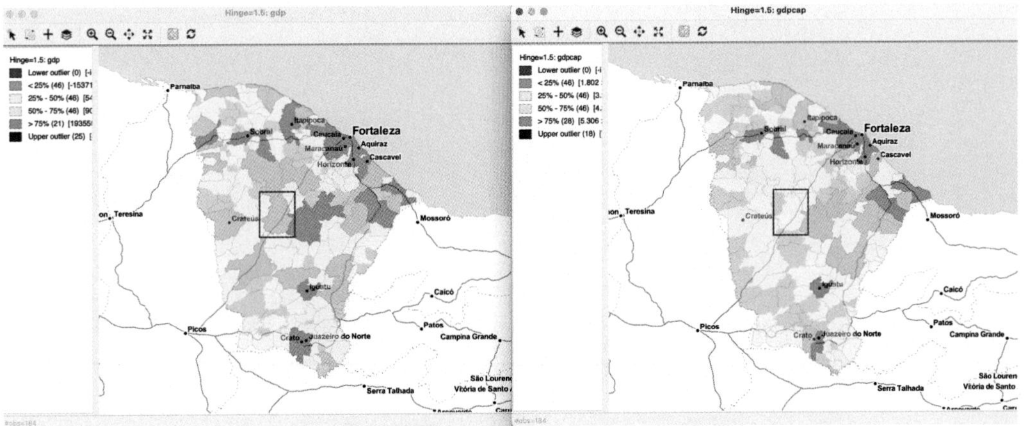

Figure 6.2: GDP and GDP per capita

(**gdpcap**) for the municipios in the state of Ceará. The left panel pertains to the spatially extensive variable, while the right-hand panel shows the corresponding spatially intensive variable.

Whereas the population box maps suggests 10 outliers, several of which are spread throughout the state, the population density map has 13 outliers, mostly concentrated around the urban area of Fortaleza. Also, as highlighted in the two maps, several of the larger areas that rank in the upper quartile for total population, drop to the first quartile in terms of population density, i.e., after the area of the municipios is corrected for (their colors changes from browns to blues). In addition, most of the outliers in the larger (rural) areas, are no longer characterized as such in the population density map.

A similar phenomenon occurs in the maps for total GDP and GDP per capita. The number of outliers for GDP (a highly skewed variable) drops from 25 to 18 in the per capita map. Again, most of the larger (and rural) areas drop in the ranking, highlighted by their change in color from browns to blues.

In sum, a proper indication of the variability in the spatial distribution of the variable of interest is only obtained when the latter is converted to a spatially intensive form.

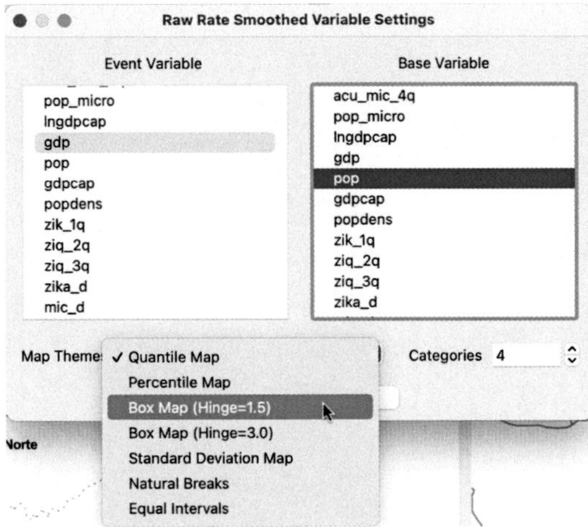

Figure 6.3: Rate map interface

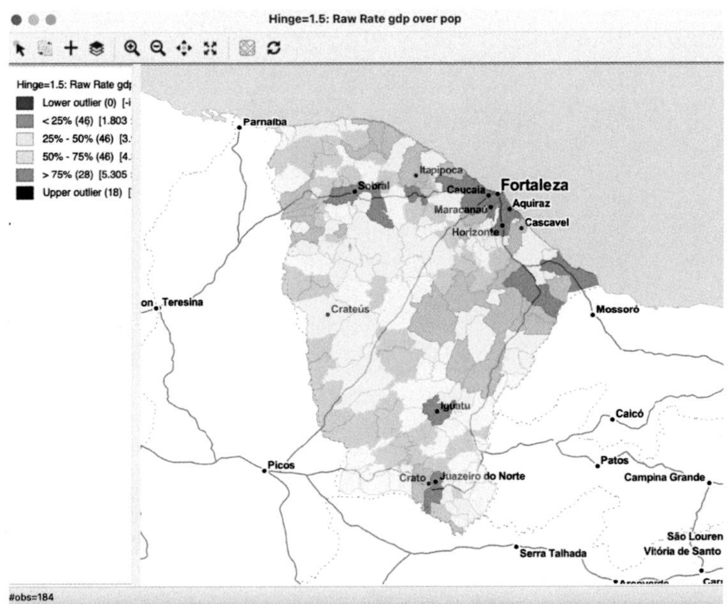

Figure 6.4: GDP per capita as a rate map

6.2.2 Raw rate map

A **Raw Rate** or crude rate map is invoked from the map menu or toolbar icon in the usual fashion, by selecting the bottom item in the interface from Figure 4.2: **Map > Rates-Calculated Maps > Raw Rate**. Alternatively, it can be created from an existing map window by selecting **Rates** in the list of map options (Figure 4.8).

In contrast to the previous maps, there is not a query for a single variable, but instead both the numerator (**Event Variable**) and denominator (**Base Variable**) of the ratio need to be specified, as shown in Figure 6.3. In the example, the respective variables are **gdp** and **pop**. The drop-down list provides all the available map types. Here, a **Box Map** with hinge 1.5 is selected, resulting in the map shown in Figure 6.4.

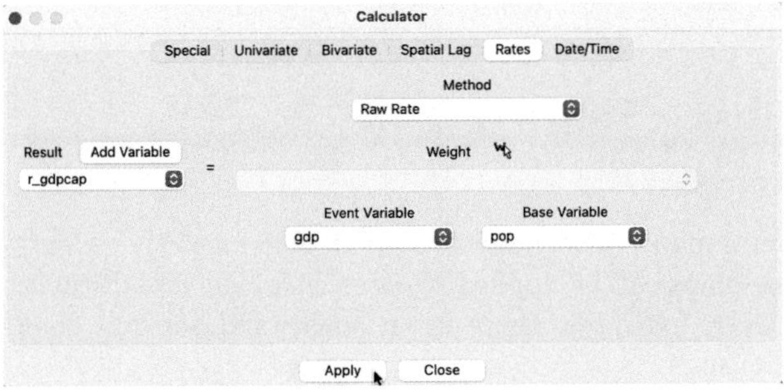

Figure 6.5: Rate calculation in table

Table - zika_ceara

		mun_name	gdpcap	R_RAW_RT	r_gdpcap
1	0	Abaiara	3.427000	3.427401	3.427401
2	1	Acarape	4.454000	4.453905	4.453905
3	0	Acaraú	5.378000	5.377665	5.377665
4	0	Acopiara	4.027000	4.026818	4.026818
5	0	Aiuaba	3.349000	3.348577	3.348577
6	1	Alcântaras	3.507000	3.507288	3.507288
7	1	Altaneira	3.588000	3.587806	3.587806
8	0	Alto Santo	4.713000	4.712758	4.712758
9	0	Amontada	4.669000	4.669148	4.669148
10	0	Antonina do Norte	4.058000	4.058276	4.058276

Figure 6.6: GDP per capita in table

Everything is the same as before, except that the map heading spells out both the numerator and denominator variables: **Raw Rate gdp over pop**. Specifically, the map is identical to the right-hand panel in Figure 6.2, which uses the **gdpcap** variable from the data table, rather than computing the ratio explicitly.

A rate map retains all the standard options of a map in `GeoDa` (see Section 4.5). Two additional features are **Rates** and **Save Rates**. **Rates** brings up a list of the different rate map options, since these are not available in the **Change Current Map Type** function in the main options menu. The **Save Rates** option allows for the calculated rates to be added to the data table. The default variable name for a raw rate is **R_RAW_RT**). As before, the new variable only becomes a permanent addition after a **Save** or **Save As** operation.

6.2.2.1 Rates in the table

Rates can also be computed directly in the table as part of the **Calculator** functionality, by selecting the **Rates** tab in the interface, shown in Figure 6.5 (see also Section 2.4.2 in Chapter 2). The particular type of rate is selected from the **Method** drop-down list. In this example, the **Raw Rate** is used. As usual, a new variable must be added (here, **r_gdpcap**) and both the **Event Variable** (**gdp**) and **Base Variable** (**pop**) must be spelled out.

Figure 6.6 illustrates the result. It shows side by side the original per capita GDP variable (**gdpcap**), the rate saved from the map window (**R_RAW_RT**), and the rate computed in the table (**r_gdpcap**). Rounded to the third decimal of precision, the values are identical.

6.3 Excess Risk – SMR – LQ

6.3.1 Relative risk

In Chapter 5, several map types were introduced that focus the attention on extreme values. Clearly, these same maps can be applied to rates, as in the illustration in Figure 6.4, showing a box map. However, for rates, there is an additional concept that helps in identifying extreme observations.

The concept is that of a relative rate or a *relative risk*, also referred to as *excess risk*. In essence, this boils down to the ratio of the observed rate at a location to some reference rate. In demography and public health analysis, such a ratio is referred to as a *standardized mortality rate* (SMR). In regional economics, when applied to the fraction of employment in a given sector, the corresponding concept is that of a *location quotient* (LQ).

The idea is to compare the observed rate at a small area to a national (or regional) standard. More specifically, the observed number of events (O_i) is compared to the number of events that would be expected (E_i), had a reference risk been applied.

In most applications, the reference risk is estimated from the aggregate of all the observations under consideration, as the ratio of the sum of numerators over the sum of denominators. Formally, this is expressed as:

$$\tilde{\pi} = \frac{\sum_{i=1}^{i=n} O_i}{\sum_{i=1}^{i=n} P_i},$$

with O_i and P_i as before. Note that this is different from a simple average of the rates. In fact, it is a population weighted average that properly assigns the contribution of each area to the overall total.

If the ratio $\tilde{\pi}$ would apply equally to each observation, the result would be the expected value for that observation. In demography and public health, this would be an expected number of deaths or incidence of a disease. In regional economics, it would be the expected share of employment in a sector (equal to the reference share). In general, this can be expressed as:

$$E_i = \tilde{\pi} \times P_i.$$

The relative risk then follows as the ratio of the observed rate over the reference rate, or, equivalently, of the observed number of events over the expected number of events:

$$RR_i = \frac{r_i}{\tilde{\pi}_i} = \frac{O_i/P_i}{E_i/P_i} = \frac{O_i}{E_i}$$

.

If an area matches the (regional) reference rate, the corresponding relative risk is one. Values greater than one suggest an excess, whereas values smaller than one suggest a shortfall. The interpretation depends on the context. For example, in disease analysis, a relative risk larger than one would indicate an area where the prevalence of the disease is *greater* than would be expected. In regional economics, a location quotient greater than one, suggests employment in a sector that exceeds the local needs, implying an *export* sector.

6.3.2 Excess risk map

An **Excess Risk** map is a special rate map that uses a custom classification to indicate observations that fall below or exceed the reference relative risk rate of one. The legend is

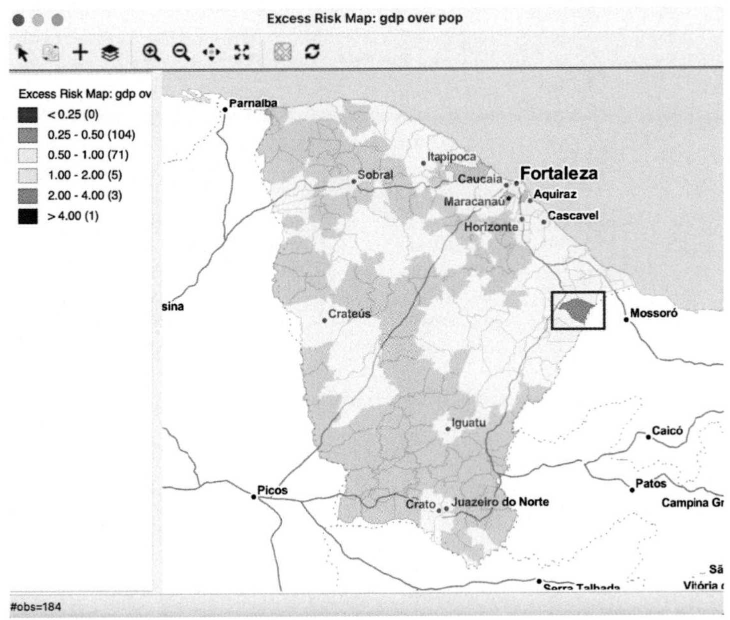

Figure 6.7: GDP per capita excess risk

divergent, using blue hues to indicate values smaller than one and brown hues for values larger than one. The intervals are 1–2, 2–4 and greater than 4 on the high end, and 0.5–1, 0.25–0.5, and <0.25 on the low end.

The map is invoked as **Maps > Rates-Calculated Maps > Excess Risk** in the usual fashion. Again, both the **Event Variable** and the **Base Variable** must be specified. In Figure 6.7, the resulting map is shown for **gdp** and **pop** as the numerator and denominator. The great majority of municipalities do not reach the state GDP per capita, yielding a blue hue in the map. Only nine observations have a GDP per capita that exceeds the state average by a multiple: five between 1 and 2, three between 2 and 4, and one observation has an excess risk rate greater than 4, highlighted in the map.

As it turns out, the identified observation is the small rural municipality of Quixeré, which specializes in high-tech agriculture. Due to the high capital intensive nature of that production and its small population, it has an extremely high GDP per capita from that activity (see also Figure 6.8). However, this extreme value may be an artifact of the particular data set.[1]

All standard map options also apply to the excess risk map.

6.3.2.1 Saving and calculating excess risk

As shown in Section 6.2.2.1, the rates can be saved to the table, as well as calculated directly in the table. This works in the same way for the excess risk ratios. The default variable name when saving the rate is **R_EXCESS**.

Figure 6.8 illustrates this for the GDP per capita. The original **gdpcap** is shown, together with the excess rate. The observations are ranked by the excess rate. The highest rate of 4.34 is obtained by Quixeré, with a GDP per capita of 40. The total GDP for the state of Ceará

[1]There may be a problem with the GDP data for that municipality in the Ceará data set. The figure for GDP is from the 2013 IBGE publication. As it turns out, in later years, this figure was revised downward.

Table - zika_ceara

	RT	mun_name	gdpcap	R_EXCESS
152	'979	Quixeré	40.018000	4.343996
56	.726	Eusébio	27.625000	2.998695
162	3910	São Gonçalo do Amar.	25.464000	2.764136
104	493	Maracanaú	19.613000	2.129066
71	1912	Horizonte	18.042000	1.958470
59	952	Fortaleza	15.132000	1.642590
167	;007	Sobral	12.475000	1.354176
75	235	Icapuí	10.415000	1.130585
12	3180	Aquiraz	9.398000	1.020183
87	)021	Itapajé	8.740000	0.948739
127	)515	Pacajus	8.321000	0.903201

Figure 6.8: GDP per capita excess risk in table

is 77,865,442 rais (i.e., the sum of the GDP in all municipalities), with a matching total population of 8,452,380. Consequently, the state GDP per capital is 9.21. Clearly, the GDP per capita of 40 for Quixeré exceeds this more than four-fold. As pointed out in the review of the map, this is actually an exception, with most of the municipalities not matching the state average. Figure 6.8 shows the top nine municipalities with excess rates greater than one. More importantly, the spatial distribution shown in the map in Figure 6.7 highlights the importance of *spatial heterogeneity*. Specifically, assuming that the economies of the municipalities in the state follow the state GDP per capita is highly misleading, as only a few such locations meet or exceed that standard.

6.4 Rate Smoothing

In many empirical investigations, the real interest is in the spatial distribution of the underlying *risk* of an event. While the concept of risk has many meanings, in this context it represents the *probability* of an event occurring. To the extent that the risk is not constant across space, its spatial distribution is represented by a so-called *risk surface*. The problem is that such a surface is not observed and thus must be estimated.

As it turns out, the crude rate is a good estimate for the unknown risk in that it is unbiased (on average, it is on target). However, it suffers from the problem that its precision (or, alternatively, its variance) is not constant and depends on the size of the corresponding population at risk (the denominator). This intrinsic *variance instability* of rates causes problems for the interpretation of maps, especially in terms of the identification of outliers. In addition, there are challenges to create meaningful maps when dealing with many locations that have no observed events.

In response, a number of procedures have been suggested that trade off some bias in order to increase overall precision, by *borrowing strength.*

In this section, I first present more formally some important concepts related to variance instability, as well as ways to improve the risk estimates by means of *smoothing rates.* In the current chapter, the discussion is limited to one particular approach, i.e., *Empirical Bayes* smoothing. Spatial smoothing methods are covered in Chapter 12.

6.4.1 Variance instability

The point of departure in the consideration of rate smoothing is to view the count of events that occur in a location i, O_i, as the number of *successes* in P_i draws from a probability distribution driven by a risk parameter π_i. The textbook analog is to draw n balls from an urn filled with red and white balls, say with a proportion of p red balls. Given n such draws, what is the probability that q balls will be red?

The formal mathematical framework for this type of situation is the binomial distribution, $B(n, p)$. The probability function for this distribution gives the probability of k successes in n successive draws from a population with risk parameter p:

$$\text{Prob}[k] = \binom{n}{k} p^k (1-p)^{n-k},$$

where $\binom{n}{k}$ is the binomial coefficient.[2] Translated into the context of O_i events out of a population of P_i, the corresponding probability is:

$$\text{Prob}[O_i] = \binom{P_i}{O_i} \pi_i^{O_i} (1-\pi_i)^{P_i - O_i},$$

with π_i as the underlying risk parameter. The mean of this binomial distribution is $\pi_i P_i$, and the variance is $\pi_i(1-\pi_i)P_i$.

Returning to the crude rate $r_i = O_i/P_i$, it can be readily seen that its mean corresponds to the underlying risk:[3]

$$E[r_i] = E[\frac{O_i}{P_i}] = \frac{E[O_i]}{P_i} = \frac{\pi_i P_i}{P_i} = \pi_i.$$

Consequently, the crude rate is an *unbiased* estimator for the underlying risk.

However, the variance has some undesirable properties. A little algebra yields the variance as:

$$Var[r_i] = \frac{\pi_i(1-\pi_i)P_i}{P_i^2} = \frac{\pi_i(1-\pi_i)}{P_i}.$$

This result implies that the variance depends on the mean, a non-standard situation and an additional degree of complexity. More importantly, it implies that the larger the population of an area (P_i in the denominator), the smaller the variance for the estimator, or, in other words, the greater the precision.

The flip side of this result is that for areas with sparse populations (small P_i), the estimate for the risk will be *imprecise* (large variance). Moreover, since the population typically varies across the areas under consideration, the precision of each rate will vary as well. This *variance instability* needs to somehow be reflected in the map, or corrected for, to avoid a spurious representation of the spatial distribution of the underlying risk. This is the main motivation for *smoothing* rates, considered next.

6.4.2 Borrowing strength

Approaches to smooth rates, also called shrinkage estimators, improve on the precision of the crude rate by *borrowing strength* from the other observations. This idea goes back to the fundamental contributions of James and Stein (the so-called James-Stein paradox),

[2] $\binom{n}{k} = n!/k!(n-k)!$, or the number of different combinations of k observations out of n.
[3] Only the number of observed events O_i is random, the population size P_i is not.

who showed that in some instances biased estimators may have better precision in a mean squared error sense (James and Stein, 1961).

Formally, the mean squared error or MSE is the sum of the variance and the square of the bias. For an unbiased estimator, the latter term is zero, so then MSE and variance are the same. The idea of borrowing strength is to trade off a (small) increase in bias for a (large) reduction in the variance component of the MSE. While the resulting estimator is biased, it is more precise in a MSE sense. In practice, this means that the chance is much smaller to be far away from the true value of the parameter.

The implementation of this idea is based on principles of Bayesian statistics, which are briefly reviewed next.

6.4.2.1 Bayes law

The formal logic behind the idea of smoothing is situated in a Bayesian framework, in which the distribution of a random variable is updated after observing data.[4] The principle behind this is the so-called **Bayes Law**, which follows from the decomposition of a joint probability (or density) of A and B into two conditional probabilities:

$$P[AB] = P[A|B] \times P[B] = P[B|A] \times P[A],$$

where A and B are random events, and $|$ stands for the conditional probability of one event, given a value for the other. The second equality yields the formal expression of Bayes law as:

$$P[A|B] = \frac{P[B|A] \times P[A]}{P[B]}.$$

In most instances in practice, the denominator in this expression can be ignored, and the equality sign is replaced by a proportionality sign:

$$P[A|B] \propto P[B|A] \times P[A].$$

In the context of estimation and inference, the A typically stands for a parameter (or a set of parameters) and B stands for the data. The general strategy is to update what is known about the parameter A *a priori* (reflected in the **prior distribution** $P[A]$), after observing the data B, to yield a **posterior distribution**, $P[A|B]$, i.e., what is known about the parameter after observing the data. The link between the prior and posterior distribution is established through the **likelihood**, $P[B|A]$. Using a more conventional notation with π as the parameter and y as the observations, this gives:[5]

$$P[\pi|y] \propto P[y|\pi] \times P[\pi].$$

For each particular estimation problem, a distribution must be specified for both the prior and the likelihood. This must be carried out in such a way that a proper posterior distribution results. Of particular interest are so-called conjugate priors, which result in a closed form expression for the combination of likelihood and prior distribution. In the context of rate smoothing, there are a few commonly used priors, such as the Gamma and the Gaussian

[4]There are several excellent books and articles on Bayesian statistics, with Gelman et al. (2014) as a classic reference.

[5]Note that in a Bayesian approach, the likelihood is expressed as a probability of the data conditional upon a value (or distribution) of the parameters. In classical statistics, it is the other way around.

(normal) distribution.[6] A formal mathematical treatment is beyond the current scope, but it is useful to get a sense of the intuition behind smoothing approaches.

In essence, it means that the estimate from the data (i.e., the crude rate) is adjusted with some prior information, such as the reference rate for a larger region (e.g., the state or country). Unreliable small area estimates are then *shrunk* toward this reference rate. For example, if a small area is observed with zero occurrences of an event, does that mean that the risk is zero as well? Typically, the answer will be no, and the *smoothed* risk will be computed by borrowing information from the reference rate.

6.4.2.2 The Empirical Bayes approach

The Empirical Bayes approach is based on the so-called *Poisson-Gamma* model, where the observed count of events (O) is viewed as the outcome of a Poisson distribution with a random intensity (mean), for example expressed as πP. The Bayesian aspect comes in the form of a prior Gamma distribution for the risk parameter π. This results in a particular expression for the *posterior* distribution, which is also Gamma, with mean and variance:

$$E[\pi] = \frac{O + \alpha}{P + \beta},$$

and

$$Var[\pi] = \frac{O + \alpha}{(P + \beta)^2}$$

where α and β are the the shape and scale parameters of the prior (Gamma) distribution.[7]

In the Empirical Bayes approach, values for α and β are *estimated* from the actual data. The smoothed rate is then expressed as a weighted average of the crude rate, say r, and the prior estimate, say θ. The latter is estimated as a reference rate, typically the overall statewide average or some other standard.

In essence, the EB technique consists of computing a weighted average between the raw rate for each small area and the reference rate, with weights proportional to the underlying population at risk. Simply put, small areas (i.e., with a small population at risk) will tend to have their rates adjusted considerably, whereas for larger areas the rates will barely change.[8]

More formally, the EB estimate for the risk in location i is:

$$\pi_i^{EB} = w_i r_i + (1 - w_i)\theta.$$

In this expression, the weights are:

$$w_i = \frac{\sigma^2}{(\sigma^2 + \mu/P_i)},$$

with P_i as the population at risk in area i, and μ and σ^2 as the mean and variance of the prior distribution. [9]

---

[6]For an extensive discussion, see, for example, the classic papers by Clayton and Kaldor (1987) and Marshall (1991).

[7]A Gamma(α, β) distribution with shape and scale parameters α and β has mean $E[\pi] = \alpha/\beta$ and variance $Var[\pi] = \alpha/\beta^2$.

[8]For an extensive technical discussion, see also Anselin et al. (2006a).

[9]In the Poisson-Gamma model, this turns out to be the same as $w_i = P_i/(P_i + \beta)$.

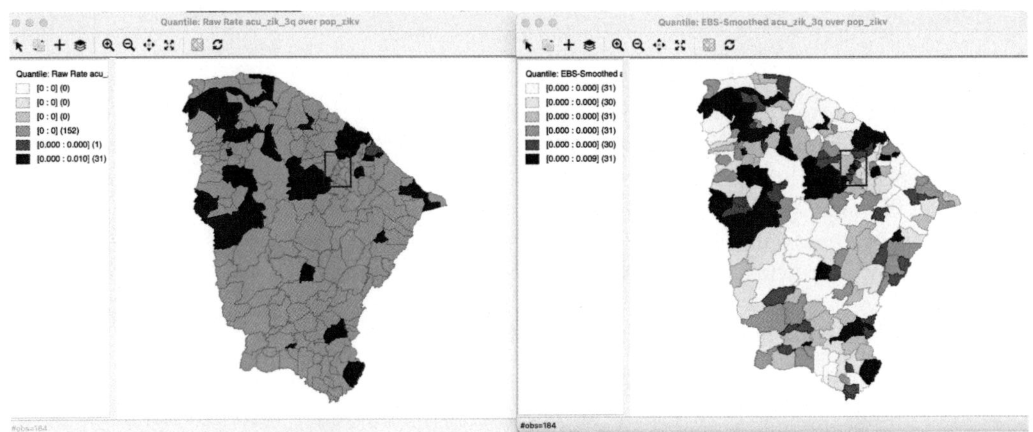

Figure 6.9: Crude and EB rate map

In the empirical Bayes approach, the mean μ and variance σ^2 of the prior (which determine the scale and shape parameters of the Gamma distribution) are estimated from the data.

For μ this estimate is simply the reference rate (the same reference used in the computation of the SMR), $\sum_{i=1}^{i=n} O_i / \sum_{i=1}^{i=n} P_i$. The estimate of the variance is a bit more complex:

$$\sigma^2 = \frac{\sum_{i=1}^{i=n} P_i (r_i - \mu)^2}{\sum_{i=1}^{i=n} P_i} - \frac{\mu}{\sum_{i=1}^{i=n} P_i / n}.$$

While easy to calculate, the estimate for the variance can yield negative values. In such instances, the conventional approach is to set σ^2 to zero. As a result, the weight w_i becomes zero, which in essence equates the smoothed rate estimate to the reference rate.

6.4.3 Empirical Bayes smoothed rate map

In Chapter 5, a unique values map was shown for the cumulative occurrence of Zika in the state of Ceará in 2016. The map in Figure 5.5 indicated that only 32 of the 184 municipios had an incidence of Zika during the period under consideration, meaning that 152 municipalities had no cases. Suggesting that these municipalities therefore had zero risk of exposure to Zika does not seem realistic. Hence the need to borrow some strength from other information.

A crude rate map that uses the cumulative Zika cases (**acu_zik_3q**) as the **Event Variable** and the population at risk (**pop_zikv**) as the **Base Variable** is shown in the left panel of Figure 6.9 as a quantile map with six categories. Clearly, all but one of the 32 non-zero observations are grouped in the top category, one is the next lower category and everything else is zero.

An Empirical Bayes smoothed map is invoked as **Maps > Rates-Calculated Maps > Empirical Bayes** in the usual fashion. The same **Event Variable** and **Base Variable** are specified and the map type is selected from the same drop-down list as in Figure 6.3. The resulting six category quantile map is shown in the right panel of Figure 6.9. The contrast with the crude rate map is striking. Each category has the required 31 or 32 observations, and the map shows much more spatial variety than before. Areas that had zero observations are *shrunk* toward an average rate, either below or above the median. The highlighted

Table - zika_ceara

	I	mun_name	pop_zikv >	R_RAW_RT	R_EBS
69	)	Guaramiranga	3720	0.000000	0.000055
65	)	Granjeiro	4494	0.000000	0.000048
130	)	Pacujá	6168	0.000162	0.000175
21	)	Baixio	6198	0.000000	0.000036
147	)	Potiretama	6318	0.000000	0.000036
62	1	General Sampaio	6763	0.000000	0.000034
55	)	Ereré	7104	0.000000	0.000032
10	)	Antonina do Norte	7227	0.000000	0.000032
7	1	Altaneira	7344	0.000272	0.000270
166	)	Senador Sá	7367	0.000000	0.000031

Figure 6.10: Effect of smoothing – small populations

Table - zika_ceara

	I	mun_name	pop_zikv <	R_RAW_RT	R_EBS
59	1	Fortaleza	2591188	0.000503	0.000503
44	1	Caucaia	353932	0.001238	0.001235
99	1	Juazeiro do Norte	266022	0.000000	0.000001
104	1	Maracanaú	221504	0.000072	0.000073
167	1	Sobral	201756	0.000045	0.000046
51	1	Crato	128680	0.000000	0.000002
88	)	Itapipoca	124950	0.000000	0.000002
105	1	Maranguape	123570	0.000032	0.000034
77	1	Iguatu	101386	0.000000	0.000003
149	1	Quixadá	85351	0.000000	0.000003

Figure 6.11: Effect of smoothing – large populations

municipalities in the map (blue rectangle) shown three small areas whose smoothed rates pulled the original zero value above the median.

6.4.3.1 The effect of smoothing

In the Ceará example, the total cumulative number of Zika cases was 2,262, for a total population of 8,904,459. This yields a reference rate of 0.000254, or 25.4 per 100,000. The effect of Empirical Bayes smoothing is that the original crude rate is moved toward this overall rate as a function of its population: areas with large populations do not see much change, whereas areas with small populations can take on very different smoothed value.

In Figure 6.10, the effect on small areas is illustrated. The municipalities are ordered by population size, with the 10 smallest areas listed. This clearly illustrates how the crude rate (**R_RAW_RT**) gets moved toward the overall mean in the Empirical Bayes rate (**R_EBS**). For example, for the smallest area, the crude rate of zero becomes 5.5 per 100,000 and most of the other crude rates get pulled up toward the overall mean. For the municipio of Altaneira, the opposite happens, i.e., the crude rate of 27.2 per 100,000 becomes 27.0 per 100,000, pulling it down toward the overall mean.

For areas with large populations, there is a minimal effect of the smoothing. Figure 6.11 lists the ten largest municipalities with their crude and smoothed rates. The difference between the two is marginal.

6.4.3.2 Calculating EB smoothed rates in the table

Just as for the other rate maps, the calculated Empirical Bayes smoothed rate can be saved to the table (the default variable name is **R_EBS**). In addition, the smoothed rates can be computed directly by means of the **Calculator** option in the Table, following the same procedures as outlined in Section 6.2.2.1.

7

Univariate and Bivariate Data Exploration

In this and the next two chapters, I continue the review of data exploration, but now shift the focus to traditional *non-spatial* EDA methods. Through the use of linking and brushing, the connection with a spatial representation (a map) can always be made explicit. This idea of *spatializing* EDA is central to the perspective taken here.

In the current chapter, I focus on techniques to describe the distribution of one variable at a time (univariate), and on the relationship between two variables (bivariate) through standard statistical graphs. These include the histogram, box plot, scatter plot and scatter plot matrix, considered in turn. The chapter closes with a discussion of *spatial heterogeneity*, both for a single variable (through the averages chart) and pertaining to the relationship between two variables (brushing the scatter plot).

To illustrate the methods covered in this and the next two chapters, I introduce a new sample data set with poverty indicators and census data for 570 municipalities in the state of Oaxaca in Mexico. The poverty indicators are from CONEVAL (the National Council for the Evaluation of Social Development Policy) for 2010 and 2020. The census variables cover 2000, 2010 and 2020 and are from INEGI (National Institute of Statistics and Geography).

The data are contained in the *Oaxaca Development* sample data set.

7.1 Topics Covered

- Computing descriptive statistics and creating visualizations of the distribution of a single variable (histogram, box plot)
- Interpreting a scatter plot and scatter plot smoothing (LOWESS)
- Linking and brushing maps and statistical graphs
- Analyzing the bivariate relationship for multiple variables in a scatter plot matrix
- Assessing spatial heterogeneity by means of the Averages Chart
- Assessing spatial heterogeneity through the Chow test

GeoDa Functions

- Explore > Histogram
  - Choose Intervals option
  - Histogram Classification option
  - View > Set as Unique Value
  - View > Display Statistics option

DOI: 10.1201/9781003274919-7

- Explore > Box Plot
  - Hinge option
- Explore > Scatter Plot
  - View > Display Precision option
  - Data option
  - Smoother option
  - LOWESS parameters setting
  - Regimes Regression option and Chow test
- Explore > Scatter Plot Matrix
  - changing the variable order in a scatter plot matrix
  - smoothing and brushing the scatter plot matrix
- Explore > Averages Chart

Toolbar Icons

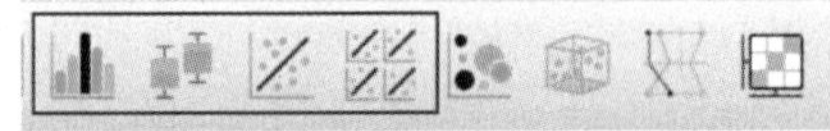

Figure 7.1: Histogram | Box Plot | Scatter Plot | Scatter Plot Matrix

7.2 Analyzing the Distribution of a Single Variable

The first step in the analysis of the distribution of a single variable is a summary of its characteristics, focusing on central tendency (mean, median), spread (variance, interquartile range) and shape (skewness, kurtosis). Here, I briefly review two familiar visualizations of a univariate distribution in the form of the histogram and the box plot.

7.2.1 Histogram

Arguably the most familiar statistical graphic is the histogram, which is a discrete representation of the density function of a continuous variable. In essence, the range of the variable (the difference between maximum and minimum) is divided into a number of equal intervals (or bins), and the number of observations that fall within each bin is depicted proportional to the height of a bar. This classification is the same as the principle underlying the equal intervals map, which was covered in Section 4.4.2. The main challenge in creating an effective visualization is to find a compromise between too much detail (many bins, containing few observations) and too much generalization (few bins, containing a broad range of observations).

7.2.1.1 Histogram basics

The histogram functionality is started by selecting **Explore > Histogram** from the menu, or by clicking on the **Histogram** toolbar icon, the left-most icon in the univariate set (within the blue rectangle) in Figure 7.1.

This brings up the **Variable Settings** dialog, which lists all the numeric variables in the data set (string variables cannot be analyzed). For example, selecting **ppov_20** from the list (percent of population living in poverty in 2020) creates the default histogram with seven bins, shown in Figure 7.2. The distribution is highly left-skewed, with a long tail toward the

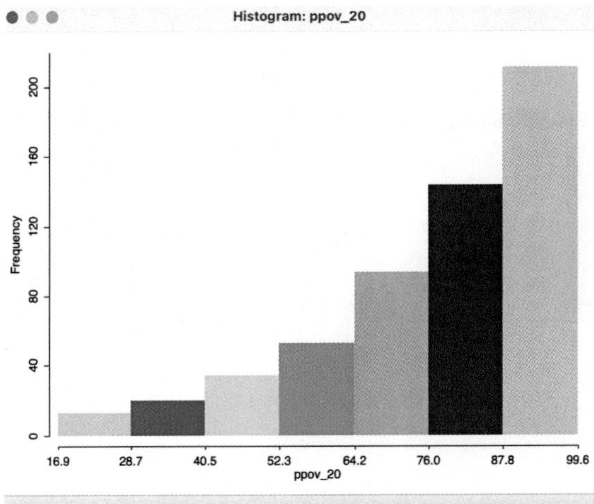

Figure 7.2: Histogram for percent poverty 2020

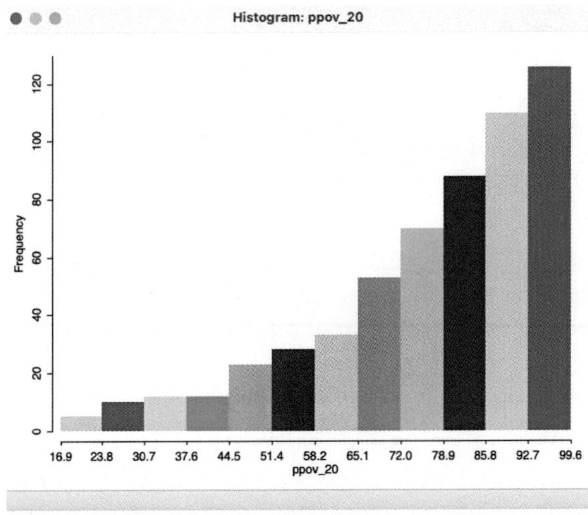

Figure 7.3: Histogram for percent poverty 2020 – 12 bins

lower percentages, highlighting the prevalence of high-poverty municipalities (on the right).

The histogram options are brought up in the usual fashion, by right-clicking on the graph. The options are grouped into three categories, the first consisting of histogram-specific items: **Choose Intervals**, **Histogram Classification** and **View**. Next is a **Color** option to set the background color (the default of white is usually best). Finally, the options to **Save Selection**, **Copy Image to Clipboard** and **Save Image As** work in the same way as for maps (see Section 4.5.5).

Of these options, the **Choose Intervals** is the most commonly used. It brings up a dialog to specify the number of bins for the histogram. For example, after changing from the default of seven to twelve bins, the histogram becomes as in Figure 7.3. In this particular instance, there is not much gained by the greater detail since the same overall pattern is maintained.

The **Histogram Classification** option allows for the selection of a custom category specification. This works in the same way as for map classification, covered in Section 4.6.

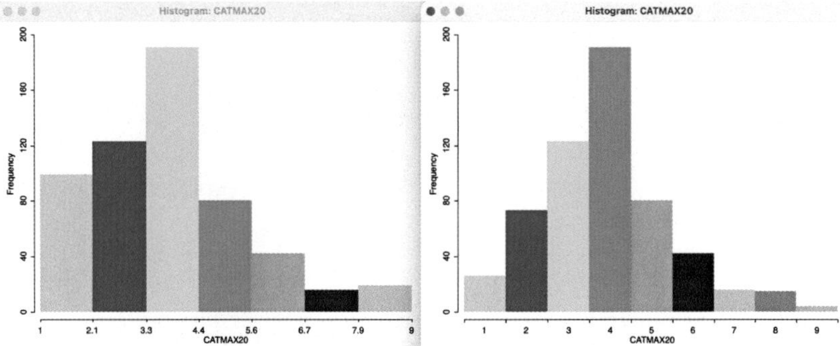

Figure 7.4: Bar chart for settlement categories

The **View** option contains six items. Four of these pertain to the overall look of the graph and are self-explanatory: **Set Display Precision** (for the variable being depicted), **Set Display Precision on Axes**, **Show Axes** and **Status Bar**. The latter two are checked by default.

7.2.1.2 Bar charts

The first item under the **View** options, **Set as Unique Value**, is useful when the variable under consideration takes on integer values, e.g., when it is categorical or ordinal. While the corresponding graph is not technically a histogram (since categories do not imply any order), the same mechanics can be used to create a *bar chart.*

Under the default settings, the bins are computed by taking the range of values into consideration, i.e., the difference between the largest and smallest integer value. This is then divided by the number of bins, leading to unrepresentative category labels for integer-valued variables. This issue is remedied by means of the **View > Set as Unique Value** option. As a result, the values taken by the variable of interest are recognized as integer and depicted as such.

For example, the left panel in Figure 7.4 shows the default histogram for the variable **CATMAX20**, a categorical variable (or, more precisely, an ordinal variable) that corresponds to the largest settlement category in each municipality (each municipality consists of one of more localities or settlements).[1]

For Oaxaca in 2020, this variable ranges from 1 to 9. As a result, the default of seven histogram bins results in a bin width of 8/7, or roughly 1.14, which does not yield meaningful intervals.

By contrast, the right-hand panel in the figure shows the result after invoking **Set as Unique Value**. Now, each bar corresponds to a distinct discrete value, matching each category. The distribution is fairly symmetric around a mode of 4 (1,000 to 2,499 inhabitants), albeit slightly right-skewed .

[1]The original variable is **tamloc**, which ranges from 1, for 1 to 249 inhabitants, to 14, for more than a million inhabitants. **CATMAX20** is the highest category obtained for a locality in the municipality. In this example, the maximum is 9, for 30,000 to 49,999 inhabitants.

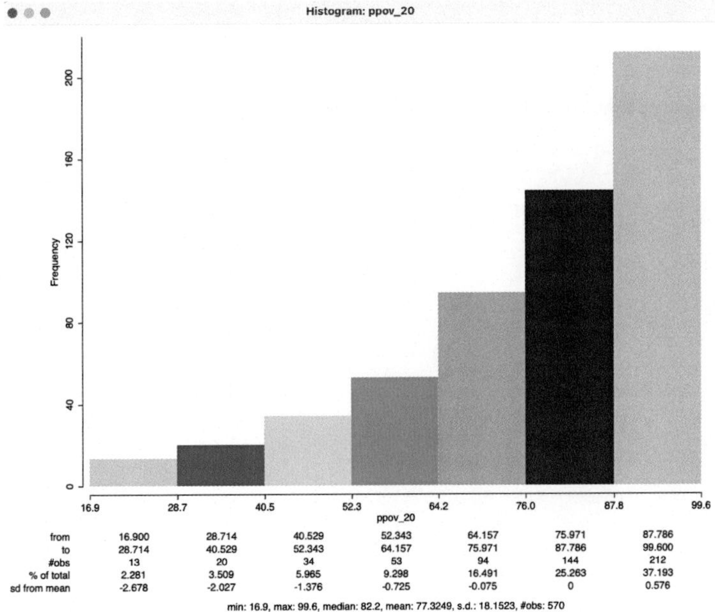

from	16.900	28.714	40.529	52.343	64.157	75.971	87.786
to	28.714	40.529	52.343	64.157	75.971	87.786	99.600
#obs	13	20	34	53	94	144	212
% of total	2.281	3.509	5.965	9.298	16.491	25.263	37.193
sd from mean	-2.678	-2.027	-1.376	-0.725	-0.075	0	0.576

Figure 7.5: Descriptive statistics

7.2.1.3 Descriptive statistics

The **View > Display Statistics** option brings up a number of descriptive statistics, shown below the histogram, as in Figure 7.5 for the **ppov_20** variable.

The overall summary statistics are listed at the bottom: minimum of 16.9, maximum of 99.6, median of 82.2, mean of 77.3, standard deviation of 18.15, for 570 observations. In addition, each histogram bar is associated with a specific range, the number of observations it contains, % of total in that bin and the number of standard deviations from the mean. For the latter, results greater than two in absolute value can be classified as outliers. In the example, this is the case for the two lowest categories.

7.2.1.4 Linking map and histogram

The concepts of linking and brushing were already briefly introduced in Chapter 4. They are most powerful when connecting a-spatial EDA graphs with a map as a visualization of the spatial distribution of a variable. The map in question can pertain to the same variable (or be themeless), allowing to visually explore the extent to which observations in the same histogram categories also occur in similar locations (a pre-cursor to the more formal notion of spatial autocorrelation). Alternatively, the map can be for a different variable, providing a way to explore the association between observations in histogram bins to the spatial distribution of a different variable.

For example, in Figure 7.6, the map on the left is a six category quantile map for **ALTID**, a representative altitude for the municipality, in meters.[2] When selecting the 73 observations in the lowest two histogram bins on the right (the largest settlement having respectively 1 to 249 or 250 to 499 inhabitants), the process of *linking* simultaneous selects the corresponding

[2]The **Basemap** is **ESRI > WorldTopoMap**. As is to be expected, the darker colors are associated with higher elevations on the topographical map.

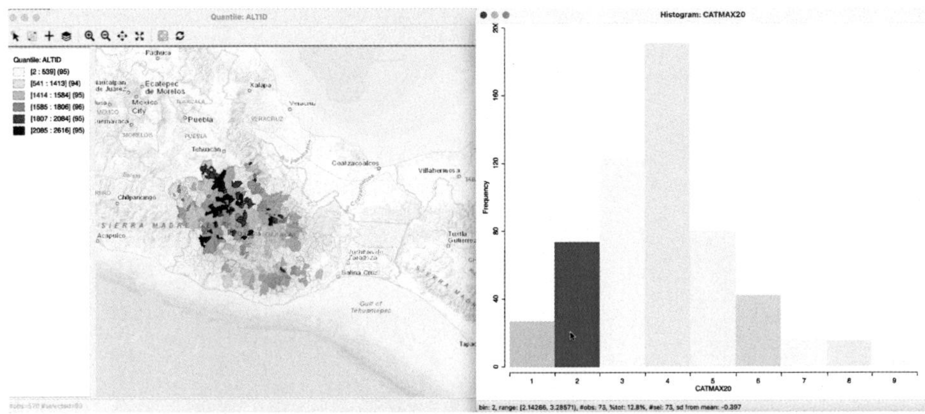

Figure 7.6: Linking between histogram and map

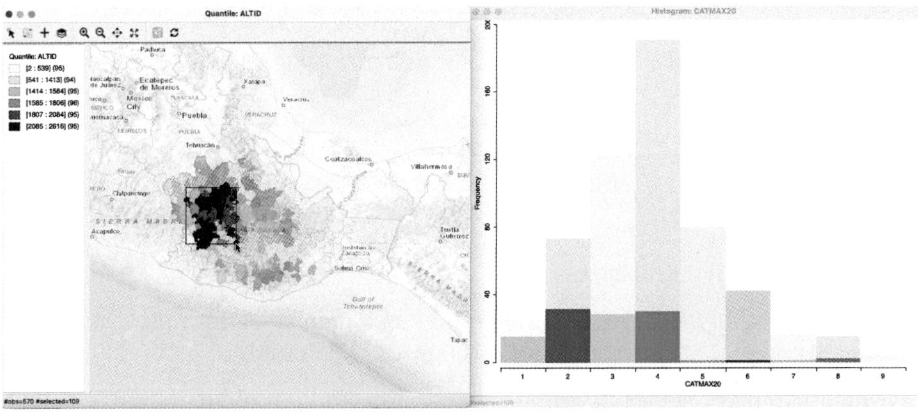

Figure 7.7: Linking between map and histogram

locations in the elevation map. The resulting configuration does not appear to be *random* (our prior expectation, or null hypothesis), but the respective municipalities are concentrated in the top elevation categories. In other words, settlements with smaller populations tend to be located in higher elevations (not an unexpected result, but still worthy of checking).

The reverse linking between the map and the histogram is illustrated in Figure 7.7. The selection rectangle in the map results in the corresponding observations to be highlighted in the histogram. Here, the focus is on *spatial heterogeneity* (further elaborated upon in Section 7.5). Everything else being the same, the expectation (null hypothesis) would be that the distribution of the selected observations largely follows that of the whole. In other words, in the histogram, the heights of the bars of the selected observations should roughly be proportional to the corresponding height in the full data set. In the example, this clearly is not the case, suggesting a difference between the distribution in the *spatial subset* and the overall distribution. In other words, this suggests the presence of spatial heterogeneity.

The process can be made dynamic through brushing, i.e., by moving the selection rectangle over the map. This results in an immediate adjustment of the selected observations in the histogram, allowing for an assessment of spatial heterogeneity *by eye*. A more formal assessment is pursued in Section 7.5.

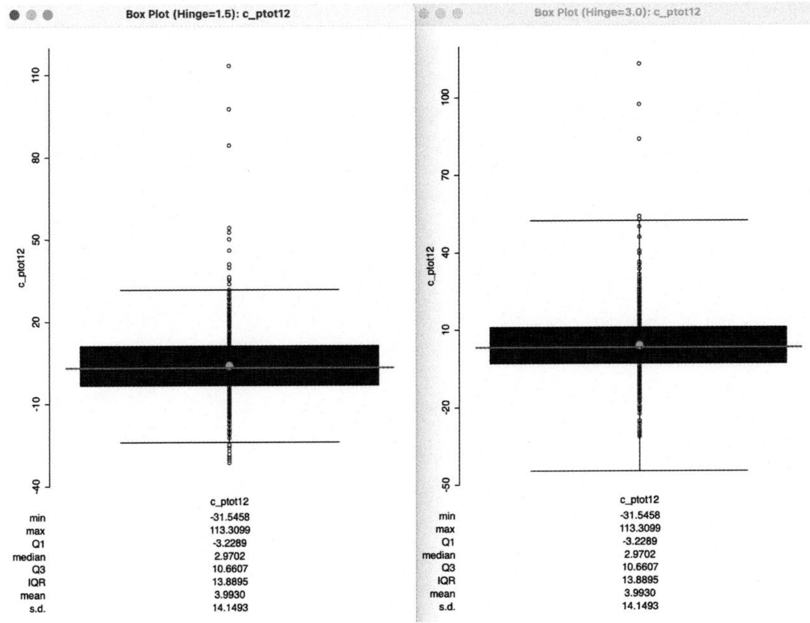

Figure 7.8: Box plot for population change 2020–2010

7.2.2 Box plot

The box plot is an alternative visualization of the distribution of a single variable that focuses on *quartiles* (see also Section 5.2.2 for a discussion of the associated *box map*). The observations are ranked from lowest to highest and (optionally) represented as dots on a vertical or horizontal line.[3] The *box* consists of a rectangle, with the lower bound drawn at the first quartile (25%) and the higher bound at the third quartile (75%). Typically, a line is also drawn at the location of the *median* (50%).

The box plot introduces the notion of *outliers*, i.e., observations that are far from the central tendency (the median), suggesting they may not belong to the same distribution as the rest. The key concept in this regard is that of the *inter-quartile range* or IQR, the difference between the values at the third and first quartile. This is a measure of the spread of the distribution around the median, a counterpart to the variance. The IQR is used to compute a *hinge* (or fence), i.e., a value that is 1.5 times (or, alternatively, 3 times) the IQR above the third quartile or below the first quartile. Observations that lie outside the hinges are designated as outliers.

7.2.2.1 Implementation

The box map is invoked as **Explore > Box Plot** from the menu, or by selecting the **Box Plot** as the second icon from the left in the toolbar in Figure 7.1. Identical to the approach followed for the histogram, next appears a **Variable Settings** dialog to select the variable.[4]

To illustrate this graph, we select the variable **c_ptot12**, the percentage population change between 2020 and 2010 (positive values are population growth). The default box plot, with

[3] `GeoDa` takes the vertical approach.

[4] In `GeoDa`, the default is that the variable from any previous analysis is already selected.

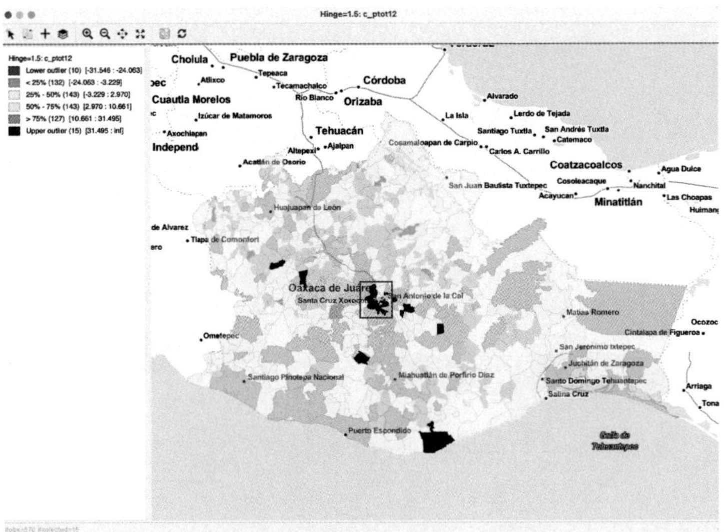

Figure 7.9: Box map for population change 2020–2010 (upper outliers selected)

a hinge of 1.5, is shown in the left-hand panel of Figure 7.8. The descriptive statistics are listed at the bottom.[5]

The observations range from −31.2% to 113.3%, with a slightly positive median of 3.0% (the mean is 4.0%, reflecting the influence of upper outliers). The interquartile range is 13.9. Consequently, the upper hinge is roughly 10.7 (Q3) + 1.5 × 13.9, or 37.9%. Fifteen observations take on values that are larger than this upper hinge and are designated as *upper outliers*. The lower hinge is −3.2 − 1.5 × 13.9, or −21.1%. Ten observations have population decreases that are even larger, hence these are *lower outliers*.

7.2.2.2 Box plot and box map

Selections in the box plot can be linked to a map, and vice versa. However, to a large extent this is already accomplished by means of the box map, discussed in Section 5.2.2.

The interesting research question is the extent to which *a-spatial* outlying observations, such as the fifteen outliers identified in the box plot in Figure 7.8 show a spatial pattern that may suggest some structure, rather than randomness. The box map for **c_ptot12** in Figure 7.9 shows the outlying observations highlighted in dark red (selected in the box plot).[6] The map indicates that about ten of the outliers are located in the center of the country (inside the blue rectangle). The research question is then whether this is purely due to chance, or, instead, suggests a pattern of clustering. This is pursued more formally in Parts IV and V.

7.2.2.3 Box plot options

Several of the box plot options are shared with the other graphs (and maps). There are seven categories. The last three are **Save Selection**, **Copy Image To Clipboard** and **Save Image As**, which work as before. The **Show Status Bar** is listed as a main option (checked by default) and the **Color** option also provides a way to change the color of the points (in addition to the background color).

[5] For the box plot, this is the default. It can be turned off by unchecking **View > Display Statistics** in the options.

[6] The basemap is **Stamen > TonerLite**.

The **View** option contains four items, all self-explanatory: **Set Display Precision**, **Set Display Precision on Axes**, **Display Statistics** and **Show Vertical Axis**.

The typical multiplier for the IQR to determine outliers is 1.5 (roughly equivalent to the practice of using two standard deviations in a parametric setting). However, a value of 3.0 is fairly common as well, which considers only truly extreme observations as outliers. The multiplier to determine the fence can be changed with the **Hinge > 3.0** option (obtained by right clicking in the plot to select the options menu, and then choosing the hinge value). This yields the box plot shown in the right-hand panel of Figure 7.8. The new hinge no longer yields lower outliers, and the upper outliers are reduced to five extreme observations.

The main purpose of the box plot in an exploratory strategy is to identify outlier observations in an a-spatial sense. Its spatial counterpart is the box map.

7.3 Bivariate Analysis – The Scatter Plot

The most commonly used tool to assess the relationship between two variables is the scatter plot, a diagram with two orthogonal axes, each corresponding to one of the variables. The observation (X, Y) coordinate pairs are plotted as points in the diagram.

Typically, the relationship between the two variables is summarized by means of a linear regression fit through the points. The regression is:

$$y = a + bx + \epsilon,$$

where a is the intercept, b is the slope and ϵ is a random error term. The coefficients are estimated by minimizing the sum of squared residuals, a so-called least squares fit, or *ordinary least squares* (OLS).

The intercept a is the average of the dependent variable (y) when the explanatory variable (x) is zero. The slope shows how much the dependent variable changes on average (Δy) for a one unit change in the explanatory variable (Δx). It is important to keep in mind that the regression applied to the scatter plot pertains to a *linear* relationship. It may not be appropriate when the variables are non-linearly related (e.g., a U shape).

When y and x are standardized (mean zero and variance one), then the slope of the regression line is the same as the correlation between the two variables (the intercept is zero). Note that while correlation is a symmetric relationship, regression is not, and the slope will be different when the explanatory variable takes on the role of dependent variable, or vice versa. Just as a linear regression fit may not be appropriate when the variables are related in a non-linear way, so is the correlation coefficient. It only measures a *linear* relationship between two variables, which is clearly demonstrated by its equivalence to the regression slope between standardized variables.

7.3.1 Scatter plot basics

A scatter plot is created by selecting **Explore > Scatter Plot** from the menu, or by clicking on the matching toolbar icon, the third from the left in Figure 7.1. This brings up the **Scatter Plot Variables** dialog, where two variables need to be selected: one for the X axis (**Independent Var X**) and one for the Y axis (**Dependent Var Y**).

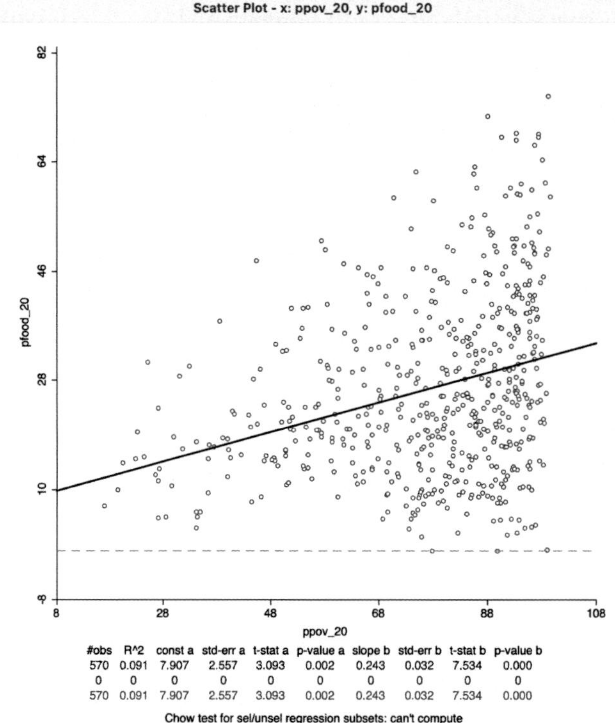

Figure 7.10: Scatter plot of food insecurity on poverty

To illustrate this graph, the X variable is **ppov_20** (percent population living in poverty in 2020) and the Y variable **pfood_20** (percent population with food insecurity in 2020). The resulting scatter plot, with all the default settings, is as in Figure 7.10.

The top of the graph spells out the X and Y variables, which are also listed along the axes. The linear regression fit is drawn over the points, with summary statistics listed below the graph. These included the fit (R^2) and the estimates for constant and slope, with estimated standard error, t-statistic and associated p-value. In the example, the fit is a low 0.091, which is not surprising, given the cone shape of the scatter plot, rather than the *ideal* cigar shape. Nevertheless, both intercept and slope coefficients are highly significant (rejecting the null hypothesis that they are zero) with p-values of 0.002 and 0.000 respectively.

The positive relationship suggests that as poverty goes up by one percent, the share in food insecurity increases by 0.24 percent.

In the current setup, no observations are selected, so that the second line in the statistical summary in Figure 7.10 (all red zeros) has no values. This line pertains to the selected observations. The blue line at the bottom relates to the unselected observation. The sum of the number of observations in each of the two subsets always equals the total number of observations, listed on the top line. The three lines are included because of the default **View** setting of **Regimes Regression**, even though there is currently no active selection (see Section 7.3.1.1).

The scatter plot has several options, invoked in the customary fashion by right clicking on the graph. They are grouped into eight categories:

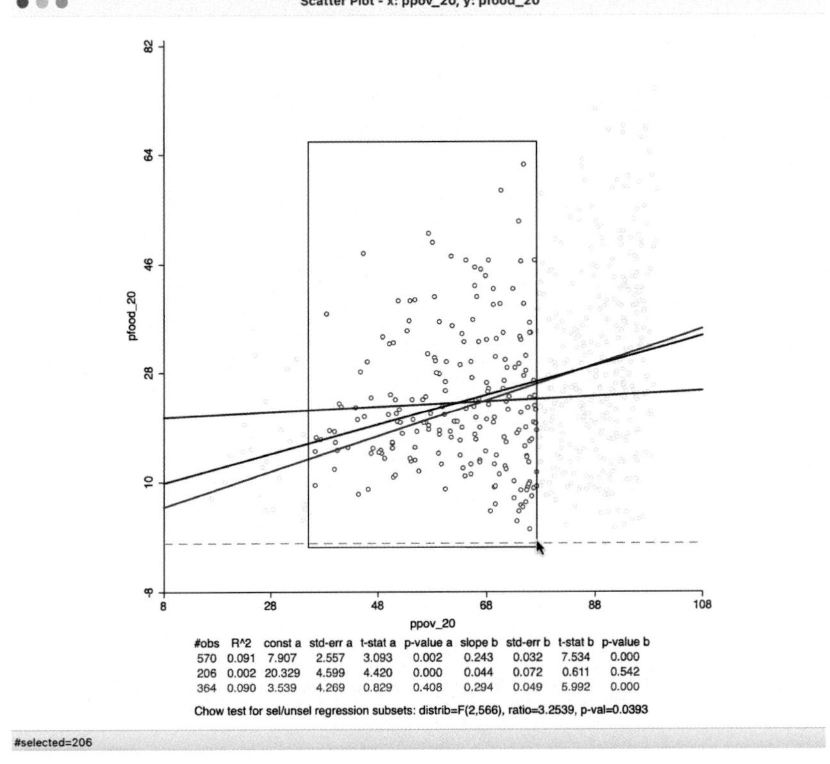

Figure 7.11: Regime regression

- **Selection Shape**
- **Data**
- **Smoother**
- **View**
- **Color**
- **Save Selection**
- **Copy Image to Clipboard**
- **Save Image As**

Selection Shape works as in the other graphs and maps (see Section 2.5.4), and so do the three last options. The **Color** options allow for the **Regression Line Color**, **Point Color** and **Background Color** to be set. The remaining options are discussed in more detail below.

7.3.1.1 View options

The **View** option manages seven settings, all but the first two checked by default:

- **Set Display Precision**
- **Set Display Precision on Axes**
- **Fixed Aspect Ratio Mode**
- **Regimes Regression**
- **Statistics**
- **Axes Through Origin**
- **Status Bar**

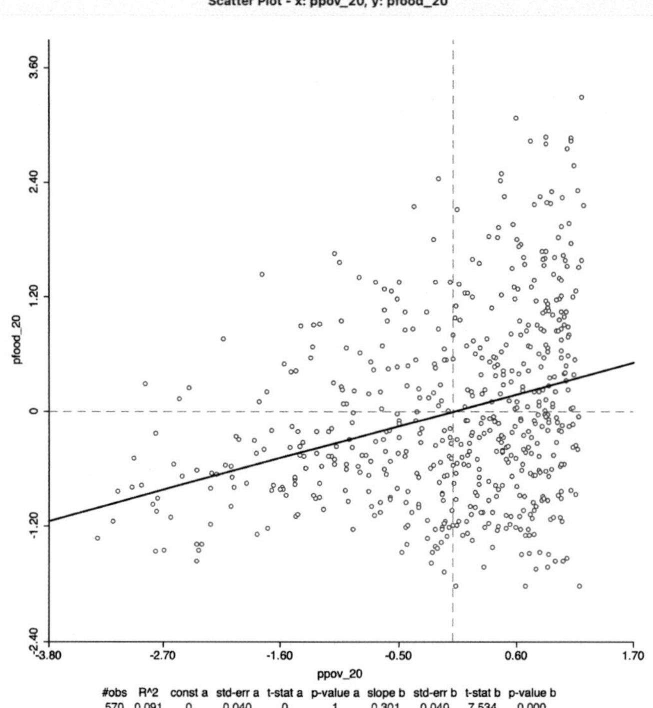

Figure 7.12: Correlation

These options are mostly self-explanatory, or work in the same way as before (such as **Statistics**), except for the **Regimes Regression**.

The latter allows for a dynamic assessment of *structural stability* by comparing the regression line for a *selected* subset of the data to its complement (unselected). For example, in Figure 7.11, 206 observations have been selected (inside the selection rectangle). This yields three regression lines: a purple one for the full data set (same results as in Figure 7.10), a red one for the *selected* observations and a blue one for its complement (384 observations). With each regression line corresponds a line in the statistics summary. In Figure 7.11, the selected observations show no significant linear relationship between the two variables. In other words, for a range of poverty roughly between 30% and 70%, there is no linear relationship with food insecurity (p-value of 0.542), whereas for the whole and for the complement, there is. This is an indication of structural instability.

A formal assessment of structural stability is based on the **Chow test** (Chow, 1960). This statistic is computed from a comparison of the fit in the overall regression to the combination of the fits of the separate regressions, while taking into account the number of regressors (k). In our simple example, there is only the intercept and the slope, so k = 2. The residual sum of squares can be computed for the full regression (RSS) and for the two subsets, say RSS_1 and RSS_2. Then, the Chow test follows as:

$$C = \frac{(\text{RSS} - (\text{RSS}_1 + \text{RSS}_2))/k}{(\text{RSS}_1 + \text{RSS}_2)/(n - 2k)},$$

distributed as an F statistic with k and $n - 2k$ degrees of freedom. Alternatively, the statistic can also be expressed as having a chi-squared distribution, which is more appropriate when

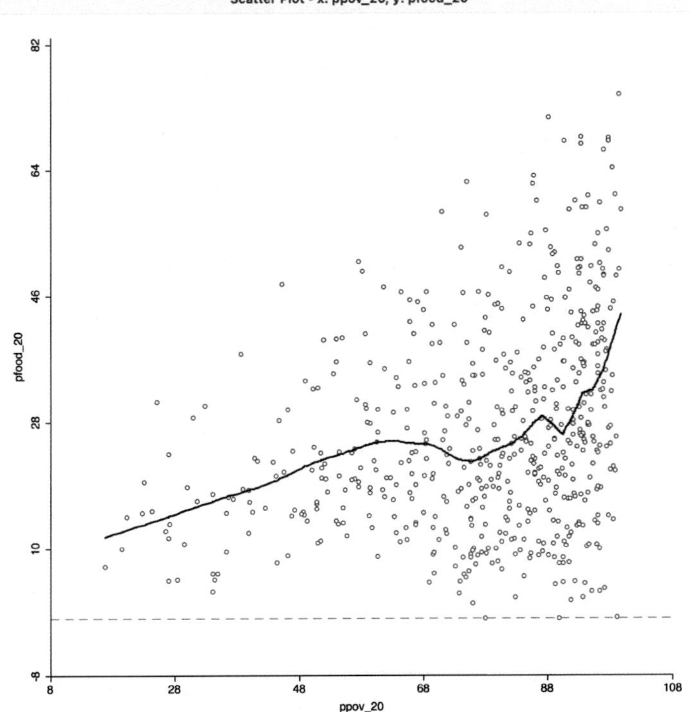

Figure 7.13: Default LOWESS local regression fit

multiple breaks and several coefficients are considered.[7] In our example, the p-value of the Chow test is 0.039, which is only weakly significant (in part this is due to the fact that the overall regression coefficient of 0.243 is not that different from zero itself). The assessment of structural stability is considered further in the context of spatial heterogeneity in Section 7.5.2.

The **Regimes Regression** option is on by default. Turning it off disables the effect of any selection and reduces the **Statistics** to a single line.

7.3.1.2 Data options

The default setting for the **Data** option is to **View Original Data**. In other words, the points in the scatter plot correspond to the respective entries for the variables in the data table. With this option turned to **View Standardized Data**, each variable is used in standardized form. As a result, the axes can be interpreted as expressing standard deviational units. Values larger than two in absolute value indicate potential outliers.

The slope of the regression line applied to the standardized variables corresponds to the *correlation coefficient*. The intercept will be zero.

In Figure 7.12, some outliers can be detected at the lower end for **ppov_20**, but not at the upper end. For, **pfood_20**, the reverse is the case, with several upper outliers, but no lower outliers. The correlation coefficient is 0.301.

[7]Technical details can be found in in Anselin and Rey (2014), pp. 287–289.

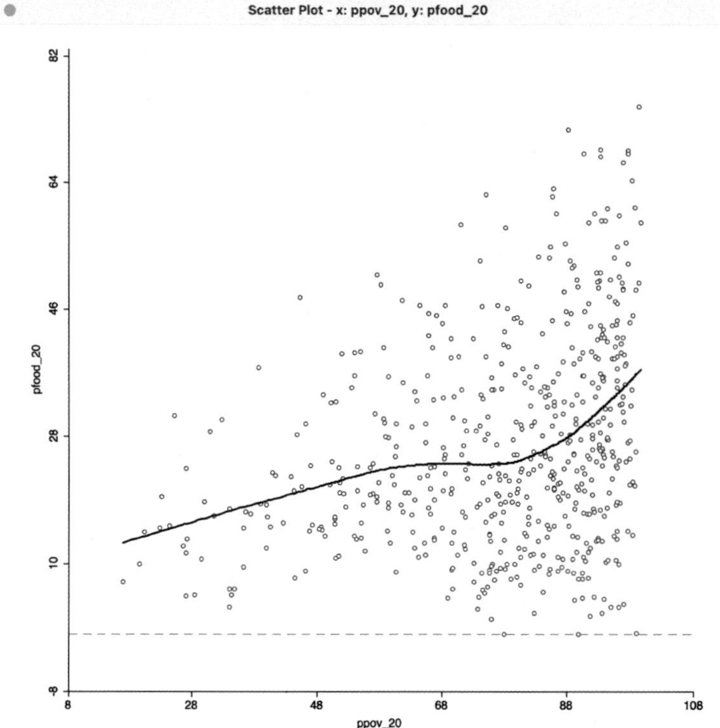

Figure 7.14: Default LOWESS local regression fit with bandwidth 0.6

7.3.2 Smoother option – local regression

The default fit to the scatter plot points is a *global* linear regression. This can be turned off by means of the **Smoother > Show Linear Smoother** option.

When the relationship between the two variables in the scatter plot is not linear, or shows complex sub-patterns in the data, a linear fit may not be appropriate. An alternative consists of a so-called *local* regression fit. A local regression fit is a nonlinear procedure that computes the slope from a subset of observations on a small interval on the X-axis. In addition, values further removed from the center of the interval may be weighted down (similar to a kernel smoother). As the interval (the bandwidth) moves over the full range of observations, the different slopes are combined into a smooth curve (for a detailed overview of the methodological issues, see, e.g., Cleveland, 1979; and Loader, 1999; Loader, 2004).

A local regression fit reveals potential nonlinearities in the bivariate relationship or may suggest the presence of structural breaks. Two common implementations are LOWESS, locally weighted scatter plot smoother, and LOESS, a local polynomial regression. The two are often confused, but implement different fitting algorithms.

The option **Smoother > Show LOWESS Smoother** calculates a local regression fit based on the LOWESS algorithm. Continuing with the same two variables, the result is as in Figure 7.13, using the default settings.

The smoothness of the local fit depends on a few parameters, considered next. Note that the **Regimes Regression** setting has no effect on the LOWESS fit. Also, in Figure 7.13, the

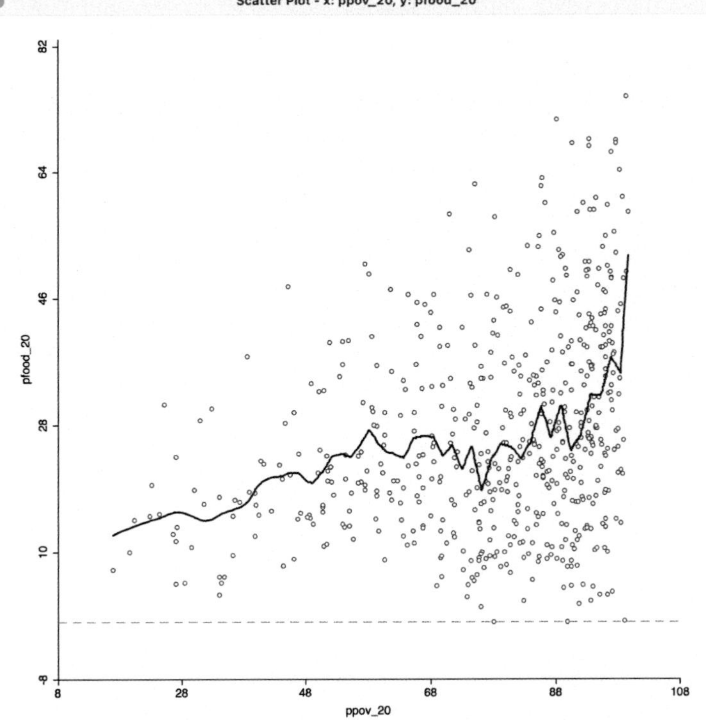

Figure 7.15: Default LOWESS local regression fit with bandwidth 0.05

linear fit has been turned off. However, in some instances, it may be insightful to have both fits showing in the scatter plot.

In the example, the curve shows a fairly regular, near linear increase up to a poverty level of about 60%, but a much more irregular pattern at higher poverty rates, with a very steep positive relationship at the very high end. In the middle range (the same range of **ppov_20** as selected in Figure 7.11), the curve is almost flat, in rough agreement with what was found in the regime regression example.

7.3.2.1 LOWESS parameters

The nonlinear fit is driven by a number of parameters, the most important of which is the *bandwidth*, i.e., the range of values on the X-axis on which the local slope estimate is based.

The smoothing parameters can be changed in the options by selecting **Edit LOWESS Parameters** in the **Smoother** option. There are three parameters: **Bandwidth** (default setting 0.20), **Iterations** and **Delta Factor**.

The bandwidth determines the smoothness of the curve and is given as a fraction of the total range in X values. In other words, the default bandwidth of 0.20 implies that for each local fit (centered on a value for X), about one fifth of the range of X-values is taken into account. The other options are technical and are best left to their default settings.[8]

[8]The LOWESS algorithm is complex and uses a weighted local polynomial fit. The **Iterations** setting determines how many times the fit is adjusted by refining the weights. A smaller value for this option will speed up computation, but result in a less robust fit. The **Delta Factor** drops points from the calculation of the local fit if they are too close (within Delta) to speed up the computations. Technical details are covered in Cleveland (1979).

A higher value of the bandwidth results in a smoother curve. For example, with the bandwidth set to 0.6, the curve in Figure 7.14 results. This more clearly suggests different patterns for three subsets of poverty rates: a slowly increasing slope at the lower end, a near flat slope in the middle and a much steeper slope at the upper end. In a data exploration, one would be interested in finding out whether these a-spatial subsets have an interesting spatial counterpart, for example, through linking with a map.

The opposite effect is obtained when the bandwidth is made smaller. For example, with a value of 0.05, the resulting curve in Figure 7.15 is much more jagged and less informative.

The literature contains many discussions of the notion of an optimal bandwidth, but in practice a trial and error approach is often more effective. In any case, a value for the bandwidth that follows one of these rules of thumb can be entered in the dialog. Currently, `GeoDa` does not compute optimal bandwidth values.

Finally, while one might expect the LOWESS fit and a linear fit to coincide with a bandwidth of 1.0, this is not the case. The LOWESS fit will be near-linear, but slightly different from a standard least squares result due to the locally weighted nature of the algorithm.

7.4 Scatter Plot Matrix

A scatter plot matrix visualizes the *bivariate relationships* among several pairs of variables. The individual scatter plots are stacked such that each variable is in turn on the X-axis and on the Y-axis.

When applied to standardized variables (with mean zero and variance one), it is the visual counterpart of a correlation matrix. In general, however, while the scatter plot matrix shows the linear association between variables, the graphs are not symmetric. In addition, in some instances a variable may not be appropriate to be both a dependent (Y-axis) and an explanatory variable (X-axis). For example, a variable such as elevation may affect socio-economic outcomes, but it is hard to imagine that socio-economic factors could in turn affect elevation.

The main interest is in the magnitude and sign of the slope in each of the scatter plots, and the extent to which this points to a significant bivariate relationship, similar to the insight provided by a correlation matrix. In `GeoDa`, the diagonal elements also contain a histogram for the variable in the corresponding row/column to provide a sense of the univariate distribution.

Even though this method deals with multiple variables, it is not really a *multivariate* exploration. Rather, it consists of a collection of *bivariate* explorations for multiple variables.

7.4.1 Implementation

The scatter plot matrix operation is started by selecting the fourth icon on the EDA toolbar (Figure 7.1), or by choosing **Explore > Scatter Plot Matrix** from the menu.

This brings up a **Scatter Plot Matrix Variables Add/Remove** dialog, through which the variables are selected. The design of the interface is such that one selects a variable from the **Variables** list on the left and clicks on the right arrow > to include it in the **Include**

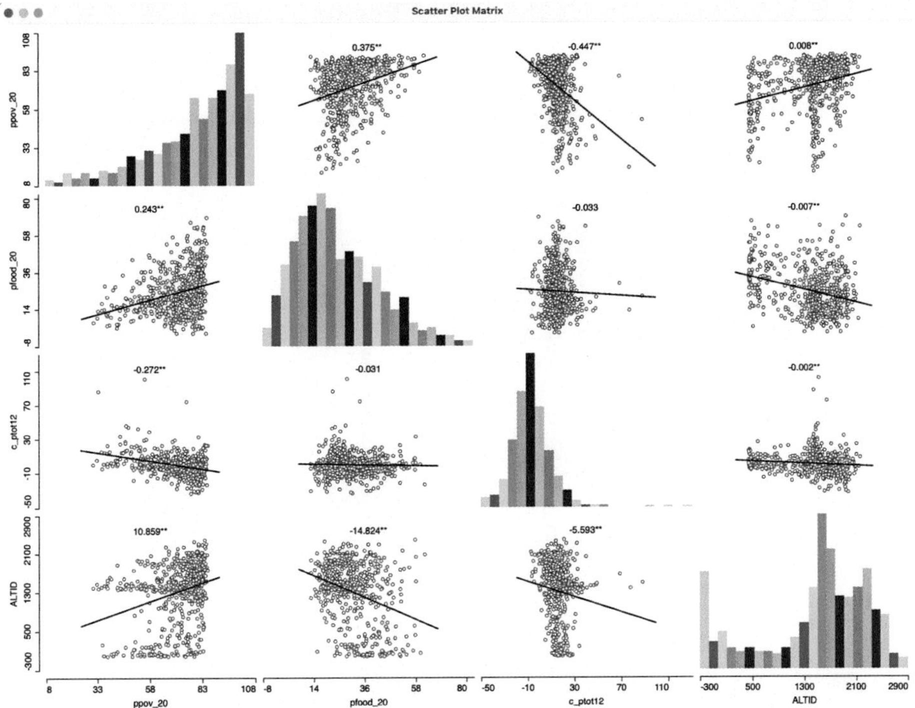

Figure 7.16: Scatter plot matrix

list on the right. Alternatively, one can double click on the variable name to move it to the right-hand column. The left arrow < is there to remove a variable from the **Include** list.

As soon as two variables are selected, the scatter plot matrix is rendered in the background. As new variables are added to the **Include** list, the matrix in the background is updated with the additional scatter plots.

Figure 7.16 shows an example with four variables that were already used in the illustrations above: **ppov_20**, **pfood_20**, **c_ptot12** and **ALTID**. Note that the latter really should not be portrayed on the Y-axis, although by construction this always happens for every variable in the scatter plot matrix.

With the default setting, each scatter plot shows the slope coefficient at the top, together with a designation of significance (** is $p < 0.01$, * is $p < 0.05$). Ignoring the bottom row (with ALTID on the Y-axis), significant coefficients are obtained in all instances, except between **pfood_20** and **c_ptot12**. In other words, food insecurity and population change do not seem to be related. On the other hand, there is a strong and negative relationship between **ppov_20** and **c_ptot12**. Altitude is significantly related with all three socio-economic outcomes, positively with **ppov_20** (greater poverty at higher altitudes), but negatively with the other two. A substantive interpretation is beyond the scope.

7.4.1.1 Scatter plot matrix options

The scatter plot matrix options consist of seven items:

- **Add/Remove Variables**
- **Selection Shape**

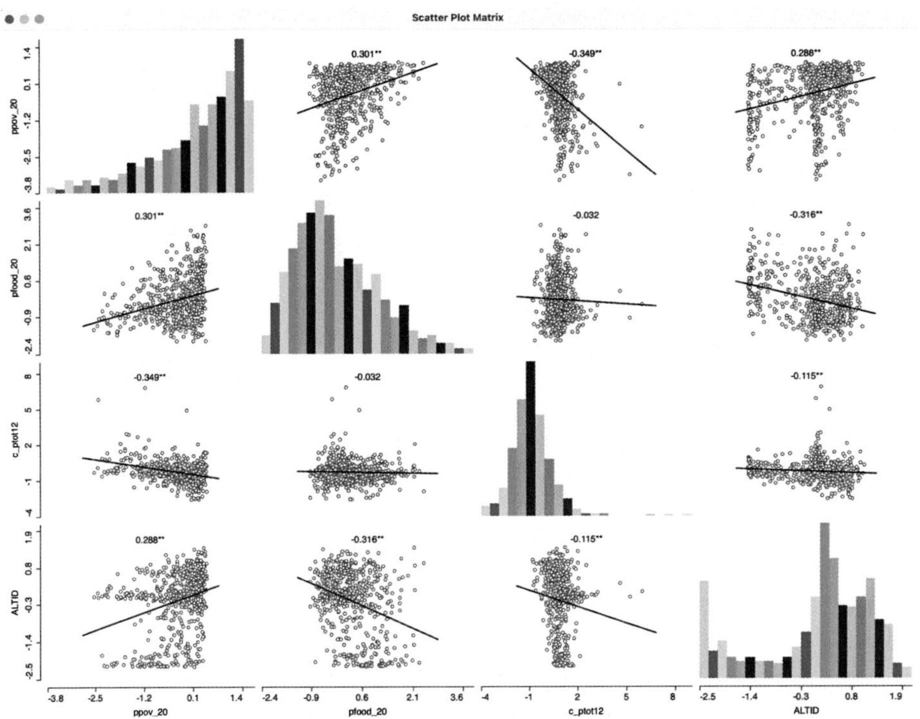

Figure 7.17: Scatter plot correlation matrix

- **Data**
- **Smoother**
- **View**
- **Color**
- **Save Image As**

The first option invokes the **Scatter Plot Matrix Variables Add/Remove** dialog to allow for changes in the variable selection. **Selection Shape** and **Save Image As** work in the usual fashion.

View has options to **Set Display Precision on Axes**, **Regimes Regression** and **Display Slope Values**. The latter is checked by default. In contrast to the standalone scatter plot, **Regimes Regression** is turned off by default in the scatter plot matrix. However, it can readily be invoked, allowing for all features of linking and brushing, e.g., to explore spatial heterogeneity across all variables (see Section 7.5.2).

The **Color** option provides a way to change the **Regression Line Color** and **Point Color**. **Data** and **Smoother** are separately considered next.

7.4.1.2 Correlation matrix

The **Data** option contains the same two items as for the standalone scatter plot. With **Data > View Standardized Data**, the scatter plot matrix turns into a correlation matrix. As pointed out, in contrast to the regression slope, the correlation coefficient is fully symmetric and the results above the diagonal are the mirror image of the results below the diagonal. On the graph, they may look slightly different due to the use of different scales on the axes, but the coefficients are identical.

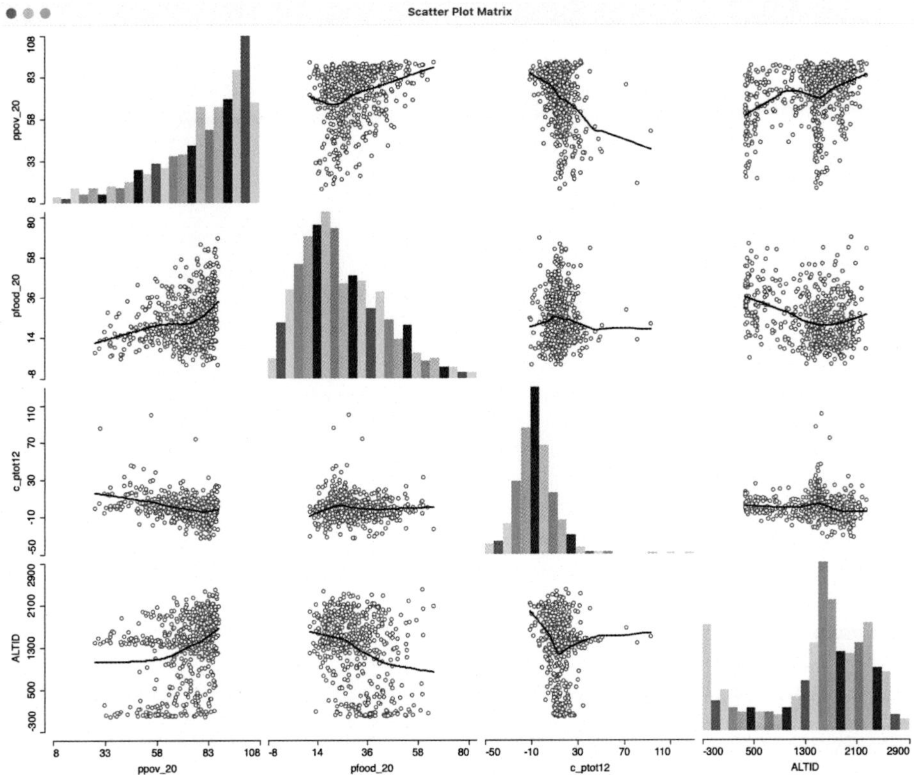

Figure 7.18: Scatter plot matrix with LOWESS fit

The result for the four variables considered before is as in Figure 7.17.

7.4.1.3 LOWESS scatter plot matrix

Finally, the **Smoother** option allows for local regression, as in the standalone scatter plot, but now applied to all bivariate relationships. This provides a quick overview of the extent to which these relationships are linear, or instead show local instability and nonlinear patterns. The LOWESS parameters can be edited in the same way as before.

The result of a LOWESS fit for a bandwidth of 0.6 is illustrated in Figure 7.18. In order to make sure that the regression coefficient estimates are no longer shown, the option **Display Slope Values** must be turned off (it remains on by default). Also, in the figure, the linear fits are not shown.

The result shows some interesting patterns, especially for the relationship between altitude and respectively poverty and food insecurity. Instead of the strong overall positive and negative slopes, the evidence is much more nuanced. In fact, especially for the altitude-food insecurity graph, the negative slope for lower altitudes turns into a positive slope for the higher altitudes. Such patterns cannot be distinguished in a purely linear approach.

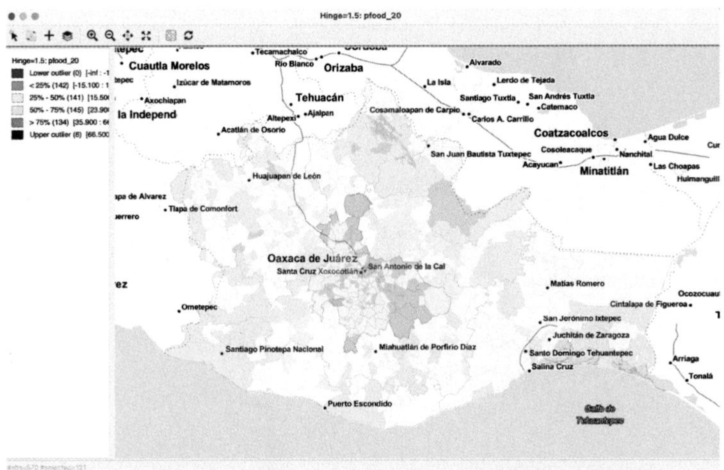

Figure 7.19: Food insecurity 2020 in Valles Centrales

7.5 Spatial Heterogeneity

Together with spatial dependence, *spatial heterogeneity* is the other critical aspect in any spatial data analysis. In essence, its presence suggests that more than one underlying distribution may be responsible for generating the observed sample. This is evidenced in the form of structural breaks, such as different mean or median values in different subsets of the data, or different slopes in a bivariate regression over observations in those subsets (see Anselin, 1988, for a more formal treatment).

The *spatial* aspect in the heterogeneity comes from the nature of the subsets in the data, which are spatially defined. Examples are the difference between center and periphery, east and west and north and south. In an exploratory data analysis, spatial heterogeneity can be assessed by selection in a map *linked* to a statistical graph. An application of this idea was illustrated in the context of a linked map and histogram (Section 7.2.1.4). Next, two further implementations of this approach are considered. One is the analysis of differences in means through the **Averages Chart**, the other the investigation of structural stability in a scatter plot by means of brushing a map linked to the scatter plot.

7.5.1 Averages chart

The **Averages Chart** is an implementation of a simple test on the difference in means between selected and unselected observations. Its most meaningful use is in the context of observations at different points in time (see Chapter 9), but it is equally applicable in a cross-sectional setting, illustrated here.

The core functionality of this chart is to illustrate and quantify the difference between the mean of a variable for *selected* observations and *unselected* observations (the complement). In GeoDa, this is not implemented as a traditional t-test, but rather as an F-statistic for a regression that includes an indicator variable for the selection (i.e., value = 1 for selected, and zero otherwise). The F-statistic on the significance of the joint slopes in that regression

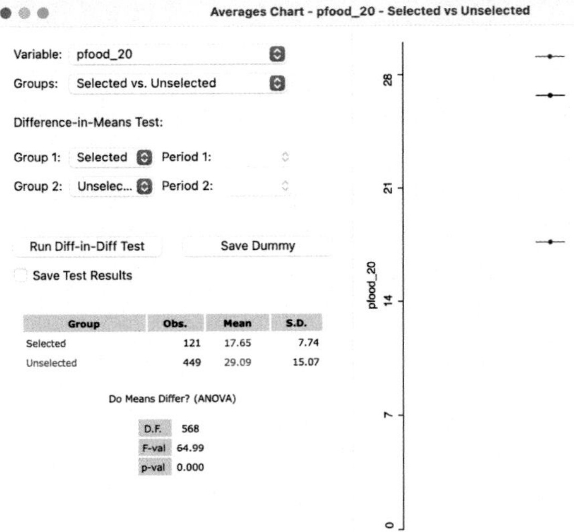

Figure 7.20: Averages Chart – Valles Centrales

is equivalent to a t-test on the coefficient of the indicator variable, since there is only one slope.

This F-statistic is basically a test on whether there is a significant gain in explanation in the regression beyond the overall mean (i.e., the constant term). Formally, the statistic uses the sum of squared residuals in the regression RSS and the sum of squared deviations from the mean for the dependent variable RSY. The statistic follows as:

$$F = \frac{RSS - RSY}{k-1} / \frac{RSY}{n-k},$$

with k as the number of explanatory variables. In our simple dummy variable regression, $k = 2$, so that the degrees of freedom for the F-statistic are $1, n-2$ (see also Anselin and Rey, 2014, pp. 98–99).

7.5.1.1 Selected and unselected observations

The averages chart is invoked from the menu as **Explore > Averages Chart**. However, its toolbar icon is not part of the EDA group, but instead is included as the right-most icon in the time management toolbar in Figure 9.1. Its most effective application is in a space-time context, but here its use is illustrated in a cross-sectional setup.

The example explores the difference in mean food insecurity in 2020 between the Oaxaca Central Valleys and the rest of the country. The so-called Valles Centrales are the location of the original Zapotec civilization and are still characterized by a large indigenous population. They also contain the state capital of Oaxaca.

The overall spatial distribution of **pfood_20** is shown in the box map in Figure 7.19, with the central valley municipalities highlighted. The selection is obtained by setting **region = 8** in the selection tool (see Section 2.5.1).

The selection is used in the averages chart to define **selected** and **unselected** as the **Groups**. In Figure 7.20, the **Variable** is specified as **pfood_20**. In the current context,

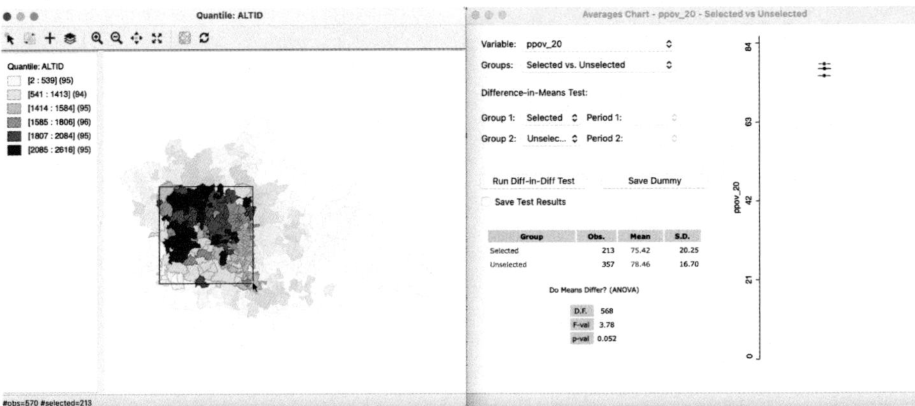

Figure 7.21: Map brushing and the averages chart – 1

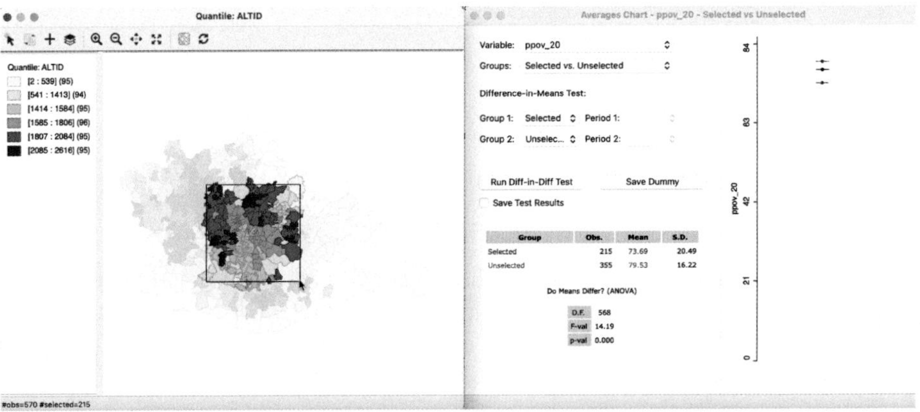

Figure 7.22: Map brushing and the averages chart – 2

the time settings can be ignored. The statistics for the two groups are listed in a small table. **Selected** has 121 observations with a **Mean** of 17.65 and **S.D.** of 7.74. In contrast, the **Unselected** group contains 449 observations, with **Mean** of 29.09 and **S.D.** of 15.07.

The formal test on equality of means results in an F-statistic of 64.99, which yields a very small p-value (essentially zero) for 568 degrees of freedom.

In the right-hand panel of the window, the overall mean (black), selected mean (red) and unselected mean (blue) are represented graphically.

In this case, there is strong evidence that food insecurity is less severe in the Central Valleys compared to the rest of the state.

7.5.1.2 Map brushing and the averages chart

A more comprehensive exploration of spatial heterogeneity is achieved by *brushing* a map linked to the averages chart. As the selection in the map changes, the statistics in the averages chart are updated. As a result, one can assess the extent to which subregions of the data have a different mean for the variable under consideration.

For example, Figure 7.21 has a six category quantile map for elevation on the left-hand side. As the *brush* is moved from west to east, the effect on the test of means can be assessed in

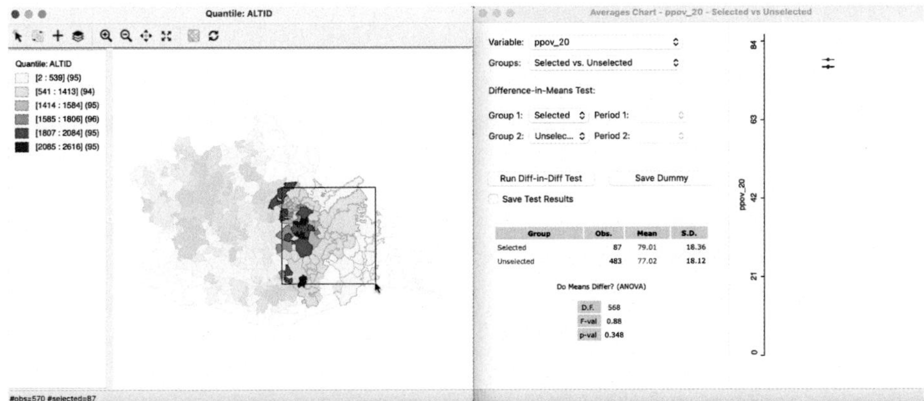

Figure 7.23: Map brushing and the averages chart – 3

the averages chart on the right. Note that the brushing is not based on the altitude variable, but is purely a move over space. However, by including a different variable, one may be able to discover what lies behind the spatial structural differences.

The variable under consideration in the averages chart is **ppov_20**. The first selection contains 213 observations with a mean of 75.42, compared to a mean of 78.46 for the rest. The test on the difference between the means is not significant at p = 0.052. Thus this initial selection of higher elevation locations is not substantially different from the rest of the state.

In Figure 7.22, the selection is moved to the center of the state, with 215 selected observations, yielding a mean of 73.69. The unselected mean is 79.53. For this selection, the test rejects the null hypothesis with a p-value of 0.000, suggesting strong spatial heterogeneity.

Finally, in Figure 7.23, the selection is moved even further to the east. The 87 selected observations have a mean of 79.01, whereas the complement has a mean of 77.02. The test on difference between the means is not significant at p = 0.348.

By moving the brush over different regions of the map, the extent of spatial structural instability can be assessed. In addition, by using a map for a different variable (such as altitude here), one can possibly gain insight into factors that may be behind the spatial structural instability. However, for this to be meaningful, one has to make sure that sufficient observations are contained in each selection.

7.5.1.3 Averages chart options

The default setting for the averages chart is to have a **Fixed scale over time**. This means that the minimum and maximum tick marks on the left side axis remain the same as the selection changes. In some instances, the respective means may no longer fall within this range. As a result, they will not be shown in the graph. To remedy this, the **Axis Option** can be set to **Enable User Defined Value Range of Y-Axis**, through which the range can be customized.

A final option is to **Set Display Precision of Y-Axis**.

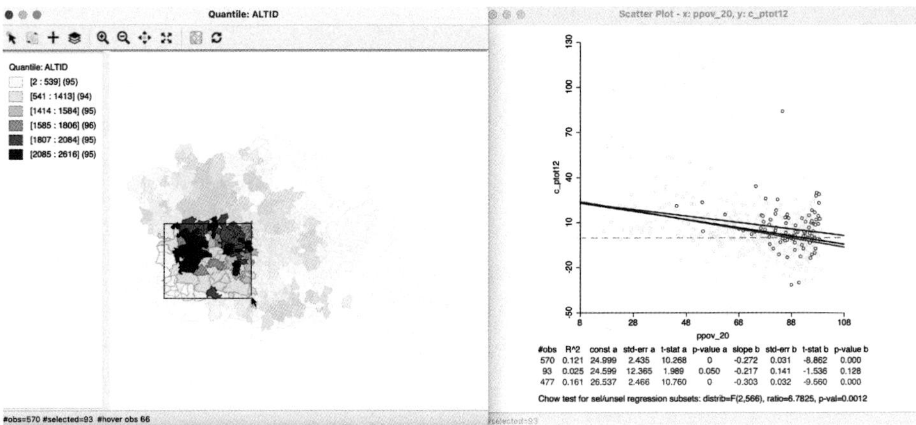

Figure 7.24: Map brushing and the scatter plot – 1

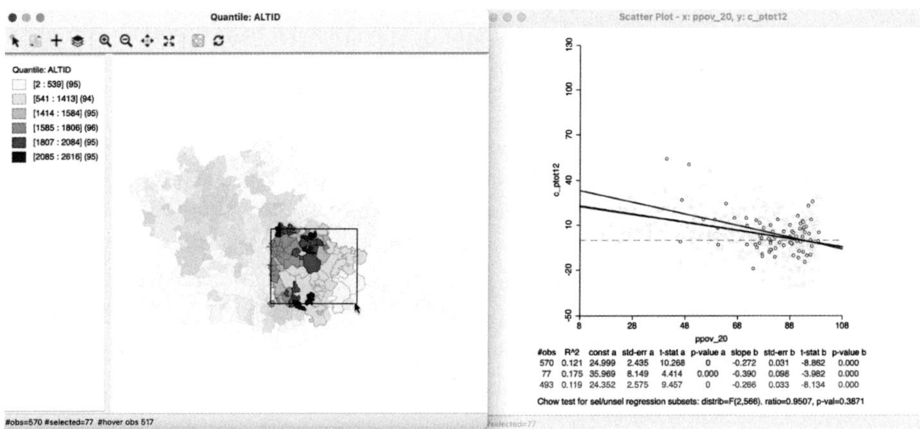

Figure 7.25: Map brushing and the scatter plot – 2

7.5.2 Brushing the scatter plot

7.5.2.1 Scatter plot brushing

The concept of scatter plot *brushing* was initially suggested by Stuetzle (1987), and extended to the map context by Monmonier (1989). The idea is to dynamically adjust the selection of observations in the scatter plot within a selection *brush* (typically a rectangle). As the brush moves over the plot, observations are added to and dropped from the selection and the slope of the linear fit is adjusted. As discussed in Section 7.3.1.1, this is implemented in GeoDa through the **Regimes Regression** option (see also Figure 7.11).

Since all windows are simultaneously linked in the basic architecture of the GeoDa software (Anselin et al., 2006b), the dynamic selection through brushing can be initiated in any open window. This results in an immediate adjustment of the slope of the linear fit in the scatter plot and an updated computation of the Chow test. As in the averages chart, *spatial heterogeneity* can be assessed by initiating a *spatial* selection through the brushing operation in a map.

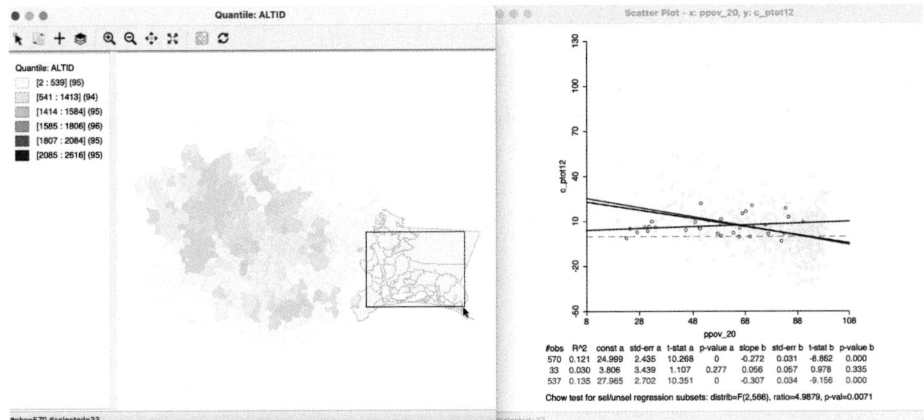

Figure 7.26: Map brushing and the scatter plot – 3

7.5.2.2 Map brushing and scatter plot

To illustrate the dynamic map brushing operation, a scatter plot of **c_ptot12** on **ppov_20** is considered jointly with the elevation map used before. The global linear fit yields a slope coefficient estimate of −0.272, with an R^2 of 0.121. The brush moves from west to east.

In Figure 7.24, the brush is initiated by selecting 93 observations in the higher altitude region in the western part of the state. The regression slope of the selected observations is −0.217, compared to −0.303 for the unselected, yielding a significant Chow test result at p = 0.0012. Note that the coefficient for the selected observations is not significant (p = 0.128).

As the brush is moved to the right, the selection is adjusted. In Figure 7.25, the slope of the 77 selected observations has changed to −0.390, compared to −0.266 for the unselected. Even though the estimates are both significant, they cannot be deemed to be different, with the Chow test yielding a p-value of 0.387.

In Figure 7.26, the brush is moved even further to the east, yielding a selection of 33 observations. Note that since the brush size is fixed, but the spatial extent of the municipalities varies, the number of observations contained in each spatial selection will not be constant. The municipalities in this part of the state tend to be larger, resulting in fewer observations in the selection window. This will affect the precision of the estimates in the subset, and, indirectly, also the Chow test statistic.

At this point, the selected observations show no significant relationship, which a slightly positive coefficient of 0.056 (p-value of 0.335). In contrast, the coefficient in the unselected observations is negative at −0.307, and highly significant.

A careful assessment of the effect of different spatial selections can provide insight into spatially defined structural instabilities in the relationship between any two variables. The inclusion of a Chow test provides some guidance, even though its results need to be interpreted with caution due to the problem of multiple comparisons (many tests carried out with the same data). However, it allows for a more quantitative measure of the spatial heterogeneity, instead of relying on a purely visual assessment, which can be misleading. This is in the spirit of the more recently developed perspectives on EDA, e.g., as discussed in Section 4.2.1.

8

Multivariate Data Exploration

In this chapter, I switch to a full multivariate exploration of spatial data, where the focus is on the potential *interaction* among multiple variables. This is distinct from a univariate (or even bivariate) analysis of multiple variables, where the properties of a distribution are considered in isolation. The goal is to discover potential pathways of interaction. For example, in a bivariate analysis, one may have found a strong correlation between lung cancer and socio-economic factors, but after *controlling for* smoking behavior (itself strongly correlated with SES), this relationship disappears.

The methods considered share the same objective, i.e., how to represent relationships in higher dimensions on a two-dimensional screen (or piece of paper). Three techniques, the bubble chart, three-dimensional scatter plot and conditional plots are limited to the analysis of three to four variables. True multivariate analyses for several variables (more than four) can be carried out by means of the parallel coordinate plot (PCP). Each is considered in turn. I continue to use the *Oaxaca Development* data set for illustrations.

As in the previous chapter, the methods covered are inherently non-spatial, but by means of linking and brushing with one or more maps, they can be *spatialized.*

Before focusing on the specific techniques, some special features of multivariate analysis are outlined in a brief discussion of the *curse of dimensionality.*

8.1 Topics Covered

- Illustrate the curse of dimensionality
- Interpreting a bubble chart
- Interpreting and manipulating a 3D scatter plot
- Interpreting and manipulating conditional plots
- Interpreting a parallel coordinate plot

GeoDa Functions

- Explore > Bubble Chart
  - bubble chart classification schemes
  - bubble size
- Explore > 3D Scatter Plot
  - rotating and zooming the 3D scatter plot
  - projecting onto one axis
  - selection in the 3D scatter plot
- Explore > Conditional Plot
  - conditional scatter plot

DOI: 10.1201/9781003274919-8

  - conditional histogram
  - conditional box plot
  - conditional map
  - conditional plot option
  - changing the condition breakpoints
- Explore > Parallel Coordinate Plot
  - changing the classification theme for the PCP
  - changing the order of the axes
  - brushing the PCP

Toolbar Icons

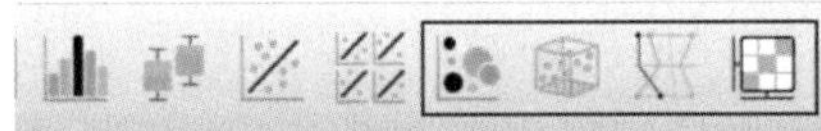

Figure 8.1: Bubble Chart | 3D Scatter Plot | Parallel Coordinate Plot | Conditional Plot

8.2 The Curse of Dimensionality

Before considering specific techniques, it is useful to focus attention on the so-called *curse of dimensionality*, a concept initially conceived by Bellman (1961). In a nutshell, the curse of dimensionality means that techniques that work well in small dimensions (e.g., k-nearest neighbor searches), either break down or become too complex or computationally impractical in higher dimensions.

In the original discussion by Bellman (1961), the problem was situated in the context of optimization. To illustrate this, consider the simple example of the search for the maximum of a function in one variable dimension, as in Figure 8.2.

One simple way to try to find the maximum is to evaluate the function at a number of discrete values for x and find the point(s) where the function value is highest. For example, in Figure 8.2, there are 10 evaluation points, equally dividing the x-range of 0–10. Here, the maximum is between 5 and 6. In a naive one-dimensional numerical optimization routine, the interval between 5 and 6 could then be further divided into ten more sub-intervals and so on. However, what happens if the optimization was with respect to two variables? At this stage, the attribute domain becomes two-dimensional. Using the same approach and dividing the range for each dimension into 10 discrete values now results in 100 evaluation points, as shown in Figure 8.3.

Similarly, for three dimensions, there would be 10^3 or 1,000 evaluation points, as in Figure 8.4.

In general, and keeping the principle of 10 discrete values per dimension, the required number of evaluation points is 10^k, with k as the number of variables. This very quickly becomes unwieldy, e.g., for 10 dimensions, there would need to be 10^{10}, or 10 billion evaluation points, and for 20 dimensions 10^{20} or 100,000,000,000,000,000,000, hence the *curse of dimensionality*. Some further illustrations of methods that break down or become impractical in high-dimensional spaces are given in Chapter 2 of Hastie et al. (2009).

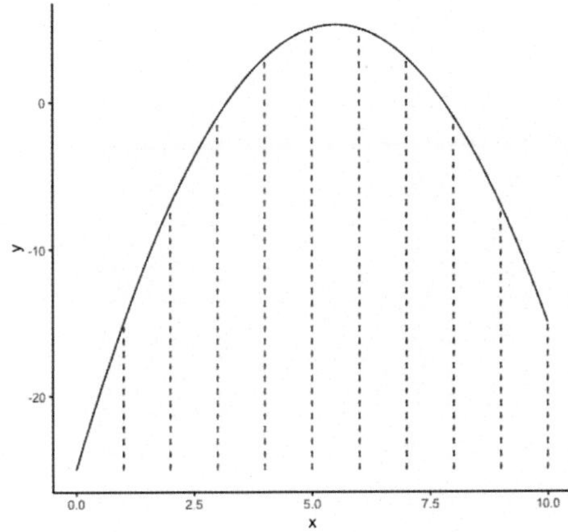

Figure 8.2: One dimension

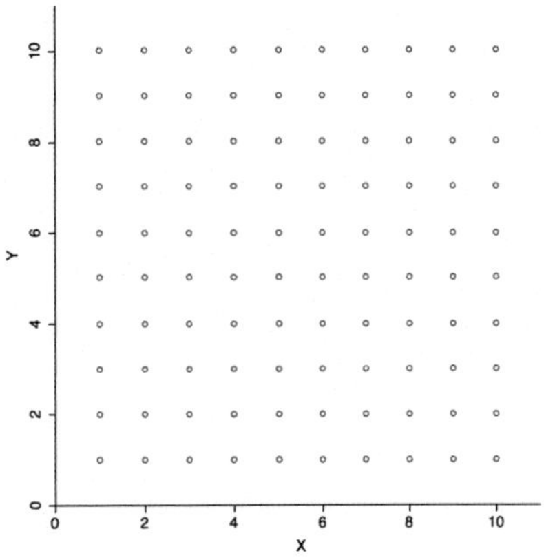

Figure 8.3: Two dimensions

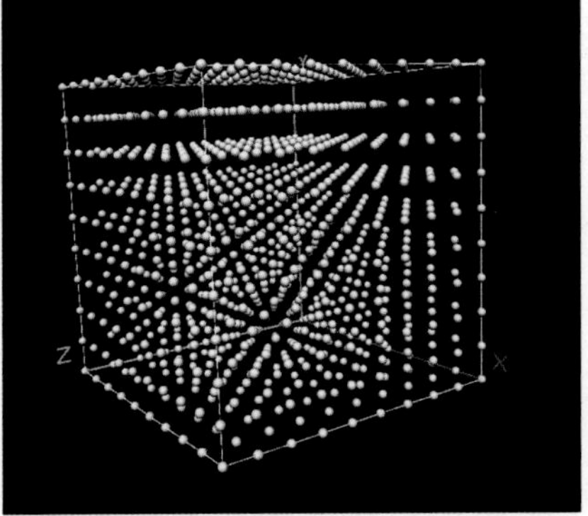

Figure 8.4: Discrete evaluation points in three variable dimensions

8.2.1 The empty space phenomenon

A related and strange property of high-dimensional analysis is the so-called *empty space phenomenon.* In essence, the same number of data points are much farther apart in higher dimensional spaces than in lower dimensional ones. This means that in order to carry out effective analysis in high dimensions, (very) large data sets are needed.

To illustrate this point, consider three variables, say x, y and z, each generated as 10 uniform random variates in the unit interval (0–1). In one dimension, the observations on these variables are fairly close together, since their range will be less than one by construction.

However, when represented in two dimensions, say in the form of a scatter plot, large empty spaces appear in the unit quadrant, as in Figure 8.5 for x and y (similar graphs are obtained for the other pairs). The sparsity of the attribute space gets even more pronounced in three dimensions, as in Figure 8.6.

To quantify this phenomenon, consider the nearest neighbor distances between the observation points in each of the dimensions.[1] In the example (the results will depend on the actual

[1]In `GeoDa`, this can be accomplished in the weights manager, using the **variables** option for distance weights, and making sure the transformation is set to **raw**. The nearest neighbor distances are the third column in a weights file created for $k = 1$. See Chapter 11 for a detailed discussion.

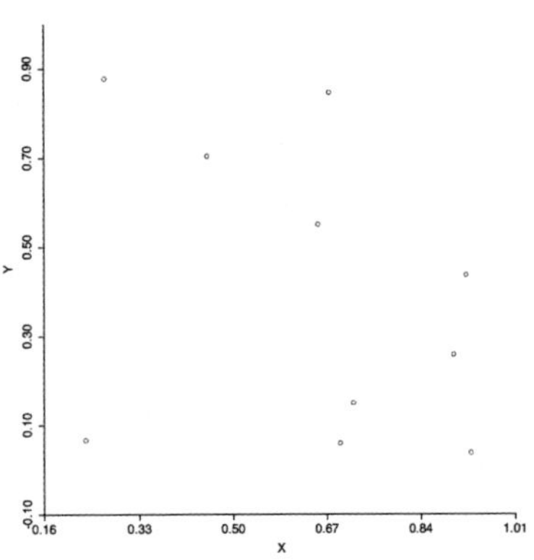

Figure 8.5: Two dimensions

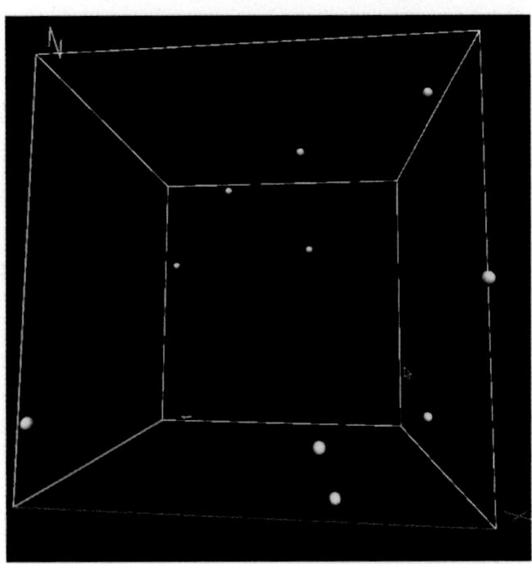

Figure 8.6: Three dimensions

random numbers generated), the smallest nearest neighbor distance in the x dimension is 0.009. In two dimensions, between x and y, the smallest distance is 0.094, and in three dimensions, it is 0.146. All distances are in the same units, since the observations fall within the unit interval for each variable.

Two insights follow from this small experiment. One is that the nearest neighbors in a lower dimension are not necessarily also nearest neighbors in higher dimensions. For example, in the x dimension, the shortest distance is between points 5 and 8, whereas in the x-y space, it is between 1 and 4. The other property is that the nearest neighbor distance increases with the dimensionality. In other words, more of the attribute space needs to be searched before neighbors are found. This becomes a critical issue for searches in high dimensions, where most lower dimensional techniques (such as k-nearest neighbors) become impractical.

Further examples of some strange aspects of data in higher dimensional spaces can be found in Chapter 1 of Lee and Verleysen (2007).

8.3 Three Variables: Bubble Chart and 3-D Scatter Plot

When the number of dimensions is at most three, it remains relatively easy to visualize the multivariate attribute space. The bubble chart implements this by augmenting a standard two-dimensional scatter plot with an additional attribute for each point, i.e., the *size* of the point or *bubble*. More directly, observations can also be visualized as points in a three-dimensional scatter plot or data cube.

8.3.1 Bubble chart

The bubble chart is an extension of the scatter plot to include a third and possibly a fourth variable into the two-dimensional chart. While the points in the scatter plot still show the association between the two variables on the axes, the size of the points, the *bubble*, is used to introduce a third variable. In addition, the color shading of the points can be used to consider a fourth variable as well, although this may stretch one's perceptual abilities.

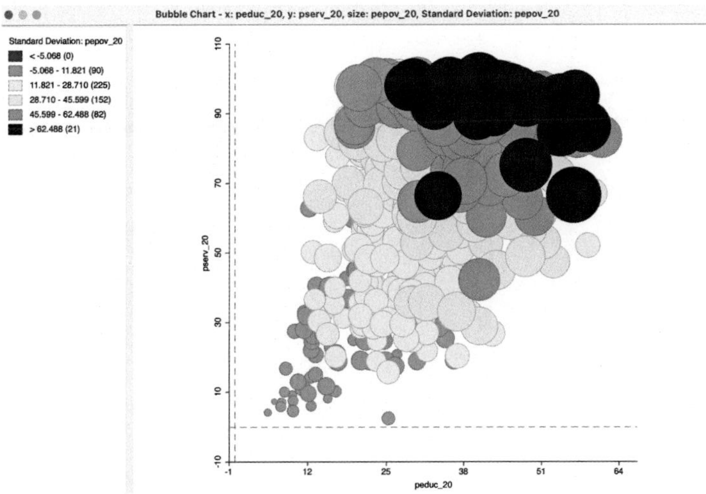

Figure 8.7: Bubble chart – default settings: education, basic services, extreme poverty

The introduction of the extra variable allows for the exploration of interaction. The point of departure (null hypothesis) is that there is no interaction. In the plot, this would be reflected by a seeming random distribution of the bubble sizes among the scatter plot points. On the other hand, if larger (or smaller) bubbles tend to be systematically located in particular subareas of the plot, this may suggest an interaction. Through the use of linking and brushing with the map, the extent to which systematic variation in the bubbles corresponds with particular spatial patterns can be readily investigated, in the same way as illustrated in the previous chapter.

The bubble chart was popularized through the well-known *Gapminder* web site, where it is used extensively, especially to show changes over time.[2] This aspect is currently not implemented in `GeoDa`.

8.3.1.1 Implementation

The bubble chart is invoked from the menu as **Explore > Bubble Chart**, and from the toolbar by selecting the left-most icon in the multivariate group of the EDA functionality, shown in Figure 8.1. This brings up a **Bubble Chart Variables** dialog to select the variables for up to four dimensions: **X-Axis**, **Y-Axis**, **Bubble Size** and **Bubble Color**.

To illustrate this method, three variables from the *Oaxaca Development* sample data set are selected: **peduc_20** (percent population with an educational gap in 2020) for the **X-Axis**, **pserv_20** (percent population without access to basic services in the dwelling in 2020) for the **Y-Axis**, and **pepov_20** (percent population living in extreme poverty in 2020) for both **Bubble Size** and **Bubble Color**. The default setting uses a standard deviational diverging legend (see also Section 5.2.3) and results in a somewhat unsightly graph, with circles that typically are too large, as in Figure 8.7.

Before going into the options in more detail, it is useful to know that the size of the circle can be readily adjusted by means of the **Adjust Bubble Size** option. With the size significantly reduced, a more appealing Figure 8.8 is the result.

[2] https://www.gapminder.org/tools/#$chart-type=bubbles&url=v1

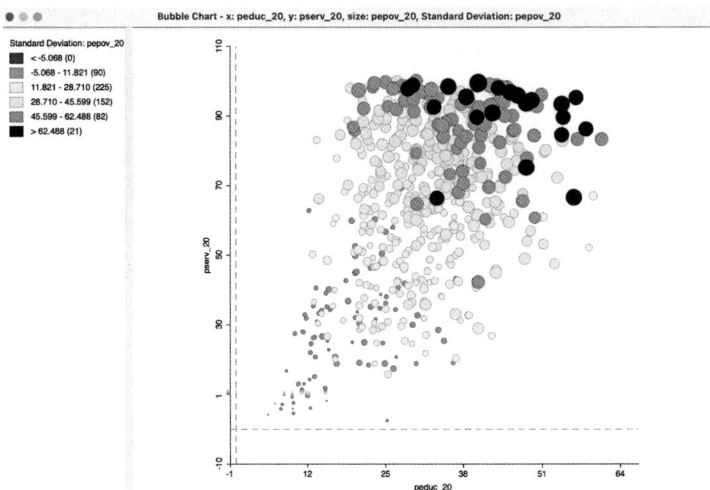

Figure 8.8: Bubble chart – bubble size adjusted: education, basic services, extreme poverty

As mentioned, the null hypothesis is that the size distribution (and matching colors) of the bubbles should be seemingly random throughout the chart. In the example, this is clearly not the case, with high values (dark brown colors and large circles) for **pepov_20** being primarily located in the upper right quadrant of the graph, corresponding to low education and low services (high values correspond with deprivation). Similarly, low values (small circles and blue colors) tend to be located in the lower left quadrant. In other words, the three variables under consideration seem to be strongly related.

Given the screen real estate taken up by the circles in the bubble chart, this is a technique that lends itself well to applications for small to medium sized data sets. For larger size data sets, this particular graph is less appropriate.

8.3.1.2 Bubble chart options

In addition to **Adjust Bubble Size**, the bubble chart has nine main options, invoked in the usual fashion by right clicking on the graph. Several of these are shared with other graphs, in particular with the scatter plot, such as **Save Selection**, **Copy Image To Clipboard**, **Save Image As**, **View**, **Show Status Bar** and **Selection Shape**. They will not be further discussed here (see, e.g., Section 7.3.1). **Classification Themes**, **Save Categories** and **Color** work the same as for any map (see Section 4.5).

Figure 8.9 provides an illustration of the flexibility that these options provide. The graph pertains to the same three variables as before, but the **Classification Themes** option has been set to **Themeless**. As such, this results in green circles, but the **Color** option applied to the legend (see Section 4.5) allows the **Opacity** for the **Fill Color for Category** to be set to zero, resulting in the empty circles.

8.3.1.3 Bubble chart with categorical variables

One particular useful feature to investigate potential structural change in the data is to use the **Unique Values** classification, in combination with setting the **Size** to a constant value.[3]

[3]A constant variable can be readily created by means of the **Calculator** option in the data table. Specifically, the **Univariate** tab with **ASSIGN** allows one to set the a variable equal to a constant.

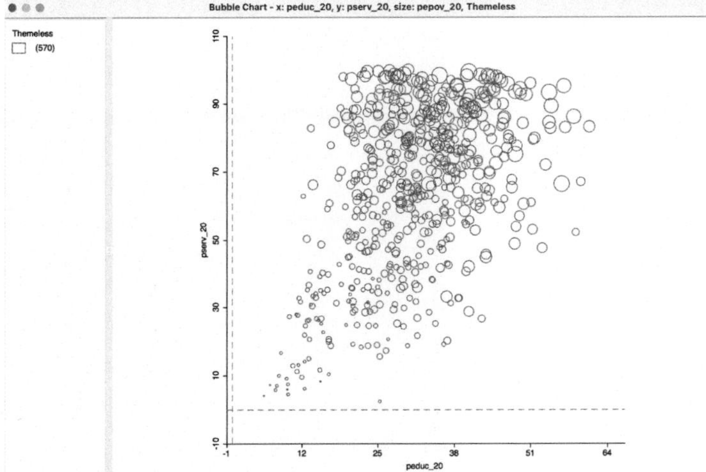

Figure 8.9: Bubble chart – no color: education, basic services, extreme poverty

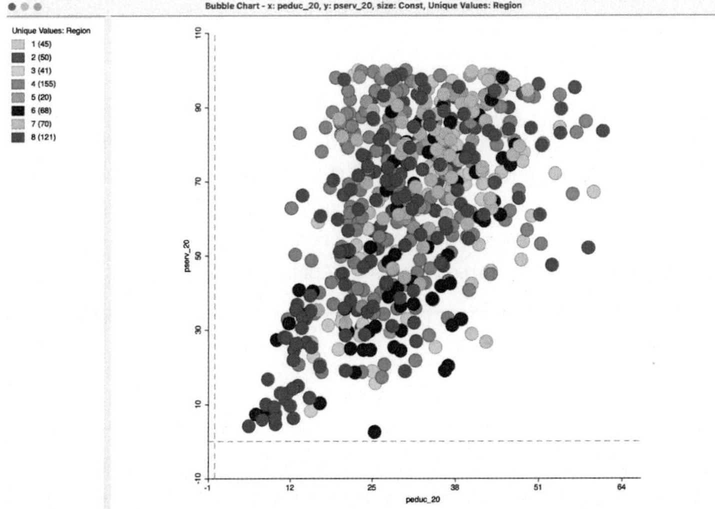

Figure 8.10: Bubble chart – categories: education, basic services, region

With **peduc_20** for the **X-Axis** and **ppov_20** (percentage population living in poverty) as the **Y-Axis**, **Bubble Size** is now set to the constant **Const**, with the categorical variable **Region** for **Bubble Color**.[4] First, this yields a rather meaningless graph, based on the default standard deviation classification. Changing **Classification Themes** to **Unique Values** results in a scatter plot of the two variables of interest, with the bubble color corresponding to the values of the categorical variable. In the example, this is **Region**, as in Figure 8.10.

In this case, there seems to be little systematic variation along the region category, in line with our starting hypothesis.

[4]The regional classifications are: (1) Canada; (2) Costa; (3) Istmo; (4) Mixteca; (5) Papaloapan; (6) Sierra Norte; (7) Sierra Sur; and (8) Valles Centrales.

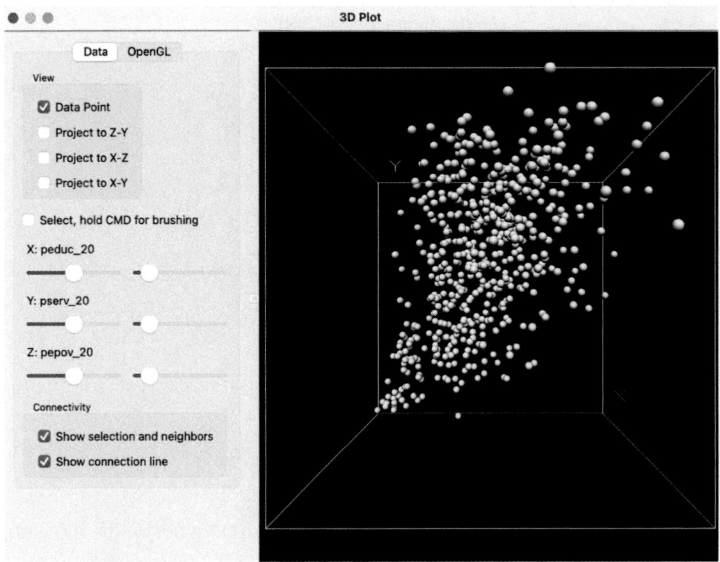

Figure 8.11: 3D scatter plot: education, basic services, extreme poverty

The use of a bubble chart to address structural change is particularly effective when more than two categories are involved. In such an instance, the binary selected/unselected logic of scatter plot brushing no longer works. Using the bubble chart in this fashion allows for an investigation of structural changes in the bivariate relationship between two variables along multiple categories, each represented by a different color bubble. This forms an alternative to the conditional scatter plot, considered in Section 8.4.2.

8.3.2 3-D scatter plot

An explicit visualization of the relationship between three variables is possible in a three-dimensional scatter plot, the direct extension of principles used in two dimensions to a three-dimensional data cube. Each of the axes of the cube corresponds to a variable, and the observations are shown as a point cloud in three dimensions (of course, rendered as a perspective plot onto the two-dimensional plane of the screen).

The data cube can be manipulated by zooming in and out, in combination with rotation, to get a better sense of the alignment of the points in three-dimensional space. This takes some practice and is not necessarily that intuitive to many users. The main challenge is that points that seem close in the two-dimensional rendering on the screen may in fact be far apart in the actual data cube. Only by careful interaction can one get a proper sense of the alignment of the points.

8.3.2.1 Implementation

The three-dimensional scatter plot method is invoked as **Explore > 3D Scatter Plot** from the menu, or by selecting the second icon from the left on the toolbar depicted in Figure 8.1. This brings up a **3D Scatter Plot Variables** selection dialog for the variables corresponding to the **X**, **Y** and **Z** dimensions. Again, **peduc_20**, **pserv_20** and **pepov_20** are used.

The corresponding initial default data cube is as in Figure 8.11, with the Y-axis (**pserv_20**) as vertical, and the X (**peduc_20**) and Z-axes (**pepov_20**) as horizontal. Note that the

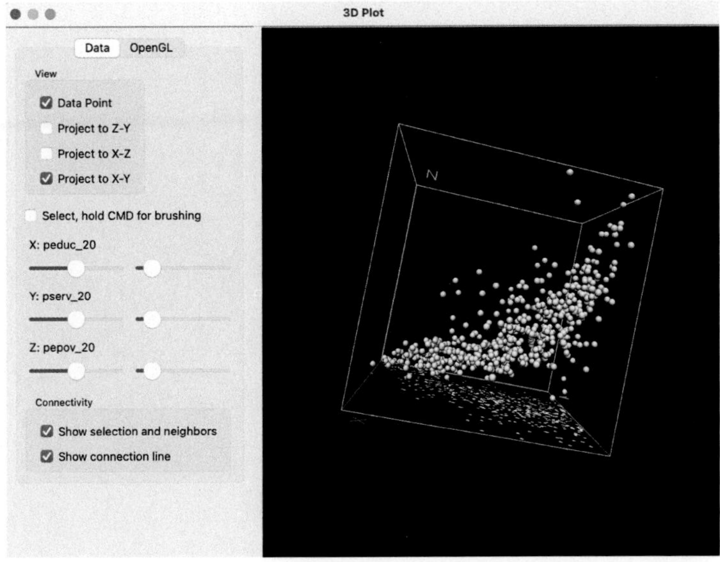

Figure 8.12: Interacting with 3D scatter plot

axis marker (e.g., the X etc.) is at the end of the axis, so that the origin is at the unmarked side, i.e., the lower left corner where the green, blue and purple axes meet.

8.3.2.2 Interacting with the 3-D scatter plot

The data cube can be re-sized by zooming in and out. It is sometimes a bit ambiguous what is meant by these terms. For the purposes of this illustration, zooming out refers to making the cube smaller, and zooming into making the cube larger (moving into the cube, so to speak).

The zoom functionality is carried out by pressing down on the track pad with two fingers and moving up (zoom out) or down (zoom in). Alternatively, one can press Control and press down on the track pad with one finger and move up or down. With a mouse, one moves the middle button up or down.

In addition, the cube can be rotated by means of the pointer: one can click anywhere in the window and *drag* the cube by moving the pointer to a different location. The controls on the left-hand side of the view allow for the projection of the point cloud onto any two-dimensional pane, and the construction of a selection box by checking the relevant boxes and/or moving the slider bar (see Section 8.3.2.3).

For example, in Figure 8.12, the cube has been zoomed out and the axes rotated such that Z is now vertical. With the **Project to X-Y** box checked on the left, a two-dimensional scatter plot is projected onto the X-Y plane, i.e., showing the relationship between **peduc_20** and **pserv_20**.

8.3.2.3 Selection in the 3-D scatter plot

Selection in the three-dimensional plot (or, rather, its two-dimensional rendering) is a little tricky and takes some practice. The selection can be done either manually, by pressing down the command key while moving the pointer, or by using the guides under the **Select, hold CMD for brushing** check box.

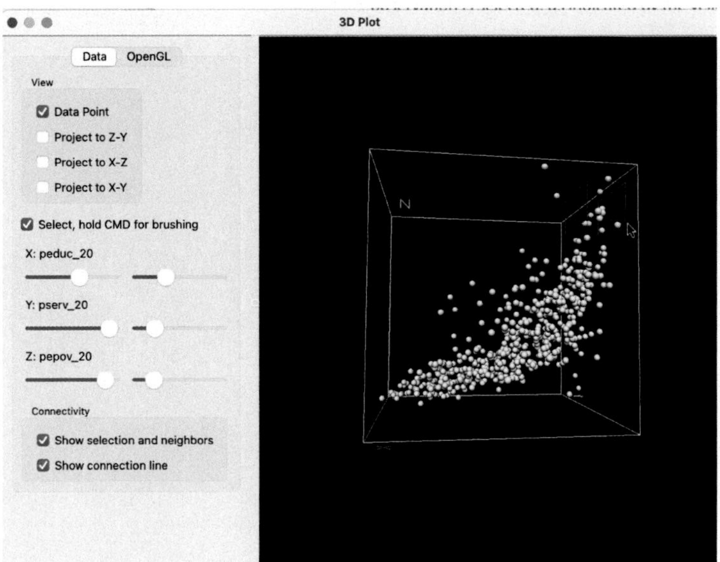

Figure 8.13: Selection in the 3D scatter plot

Checking this box creates a small red selection cube in the graph. The selection cube can be moved around with the command key pressed, or can be moved and resized by using the controls to the left, next to **X:**, **Y:** and **Z:**, with the corresponding variables listed.

The first set of controls (to the left) move the box along the matching dimension, e.g., up or down the X values for larger or smaller values of **peduc_20**, and the same for the other two variables. The slider to the right changes the size of the box in the corresponding dimension (e.g., larger along the X dimension). The combination of these controls moves the box around to select observation points, with the selected points colored yellow.

The most effective way to approach this is to combine moving around the selection box and rotating the cube. The reason for this is that the cube is in effect a perspective plot, and one cannot always judge exactly where the selection box is located in three-dimensional space.

A selection is illustrated in Figure 8.13, where only a few out of seemingly close points in the data cube are selected (yellow). By further rotating the plot, one can get a better sense of their location in the point cloud.

As in all other graphs, linking and brushing is implemented for the 3D scatter plot as well. Figure 8.14 shows an example of brushing in the six quantile map for **ALTID** and the associated selection in the point cloud. The assessment of the match between closeness in geographical space (the selection in the map) and closeness in multivariate attribute space (the point cloud) is a fundamental notion in the consideration of multivariate spatial correlation, discussed in Chapter 18.

Similar to the bubble chart, the 3D scatter plot is most useful for small to medium sized data sets. For larger numbers of observations, the point cloud quickly becomes overwhelming and is no longer effective for visualization.

8.3.2.4 3-D scatter options

The 3D scatter plot has a few specialized options, available either on the **Data** pane (which was considered exclusively so far), or on the **OpenGL** pane. The latter gives the option

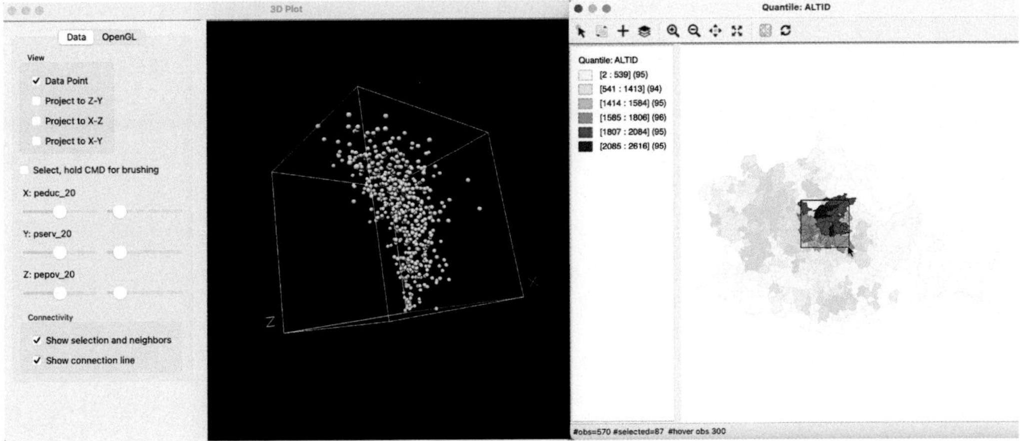

Figure 8.14: Brushing and linking with the 3D scatter plot

to adjust the **Rendering quality of points**, the **Radius of points**, as well as the **Line width/thickness** and **Line color**. These are fairly technical options that basically affect the quality of the graph and the speed by which it is updated. In most situations, the default settings are fine.

Finally, under the **Data** button, there are the **Show selection and neighbors** and **Show connection line** options. These items are relevant when a spatial weights matrix has been specified. This is covered in Chapter 10.

8.4 Conditional Plots

Conditional plots, also known as small multiples (Tufte, 1983), Trellis graphs (Becker et al., 1996), or facet graphs (Wickham, 2016), provide a means to assess interactions among up to four variables. Multiple graphs are constructed for different subsets of the observations, obtained as a result of *conditioning* on the value of one or two variables, different from the variable(s) of interest. The graphs can be any kind of statistical graph, such as a histogram, box plot, or scatter plot. The same principle can be applied to choropleth maps, resulting in so-called *micromaps* (Carr and Pickle, 2010).

With one conditioning variable, the observations are grouped into subsets according to the values taken by the conditioning variable, organized along the x-axis from low to high, e.g., below or above the median. For each of these subsets, a separate graph or map is drawn for the variable(s) of interest. The same principle is applied when there are two conditioning variables, resulting in a matrix of graphs, each corresponding to a subset of the observations that fall in the specified intervals for the conditioning variables on the x and y-axis. Of course, for this to be meaningful, one has to make sure that each of the subsets contains a sufficient number of observations.

The point of departure in the conditional plots is that the subsetting of observations should have no impact on the statistic in question. In other words, all the graphs or maps should look more or less the same, irrespective of the subset. Systematic variation across subsets indicates the presence of heterogeneity, either in the form of structural breaks (discrete

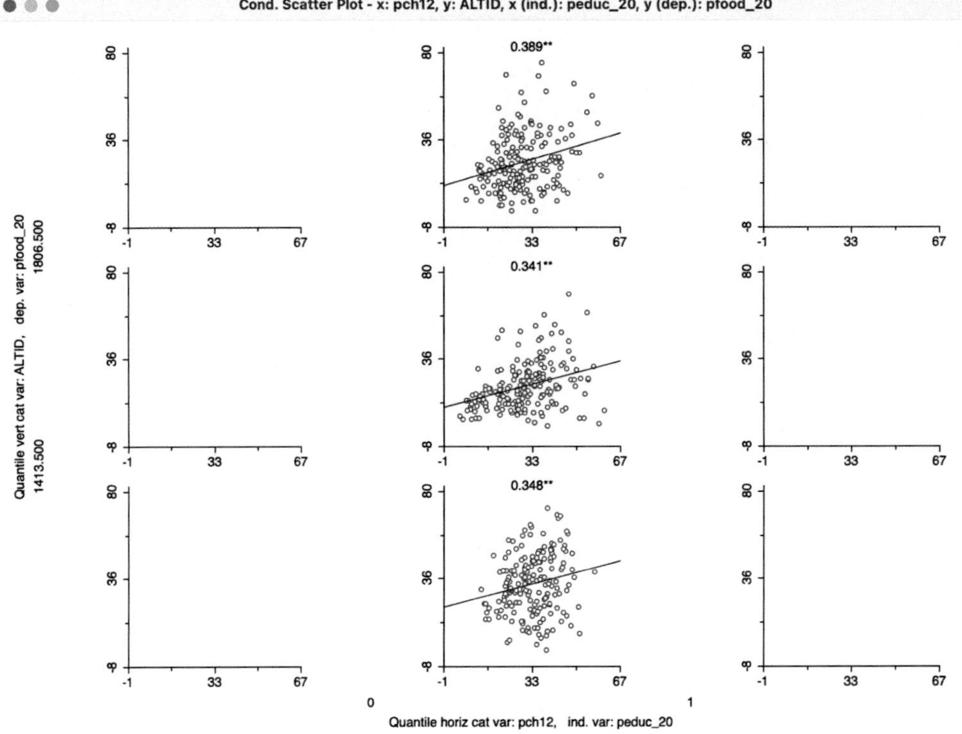

Figure 8.15: Conditional scatter plot – 3 by 3

categories), or suggesting an interaction effect between the conditioning variable(s) and the statistic under scrutiny.

In `GeoDa`, conditional graphs are implemented for the histogram, box plot, scatter plot and thematic map.

8.4.1 Implementation

A conditional plot is invoked from the menu by selecting **Explore > Conditional Plot**. This brings up a list of four options: **Map**, **Histogram**, **Scatter Plot** and **Box Plot**. The same list of four options is also created after selecting the **Conditional Plot** icon, the right-most in the multivariate EDA subset in the toolbar shown in Figure 8.1. In addition, the conditional map can also be started as the third item from the bottom in the **Map** menu.

Next follows a dialog to select the conditioning variables and the variable(s) of interest. The conditioning variables are referred to as **Horizontal Cells** for the x-axis, and **Vertical Cells** for the y-axis. They do not need to be chosen both. Conditioning can also be carried out for a single conditioning variable, on either the horizontal or on the vertical axis alone.

The remaining columns in the dialog are either for a single variable (histogram, box plot, map), or for both the **Independent Var (x-axis)** and the **Dependent Var (y-axis)** in the scatter plot option.

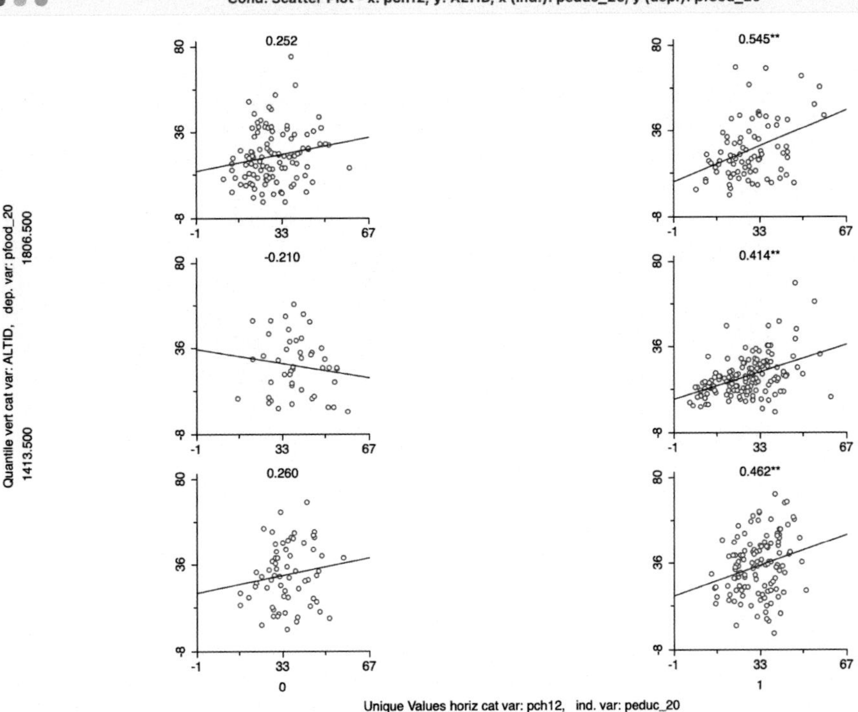

Figure 8.16: Conditional scatter plot – unique values

8.4.1.1 Conditional plot options

Two important options, special to the conditional plot, are **Vertical Bins Breaks** and **Horizontal Bins Breaks**. These options provide a way to select the observation subsets for each conditioning variable. They include all the same options used in map classification (see Figure 4.2 in Chapter 4). Of special interest are the **Unique Values** option and the **Custom Breaks** options. **Unique Values** is particularly appropriate when the conditioning variable takes on discrete values, in which case other classifications may result in meaningless subsets (see the discussion of Figure 8.15). **Custom Breaks** is useful when subclasses for the conditioning variable were determined previously, preferably contained in a project file (see Section 4.6.3).

Each graph also has its usual range of options available, with the exception of **Regimes Regression** for the conditional scatter plot. Selection of observations is implemented, but it does not result in a re-computation of the linear regression fit. On the other hand, the conditional scatter plot does include the **LOWESS Smoother** option.

8.4.2 Conditional statistical graphs

To illustrate these concepts, first, a set of conditional scatter plots is created for the relationship between **peduc_20** (x-axis) and **pfood_20** (y-axis). This is conditioned by three subcategories (the default number) for **pch12** in the horizontal dimension (an indicator variable for whether the population increased, =1, or decreased, =0, between 2010 and 2020).

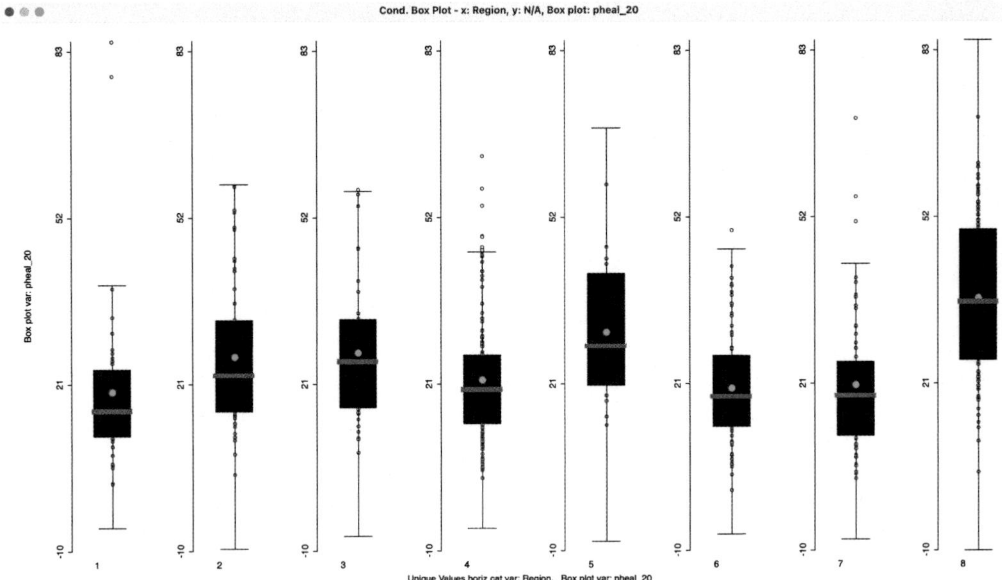

Figure 8.17: Conditional box plot – lack of health access by region

The conditioning variable in the vertical dimension is **ALTID**. The default classification is to use **Quantile** with three categories for both conditioning variables.

This yields the plot shown in Figure 8.15. Something clearly is wrong, since there turns out to be only one subcategory for the horizontal conditioning variable. This is because a 3-quantile classification of the indicator variable does not provide meaningful categories. This will be dealt with next.

However, first consider the substantive interpretation of this graph. The three plots show a strong positive and significant relationship in each case (significance indicated by **). In other words less access to education is linearly related to less food security, more or less in line with our prior expectations. The lack of difference between the graphs would suggest that the relationship is stable across all three ranges for altitude, the conditioning variable.

Figure 8.16 results after changing the **Horizontal Bins Breaks** option to **Unique Values**. At this point, the two categories for the horizontal conditioning variable correspond to the values of 0 and 1 for the population change indicator variable.

Conditioning on this variable does provide some indication of *interaction*. For **pch12 = 1**, the linear relationship between **peduc_20** and **pfood_20** is strong and positive, and not affected by altitude. However, for **pch12 = 0**, there does not seem to be a significant slope for any of the subgraphs, suggesting a lack of relationship between the two variables in those municipalities suffering from population loss. The change in sign of the slope along altitude is not meaningful since the slopes are not significant.

If only the horizontal classification had been used as a condition, the result would be the same as a **Regimes Regression** in a standard scatter plot. However, in order to consider the simultaneous conditioning on two variable,six separate selection sets would be required to create six individual scatter plots with regimes, which is not very practical. In addition, the double conditioning allows for the investigation of more complex interactions among the variables.

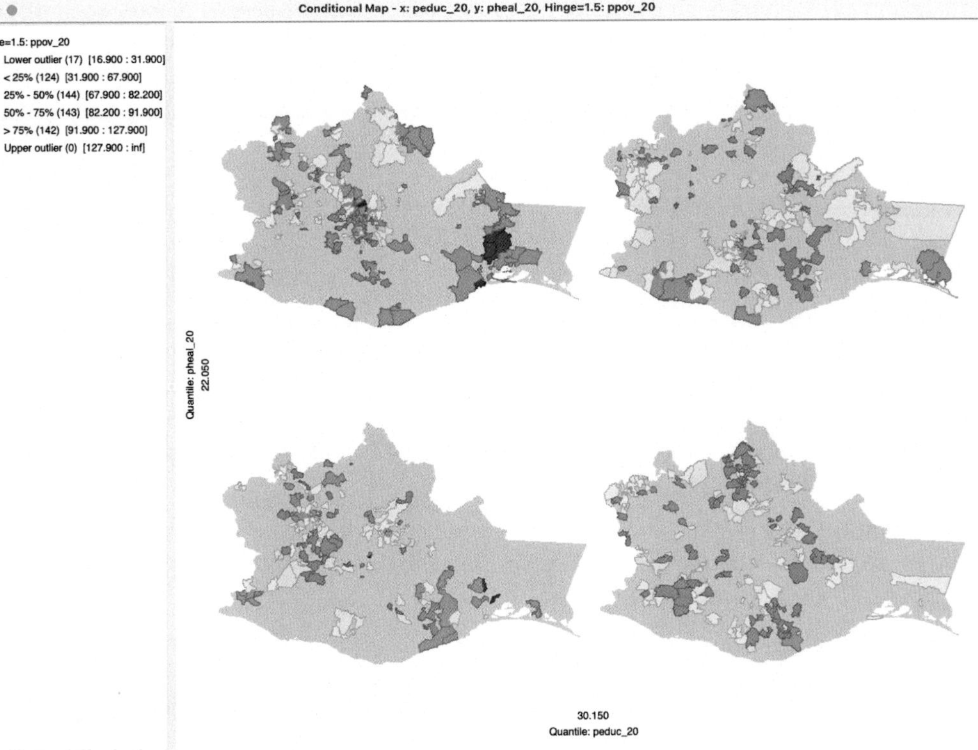

Figure 8.18: Conditional box map – 2 by 2

An alternative as well as generalization of the consideration of spatial heterogeneity by means of the **Averages Chart** (see Section 7.5.1) is the application of a **Conditional Box Plot** with a conditioning variable that corresponds to *spatial subsets* of the data. In Figure 8.17, this is illustrated for **pheal_20** (percent population that lacks access to health care), using the **Region** classification as the conditioning variable along the horizontal axis. In the graph, **Valles Centrales** (8) shows a median value that is higher than the other regions, with **Canada** (1) having the best median health outcome (lowest value for lack of access). While the conditional box plot is not an alternative to a formal difference in means (or medians) test, it provides a visual overview of the extent to which heterogeneity may be present.

8.4.3 Conditional maps

A final example is the conditional map or micromap matrix shown in Figure 8.18. Four box maps are included for **ppov_20**, conditioned by **peduc_20** on the horizontal axis and **pheal_20** on the vertical axis. The graph is obtained after changing the bin breaks to two quantiles, i.e., below and above the median for each of the conditioning variable.

As before, the interest lies in the extent to which the maps represent similar spatial patterns in each of the subcategories. The maps suggest an interaction with **peduc**, where more of the brown values (more poverty) are found for above median values for lack of education, and more blue values (less poverty) in the lower median (better education). A potential interaction with health access is less pronounced.

As in any EDA exercise, considerable experimentation may be needed before any meaningful patterns are found for the right categories of conditioning variables. Of course, this runs the danger than one ends up finding what one wants to find, an issue which revisited in Chapter 21.

8.5 Parallel Coordinate Plot

So far, the methods discussed were limited to three or four variables. Once more than three variables are considered, it is no longer possible to graphically represent the data points in a multi-dimensional data cube. The parallel coordinate plot or PCP provides an alternative that can be generalized to many dimensions. Data points are replaced by data lines, which allows a large number of variables to be considered. The only limitation is human perception and screen real estate.

This approach was originally suggested by Inselberg (1985; see also Inselberg and Dimsdale, 1990), and it has become a main feature in many visual data mining frameworks (e.g., Wegman, 1990; Wegman and Dorfman, 2003).

In a PCP, each variable is represented as a *parallel* axis, and each observation consists of a line that connects points on the axes. Additional variables are represented by new axes, so that it is easy to go beyond three variables. The only visual limitation is the number of observations, which can make the plot of lines very crowded. However, techniques exist to overcome this hurdle, such as binning, although this is currently not implemented in `GeoDa`.

A main objective in the application of PCP is the identification of clusters and outliers in multi-attribute space. *Clusters* are found as groups of lines (i.e., observations) that follow a similar path. This is equivalent to points that are close together in multidimensional variable space. *Outliers* in a PCP are lines that show a very different pattern from the rest, similar to outlying points in a multi-dimensional cloud.

8.5.1 Implementation

The PCP functionality is invoked from the menu as **Explore > Parallel Coordinate Plot**, or by means of the PCP toolbar icon, the third item from the left in Figure 8.1. This brings up a **Parallel Coordinate Plot** variable selection interface that consists of two columns: **Exclude** and **Include**. Variables are chosen by moving them to the **Include** column by means of arrow buttons.

Four variables are selected to illustrate the PCP: **peduc_20**, **pserv_20** and **pepov_20**, as well as **pss_20** (percent of population without social security). With the default settings, the result is as in Figure 8.19. Each variable is represented by a horizontal axis, with the observations as lines connecting points on each axis. The variable name is listed at the left, together with the range of the variable, its mean and standard deviation. Note that the axes are represented with equal lengths, which corresponds to the range for each individual variable. This implies that the distance between points on each axis does not necessarily correspond with the same difference in value for each variable.

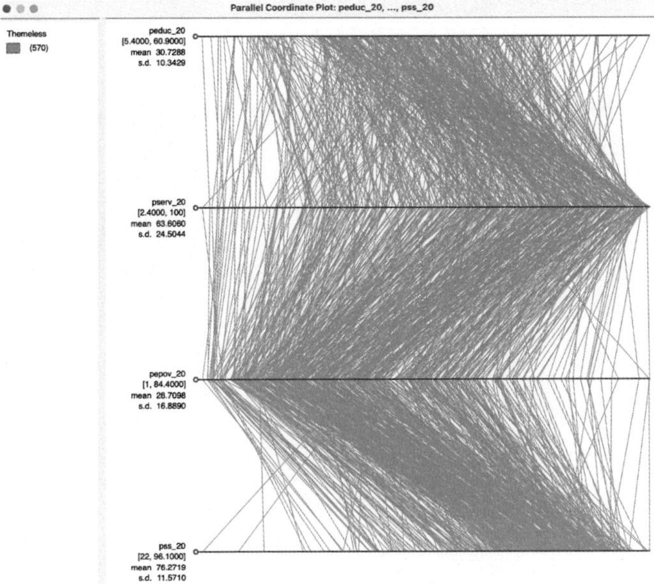

Figure 8.19: Parallel coordinate plot

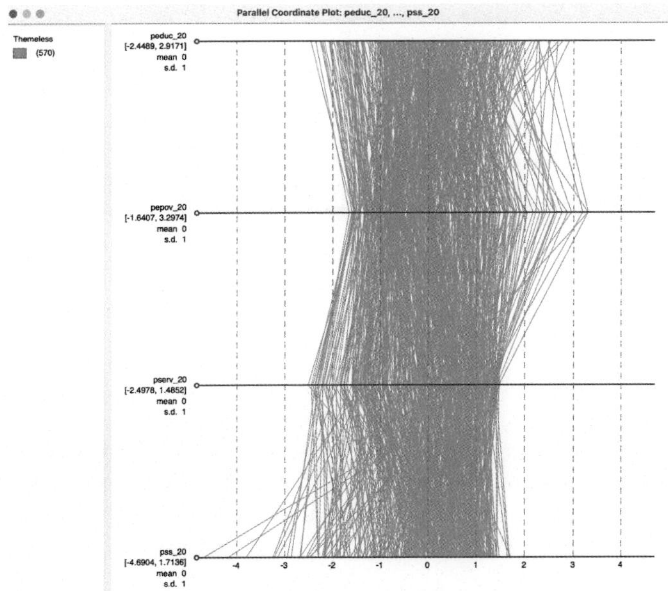

Figure 8.20: Parallel coordinate plot – standardized variables

8.5.1.1 PCP options

The PCP has nine main options, invoked by right clicking on the graph. They are the same as discussed for other graphs and are not further considered in detail.[5] The **Data** option

[5]Specifically, the options are **Classification Themes**, **Save Categories**, **Data**, **View**, **Selection Shape**, **Color**, **Save Selection**, **Copy Image to Clipboard** and **Save Image As**. Note that **Classification Themes** always pertains to the variable that was originally listed on top, even when the order of axes is later changed.

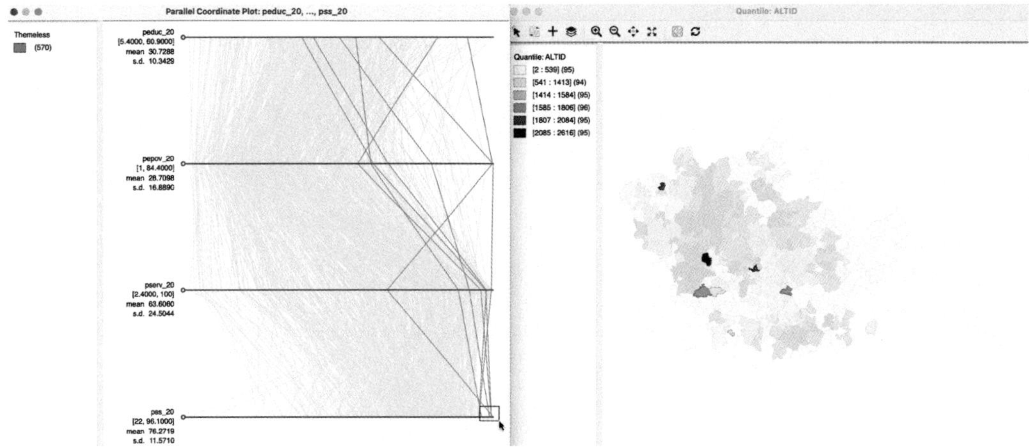

Figure 8.21: Brushing the PCP

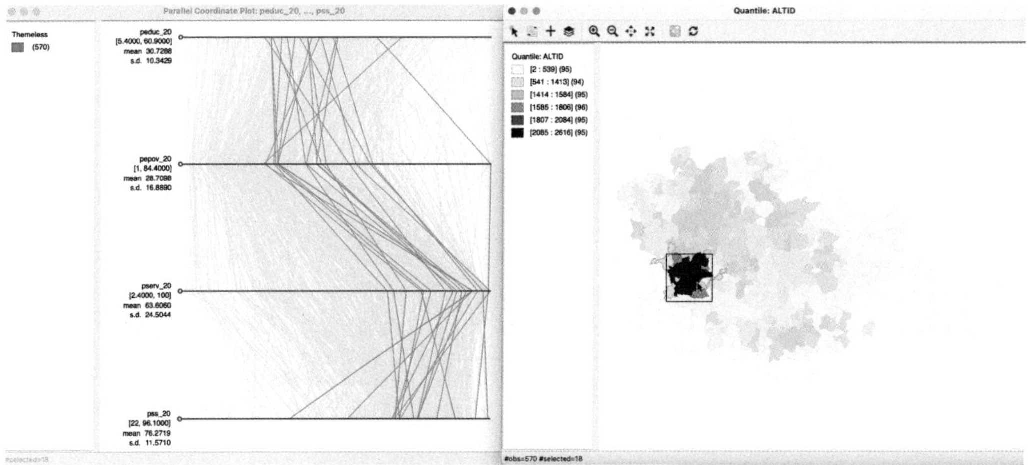

Figure 8.22: Brushing map and PCP

offers a way to make all axes lengths equivalent by selecting **View Standardized Data**. As a result, the data points on each axis are given in standard deviational units and become directly comparable between the different variables, as in Figure 8.20. Observations that are outside the −2 to +2 range can be considered to be outliers.

This graph also illustrates another feature of the PCP. By grabbing the small circular handle on the left of an axis, it can be moved up or down in the graph, changing the order of the variables. For example, in Figure 8.20, the axis for **pepov_20** has been moved up to be above **pserv_20**. Realigning the axes in this manner can often bring out patterns in the data. However, in practice, it is not always clear which is the optimal alignment.

8.5.1.2 Brushing the PCP

Selection of observations in the PCP is implemented by moving the selection shape over one of the axes. As for the other graphs, the default is a rectangle. Figure 8.21 illustrates this with a small selection rectangle situated over the highest values for **pss_20**. This highlights the lines (observations) that are selected. Of particular interest is the extent to

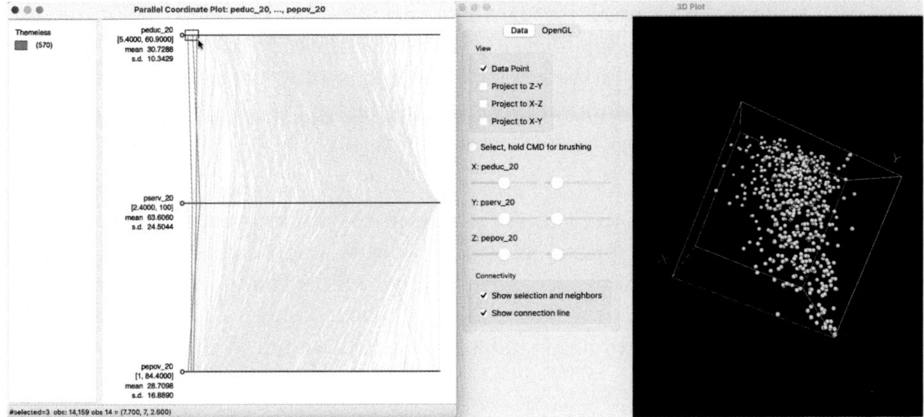

Figure 8.23: Clusters of observations in PCP

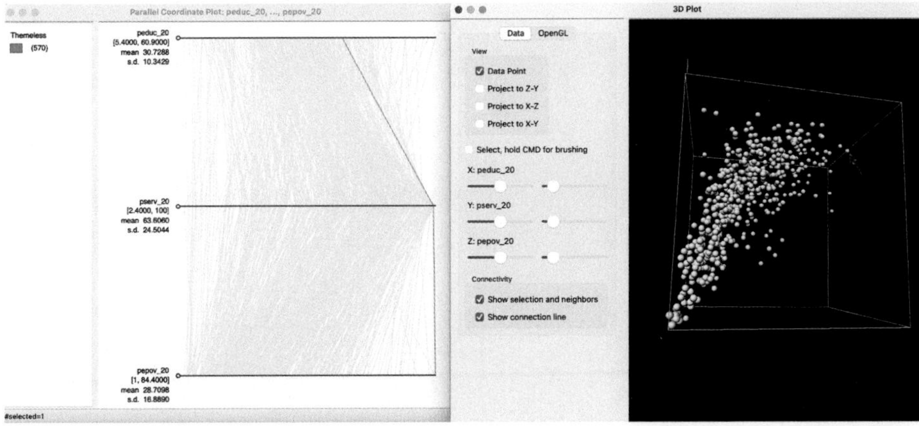

Figure 8.24: Outlier observation in PCP

which these lines move together. For example, in the example, several observations follow a similar path for **pss_20**, **pserv_20** and **pepov_20**, but not for **peduc_20**. Insight into any corresponding spatial pattern can be obtained by linking the selection with a map, as in the example, using the six quantile map for **ALTID**.

A more explicit spatial perspective is taken by brushing a map linked to the PCP. In Figure 8.22, this is again implemented with the same six quantile map. The selection in the map suggests that several locations within close geographical proximity also take on similar values for a subset of the variables (close lines), but not for all. This illustrates some of the practical difficulties associated with the concept of multivariate spatial correlation.

8.5.2 Clusters and outliers in PCP

Visual cluster detection (e.g., Wegman and Dorfman, 2003) consists of locating groups of lines in the PCP that follow a similar path. To illustrate this feature, Figure 8.23 shows a three-variable PCP side by side with a three-dimensional data cube for the variables **peduc_20**, **pserv_20** and **pepov_20**. The selection rectangle is centered on the three lowest values of **peduc_20** in the top axis (i.e., the best educational access). In the PCP, the observation lines closely track for the three variables. In the data cube on the right, the same three observations are highlighted in yellow (with the red arrow pointing to them), very close together near the origin of the axes. This demonstrates the equivalence of lines being close together in the PCP and observation points being close together in multi-attribute variable space. Clearly, for more than three variables, only the PCP remains a viable method to assess such closeness.

Figure 8.24 illustrates the concept of an outlier in the PCP. Now, the selection rectangle is over the highest value for **pepov_20** (in the lowest axis), an observation with extreme poverty. As the graph shows, this also corresponds to an extreme value for **pserv_20**, but much less so for **peduc_20**. In the data cube, the selected observation, highlighted in yellow (with the red arrow pointing to it), is indeed somewhat removed from the rest of the data cloud.

The number of variables that can be visualized with a PCP is only constrained by screen real estate. However, as pointed out, for larger data sets, the graph can become quite cluttered, making it less practical.

9

Space-Time Exploration

While the primary focus of `GeoDa` is on the exploration of cross-sectional or static data, it also includes limited functionality to investigate space-time dynamics through the use of the **Time Editor**, **Time Player** and **Averages Chart**.

This allows for the evolution of a variable to be shown over time through maps and graphs, in a form of *comparative statics*. One limitation of this approach is that it is cross-section oriented. More precisely, the different time periods are considered as separate cross-sectional variables (one cross-section for each time period). While this results in quite a bit of flexibility when it comes to *grouping* variables, it also means there is no inherent time-awareness, nor a dedicated panel data structure.

After explaining the approach to space-time exploration through the **Time Editor** and **Time Player**, I revisit the use of the **Averages Chart** for treatment effect analysis. In contrast to the discussion in Chapter 7, here a dynamic analysis is possible, in that spatial structural change can be assessed at two points in time through a *difference in difference* approach.

To illustrate these methods, I continue to use the *Oaxaca Development* sample data set, particularly the variables from the Mexican census at three different points in time: 2000, 2010 and 2020.

9.1 Topics Covered

- Create grouped variables by means of the Time Editor
- Save group definitions in the Project File
- Assess space-time patterns by means of the Time Player
- Carry out treatment effect analysis with the Averages Chart

GeoDa Functions

- Time > Time Editor
  - grouping variables
  - ungrouping variables
  - editing the Time label of a variable
  - removing time periods from a time-enabled variable
- Time > Time Player
  - starting and pausing the player
- Box Plot
  - changing the number of time-enabled variables displayed
  - changing axes and synchronization options

DOI: 10.1201/9781003274919-9

- Scatter Plot
  - time-wise autoregressive scatter plot
- Map
  - moving a choropleth map through time using custom categories
- Averages Chart
  - difference in means test
  - difference in difference (DID) analysis
  - saving the dummy variables

Toolbar Icons

Figure 9.1: Time | Averages Chart

9.2 Time Editor

The **Time Editor** is the mechanism through which different cross-sectional variables are *grouped* into a specific temporal sequence. In general, this could be any set of variables, but it is most relevant if the variables pertain to successive time periods. This requires that the observations for each time period are included in the data table as a separate cross-section. The design is that of a *balanced panel*, with the same cross-sectional units in each time period.[1] The system is agnostic to the fact that the variables pertain to different time periods, so care must be taken to avoid counter intuitive situations (e.g., by selecting the wrong sequence of variables).

The space-time functionality is invoked from the menu as **Time > Time Editor**. The menu includes a separate item for each of the **Time Editor** and the **Time Player**. In contrast, when using the toolbar icon, i.e., the left icon in Figure 9.1, both are started.

The main purpose of the **Time Editor** dialog is to create *grouped* variables.

9.2.1 Creating grouped variables

The creation of grouped variables is a somewhat tedious process that involves identifying the variable names for each time period and combining them into a *group*. The interface, shown in Figure 9.2, consists of three major panels: **Ungrouped Variables**, **New Group Details** and **Grouped Variables**. Each has a brief help window to explain the meaning of these terms, brought up by clicking on the **?** symbol.

The left-most panel contains the variable names for all the numeric variables in the data table. They are selected by double clicking or by pressing on the arrow key (>). This moves the variable to the middle panel, where they are listed under **Name**. At the same time, the **variables to include** item is updated. The column associated with **Time** includes a period identifier, initially **time 0**, **time 1**, etc., but this can easily be edited. The **name**

[1] An *unbalanced* panel would have different cross-sectional units by time period, which cannot be supported by `GeoDa's` current data structure.

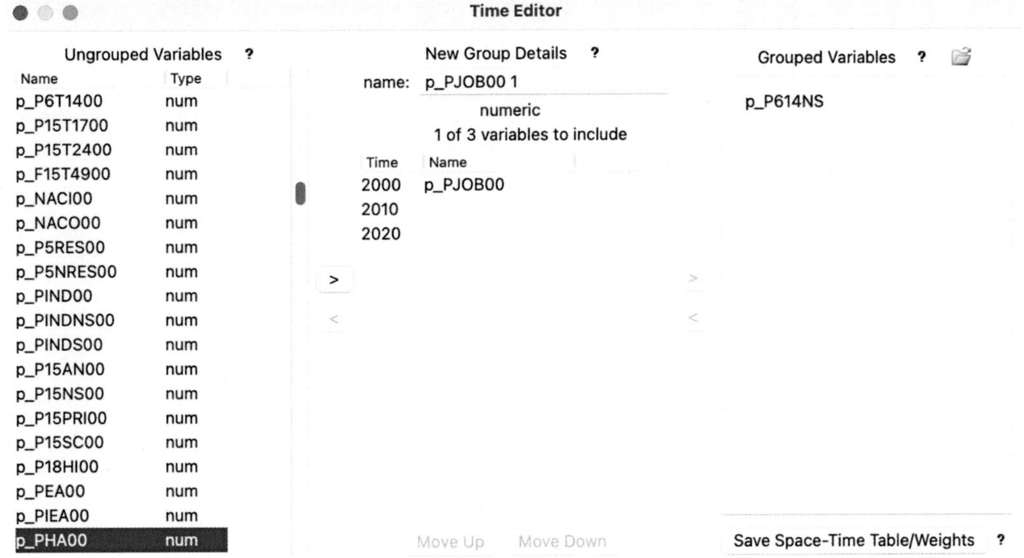

Figure 9.2: Time Editor interface

	p_PJOB (2000)	p_PHA (2000)
1	13.972056	1.103896
2	42.157900	41.806680
3	50.981308	45.930031

Figure 9.3: Grouped variables in data table

for the group is initially derived from the variable name entered but also typically requires some further editing.

Individual variables can be moved up or down in the sequence (the sequence determines the order of the comparative statics graphs and maps) by selecting the variable and using the buttons at the bottom of the panel (**Move Up** and **Move Down**). In addition, any variable can be moved out of the grouping, back to the ungrouped variables list, by selecting it and clicking on the left arrow buttons (<).

When all the variables are included under **New Group Details**, in the correct order and with the correct **Time** label, another arrow key (to the right of the variable list) adds the group under the **Grouped Variables** list with the **name** as specified in the central panel. At this point, the process can be repeated for additional variables. However, for each grouped variable, the same number of original variables must be combined. In other words, the grouping always pertains to a fixed number of time periods, set when the first grouped variable is created.

The example in Figure 9.2 is for the *Oaxaca* data set. It shows the situation after the variable **p_P614NS** (percentage of the population between 6 and 14 years of age that received no schooling) has been created by grouping **p_P614NS00** (for 2000), **p_P614NS10** (for 2010) and **p_P614NS20** (for 2020). The **Time** labels have been set to **2000**, **2010** and **2020**. A new grouped variable is in the process of being created using **p_PJOB00** for **2000** (percentage of the population over 12 years of age with a job). Note how the entry for **name** is not very useful at this point, and should be edited to **p_PJOB**, for example. The next variable to be selected should be **p_PJOB10**, for **2010**.

To support a range of illustrations, two more groups are created: **p_PHA** (percentage of the population with access to health services), and **p_CAR** (percentage private inhabited dwellings with a car or van).

The grouped variables are shown in the data table with their new name, followed by the first time period in parenthesis. For example, in Figure 9.3, the entries are shown for **p_PJOB** and **p_PHA**.

Note that the grouped variables have not been dropped as individual entries from the table, they are just not displayed as such. The new label in parentheses indicates that they have been grouped.

Finally, as mentioned, `GeoDa` is agnostic to what variables are grouped and in which order they are arranged, so this procedure can also be used to group variables that do not necessarily correspond to different time periods. For example, this may be useful when constructing a box plot graph that shows the box plots for multiple variables in one window.

9.2.1.1 Grouped variables in Project File

As discussed in the case of custom map categories (Section 4.6.3), any special definition such as a grouped variable will be lost if it is not saved in a project file (accomplished through **File > Save Project**). Once contained in a project file, the grouped variables can be re-used in later analyses. There can be several different project files associated with the same data layer, each focused on a special aspect of time dynamics.

The project file contains the **time_ids** specified in the time editor, and the new variable definitions under the **group** entry. For example, in Figure 9.4, the three years for the *Oaxaca* data set are included under the **time_ids** listing below the start of **variable_order**. Further down in the file, the group definitions for the **p_PJOB** and **p_PHA** variables are shown, together with some of the non-grouped variable definitions.

Note that the grouping of variables can only use one set of time periods, due to the synchronization mechanism that underlies the time player. So, in order to look at different time periods, such as two periods for one variable and three for another, two separate project files would need to be created. More precisely, any active grouping has to pertain to the same time labels for all the grouped variable. For example, in the *Oaxaca* data set, the poverty indicators pertain only to 2010 and 2020, whereas the census variables are for three years. In order to carry out any type of space-time analysis for the poverty indicators, a separate **Time Editor** operation would need to be carried out on a file that does not contain any other grouped variables (more precisely, one would need to start over). The new definitions should be saved in a different dedicated project file. Consequently, in the *Oaxaca* situation, there could need to be separate project files for a two-period analysis (2010 and 2020) and for a three period analysis (including 2000).

As seen before, instead of loading a spatial layer in the **Connect to Data Source** of the file open operation, one should load a project file with the **gda** extension. This will retain all grouped variable definitions contained in the project file.

9.2.1.2 Saving the space-time data table

A final option in the lower right hand side of the interface shown in Figure 9.2 is invoked by the **Save Space-Time Table/Weights** button. This creates a data file (typically in csv format) that contains a new space-time identifier, **STID**, as well as identifiers for the

```
<?xml version="1.0" encoding="utf-8"?>
<project>
    <title>oaxaca</title>
    <layers>
        <layer>
            <datasource>
                <type>ESRI Shapefile</type>
                <path>oaxaca.shp</path>
            </datasource>
            <layername>oaxaca</layername>
            <title/>
            <variable_order>
                <time_ids>
                    <id>2000</id>
                    <id>2010</id>
                    <id>2020</id>
                </time_ids>
                <var>ID</var>
                <var>
                    Shape_Leng
                    <displayed_decimals>6</displayed_decimals>
                </var>
                        .
                        .
                        .

                <var>
                    p_PIEA00
                    <displayed_decimals>6</displayed_decimals>
                </var>
                <group>
                    <name>p_PJOB</name>
                    <displayed_decimals>6</displayed_decimals>
                    <var>p_PJOB00</var>
                    <var>p_PJOB10</var>
                    <var>p_PJOB20</var>
                </group>
                <group>
                    <name>p_PHA</name>
                    <displayed_decimals>6</displayed_decimals>
                    <var>p_PHA00</var>
                    <var>p_PHA10</var>
                    <var>p_PHA20</var>
                </group>
                <var>
                    p_NOH00
                    <displayed_decimals>6</displayed_decimals>
                </var>
```

Figure 9.4: Grouped variables in project file

cross-sectional observation, the time period and time period dummy variables.[2] The time grouped variables are stacked as cross-sections for each successive time period.

The space-time data file allows for pooled cross-section and time series analyses, but does not work with a corresponding map. In other words, the data can be loaded into a table for analysis (and, if appropriate, spatial weights included in the weights manager), but no map is available.

In addition, if a spatial weights file is active, a corresponding space-time weights GAL file is created as well, as detailed in Section 11.5.4.

[2]When a spatial weights file is active, its ID variable is used as the cross-sectional identifier. In the absence of an active spatial weights file, a sequence of integer values is used.

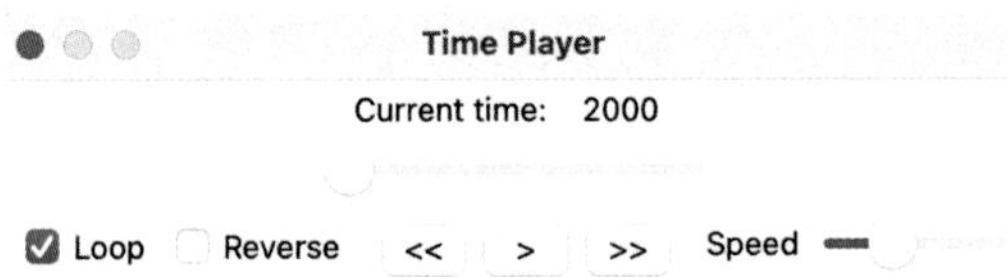

Figure 9.5: Time Player dialog

9.3 Time Player – Comparative Statics

The **Time Player** is the mechanism through which the space-time dynamics are implemented. It is automatically started when selecting the **Time** icon from the toolbar, but not from the menu, where it is a separate item from **Time Editor**. From the menu, it is invoked as **Time > Time Player**.

The dialog, shown in Figure 9.5, shares several aspects with the design of the **Animation** interface (see Section 5.5). The **Current time** is shown at the top of the interface (**2000** in the example). The default is to **Loop** through the different time periods, which is started by pressing the > button (the loop can also be run in **Reverse** by checking the corresponding box). Single step forward or backward is implemented through the double arrows, » and «. Finally, the **Speed** of the loop can be adjusted using the slider, just as in the animation tool.

Whenever a grouped (time-enabled) variable is selected for a graph or map, the **Time Player** becomes active. The starting point (time) is determined from a dropdown list below the variable name, which includes all available periods (for an example, see Figure 9.8 for a scatter plot). After the initial period is selected, one can move through discrete time steps by means of the player. While such a process does not assess true space-time dynamics, it allows one to carry out comparative statics, i.e., to compare cross-sectional patterns at different discrete time periods.

The comparative statics approach is illustrated with three examples that highlight some additional features of space-time exploration: the evolution of a box plot over time; a scatter plot with a time lagged variable; and a thematic map over time.

9.3.1 Box plot over time

To assess the evolution of the distribution of a variable over time, it is often useful to plot the corresponding box plots side-by-side. This is the default behavior when selecting the box plot functionality for a time-enabled (grouped) variable.

For example, in Figure 9.6, this is implemented for **p_P614NS**. The three graphs are shown side-by-side, with summary statistics below. It is clear that the schooling situation improved considerably between 2000 and 2010, but less so from 2010 to 2020. The median percent unschooled 6 to 14 year olds decreases from 9.13% to 4.53% in the first decade, and to 4.51% in the ten years afterward. However, the distribution is slightly more compact in 2020, with an IQR of 3.85 compared to 4.27 in 2010 (in 2020, the IQR was 6.8). Interestingly, this resulted in more upper outliers (i.e., municipalities that do much worse than the bulk), 22 in 2020 vs. 13 in 2010. The combination of the descriptive statistics with a visual representation of the distribution provides an effective way to assess broad changes over time.

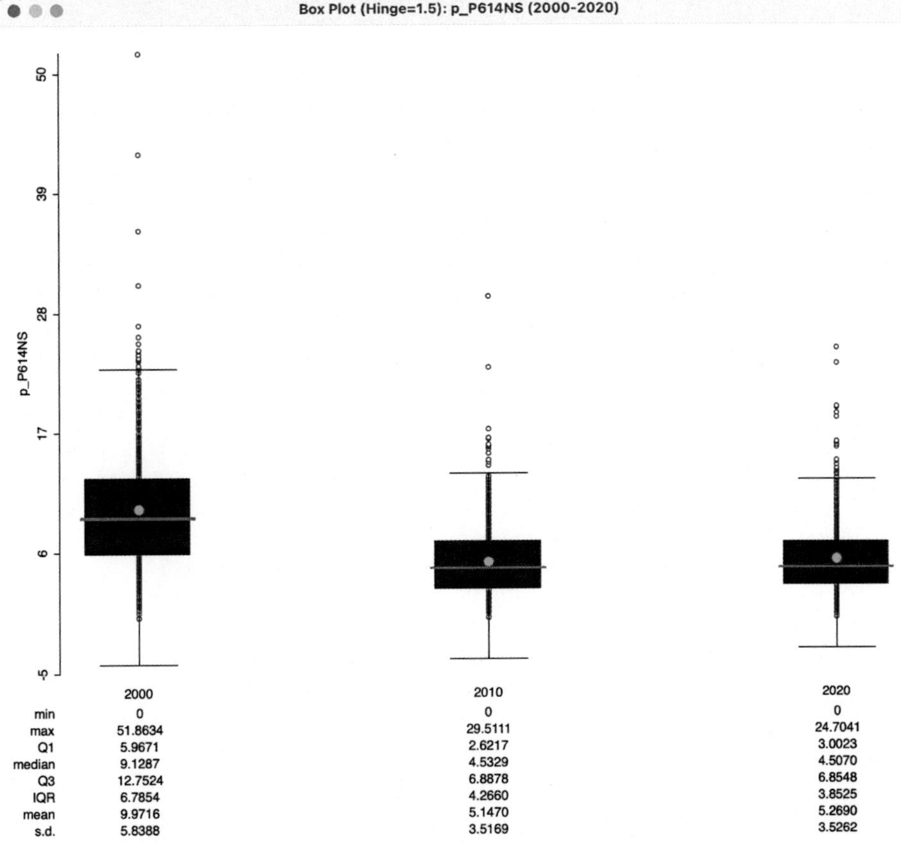

Figure 9.6: Box plot over time

9.3.1.1 Dynamic box plot options

The dynamic box plots share all the standard options with the regular box plot, see Section 7.2.2.3. In addition, there are three sets of special options: **Scale Options**, **Number of Box Plots** and **Time Variable Options**. The latter is by default checked to allow synchronization with the **Time Control**. This can be specified for each individual graph. Turning this check off disables the forward and backward movement of time periods through the **Time Player**, i.e., it makes the graph a static graph.

The **Scale Options** have by default **Fixed scale over time**. Turning this off adjusts the scale for each time period, in function of the relevant variable range. This is immaterial when all three box plots are shown, but matters when this is not the case.

Finally, **Number of Box Plots** is set by default to **All**, resulting in the side-by-side plots as in Figure 9.6. Other options are all values up to the number of periods in the group, in this example, **1** and **2**. Figure 9.7 illustrates the case with two plots, for 2000 and 2010.

9.3.2 Scatter plot with a time-lagged variable

With time-enabled variables, it becomes possible to create a scatter plot of the same variable at two points in time. The slope of the linear fit then corresponds to the serial autoregressive

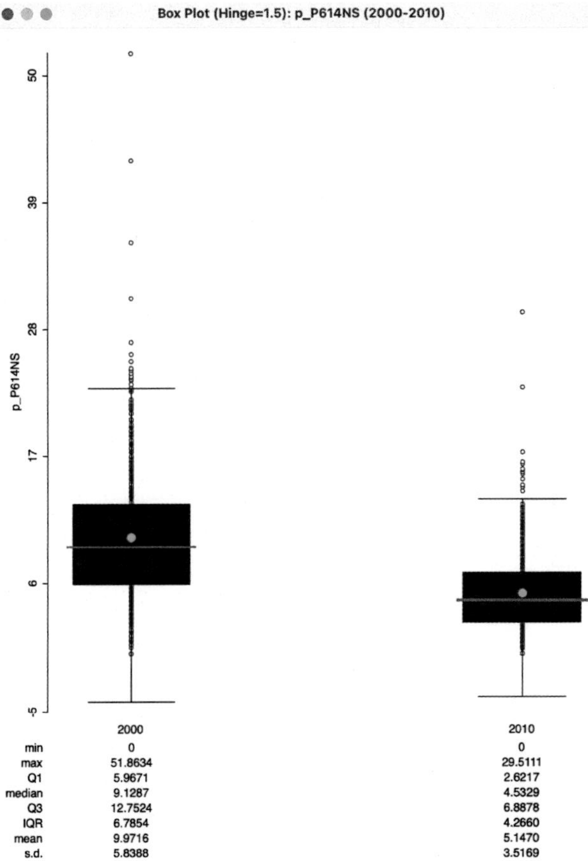

Figure 9.7: Box plot for two time periods

coefficient, i.e., how the variable in one period is related to its values in a previous period. More formally, this is the slope coefficient in a linear regression of y_t on y_{t-1}.

Since the time periods for the variables are available from a drop-down list in the variable selection dialog, it is easy to change the time selection for both X and Y variables. In Figure 9.8, this is illustrated for **p_PJOB**. The **Independent Var X** is set to **Time 2000**, and the drop-down list for **Dependent Var Y** is checked for **Time 2010**.

This results in the scatter plot shown in the left-hand panel of Figure 9.9, including both a linear fit and a LOWESS smoother (with the bandwidth changed to 0.4). The slope of 0.288, while highly significant, only results in an overall fit of 10%. The LOWESS smoother illustrates why this is the case. There is a clear break in the slope, with a positive slope (steeper than the overall 0.288) for values of **p_PJOB** in 2000 up to roughly 52%, but beyond that it turns to a negative slope. So, for municipalities that already had a high percentages employed people, the relationship between 2000 and 2010 is negative, suggesting a loss of employment for those municipalities. In an actual application, this structural break should be further investigated by means of a Chow test. Also, its spatial footprint can be assessed by means of linking and brushing with a map (see Sections 7.3.1.1 and 7.5.2 in Chapter 7).

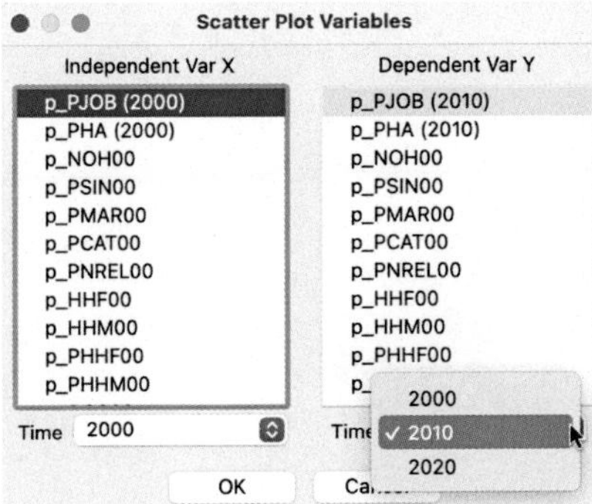

Figure 9.8: Time lagged scatter plot variables

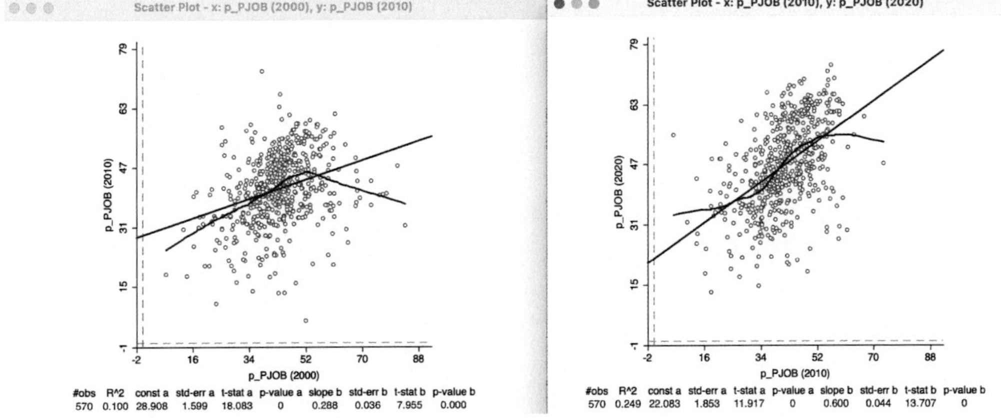

Figure 9.9: Scatter plots with time lagged variables

After adjusting the **Time** for the X-variable to 2010 and for the Y-variable to 2020, the scatter plot in the right-hand panel of Figure 9.9 results. Now, the slope is much steeper, and the fit quite a bit better (R^2 of 24.9%). However, there is still considerable evidence of structural instability, as emphasized by the S-shaped LOWESS fit (again with bandwidth at 0.4). In an actual application, this would need to be investigated further.

Unlike what is the case for univariate graphs (and maps), the **Time Player** does not necessarily have the desired effect. It is essentially a univariate tool, so it only changes the **Time** period for the dependent variable (Y). The X-variable remains unchanged, rather than updating both axes to move the time lag forward or backward.

9.3.3 Thematic map over time

Since a thematic map shows the spatial distribution of a single variable, it can be readily manipulated by means of the **Time Player** when it pertains to a time enabled variable. However, since any map classification is by design *relative*, this may lead to an unsatisfactory outcome.

For example, the left-hand panel of Figures 9.10–9.12 contains a six quantile map for the percentage car ownership, **p_CAR**, in 2000, 2010 and 2020. These maps show an overall spatial distribution that seems to undergo little change. However, unless one carefully inspects

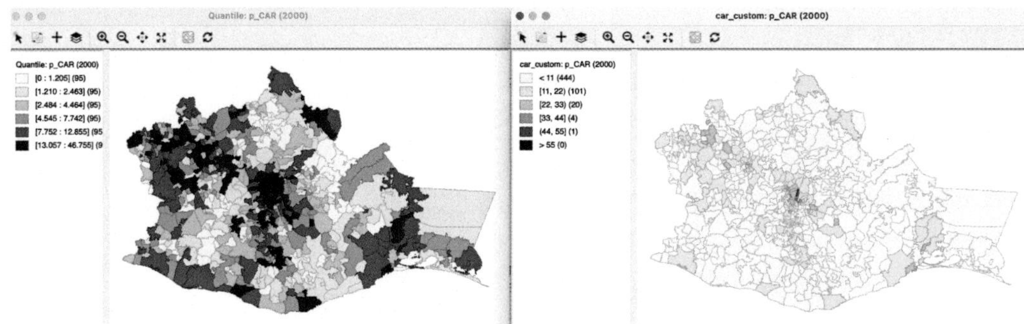

Figure 9.10: Thematic maps for car ownership in 2000

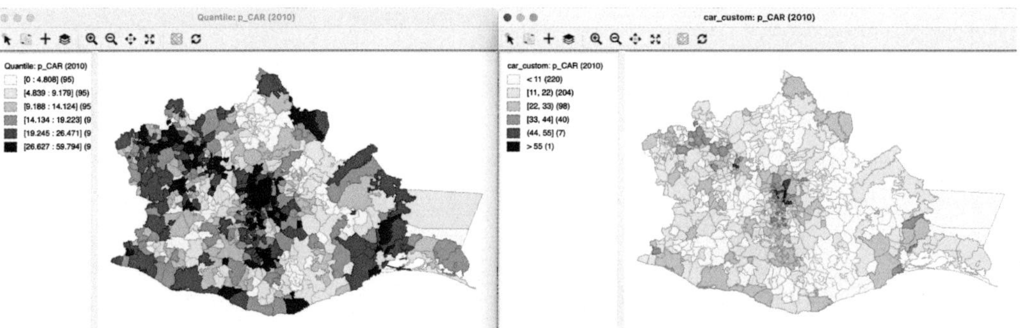

Figure 9.11: Thematic maps for car ownership in 2010

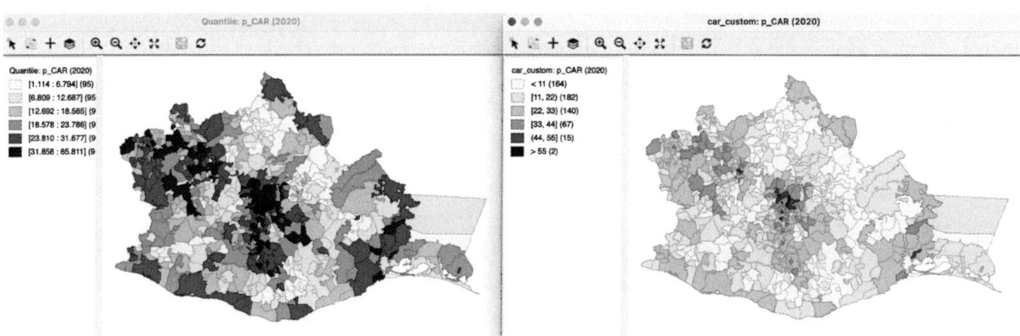

Figure 9.12: Thematic maps for car ownership in 2020

each map legend, the maps represent very different realities. In 2000, the percentage car ownership ranged from 0% to 46.8%, whereas in 2020, it varied between 1.1% and 65.8%, a 20% difference for the maximum.

In order to better represent these absolute changes, a set of custom categories must be created by means of the **Category Editor** (see Section 4.6). In the right-hand panel of Figures 9.10–9.12, a six-category equal interval classification is employed, going from 0 to 66 (the map legend lists **car_map** as the classification method).

The result is very different. The paucity of car ownership in the first period is emphasized (there are only five observations over 33%) and the maps show a gradual increase over time, with growing prevalence of the darker colors (from the center outward and to the east).

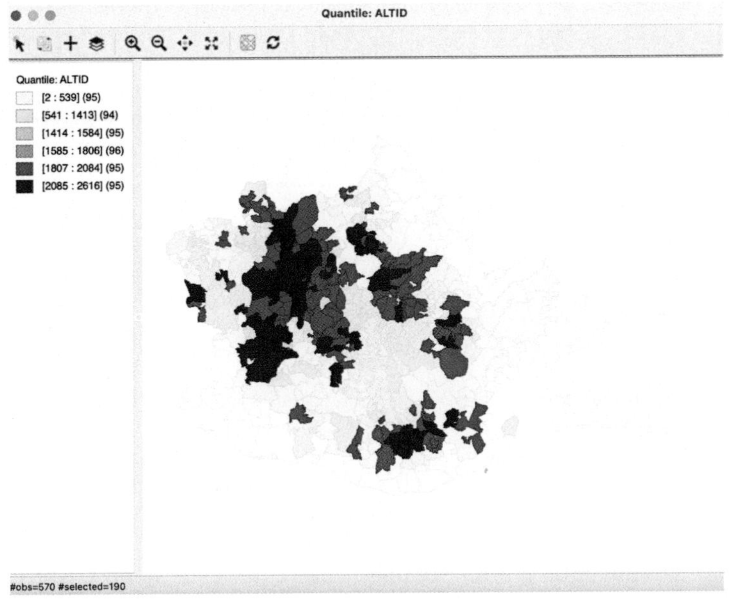

Figure 9.13: Selected municipalities with altitude in two upper quantiles

9.4 Treatment Effect Analysis – Averages Chart

As pointed out in Section 7.5.1, the real power of the **Averages Chart** lies in the comparison *over time* of the differences in a variable distribution between two subsets of the data. This is particularly relevant in so-called *treatment effect analysis*, where one group of observations (the **Selected**) is subject to a policy (treatment) between a beginning point and an endpoint. A target variable (outcome) is observed for both time periods, i.e., before and after the policy is implemented. The objective is to assess whether the treatment had a significant effect on the target variable by comparing its change over time to that for a control group (the **Unselected**).

In the example used here, the target variable is **p_PHA**, the percentage population with access to health care. Consider an imaginary policy experiment, where between 2000 and 2020 a targeted set of policies would have been applied to improve access to health care in higher elevation municipalities, specifically municipalities in the upper two quantiles of a six quantile distribution for altitude (**ALTID**). Note that this example is used only to illustrate the logic and implementation of a treatment effect analysis. It does not in fact correspond to an actual policy carried out in Oaxaca.

The *treated* municipalities (**Selected**) are depicted in the familiar quantile map for **ALTID**, with their selection highlighted in Figure 9.13.

The **Averages Charts** is invoked in the same manner as in the cross-sectional case, either from the menu, as **Explore > Averages Chart**, or by selecting the right icon in the toolbar shown in Figure 9.1. Unlike what was the case in the cross-sectional context, once the **Averages Chart** is applied to a time enabled variable, it becomes possible to specify **Period 1** and **Period 2** in the interface.

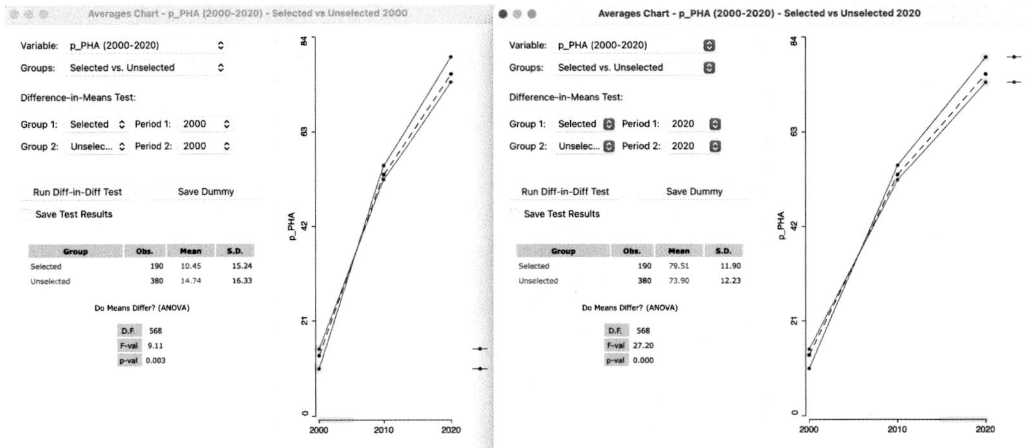

Figure 9.14: Difference in means, selected and unselected, static, in 2000 and 2020

This allows for two different types of applications. One is the same **Difference-in-Means Test** as considered before, applied statically to the variable at different points in time. A second is a more complete treatment effect analysis, implemented through the **Run Diff-in-Diff Test** option. Each is considered in turn.

9.4.1 Difference in means

To carry out a simple static difference in means test, the time-enabled variable **p_PHA** must be specified in the **Variable** drop-down list, with the **Groups** set as **Selected vs. Unselected**. The time-enabled property of the variable is revealed by the beginning and end period in parentheses, **(2000–2020)**, following the variable name.

As before, the **Difference-in-Means Test** requires **Selected** for **Group 1**, and **Unselected** for **Group 2**, but now the **Period** selection allows a choice among three time periods for **1** and for **2** (the period for **1** must always be before or equal to that for **2**). A static analysis requires both periods to be the same. In the left-hand panel of Figure 9.14, the date is set to **2000**, whereas in the right-hand panel, it pertains to **2020**.

The graph in each panel shows the values for all three time periods. However, the relevant period is highlighted with a yellow bar and the statistics pertain only to that period.

In the illustration, in 2000, the mean for **Selected** is 10.45, compared to 14.74 for **Unselected**, for a difference of **−4.29**. In 2020, the situation is reversed, with a mean of 79.51 for **Selected**, compared to 73.90 for **Unselected**, now yielding a positive difference of **5.61**. In both cases the static difference in means test is highly significant.

To what extent might the difference of between the mean of 79.51 in 2020 and 10.45 in 2000 be due to our imaginary policy? This is answered by the **Diff-in-Diff Test**.

9.4.2 Difference in difference

The principle behind treatment effect analysis is to compare the evolution of the target variable in the treated group *before and after* the policy implementation (respectively **Period 1** and **Period 2**) to a *counterfactual*, i.e., what would have happened to that variable had the treatment *not* been applied. Rather than a simple before and after difference in means

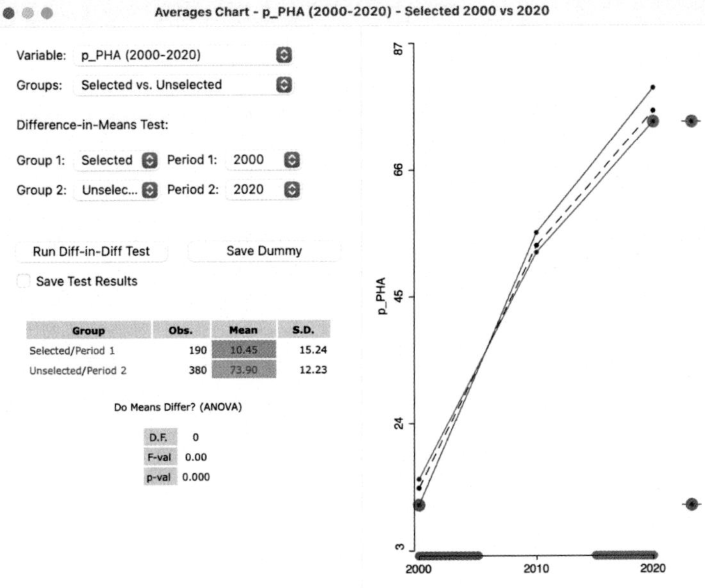

Figure 9.15: Difference in difference setup, 2000 to 2020

test, the evolution of the target variable for a control group is taken into account as well. This is the basis for a difference in difference analysis. The problem is that the counterfactual is not actually observed. Its behavior is inferred from what happens to the control group.

A critical assumption is that in the absence of the policy, the target variable follows parallel paths over time for the treatment and control group. In other words, the difference between the treatment and control group at period 1 would be the same in period 2, in the absence of a treatment effect.

The counterfactual is thus a simple trend extrapolation of the treated group. In the absence of a treatment effect, the value of the target variable in period 2 should be equal to the difference between treated and control in period 1 added to the change for the control between period 1 and period 2, i.e., the trend extrapolation. The difference between that extrapolated value for the counterfactual and the actual target variable at period 2 is then the estimate of the treatment effect.

Formally, this can be expressed as a simple linear regression of the target variable stacked over the two periods, using a dummy variable for the treatment-control dichotomy (the space dummy, S, equal to 1 for the **Selected**), a second one for the before and after dichotomy (the time dummy, T, equal to 1 for **Period 2**) and a third dummy for the interaction between the two (i.e., treatment in the second period, $S \times T$):[3]

$$y_t = \beta_0 + \beta_1 S + \beta_2 T + \beta_3 (S \times T) + \epsilon,$$

with β as the estimated coefficients, and ϵ as a random error term. The treatment effect is the coefficient β_3. The coefficient β_0 corresponds to the mean in the control group (**Unselected**) in 2000 (14.74 in Figure 9.14). β_1 is the pure space effect, the difference in means between **Selected** and **Unselected** in 2000 (-4.29 in Figure 9.14). Finally, β_2 is the time trend in

[3]A full discussion of the econometrics of difference in difference analysis is beyond the scope of this chapter and can be found in Angrist and Pischke (2015), among others.

```
REGRESSION (DIFF-IN-DIFF, COMPARE REGIMES AND TIME PERIOD)
----------
SUMMARY OF OUTPUT: ORDINARY LEAST SQUARES ESTIMATION
Data Set            :  oaxaca_master
Dependent Variable  :  p_PHA (2000,2020)
Number of Observations: 1140
Mean dependent var  :      44.5403  Number of Variables    :    4
S.D. dependent var  :      34.3691  Degrees of Freedom     : 1136

R-squared             :   0.830374  F-statistic            :        1853.7
Adjusted R-squared  :   0.829926  Prob(F-statistic)      :             0
Sum squared residual:      228420  Log likelihood         :      -4638.68
Sigma-square          :    201.074  Akaike info criterion :       9285.36
S.E. of regression  :       14.18  Schwarz criterion      :       9305.51
Sigma-square ML       :    200.368
S.E of regression ML:     14.1551

------------------------------------------------------------------------------
       Variable       Coefficient        Std.Error     t-Statistic   Probability
------------------------------------------------------------------------------
          CONSTANT        14.7386         0.727421         20.2615       0.00000
             SPACE       -4.28552          1.25993         -3.4014       0.00069
        T2000_2020        59.1596          1.02873         57.5075       0.00000
          INTERACT        9.90215          1.78181         5.55735       0.00000
------------------------------------------------------------------------------
```

Figure 9.16: Difference in difference results

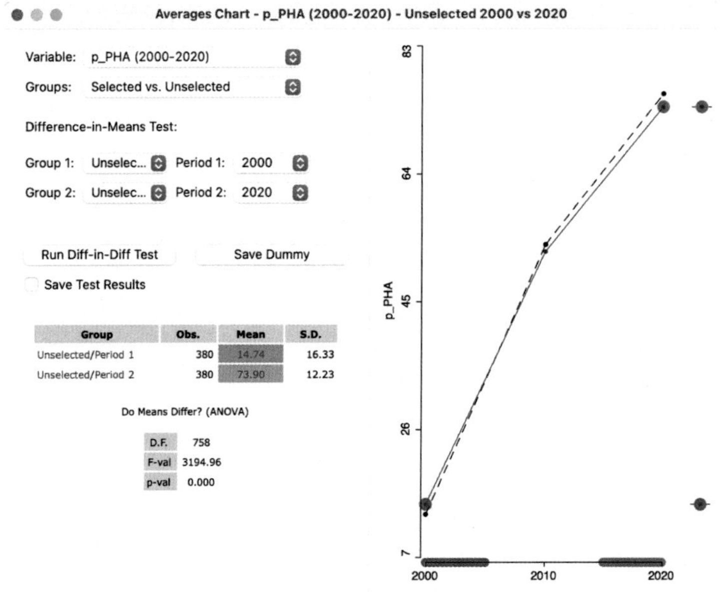

Figure 9.17: Difference in means, unselected, dynamic, 2000 to 2020

the control group, which has not yet been computed in the cross-sectional analysis (but see Figure 9.17).

9.4.2.1 Implementation

The difference in difference implementation uses the same **Averages Chart** interface as before, but with settings that disable the difference in means test. As shown in Figure 9.15, the two groups are **Selected** and **Unselected**, but **Period 1** is **2000**, with **Period 2** as **2020**. Even though the two means are listed and the graph is shown in the right hand panel, the results of the difference in means test are listed as **0**.

The analysis is carried out by selecting the **Run Diff-in-Diff Test** button. This yields the standard GeoDa regression output, shown in Figure 9.16.[4] The overall fit is quite good, with an R^2 of 83%, and all coefficients are highly significant.

9.4.2.2 Interpretation and options of Diff-in-Diff results

The interpretation of the Diff-in-Diff results centers on the regression coefficients in Figure 9.16. The estimate of the treatment effect is the coefficient of **INTERACT**, **9.90** and highly significant. This would suggest that our imaginary high altitude policies indeed had a significant positive effect on health care access.

The other pieces of the puzzle can be found in the remaining coefficients. The **CONSTANT** is 14.7386, which is the value shown for the analysis in 2000 in Figure 9.14 as the mean of the **Unselected**. The **SPACE** coefficient estimate of -4.2855 corresponds to the difference in means between **Selected** and **Unselected** in 2000 in Figure 9.14. The trend coefficient (**T2000_2020**) is 59.16. In Figure 9.17, the means for **Unselected** are shown for **Period 1** (14.74) and **Period 2** (73.90). The difference between the two is 59.16, the same as the coefficient for the time trend, **T2000_2020**.

The interface offers two options to **Save Dummy** and **Save Test Results**. Selecting those options brings up a **File > Save** dialog to specify the file type and file name to save a panel data listing of the observations and dummy variables. This can be used for more sophisticated regression analysis in other software (including GeoDa), e.g., to include relevant control variables, and correct for serial and spatial correlation. This is beyond the current scope.[5]

In spite of the limitations of the simple difference in difference analysis, it provides a way to more formally assess the potential impact of *spatial* policy initiatives.

[4]The regression functionality of GeoDa in not considered in this book. For a detailed treatment, see Anselin and Rey (2014).

[5]For recent reviews of some of the methodological issues associated with incorporating spatial effects in treatment effects analysis, see, among others, Kolak and Anselin (2020), Reich et al. (2021) and Akbari et al. (2023).

Part III

Spatial Weights

10

Contiguity-Based Spatial Weights

A core concept in the exploration of spatial patterns is *spatial autocorrelation*. It is considered in depth in Parts IV and V. At this point, it suffices to recognize that spatial autocorrelation is essentially a compromise between *attribute similarity* (the usual notion of correlation) and *spatial similarity*.

Spatial similarity in this context is typically interpreted as locational or geographical similarity, i.e., the presence of a neighbor relationship between a pair of observations. However, the concept is perfectly general, and pertains to any network structure that represents observations as *nodes*, connected by *edges*. The existence of an edge between a pair of observations then becomes the neighbor structure.

The matrix representation of this network structure is the so-called *spatial weights matrix*. In the current and the next two chapters, this concept is discussed in more detail, although with a particular focus on its implementation in `GeoDa`. Consequently, only a limited set of aspects of this broad ranging topic will be covered. More extensive and technical discussions as well as further references can be found in Anselin (1988), Bavaud (1998), Getis (2009), Harris et al. (2011), Anselin and Rey (2014) and Anselin (2021), among others.

This chapter deals with spatial weights derived from the presence of a common border between two observations (typically polygons), or *contiguity*-based weights. It begins with a more formal introduction of the concept of spatial weights, followed by a review of the creation of so-called rook and queen weights, as well as higher order contiguity weights and space-time weights. The chapter closes with a discussion of a range of methods that can be used to analyze and visualize characteristics of the spatial weights.

To illustrate these concepts, I will mostly use the *Ceará Zika* sample data set, except in the discussion of space-time weights, where *Oaxaca Development* is employed. Note that to deal efficiently with the resulting spatial weights files, it is preferable to save the sample data sets onto a designated working directory before constructing the weights.[1]

10.1 Topics Covered

- Understand the concept of spatial weight
- Construct rook and queen contiguity-based spatial weights
- Compute higher order contiguity weights
- Create space-time weights
- Store the weights information in a Project File

[1]The sample data sets are contained within a special file attached to the `GeoDa` software and cannot be changed.

DOI: 10.1201/9781003274919-10

- Assess the characteristics of spatial weights
- Visualize the graph structure of spatial weights
- Identify the neighbors of selected observations

GeoDa Functions

- Weights Manager (Tools > Weights Manager)
  - weights file creation interface
  - ID variable
  - rook and queen contiguity
  - precision threshold option
  - spatial weights file name
  - weights properties in the weights manager
  - loading weights from a file
- Time Editor > Save Space-Time Table/Weights
- Structure of a GAL file
- Connectivity histogram
- Connectivity map
- Connectivity graph
- Select neighbors of selected
  - Map > Connectivity
  - Table > Add Neighbors To Selection

Toolbar Icons

Figure 10.1: Weights Manager

10.2 The Concept of Spatial Weights

Spatial weights are a key component in any cross-sectional analysis of spatial dependence. As mentioned in the introduction, they are an essential element in the construction of spatial autocorrelation statistics. In addition, spatial weights provide the means to create *spatially explicit* variables, such as spatially lagged variables and spatially smoothed rates (see Chapter 12).

Spatial weights are a necessary evil. In principle, one would like to have the data determine the strength of the correlation between any pair of observations. However, with only n observations in a cross-section, it is impossible to estimate all $n \times (n-1)/2$ pairwise correlations. In this situation, growing the sample, i.e., so-called *asymptotic* analysis, does not help, since the number of correlation parameters grows with n^2, while the sample only increases linearly. In the literature, this is referred to as the *incidental parameter problem*.[2] It has no solution other than imposing some structure that simplifies the problem.

The spatial weights accomplish such a simplification by *constraining* the number of possible neighbor relations. In addition, the same correlation is imposed for all pairs of observations,

[2]See Lancaster (2000) for a technical review.

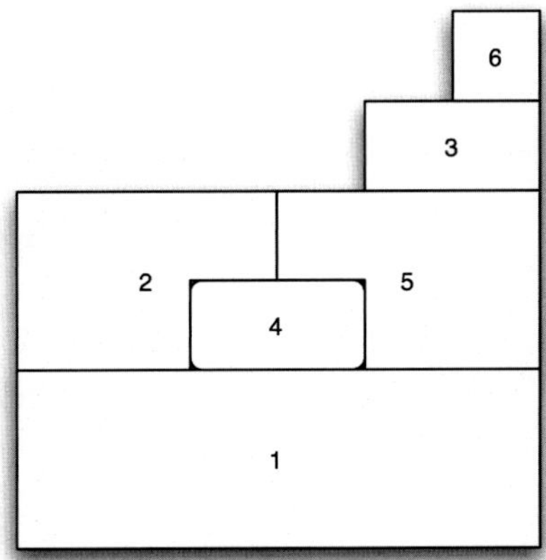

Figure 10.2: Example spatial layout

but this is only relevant in a discussion of global spatial autocorrelation statistics, which is postponed until Part IV.

There are different criteria that can be used to reduce the number of neighbors for each observation. In this chapter, approaches based on the notion of *contiguity* are considered, i.e., sharing a common border. In spatial analysis, this is typically based on geographic borders, although as discussed in Chapter 11, the concept is perfectly general.

In the remainder of this section, the concept is introduced more formally first, followed by a brief discussion of the two most common forms, i.e., rook and queen contiguity. This is followed by a treatment of the concept of higher order contiguity and some practical considerations.

10.2.1 Spatial weights matrix

A spatial weights matrix expresses the neighbor structure between the n observations in a cross-section as a $n \times n$ matrix $\mathbf{W}$. For each pair $i - j$, where i is the matrix row and j the matrix column, there is a spatial weight element w_{ij}:

$$\mathbf{W} = \begin{bmatrix} w_{11} & w_{12} & \dots & w_{1n} \\ w_{21} & w_{22} & \dots & w_{2n} \\ \vdots & \vdots & \ddots & \vdots \\ w_{n1} & w_{n2} & \dots & w_{nn} \end{bmatrix}.$$

The spatial weights w_{ij} are non-zero when i and j are neighbors, and zero otherwise. By convention, the self-neighbor relation is excluded, so that the diagonal elements of $\mathbf{W}$ are zero, $w_{ii} = 0$.[3]

To make the concept of a spatial weights matrix more concrete, a spatial layout is used with six polygons, shown in Figure 10.2. Two spatial units are defined as neighbors when they share a common border, i.e., when they are *contiguous*. For example, in the figure, unit 1 shares a border with units 2, 4 and 5, and hence they are neighbors.

[3]An exception to this rule are the diagonal elements in kernel-based weights, which are considered in Chapter 12.

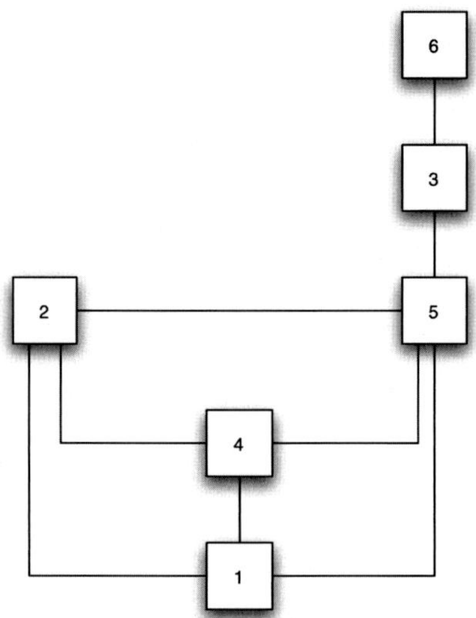

Figure 10.3: Contiguity as a network

The same neighbor structure can also be represented in the form of a network or *graph*, as in Figure 10.3. Here, each spatial unit becomes a node in the network, and the existence of a neighbor relation is represented by an edge or link connecting the respective nodes. Again, node 1 is connected with a link to nodes 2, 4 and 5. It is very important to keep in mind the generality of this network representation, since it applies well beyond purely geographic concepts of contiguity (see Chapter 11).

In its simplest form, the spatial weights matrix expresses the existence of a neighbor relation in binary form, with weights 1 and 0. For the layout in Figure 10.2, this yields a 6×6 matrix:

$$\mathbf{W} = \begin{bmatrix} 0 & 1 & 0 & 1 & 1 & 0 \\ 1 & 0 & 0 & 1 & 1 & 0 \\ 0 & 0 & 0 & 0 & 1 & 1 \\ 1 & 1 & 0 & 0 & 1 & 0 \\ 1 & 1 & 1 & 1 & 0 & 0 \\ 0 & 0 & 1 & 0 & 0 & 0 \end{bmatrix}. \tag{10.1}$$

In the first row of this matrix, matching unit 1, the non-zero elements correspond to columns (units) 2, 4 and 5.

In sum, the polygon boundaries in the map layout, the presence of edges in the network and the non-zero weights in the matrix are all equivalent representations of the topology or spatial arrangement of the data.

While naturally associated with areal units represented as polygons, the notion of contiguity can also be applied in situations where the observations are represented as points, see Section 11.5.1.

Finally, it is important to note that even though it is referred to as a spatial weights *matrix*, no such matrix is actually used in actual software operations. Spatial weights are typically very sparse matrices, and this sparsity is exploited by using specialized data structures (there is no point in storing lots and lots of zeros).

10.2.1.1 Row-standardizing spatial weights

With a few exceptions, the analyses in GeoDa that employ spatial weights use them in *row-standardized* form. Row-standardization takes the given weights w_{ij} (e.g., the binary zero-one weights) and divides them by the row sum:

$$w_{ij(s)} = w_{ij} / \sum_j w_{ij}.$$

As a result, each row sum of the new matrix equals one. Also, the sum of all weights, $S_0 = \sum_i \sum_j w_{ij}$, equals n, the total number of observations.[4]

For the layout in Figure 10.2, the matching row-standardized weights matrix is:

$$\mathbf{W_s} = \begin{bmatrix} 0 & 1/3 & 0 & 1/3 & 1/3 & 0 \\ 1/3 & 0 & 0 & 1/3 & 1/3 & 0 \\ 0 & 0 & 0 & 0 & 1/2 & 1/2 \\ 1/3 & 1/3 & 0 & 0 & 1/3 & 0 \\ 1/4 & 1/4 & 1/4 & 1/4 & 0 & 0 \\ 0 & 0 & 1 & 0 & 0 & 0 \end{bmatrix}. \qquad (10.2)$$

Importantly, whereas the original binary weights matrix was symmetric, this is no longer the case with the row-standardized weights. For example, the element in row 1-column 5 equals 1/3, whereas the element in row 5-column 1 equals 1/4. This has important consequences for some operations in spatial regression analysis, but less so in an exploratory context (but see Section 12.3.1). Also, the weights are no longer equal. In fact, they are inversely related to the number of neighbors. With more neighbors (say, k), the weight given to each neighboring observation ($1/k$) becomes smaller. More importantly, when there is only one neighbor (as in row 6, column 3 in the example), the weights remains 1. This may lead to counter intuitive results when computing spatial transformations, as in Chapter 12.

10.2.2 Types of contiguity

Sharing a common border may seem a straightforward definition of contiguity. However, in practice, it is not always clear what constitutes a border of non-zero length. For example, consider the regular three by three grid layout shown in Figure 10.4

In this layout, contiguity can be defined in three ways, each yielding a different spatial weights matrix. The three criteria are referred to as *rook*, *bishop* and *queen*, in analogy to the moves allowed by matching pieces on a chess board.

The rook criterion defines neighbors as those cells to the east, west, north and south, yielding four neighbors for each spatial unit that is not on the boundary of the spatial layout. In the example, this yields 2, 4, 6 and 8 as the neighbors of cell 5, as shown in Figure 10.5.

The matching binary contiguity weights matrix for the nine spatial units is then the 9×9

---

[4]Strictly speaking, this is only correct in the absence of so-called isolates, i.e., observations without neighbors. With q isolates, the sum $S_0 = n - q$. See Section 11.4.2.

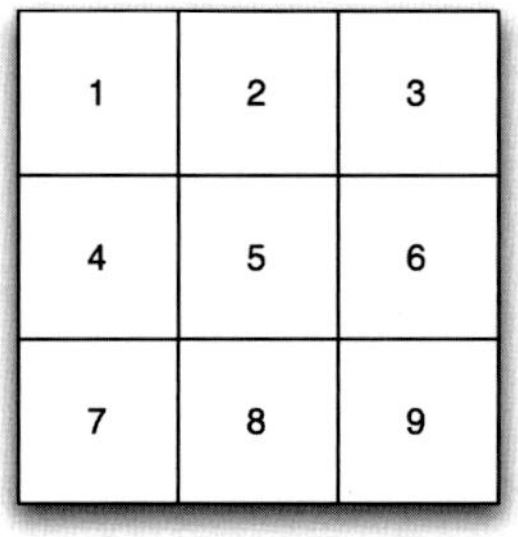

Figure 10.4: Regular grid layout

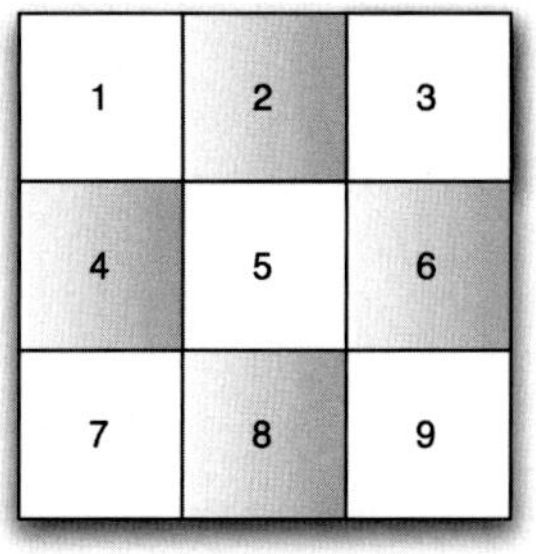

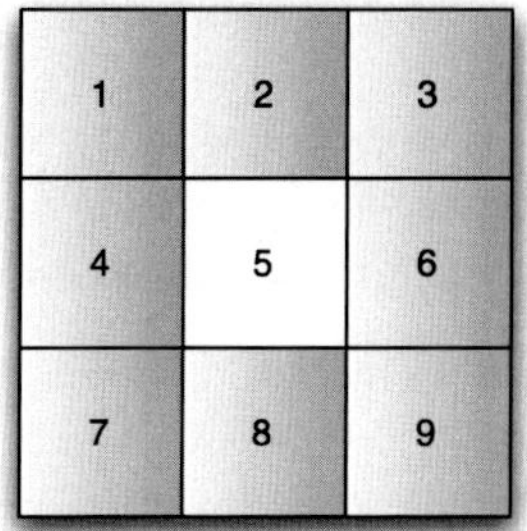

Figure 10.5: Rook contiguity

Figure 10.6: Queen contiguity

matrix:

$$\mathbf{W} = \begin{bmatrix} 0 & 1 & 0 & 1 & 0 & 0 & 0 & 0 & 0 \\ 1 & 0 & 1 & 0 & 1 & 0 & 0 & 0 & 0 \\ 0 & 1 & 0 & 0 & 0 & 1 & 0 & 0 & 0 \\ 1 & 0 & 0 & 0 & 1 & 0 & 1 & 0 & 0 \\ 0 & 1 & 0 & 1 & 0 & 1 & 0 & 1 & 0 \\ 0 & 0 & 1 & 0 & 1 & 0 & 0 & 0 & 1 \\ 0 & 0 & 0 & 1 & 0 & 0 & 0 & 1 & 0 \\ 0 & 0 & 0 & 0 & 1 & 0 & 1 & 0 & 1 \\ 0 & 0 & 0 & 0 & 0 & 1 & 0 & 1 & 0 \end{bmatrix}. \tag{10.3}$$

The structure of the weights matrix is characterized by a band of ones. In addition, there is a strong effect of boundary units in this small data set. Internal units have four neighbors, but corner units only have two, and the other boundary units only have three.

A second criterion to define contiguity, less frequently used, is based on common vertices (corners) as the convention to identify neighbors. This is referred to as bishop contiguity. Due to its lack of adoption in practice, it is not further considered here.

Finally, the queen criterion combines rook and bishop. As a result, neighbors share either a common edge or a common vertex. This yields eight neighbors for the non-boundary units. For example, as shown in Figure 10.6, the central cell 5 has all the other cells being neighbors.

The matching 9×9 binary contiguity matrix is much denser than before:

$$\mathbf{W} = \begin{bmatrix} 0 & 1 & 0 & 1 & 1 & 0 & 0 & 0 & 0 \\ 1 & 0 & 1 & 1 & 1 & 1 & 0 & 0 & 0 \\ 0 & 1 & 0 & 0 & 1 & 1 & 0 & 0 & 0 \\ 1 & 1 & 0 & 0 & 1 & 0 & 1 & 1 & 0 \\ 1 & 1 & 1 & 1 & 0 & 1 & 1 & 1 & 1 \\ 0 & 1 & 1 & 0 & 1 & 0 & 0 & 1 & 1 \\ 0 & 0 & 0 & 1 & 1 & 0 & 0 & 1 & 0 \\ 0 & 0 & 0 & 1 & 1 & 1 & 1 & 0 & 1 \\ 0 & 0 & 0 & 0 & 1 & 1 & 0 & 1 & 0 \end{bmatrix}$$

Now, the corner cells have three neighbors, and the other boundary cells have five.

The concepts of rook and queen contiguity can easily be extended to other regular tessellations, such as hexagons.[5] In addition, they apply to irregular lattice layouts as well. In this context, rook neighbors are those spatial units that share a common edge, and queen neighbors those that share either a common edge or a common vertex. As a result, spatial weights based on the queen criterion will always have at least as many neighbors as a corresponding rook-based weights matrix.

In practice, whether a common segment is categorized as a point (vertex) or line segment (edge) depends on the precision of the geocoding in a GIS. Due to *generalization* (a process of simplifying the detail in a spatial layer), some small line segments may be represented as points in one GIS and not in another. In addition, similar quirks in the representation of polygon boundaries may result in (very) small empty spaces between them. In a naive approach to defining common boundaries, such instances would fail to identify neighbors. Therefore, some tolerance may be required to make the weights creation robust to such characteristics.

10.2.3 Higher-order contiguity

So far, the focus has been on the notion of a direct neighbor, or, more precisely, a *first-order* neighbor. The concept of contiguity can be generalized to allow for higher order neighbors, similar to the time series context, where a time shift can pertain to multiple periods.

A higher order neighbor is defined in a *recursive* fashion, as a first-order neighbor to a lower order neighbor. More formally, j is a neighbor of order k to i if:

- j is a first-order neighbor to h,
- h and i are neighbors of order $k - 1$ and
- j is not already a lower order neighbor to i.

[5] In the literature pertaining to cellular automata, counterparts to the rook and queen contiguity are the so-called von Neumann and Moore neighborhoods. The main difference is that here, the central cell is not included in the neighborhood concept, whereas it is in cellular automata.

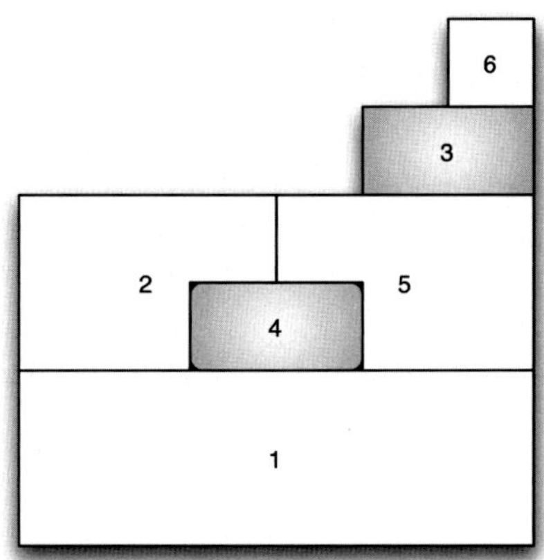

Figure 10.7: Second-order contiguity

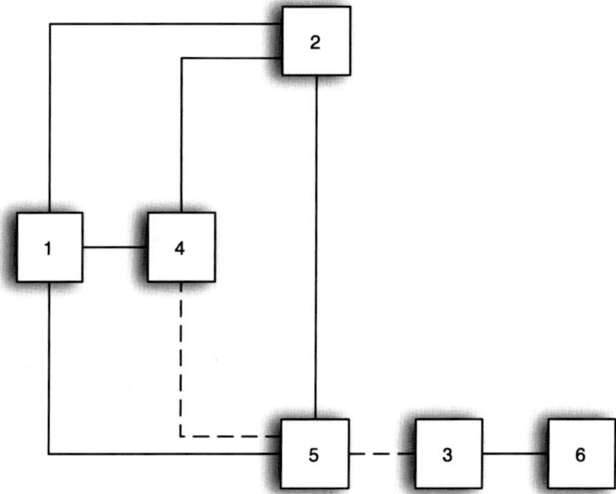

Figure 10.8: Higher order contiguity as multiple steps in the graph

Using this logic, candidates to be a second-order neighbor would be any first-order neighbor to another observation that is already a first-order neighbor. However, this would be limited to only those locations that are not already first-order neighbors. Parenthetically, it should be noted that the concept of higher order contiguity is similar to the notion of reachability in social networks (e.g., Newman, 2018).

In the layout shown in Figure 10.7, cells 4 and 3 are highlighted. Cell 4 has three first-order neighbors, immediately surrounding it: 1, 2 and 5. Cell 3 is a first-order neighbor to 5, hence 3 is a second-order neighbor to 4.

In the graph representation of the contiguity structure in Figure 10.8, second-order contiguity corresponds to of a *path* of length 2 from node 4 to node 3, illustrated by the dashed lines. This path requires the traversal of two edges in the graph, [4-5] and [5-3] and is thus of length 2. As it turns out, the number of edges separating a pair of nodes corresponds to the order of contiguity.

As such, identifying paths of length 2 between nodes is not sufficient to find the correct second-order neighbors, since it does not preclude redundant or circular paths. For example, a closer examination of Figure 10.8 reveals that 1 and 5 can be reached in two steps from 4 as well (e.g., 4-1-2, 4-5-1). In fact, multiple paths of length 2 can exist between two nodes (e.g., 4-1-2 and 4-5-2).

These paths illustrate the complexity of the problem of finding higher order neighbors. For example, nodes 2 and 5 cannot be both first-order and second-order neighbors of node 4, so a strict application of the graph-theoretical approach is not sufficient in the spatial weights case. Ways must be found to eliminate these redundant paths in order to yield the proper second-order neighbor, node 3.

10.2.3.1 Circular and redundant paths

The problem of circular and redundant paths was pointed out early on by Blommestein (1985), in the context of the specification of spatial autoregressive models. For such models, the inclusion of the same observation as a neighbor at different orders of contiguity (e.g., both first- and second-order neighbor) constitutes a so-called identification problem (it is impossible to tell certain parameters apart). An efficient algorithm to remove such redundancies is developed by Anselin and Smirnov (1996), based on shortest paths in the network structure.

However, in an exploratory context, it may be useful to include lower order neighbors in order to mimic the effect of expanding bands of neighbors, similar to an expanding distance band (see Section 11.3.1). `GeoDa` includes both concepts of higher order weights (Section 10.3.4).

10.2.4 Practical considerations

In practice, the construction of the spatial weights from the geometry of the data cannot be done by visual inspection or manual calculation, except in the most trivial of situations. Specialized software is needed to compute distances between points from their coordinates or to derive the presence of common borders from the boundary information for polygons.

To assess whether two polygons are contiguous is quite demanding. This requires the use of explicit spatial data structures to deal with the location and arrangement of the polygons, a functionality that is lacking in most econometric and statistical software. The specialized spatial data representation that differentiates a GIS from a standard data base system, and especially the use of a *spatial index* are crucial in this respect.

Many commonly used spatial data structures are in a so-called spaghetti representation. While this approach is more efficient for rapid rendering of maps, it lacks indexing and topology. It is popular in many cartographic packages and map-oriented GIS. A well known example is ESRI's shape file format.

`GeoDa` uses its own internal data structure to represent polygons loaded from different sources (files as well as spatial data bases), so that it can take advantage of spatial indexing to optimize the generation of the weights information.

It is important to keep in mind that the spatial weights are critically dependent on the quality of the spatial data source (GIS) from which they are constructed. Problems with the topology in the GIS (e.g., slivers) will result in inaccuracies for the neighbor relations included in the spatial weights. In practice, it is essential to check the characteristics of the weights for any evidence of problems. When problems are detected, the solution is to go

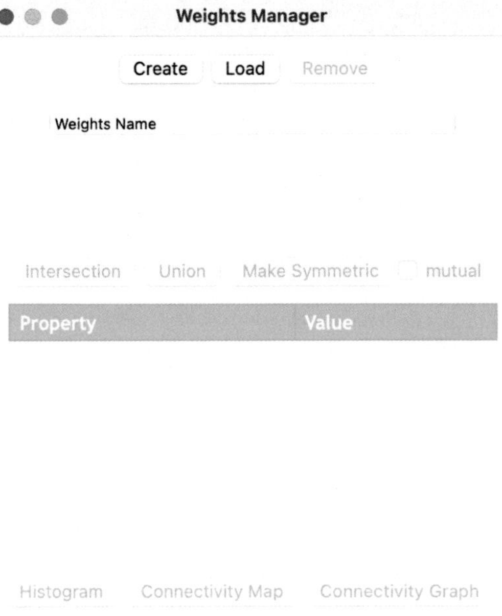

Figure 10.9: Weights Manager interface

back to the GIS and fix or clean the topology of the data set. Editing of spatial layers is not implemented in `GeoDa`, but this is a routine operation in most GIS software.

10.3 Creating Contiguity Weigths

In `GeoDa`, the creation and manipulation of spatial weights is carried out by means of a **Weights Manager**. This dialog is invoked from the menu as **Tools > Weights Manager**. Alternatively, the large **W** icon on the toolbar can be used, highlighted in Figure 10.1.

The main interface is shown in Figure 10.9. It contains buttons to **Create**, **Load** and **Remove** weights that are listed under **Weights Name**. The highlighted weight is the currently active one (for example, see Figure 10.11). Its summary characteristics are listed in the table with columns **Property** and **Value**.

A second row of buttons allows various operations on existing weights, including **Intersection**, **Union** and **Make Symmetric**. Finally, at the bottom of the interface are options to visualize the properties of the weights, in the form of a **Histogram**, **Connectivity Map** and **Connectivity Graph**.

10.3.1 Weights manager

The actual construction of the weights is implemented through the **Weights File Creation** interface, invoked by the **Create** button. It is shown in Figure 10.10 for the *Ceará Zika* sample data set. The interface provides the entry point to all the available options. Two buttons correspond to the main types, **Contiguity Weight** or **Distance Weight**. In this example, the option for **Rook contiguity** is checked, with the **Contiguity Weight** button active.

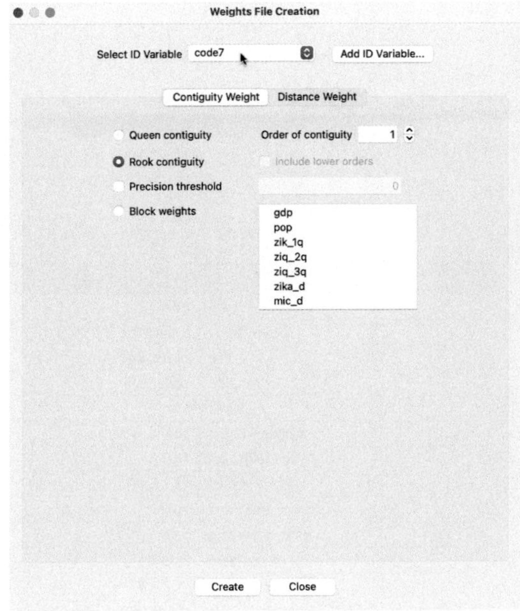

Figure 10.10: Rook contiguity in the Weights File Creation interface

10.3.1.1 ID variable

The first item to specify in the dialog is the **ID Variable**. This variable has to be *unique* for each observation. It is a critical element to make sure that the weights are connected to the correct observations in the data table. In other words, the ID variable is a so-called *key* that links the data to the weights.

In `GeoDa`, it is best to have the ID Variable be integer. In practice, this is often not the case, and even though the identifier may *look* like an integer value, it may be stored as a string. For example, the standard identifiers that come with many open data sets (such as the US census data) are typically character variables. One way to deal with this problem is to use the **Edit Variable Properties** functionality in the table to turn a string into an integer, as shown in Section 2.4.1.1.

In some instances, there is no easy way to identify an ID variable. In that case, the **Add ID Variable** button provides a solution: the added ID variable is simply an integer sequence number that is inserted into the data table (as always, there must be a **Save** operation to make the addition permanent).

For the *Ceará Zika* data set, **code7** is used as the ID variable, as indicated in Figure 10.10.

10.3.2 Rook contiguity

With the **Rook contiguity** radio button checked, as in Figure 10.10, a click on **Create** will start the weights construction process.

First, a file dialog appears in which a file name for the weights must be specified (the file extension **GAL** is added automatically). In the example, this file is **ceara_r.gal**. Since there are no real metadata in a spatial weights file, it is a good practice to make the file name something meaningful, so that one can remember what type of weight was created.

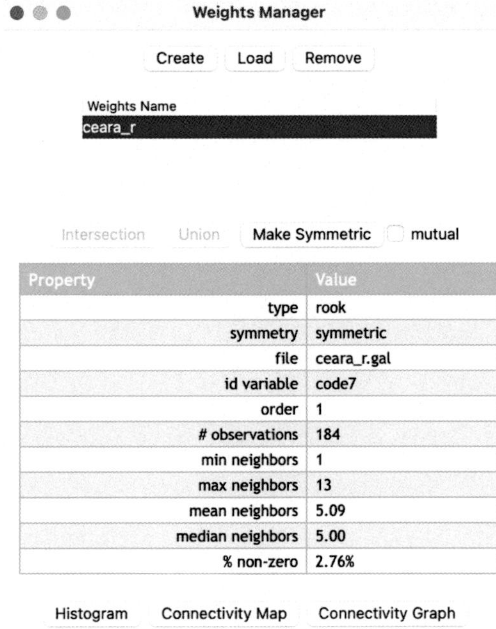

Figure 10.11: Rook weights summary interface

```
0 184 zika_ceara code7
2300101 3
2302503 2308302 2308401
2300150 4
2301950 2304954 2309607 2311603
2300200 6
2300754 2302305 2304251 2306553 2307809 2308906
2300309 10
2301505 2303600 2304269 2305506 2307403 2308500 2310902 2311355 2311900 2313005
```

Figure 10.12: GAL file contents

For example, an **_r** added to the name of the data would suggest rook weights. However, as outlined in Section 10.3.5, if a **Project File** is saved, several of the characteristics of the weights (i.e., its metadata) are stored in that file for later reuse.

The weights are immediately computed and written to the GAL file. At the end of this operation, a success message appears (or an **Error** message if something went wrong).

A useful option in the weights file creation dialog is the specification of a **Precision threshold** (situated right below the **Rook contiguity** radio button). In most cases, this is not needed, but in some instances the precision of the underlying shape file is insufficient to allow for an exact match of coordinates (to determine which polygons are neighbors). When this happens, `GeoDa` suggests a default error band to allow for a fuzzy comparison. In most cases, this will be sufficient to fix the problem.

After the weights are created, the weights manager becomes populated, as shown in Figure 10.11. The name for the file that was just created is now included under the **Weights Name**. In addition, under the item **Property**, several summary characteristics are listed. This is further examined in Section 10.4.1.

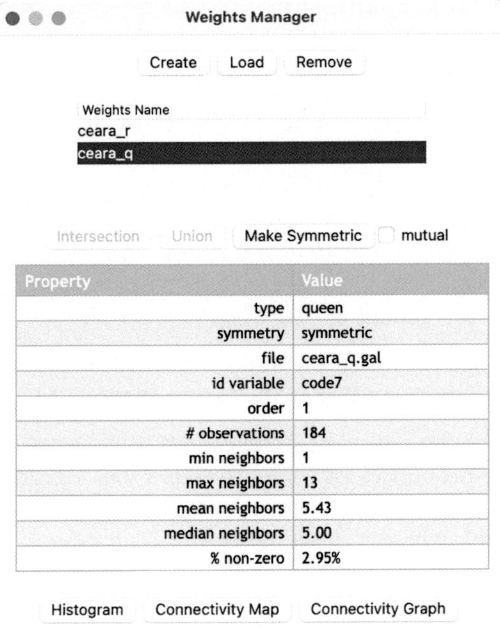

Figure 10.13: Summary properties of first-order queen contiguity

10.3.2.1 GAL file

The GAL weights file is a simple text file that contains, for each observation, the number of neighbors and their identifiers. The format was suggested in the 1980s by the Geometric Algorithms Lab at Nottingham University. It achieved widespread use after its inclusion in `SpaceStat` (Anselin, 1992), and subsequent adoption by the `R spdep` package and others.

The one innovation `SpaceStat` added to the format was the inclusion of a header line, with some metadata for the weights, such as the number of observations (**184**), the name of the shape file from which the weights were derived (**zika_ceara**), and the name of the ID variable (**code7**).[6] As illustrated in Figure 10.12, for each observation, the number of neighbors is listed after its ID. For example, for the municipality with **code7** 2300101, there are 3 neighbors, with respective IDs 2302503, 2308302 and 2308401.

The GAL file is a simple text file. It can thus be edited (e.g., change neighbors), although this is not recommended: it is easy to break the symmetry of the weights.

10.3.3 Queen contiguity

Spatial weights based on the queen contiguity criterion are created in the same way, but with the **Queen contiguity** button checked in the dialog of Figure 10.10. If needed, the **Precision threshold** option is available as well.

After the weights file is created, e.g., *ceara_q.gal*, its summary properties are listed in the weights manager, with the associated file name highlighted, as in Figure 10.13.

[6]The **0** at the start of the header line is currently not used.

10.3.4 Higher-order contiguity

Higher order contiguity weights are constructed in the same general manner as the first-order weights just considered. To accomplish this, a value larger than the default of 1 must be specified in the **Order of contiguity** box in Figure 10.10.

One important aspect of higher order contiguity weights is whether or not the lower order neighbors should be included in the weights structure. This is determined by a check box.

Importantly, as outlined in Section 10.2.3, there is a difference between the two concepts. The *pure* higher order contiguity does not include any lower order neighbors, whereas they are included when the corresponding check box is checked.

Using the second-order for the queen contiguity criterion in the Ceará data, e.g., saved into *ceara_q2.gal*, the summary properties are shown in Figure 10.14.

In contrast, an encompassing notion of second-order neighbors would include the first-order neighbors as well. The properties of such a weights file, e.g., *ceara_q2i.gal*, are listed in Figure 10.15. Note how the **include lower orders** entry is set to **true**, in contrast with the entry in Figure 10.14, where it is set to **false**.

The difference between the two concepts can be easily gathered from the descriptive statistics. As expected, the inclusive second-order weights are denser. For example, this is illustrated by the percent non-zero: 6.56% for the pure second-order and 9.51% for the inclusive second-order. Weights characteristics are considered in more detail in Section 10.4.

10.3.5 Project file

Similar to how custom classifications and grouped variable specifications were earlier saved to a **Project File**, the same can be achieved for spatial weights. This is very important, since otherwise any weights metadata are lost. They are not included in the saved weights file (see Section 10.3.6).

As before, this is accomplished by invoking **File > Save Project** and specifying a file name. A **gda** file extension will be added.

In Figure 10.16, the respective summary weights properties are listed under the **weights_entries** tag in the file.

As in the previous cases, the **gda** file must be specified as the input file when starting a new analysis, to make sure the weights are loaded properly. Their properties will immediately appear in the weights manager dialog.

10.3.6 Using existing weights

When a Project File is not used as the input, several of the summary properties of the weights are lost. For example, when using the **Load** option in the weights manager (the middle button in Figure 10.9) for the second-order queen contiguity file *ceara_q2.gal*, the resulting summary is as in Figure 10.17.

In contrast to the entries in Figure 10.14, there is no awareness of the **type** (set to **custom** compared to **queen**), the **symmetry**, or **order** of contiguity. Since the numerical properties are computed on the fly, they are still included.

In general, it is highly recommended to use project files to keep track of the spatial weights metadata.

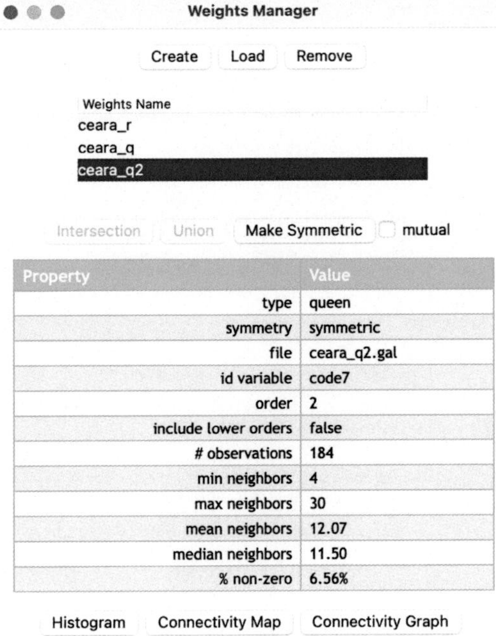

Figure 10.14: Summary properties of second-order queen contiguity

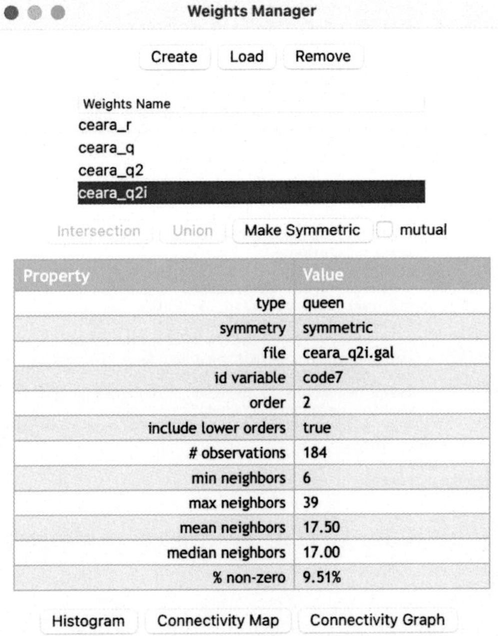

Figure 10.15: Summary properties of inclusive second-order queen contiguity

10.3.7 Space-time weights

The **Time Editor** considered in Chapter 9, as depicted in Figure 9.2, contains a button in the lower right corner to **Save Space-Time Table/Weights**.

```
<weights_entries>
    <weights>
        <title>ceara_r</title>
        <meta_info>
            <weights_type>rook</weights_type>
            <order>1</order>
            <inc_lower_orders>false</inc_lower_orders>
            <path>ceara_r.gal</path>
            <id_variable>code7</id_variable>
            <symmetry>symmetric</symmetry>
            <num_observations>184</num_observations>
            <min_neighbors>1</min_neighbors>
            <max_neighbors>13</max_neighbors>
            <mean_neighbors>5.0869565217391308</mean_neighbors>
            <median_neighbors>5</median_neighbors>
            <non_zero_perc>2.7646502835538751</non_zero_perc>
        </meta_info>
    </weights>
    <weights>
        <title>ceara_q</title>
        <meta_info>
            <weights_type>queen</weights_type>
            <order>1</order>
            <inc_lower_orders>false</inc_lower_orders>
            <path>ceara_q.gal</path>
            <id_variable>code7</id_variable>
            <symmetry>symmetric</symmetry>
            <num_observations>184</num_observations>
            <min_neighbors>1</min_neighbors>
            <max_neighbors>13</max_neighbors>
            <mean_neighbors>5.4347826086956523</mean_neighbors>
            <median_neighbors>5</median_neighbors>
            <non_zero_perc>2.9536862003780717</non_zero_perc>
        </meta_info>
    </weights>
    <weights>
        <title>ceara_q2</title>
        <meta_info>
            <weights_type>queen</weights_type>
            <order>2</order>
            <inc_lower_orders>false</inc_lower_orders>
            <path>ceara_q2.gal</path>
            <id_variable>code7</id_variable>
            <symmetry>symmetric</symmetry>
            <num_observations>184</num_observations>
            <min_neighbors>4</min_neighbors>
            <max_neighbors>30</max_neighbors>
            <mean_neighbors>12.065217391304348</mean_neighbors>
            <median_neighbors>11.5</median_neighbors>
            <non_zero_perc>6.5571833648393199</non_zero_perc>
        </meta_info>
    </weights>
    <weights>
        <title>ceara_q2i</title>
        <default/>
        <meta_info>
            <weights_type>queen</weights_type>
            <order>2</order>
            <inc_lower_orders>true</inc_lower_orders>
            <path>ceara_q2i.gal</path>
            <id_variable>code7</id_variable>
            <symmetry>symmetric</symmetry>
            <num_observations>184</num_observations>
            <min_neighbors>6</min_neighbors>
            <max_neighbors>39</max_neighbors>
            <mean_neighbors>17.5</mean_neighbors>
            <median_neighbors>17</median_neighbors>
            <non_zero_perc>9.5108695652173907</non_zero_perc>
        </meta_info>
    </weights>
</weights_entries>
```


Figure 10.16: Weights description in Project File

This is useful in order to *trick* the cross-sectional data structure of `GeoDa` to allow for simple *pooled* time series/cross section analyses.[7] The pooled aspect means that the same coefficients or model parameters hold for all time periods, which is not necessarily a realistic assumption. However, it is often the point of departure in an analysis of space-time dynamics.

[7]The most common application is to carry out a regression analysis, but this approach can also be used to compute spatial autocorrelation statistics for the pooled data.

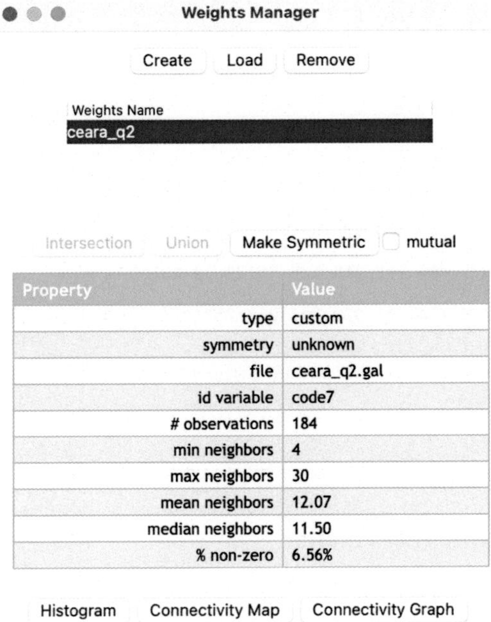

Figure 10.17: Second-order contiguity weights loaded from file

```
STID,mun,TIME,T_2000,T_2010,T_2020,p_P614NS,p_PJOB,p_PHA,p_ELEC,p_WAT,p_CAR
"1","1","2000","1","0","0",8.845208845000000,13.972055890000000,1.1038961040
"2","2","2000","1","0","0",10.554728219999999,42.157899800000003,41.80668028
"3","174","2000","1","0","0",3.508771930000000,50.981308409999997,45.9300311
"4","3","2000","1","0","0",13.445378150000000,53.374233130000000,3.272302298
"5","4","2000","1","0","0",5.701754386000000,47.487001730000003,12.5707814300
"6","5","2000","1","0","0",4.614850798000000,42.477957379999999,28.2335602500
"7","6","2000","1","0","0",5.927099842000000,52.063031310000000,17.533648599
```

Figure 10.18: Saved space-time table as csv file

The critical element in this endeavor is to create a unique space-time ID variable (**STID**) so that the stacked cross-sections (one for each time period) can be handled as if they were one single cross-sectional data set, with a unique ID for each space-time observation. Analogously, the spatial weights are block-diagonal, with the same weights structure repeated for each stacked time period.

This is illustrated for the *Oaxaca Development* data set from Chapter 9. Note that it is necessary to have a spatial weights file active in the weights manager for this to work properly. In the example, this is a first-order queen contiguity weights file using **mun** as the ID. A snapshot of the corresponding space-time data table, saved as a *csv* file, is shown in Figure 10.18. In addition to the cross-sectional identifier **mun**, there is a unique space-time identifier **STID**, as well as an indicator variable for each time period: **T_2000**, **T_2010** and **T_2020**. The remaining entries consist of the time-enabled (grouped) variables.

The contents of the corresponding *gal* file are illustrated in Figure 10.19. The header line lists **1710** as the number of space-time observations (570×3), **stoaxaca** as the name of the source file with the data (i.e., the file shown in Figure 10.18), and **STID** as the key. The remaining entries follow the same format as before.

```
0 1710 stoaxaca STID
1 6
362 298 264 218 178 177
2 3
405 281 25
3 3
409 389 85
4 5
539 457 438 400 205
5 5
479 403 184 153 38
6 11
509 475 406 306 268 126 58 39 30 29 19
7 15
466 454 404 401 392 307 267 252 228 221 219 163 98 60 52
```

Figure 10.19: Space-time GAL contiguity file

10.4 Weights Characteristics

In addition to creating and loading spatial weights, the **Weights Manager** also contains functionality to compute and visualize some important properties of the weights. This includes summary characteristics, as well as the visualization of the connectivity structure by means of a histogram, map, or graph.

10.4.1 Summary characteristics

Figures 10.11 and 10.13–10.15 show the summary properties for rook, queen, second-order queen and inclusive second-order queen weights computed for the Ceará example. These include both metadata as well as descriptive statistics computed from the weights.

The metadata include the **type** of weights, **symmetry**, the original spatial layer (**file**), the **id variable**, the **order**, and, for higher order contiguity, whether **include lower orders** is **true** or **false**.

The descriptive statistics consist of the number of **observations**, the **min**, **max**, **mean** and **median** number of neighbors, as well as an overall measure of sparsity, in the form of the **% non-zero**.

For the four weights considered so far, there is an increasing degree of density, reflected in all the descriptive statistics. For example, the mean number of neighbors increases from **5.09** for rook, **5.43** for queen, **12.07** for second-order queen, to **17.50** for inclusive second-order queen.

10.4.2 Connectivity histogram

The **Histogram** button at the left-hand side of the bottom of the weights manager (Figure 10.9) produces a connectivity histogram. This shows the number of observations for each value of the *cardinality* of neighbors, i.e., how many observations have the given number of neighbors. The graph is illustrated in Figure 10.20 for the rook contiguity associated with the municipalities of Ceará. It is a standard `GeoDa` histogram, with a number of options available, most of which are generic (see Section 7.2.1), and some specific to the connectivity histogram.

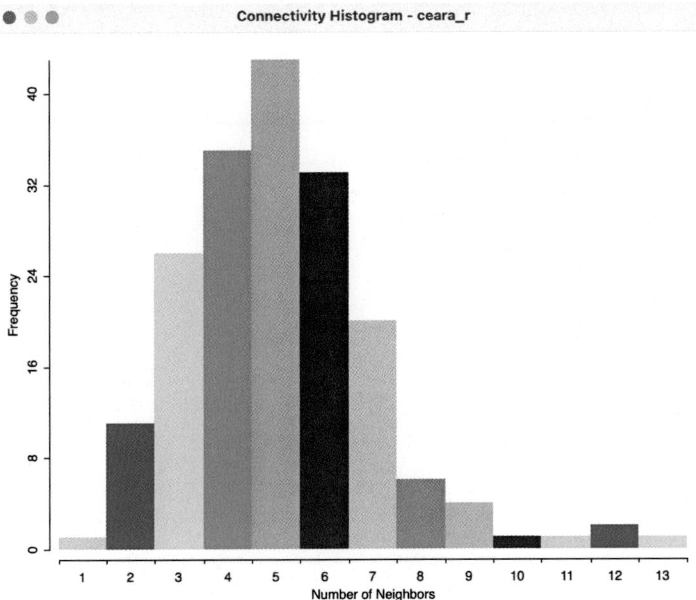

Figure 10.20: Rook contiguity histogram

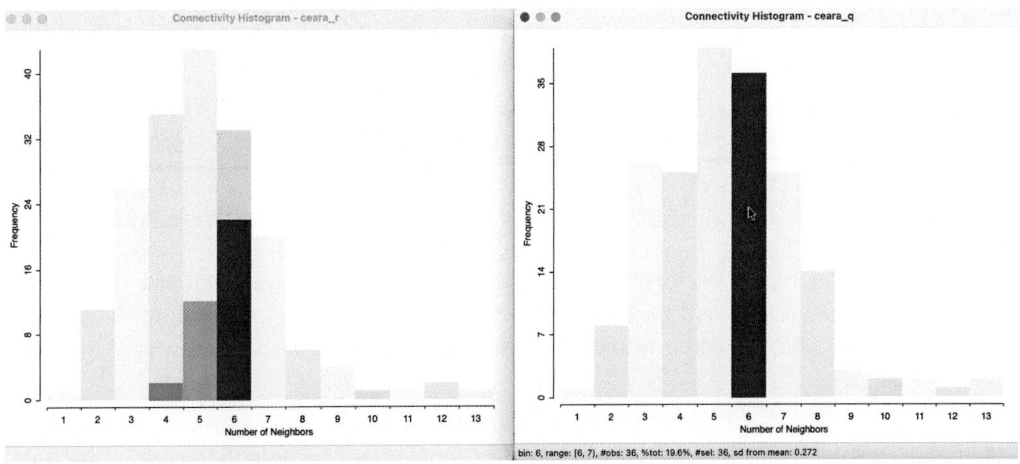

Figure 10.21: Rook and queen contiguity histograms

The overall pattern is quite symmetric, with a mode of 5 (i.e., most municipalities have 5 rook neighbors), but quite a long tail to the right (i.e., a few municipalities have a much larger number of neighbors). In addition to the visual inspection, the usual statistics of the distribution can be added to the bottom of the table by means of the **View > Display Statistics** option. The descriptive statistics are the same as listed in the weights manager summary.

Figure 10.21 illustrates the minor differences between rook and queen contiguity. In the histogram on the right, for queen contiguity, the municipalities with six neighbors are selected. In the linked contiguity histogram for rook weights on the left, the matching distribution is shown. Several of the observations with 6 neighbors using the queen criterion, have only

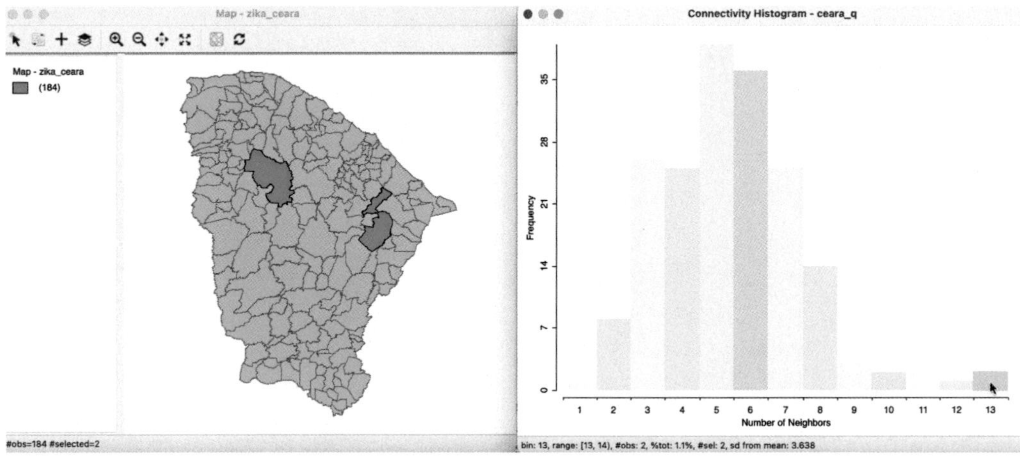

Figure 10.22: Linked contiguity histogram and map

5 or even 4 neighbors for the rook criterion. This highlights the more inclusive nature of queen contiguity.

In addition, the contiguity histogram can be linked to any other window, such as the themeless map in the left-hand panel of Figure 10.22. In the contiguity histogram on the right (queen contiguity), the two observations with 13 neighbors are selected. A closer examination of the map identifies two very large areal units. Consequently, they will have more neighbors than the average sized unit.

It is good practice to check the connectivity histogram for any strange patterns, such as observations with only one neighbor, or neighborless observations (isolates, see Section 11.4.2).

Ideally, the distribution of the cardinalities should be smooth (i.e., no major gaps) and symmetric, with a limited range. Deviations from symmetry, such as bi-modal distributions (some observations have few neighbors and some have many) require further investigation. Any such characteristics will have an immediate effect on the corresponding weights, with implications for all the computations that use these weights.

10.4.2.1 Neighbor cardinality

A useful option of the connectivity histogram is to save the neighbor cardinality to the data table as an additional column/variable. This can then be used as an input in further analysis. For example, the number of neighbors is an important input into the calculation of the significance of the local join count statistic, covered in Chapter 19.

The **Save Connectivity to Table** option is invoked in the usual way by right clicking on the connectivity histogram window. This brings up a small dialog to specify the variable name (the default is **NUM_NBRS**).

10.4.3 Connectivity map

The middle button at the bottom of the **Weights Manager** interface (Figure 10.9) brings up a **Connectivity Map**. This is a standard `GeoDa` themeless map view, with all the usual features invoked through toolbar icons, including zooming and panning. In addition, the connectivity map includes a special functionality that highlights the neighbors of any *selected*

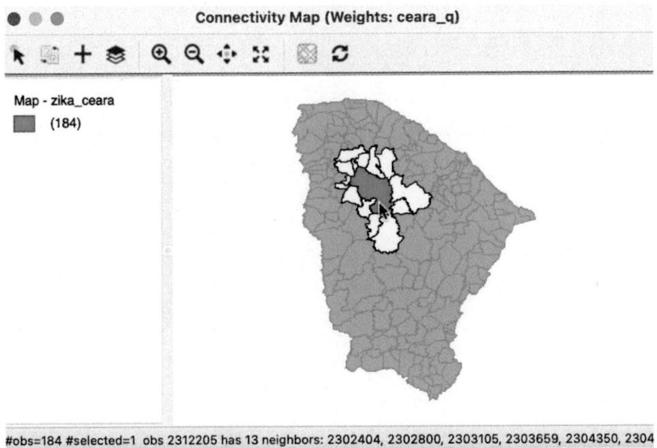

Figure 10.23: Connectivity map for queen contiguity

observation. The window header lists the weights matrix to which the connectivity structure pertains.

In Figure 10.23, this is illustrated using queen contiguity for the Ceará municipalities. As soon as the pointer is moved over one of the observations, it is selected and its neighbors are highlighted. Note that this behavior is slightly different from the standard selection in a map, since it does not require a click to select. Instead, the pointer is in so-called *hover* mode. As soon as the pointer moves outside the main map, the complete layout is shown.

In the example, the selection is over the municipality of Santa Quitéria (**code7 2312205**), one of the two observations identified as having 13 neighbors. This outcome is listed in the status bar below the map, as well as the identifying codes for the neighbors. Moving the pointer to another observation immediately updates the selection and its neighbors.

The **Connectivity Map** in the weights manager is actually a special case of the **Connectivity** option in any map. This is considered in more detail in Section 10.4.5.1.

10.4.4 Connectivity graph

The right-most button at the bottom of the **Weights Manager** interface (Figure 10.9) brings up a **Connectivity Graph**. Like the connectivity map, this is a standard `GeoDa` map view, with all the usual options invoked through toolbar icons, including zooming and panning.

On additional feature is that the graph structure that corresponds to the spatial weights connectivity is superimposed on the map. For the queen contiguity in Ceará, this yields Figure 10.24. Again, the window header lists the corresponding spatial weights file (**ceara_q**).

As in the connectivity map, there are a few options that allow for some customization of the graph structure representation. These include the color and thickness of lines, and whether the background map is shown or not.

The default edge thickness is **Normal**, but two other options are available, to make the lines thicker (**Strong**) or thinner (**Light**). It is also often useful to change the color of the graph to something different from the default black. In Figure 10.24, these options have been used to change the thickness to **Strong**, the color to **blue**, the **Opacity** of the **Fill Color** to zero and the **Outline Color for Category** to **black**.

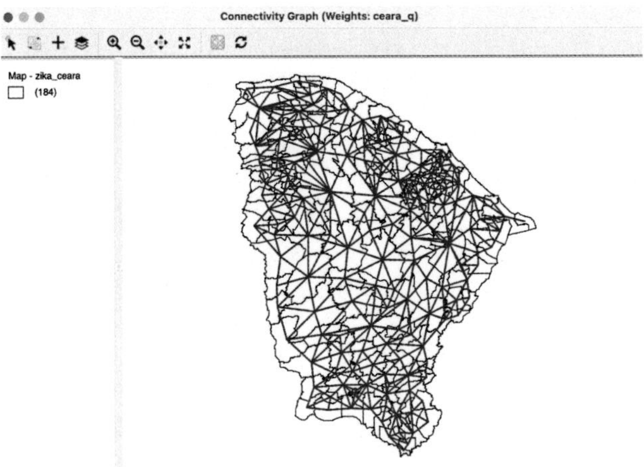

Figure 10.24: Connectivity graph for queen contiguity

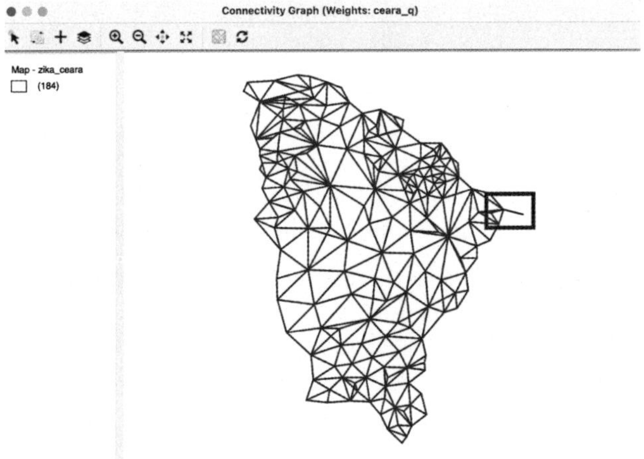

Figure 10.25: Connectivity graph for queen contiguity without map

The most dramatic of the options is the **Hide Map** feature. By checking this, the background map is removed and only the pure graph structure remains, as in Figure 10.25. This clearly brings out some interesting features of the graph, such as the observation with a single neighbor (a single edge in the graph) in the extreme north-east (highlighted in red).

10.4.5 Connectivity option

Whenever there is an active weights matrix specified in the weights manager, the **Connectivity Option** becomes available in any map window as well as in the **Selection Tool** for the table.

10.4.5.1 Connectivity option in the map

In the discussion of the various map options in Section 4.5 of Chapter 4, the **Connectivity** option was deferred to the current chapter.

This option supports operations similar to those in a **Connectivity Map**, but applied to any current map. The functionality is invoked by selecting **Connectivity > Show Selection and Neighbors** from the options menu, in the usual fashion (right click).

With the option checked, the selection feature works the same as in the **Connectivity Map**. The main difference between the functionality in the **Connectivity Map** and the **Connectivity** option in any thematic map is that the latter is updated to the current active spatial weights file. In other words, as the active weights change, the connectivity structure is adjusted. In contrast, when selecting the **Connectivity Map** from the weights manager, there is only one such map, tied directly to the weights file from which it was invoked.

With the **Show Selection and Neighbors** option checked, it becomes possible to **Change Outline Color of Neighbors**, to **Change Fill Color of Neighbors**, and implement other cosmetic adjustments to the look of the selected polygons.

10.4.5.2 Connectivity option in the table

Neighbors of selected observations can also be displayed by means of a a feature of the table **Selection Tool**, as long as a weights matrix is active. The **Add Neighbors to Selection** button is central in the middle panel of of the **Selection Tool** in Figure 2.24 of Chapter 2.

This functionality requires that there is a current selection. Also, the **Weights** box must be specified. By default, the drop-down list will show the currently active weights, e.g., **ceara_q**.

Clicking on this button will add the neighbors of a current selection to the selection, similar to the operation of the **Append to Current Selection** option. Note that once the neighbors have been added, the *current selection* has grown to include those observations, so another **Add Neighbors to Selection** operation increases the selection set even more. This is not always what one has in mind.

11

Distance-Based Spatial Weights

In this second chapter devoted to spatial weights, the focus is on weights that are derived from a notion of distance. Intrinsically, this is most appropriate for point layers, but it can easily be generalized to polygons as well, through the use of the polygon centroids.

The chapter starts with a brief overview of *distance metrics* and the fundamental difference between points expressed in projected coordinates and points in latitude-longitude decimal degrees. Whereas for the former the familiar concept of *Euclidean distance* can be applied, the latter requires the computation of *great circle distance* or arc distance. In addition, a concept of distance in multivariate attribute space is introduced.

This is followed by a review of spatial weights that are based on distance-bands, k-nearest neighbor weights and the use of generalized concepts of distance.

Next is a discussion of the implementation of the weights functionality in `GeoDa`, including a broadening of the strict concept of contiguity to apply to points.

The chapter closes with a discussion of set operations on weights, such as intersection, union and making symmetric.

The methods are illustrated with a point data set that contains performance measures for 261 Italian community banks for 2011–2017. This data set was used in an analysis of spatial spillovers in technical efficiency by Algeri et al. (2022). The data are contained in the *Italy Community Banks* sample data set. The space-time weights are illustrated with the *Oaxaca Development* data.

As mentioned in the previous chapter, to work most effectively with the spatial weights files, the data sets should first be copied (**Save As**) to a working directory.

11.1 Topics Covered

- The concept of a distance metric
- Construct distance band spatial weights
- Great circle distance
- The contents of a gwt weights file
- Assess the characteristics of distance-based weights
- Assess the effect of the max-min distance cut-off
- Identify isolates
- Construct k-nearest neighbor spatial weights
- Making a k-nearest neighbor weights matrix symmetric
- Construct contiguity weights for points and distance weights for polygons
- Block weights

DOI: 10.1201/9781003274919-11

- Space-time weights
- Create a general dissimilarity matrix based on multi-attribute distance
- Investigate commonalities among spatial weights (intersection and union)

GeoDa Functions

- Weight File Creation dialog
  - distance-band weights
  - k-nearest neighbor weights
  - great circle distance option
  - block weights option
- Weights Manager
  - Connectivity Histogram, Map and Graph
  - Intersection, Union
  - Make symmetric
- Time Editor
  - Save Space-Time Weights

11.2 Distance Metrics

Pairs of observations are considered to be neighbors based on how *close* they are to each other. To make this precise, there is a need for a formal definition of distance. Two important concepts are considered: distance between points in a Cartesian coordinate system, and distance between points located on a sphere. This can be further generalized to distance between points in multivariate attribute space, i.e., to a non-geographical notion of distance.

11.2.1 Distance in a Cartesian coordinate system

A general concept for the distance between two points i and j, with respective coordinates (x_i, y_i) and (x_j, y_j) in a Cartesian coordinate system, is the so-called Minkowski metric:

$$d_{ij}^p = (|x_i - x_j|^p + |y_i - y_j|^p)^{(1/p)},$$

with p as a general exponent. The Minkowski metric itself is not often used in spatial analysis, but there are two special cases of great interest.

Arguably the most familiar case is for $p = 2$, i.e., *Euclidean* or straight line distance, d_{ij}, *as the crow flies*:[1]

$$d_{ij} = \left(|x_i - x_j|^2 + |y_i - y_j|^2\right)^{(1/2)},$$

or, in its more familiar form:

$$d_{ij} = \sqrt{(x_i - x_j)^2 + (y_i - y_j)^2}.$$

An alternative to Euclidean distance that is sometimes preferred because it lessens the influence of outliers is the so-called *Manhattan* block distance. This metric is obtained by setting $p = 1$ in the Minkowski expression. This notion only considers movement along the

[1]Since $|x_i - x_j|^2$ is equivalent to $(x_i - x_j)^2$, the latter expression is typically used for Euclidean distance.

east-west and north-south directions, as in the city blocks of Manhattan. Formally, it is expressed as:

$$d_{ij}^{m} = |x_i - x_j| + |y_i - y_j|.$$

11.2.2 Great circle distance

Euclidean inter-point distances are only meaningful when the coordinates are recorded on a plane with a Cartesian coordinate system. This implies that any point layer must be *projected* for Euclidean distances to be appropriate.

In practice, one often works with unprojected points, expressed as *decimal degrees* of latitude and longitude. In this case, using a straight line distance measure is inappropriate, since it ignores the curvature of the earth. This is especially the case for longer distances, such as when observations span a continent. Also, decimal degrees are not Cartesian coordinates, and consequently a notion of distance between degrees is meaningless.

The proper distance measure in this case is the so-called *arc distance* or *great circle distance.* This approach takes the latitude and longitude in decimal degrees as input into a conversion formula.[2] Decimal degrees are obtained from the degree-minute-second value as degrees + minutes/60 + seconds/3600.

The latitude and longitude in decimal degrees are converted into radians as:

$$\begin{aligned}\text{Lat}_r &= (\text{Lat}_d - 90) * \pi/180\\ \text{Lon}_r &= \text{Lon}_d * \pi/180,\end{aligned}$$

where the subscripts d and r refer respectively to decimal degrees and radians, and $\pi = 3.14159\ldots$. With $\Delta\text{Lon} = \text{Lon}_{r(j)} - \text{Lon}_{r(i)}$, the expression for the arc distance is:

$$\begin{aligned}d_{ij} &= \text{R} * \arccos[\cos(\Delta\text{Lon}) * \sin\text{Lat}_{r(i)} * \sin\text{Lat}_{r(j)})\\ &\quad + \cos\text{Lat}_{r(i)} * \cos\text{Lat}_{r(j)}],\end{aligned}$$

or, equivalently:

$$\begin{aligned}d_{ij} &= \text{R} * \arccos[\cos(\Delta\text{Lon}) * \cos\text{Lat}_{r(i)} * \cos\text{Lat}_{r(j)})\\ &\quad + \sin\text{Lat}_{r(i)} * \sin\text{Lat}_{r(j)}],\end{aligned}$$

where R is the radius of the earth. In `GeoDa`, the arc distance is obtained in miles with R = 3959, and in kilometers with R = 6371.

However, it should be noted that these calculated distance values are only approximate, since the radius of the earth is taken at the equator. A more precise measure would take into account the actual latitude at which the distance is measured. In addition, the earth's shape is much more complex than a sphere. In most instances in practice, the approximation works fine.

11.2.3 General distance

The distance metric used to construct weights as discussed in Section 11.3 can be readily extended to a notion of *general distance*, familiar in regional science theory (Isard, 1969). In addition, multidimensional scaling can be applied to create *coordinates* in a multivariate

[2]The latitude is the y dimension, and the longitude the x dimension, so that the traditional reference to the pair (lat, lon) actually pertains to the coordinates as (y,x) and not (x,y).

variable space (considered in Part II of Volume 2). These artificial locations can then be used to derive spatial weights based on the distance between them.

In the empirical literature, such weights are often referred to as *economic* weights, based on on a notion of economic distance. Simply put, the distance between observation i and j expressed in terms of an economic (or social) variable z is:

$$d_{ij} = |z_i - z_j|^\alpha,$$

where α is a scaling factor, often simply set equal to one.

A general distance measure readily serves as a proxy for the extent and strength of interactions between locations (e.g., a social network among neighborhoods) by extending the one-dimensional approach to a multivariate setting. Such a distance measure includes the difference between two observations according to several dimensions, summarized as a single value. Formally, with $h = 1, \ldots, H$ as the number of variables z_h considered to characterize the socio-economic makeup of each observation, the Euclidean distance between i and j is:

$$d_{ij} = \sqrt{\sum_{h=1}^{H} (z_{ih} - z_{jh})^2}.$$

This approach is often used in economic applications.

An alternative way to derive distance-based weights in a multivariate setting is to employ multidimensional scaling, or MDS (see Volume 2). This yields a point *map* of locations in two-dimensional space, where the location and configuration of the points approximates their distance relationship in multivariate dimensions. Loosely put, the two dimensions in the MDS map correspond to the largest eigenvalues and matching eigenvectors of the correlation matrix for the variables considered. This yields a best fit solution in the sense that the relative distances between the points on the MDS map are the closest to the actual relative distances between observations. The distances in MDS space can then be used to define neighbors in multivariate attribute space.

11.3 Distance-Based Weights

With a definition of distance in hand, neighbors can be defined as those pairs of points that are within a critical cut-off distance from each other. Alternatively, the k closest points, or k-nearest neighbors can be selected. These approaches can be readily extended to more general concepts of distance.

11.3.1 Distance-band weights

The most straightforward spatial weights matrix constructed from a distance measure is obtained when i and j are considered neighbors whenever j falls within a critical *distance band* from i. More precisely, $w_{ij} = 1$ when $d_{ij} \leq \delta$, and $w_{ij} = 0$ otherwise, where δ is a preset critical distance cutoff.

In order to avoid isolates (islands) that would result from too stringent a critical distance, the distance must be chosen such that each location has at least one neighbor. Such a

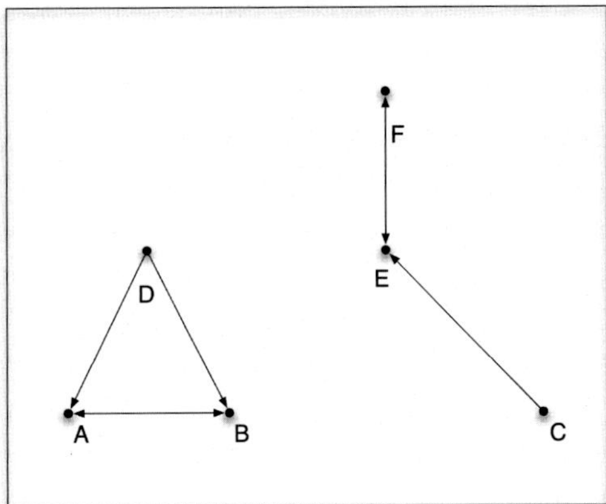

Figure 11.1: Nearest neighbor properties

distance conforms to a *max-min* criterion, i.e., it is the largest of the nearest neighbor distances.[3]

In practice, the max-min criterion often leads to too many neighbors for locations that are somewhat clustered, since the critical distance is determined by the points that are farthest apart. This problem frequently occurs when the density of the points is uneven across the data set, such as when some of the points are clustered and others are more spread out. This is examined more closely in the illustrations.

11.3.2 K-nearest neighbor weights

With a full matrix of inter-point distances D, each i-th row contains the distances between a point i and all other points j. With these distances sorted from smallest to largest, as d^*_{ij}, a concept of k-nearest neighbors can be defined.

For each i, the k-nearest neighbors are those j with d^*_{ij} for $j = 1, k$, i.e., the k closest points (this assumes there are no ties, which is considered below). This is not a symmetric relationship. Even though $d_{ij} = d_{ji}$ the sorted order of distances for i is often not the same as the sorted order for j.

Imagine the situation with three points on a line as A, B and C, with C closer to B than A is to B. Then B will be the nearest neighbor for A, but A will not be the nearest neighbor for B. This asymmetry can cause problems in analyses that depend on the intrinsic symmetry of the weights (e.g., some algorithms to estimate spatial regression models).

Another potential issue with k-nearest neighbor weights is the occurrence of ties, i.e., when more than one location j has the same distance from i. A number of solutions exist to break the tie, from randomly selecting one of the k-th order neighbors, to including all of them. In `GeoDa`, random selection is implemented.

To illustrate these properties, consider Figure 11.1, portraying six points. The nearest neighbor relation is shown by an arrow from i to j. For example, E is clearly the nearest neighbor of C, but C is not the nearest neighbor for E (that is F). The triangle D-A-B

[3]The nearest neighbor distance is the smallest distance from a given point to all the other points, or, the distance from a point to its nearest neighbor.

illustrates the problem of ties. D is equidistant from A and B, so it is impossible to pick one or the other as *the* nearest neighbor. On the other hand, the nearest neighbor relation between A and B is symmetric (arrow in both directions).

In practice, k-nearest neighbor weights often provide a solution to the problem of *isolates*, since they ensure that each observation has exactly k neighbors (see Section 11.4.2). However, this approach should not be applied uncritically. One drawback of the concept is that it warps the effect of distance.

For sparsely distributed points, the nearest neighbor (and, by extension, the k nearest neighbors) can be quite far removed, whereas for densely distributed points, many neighbors can be found within a small distance band. In terms of *interaction*, this implies that the actual distance is of less relevance than the fact that a neighbor is encountered, which is not always an appropriate assumption. For example, when considering commuting distances, a destination 40 miles away may be a reasonable *neighbor*, whereas there may be very little interaction with a nearest neighbor that is 150 miles away.

Since the k-nearest neighbor relationship is typically asymmetric, it is sometimes desired to convert this to symmetric spatial weights. One approach to accomplish this is to replace the original weights matrix $\mathbf{W}$ by $(\mathbf{W} + \mathbf{W}')/2$, which is symmetric by construction.[4] Each new weight is then $(w_{ij} + w_{ji})/2$. The resulting matrix will have more neighbors than the original knn weights.

An alternative approach is to only use the instances where $w_{ij} = w_{ji} = 1$. This is referred to as *mutual* symmetry. The result will be a much sparser weights matrix, with the possibility of isolates.

11.3.3 General distance weights

With a general concept of distance in attribute space, the implementation of distance-band and k-nearest neighbor weights is straightforward. However, when using a distance band, the meaning of the implied max-min distance is not always obvious, since it is a combination of distances among multiple variable dimensions.

An alternative weights construction is sometimes considered as the inverse of the generalized distance:

$$w_{ij} = 1/|z_i - z_j|^{\alpha} = |z_i - z_j|^{-\alpha},$$

or, in standardized form:

$$w_{ij} = \frac{|z_i - z_j|^{-\alpha}}{\sum_j |z_i - z_j|^{-\alpha}}.$$

For example, in the much cited study by Case et al. (1993), two separate weights matrices were constructed using economic weights. One was based on the difference in per capita income between states and the other on the difference in the proportion of their Black population. In another nice example, Greenbaum (2002) uses the concept of a *similarity matrix* to characterize school districts that are similar in per-student income. Instead of using those weights as such, they are combined with a nearest neighbor distance cut-off, so that the similarity weight is only computed when the school districts are also within a given order of geographic nearest neighbors (see also the discussion of weights intersection in Section 11.6.1).

Weights created as functions of distance are further considered in Chapter 12.

[4] $\mathbf{W}'$ is the *transpose* of the weights matrix $\mathbf{W}$, such that rows of the original matrix become columns in the transpose.

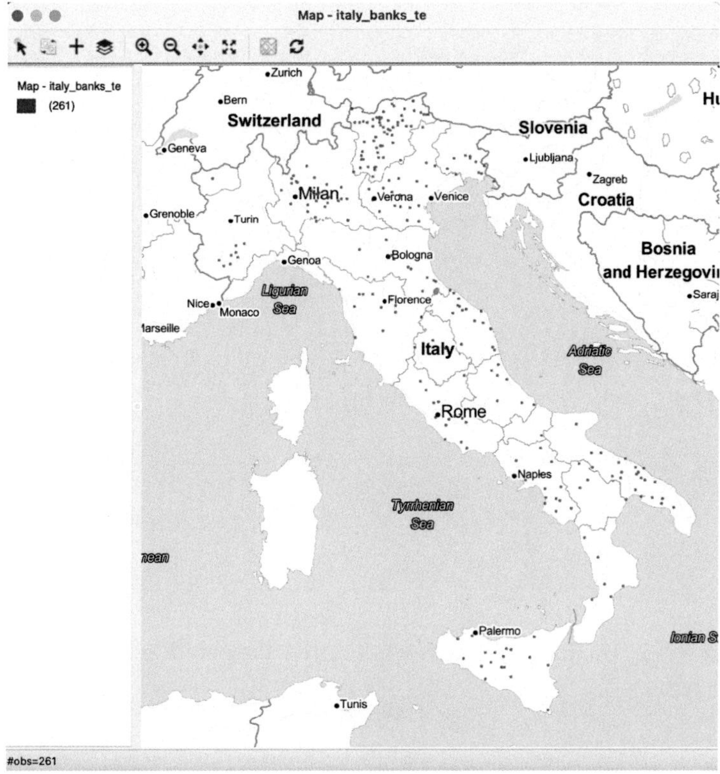

Figure 11.2: Italian community bank point layer

11.4 Implementation

As mentioned in the chapter introduction, the distance-based weights functionality is illustrated with a point layer of 261 Italian community banks, shown in Figure 11.2.[5]

Distance weights are invoked in the same way as contiguity weights, through the **Weights Manager**, as outlined in Section 10.3.1. The **Weights File Creation** dialog with the **Distance Weight** button selected is shown in Figure 11.3. The **ID Variable** is set to **idd**.

The dialog controls the various options available to create distance-based weights. These include the choice between geographic distance – the **Geometric centroids** button under **Variables** – or generalized distance – the **Variables** button (see Section 11.4.4).

The default for geographic distances is to have the **X-coordinate variable** as the internally obtained **X-Centroids**, and the **Y-coordinate variable** as the internally obtained **Y-Centroids**. However, any other pair of variables can be selected, including non-geographic ones. The resulting concept of general distance is limited to two dimensions, whereas the **Variables** tab allows for higher dimensions as well.

[5]The map is customized by selecting **Fill Color for Category** and **Outline Color for Category** as red (instead of the default green), changing the **Point Radius** to 1, and using **Stamen > Toner Lite** as the **Basemap** (with **Transparency** set to 30).

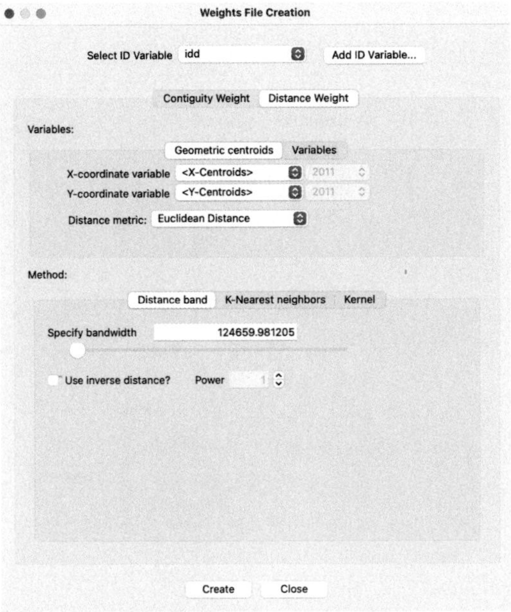

Figure 11.3: Distance-based weights in the Weights File Creation interface

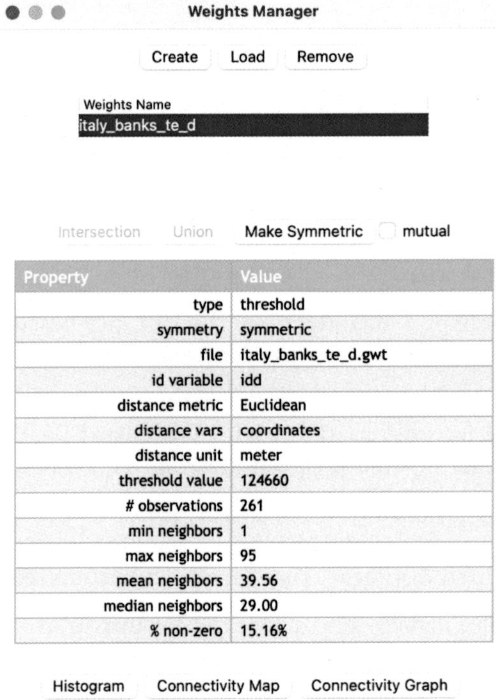

Property	Value
type	threshold
symmetry	symmetric
file	italy_banks_te_d.gwt
id variable	idd
distance metric	Euclidean
distance vars	coordinates
distance unit	meter
threshold value	124660
# observations	261
min neighbors	1
max neighbors	95
mean neighbors	39.56
median neighbors	29.00
% non-zero	15.16%

Figure 11.4: Default distance-based weights summary properties

The default **Distance metric** is **Euclidean Distance**. Other options, appropriate for unprojected coordinates, are **Arc Distance (mi)** and **Arc Distance (km)**, for great circle distance expressed as either miles or kilometers.

```
0 261 italy_banks_te idd
1 95        49615.6221
1 64        38864.5283
1 96         49580.036
1 305        61542.2186
1 249        54250.7586
1 372         74488.915
1 419        89935.5477
1 252        82430.6115
1 22        80909.9013
1 222        93241.9107
1 273        101209.558
1 286        121276.625
1 356        99131.6876
1 70        57996.3309
1 309        33939.0986
1 272        20168.8519
1 267        47765.1824
1 77        98475.6655
1 304        120019.699
1 302        71928.5133
1 266        120684.357
1 423        124601.262
1 422        78679.2664
1 277        102160.684
1 75        102252.866
1 237        121265.773
1 206        122785.456
1 38        115892.237
8 218        119612.331
8 361        104410.953
8 421        25595.0177
8 30        138.146576
```

Figure 11.5: GWT weights file format

The type of weight is set by the **Method** tab. The **Distance band** criterion is the default. Other options are **K-Nearest neighbors** and **Kernel** weights.[6] The default **bandwidth** for distance band weights is the max-min distance. In the current example, this is listed as about 125 km (the units are meters). The options for **inverse distance** and **Power** are considered in Chapter 12.

11.4.1 Distance-band weights

With the default settings as in Figure 11.3, invoking **Create** will bring up the familiar file save dialog. As was suggested for contiguity weights, it is useful to make the file name reflect the nature of the weights since there are no metadata, unless a project file is created. In this example, the file name is *italy_banks_te_d* with the **GWT** file extension added automatically.

Once the file is created, it appears in the **Weights Manager**, as shown in Figure 11.4.

The summary properties indicate the **type** as **threshold**, with **threshold value** as **124660**, **distance metric** as **Euclidean**, **distance vars** as **coordinates** and **distance unit** as **meter**. The other properties are the same as for contiguity weights (see Section 10.4.1).

Note how the range of the neighbor cardinality is much larger than for contiguity, going from 1 to 95. This reflects the impact of the max-min default cut-off distances on densely distributed points. The distance of about 125 km is much larger than the average distance that separates the community banks.

[6]Kernel weights are considered in Chapter 12.

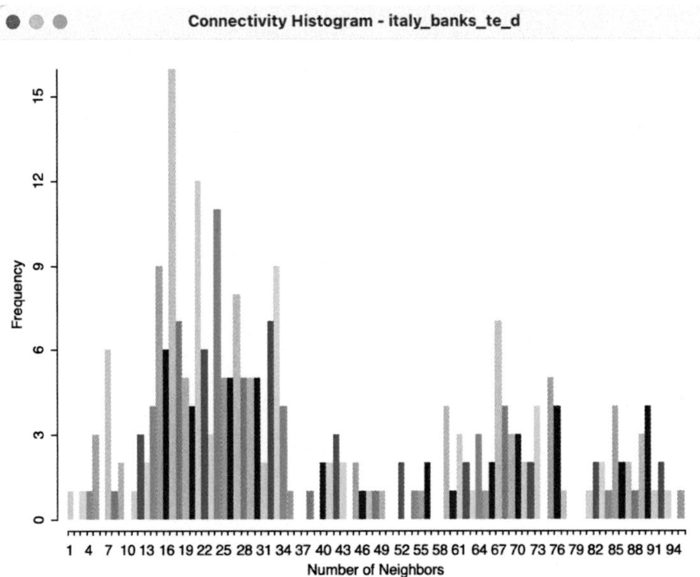

Figure 11.6: Connectivity histogram for default distance weights

11.4.1.1 GWT file

Distance-based weights are saved in files with a **GWT** file extension. This format, illustrated in Figure 11.5, is slightly different from the GAL format used for contiguity weights. It was first introduced in `SpaceStat` in 1995, and later adopted by `R spdep` and other software packages. The header line is the same as for GAL files, but each pair of neighbors is listed, with the ID of the observation, the ID of its neighbor and the distance that separates them.

The pairwise distance is currently only included for informational purposes, since `GeoDa` **does not use the actual distance value in any statistical operations**. In other words, the weights are turned into binary form and row-standardized by default. The only place where the distances are taken into account is in the computation of distance functions, covered in Chapter 12.

Note how observation with **idd=1** has no less than 28 neighbors, with inter-point distances ranging from slightly over 20 km to almost 125 km.

11.4.1.2 Weights characteristics

In the same way as for contiguity weights, the characteristics of distance-based weights can be assessed by means of the **Connectivity Histogram**, the **Connectivity Map** and the **Connectivity Graph**. These are available through the buttons at the bottom of the weights manager.

The shape of the connectivity histogram for distance-band weights is typically very different from that of contiguity-based weights. As illustrated in Figure 11.6, the range in the neighbor cardinality is much larger than for contiguity weights. The distribution also includes extremes, with one observation having only one neighbor, and one having 95.

As mentioned, the range in the number of neighbors is directly related to the density of the spatial distribution of the points. Locations that are somewhat isolated will drive the determination of the largest nearest neighbor cut-off point (their nearest neighbor distance

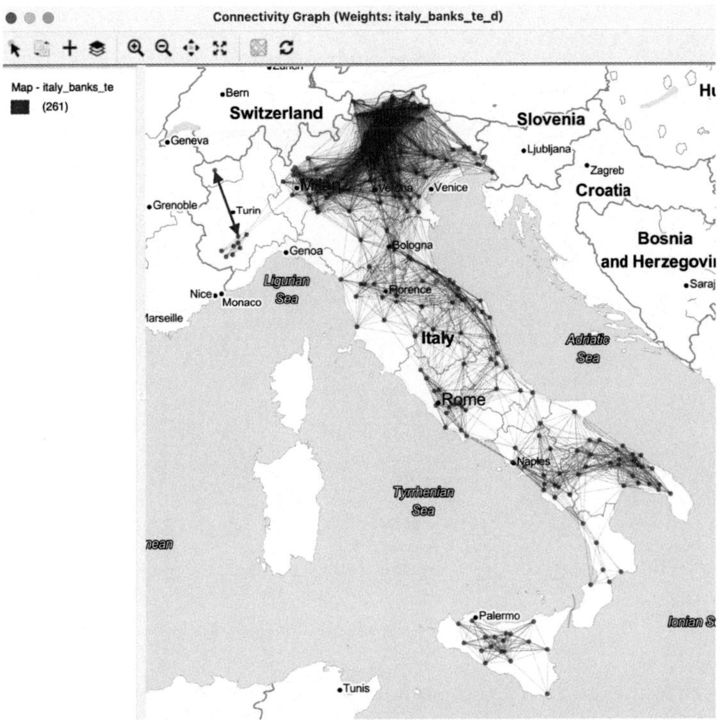

Figure 11.7: Connectivity graph for default distance weights

will be large), whereas dense clusters of locations will encompass many neighbors using this large cut-off distance.

The properties of the distance band weights can be further investigated by means of the **Connectivity Graph**. As before, this is invoked through the right-most button at the bottom of the weights manager.

The pattern shown in Figure 11.7 highlights how the connectivity is highly uneven, with very dense areas alternating with much sparser distributions. The connection highlighted in blue corresponds to the max-min distance. Clearly, this distance is much larger than the average distance among locations, leading to a highly unbalanced weights matrix.

11.4.2 Isolates

The default distance band ensures that each observation has at least one neighbor, but this has several undesirable side effects. To create a more balanced weights matrix, one can lower the distance threshold. However, this will result in observations that do not have neighbors, *isolates* or islands.

The **Weights File Creation** dialog is flexible enough that a specific distance cut-off can be entered in the box, or the movable button can be dragged to any value larger than the minimum distance, but smaller than the max-min distance. Sometimes, theoretical or policy considerations suggest a specific value for the cut-off that may be smaller than the max-min distance.

When such a threshold is chosen, a warning will appear, pointing out that the specified cut-off value is smaller than the distance needed to ensure that each observation has at least one neighbor.

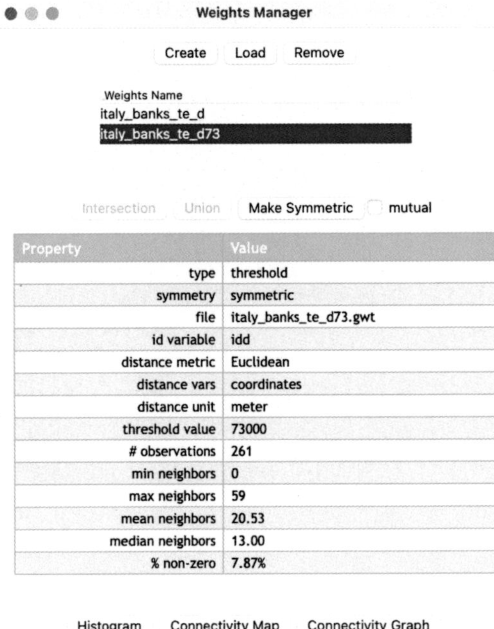

Property	Value
type	threshold
symmetry	symmetric
file	italy_banks_te_d73.gwt
id variable	idd
distance metric	Euclidean
distance vars	coordinates
distance unit	meter
threshold value	73000
# observations	261
min neighbors	0
max neighbors	59
mean neighbors	20.53
median neighbors	13.00
% non-zero	7.87%

Figure 11.8: Weights summary properties for distance 73 km

For example, using a cut-off distance of 73 km instead of the default 125 km yields a much sparser weights matrix (*italy_banks_te_d73.gwt*), as evidenced by the summary in Figure 11.8. The maximum number of neighbors has been reduced from 95 to 59 (which is still quite large), but, more importantly, the **min neighbors** is listed as **0**. Compared to the default, the density has decreased substantially, going from 15.16% non-zero elements to 7.87% non-zero.

The connectivity graph shown in Figure 11.9 illustrates the greater sparsity. As it turns out, the distance band of 73 km resulted in only two isolates, highlighted in the figure, one near Aosta, in the north-west, and a second on the island of Sicily. All other observations are connected. However, there begin to form disconnected entities, i.e., separate *components* in the graph that are connected internally, but not between them.

Further lowering the cut-off distance may result in more meaningful groupings, but at the expense of a growing number of isolates.

11.4.2.1 How to deal with isolates

Since the isolated observations are not included in the spatial weights (in effect, the corresponding row in the spatial weights matrix consists of zeros), they are not accounted for in any spatial analysis, such as tests for spatial autocorrelation, or spatial regression. For all practical purposes, they should be removed from such analysis. However, they are fine to be included in a traditional *non-spatial* data analysis.

Ignoring isolates may cause problems in the calculation of spatially lagged variables, or measures of local spatial autocorrelation. By construction, the spatially lagged variable will be zero, which may suggest spurious correlations.

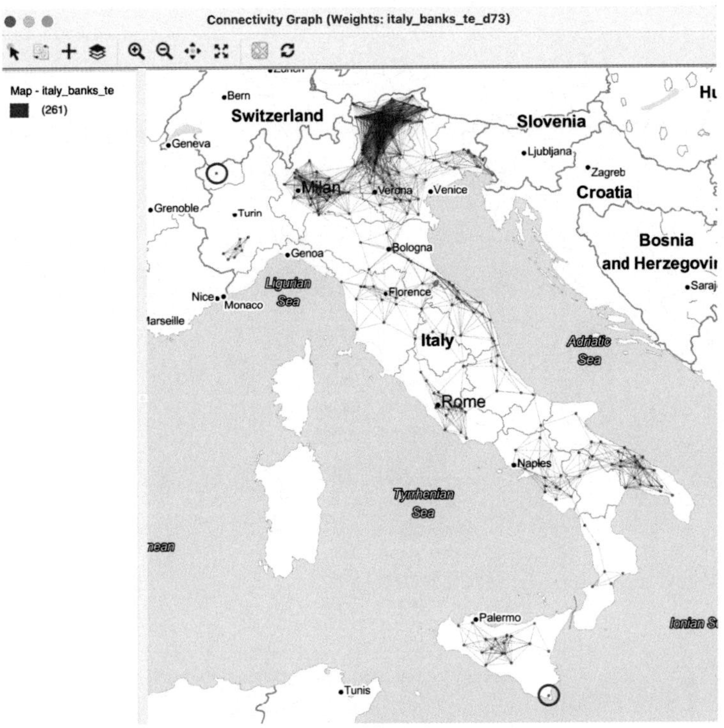

Figure 11.9: Connectivity graph for distance band 73 km

11.4.3 K-nearest neighbor weights

K-nearest neighbor (KNN) weights are computed by selecting the corresponding button in **Distance Weight** panel of the **Weights File Creation** interface. The value for the **Number of neighbors** (**k**) is the only option. The default is 4, but **6** is used in the example (as in Algeri et al., 2022).

The weights (saved as the file *italy_banks_te_k6.gwt*) are added to the collection contained in the weights manager. In addition, all its properties are listed, as illustrated in Figure 11.10. Note that the properties now include the number of **neighbors** instead of the distance threshold value, as was the case of distance-band weights. Also, **symmetry** is set to **asymmetric**, which is a fundamental difference with distance-band weights.

The properties listed in the weights manager also include the mean and median number of neighbors, which of course equal k (in the example, they equal 6). The resulting weights matrix is dramatically less dense relative to the distance-band weights (2.30% compared to 15.16%).

K-nearest neighbors weights are often an alternative when a realistic distance band results in too many isolates, since each observation has k neighbors by design. However, this approach is not without some drawbacks and may not be appropriate in all situations.

11.4.3.1 Weights characteristics

Again, the connectivity histogram, map and graph can be used to inspect the neighbor characteristics of the observations. However, in this case, the histogram doesn't make much sense, since all observations have the same number of neighbors (by construction).

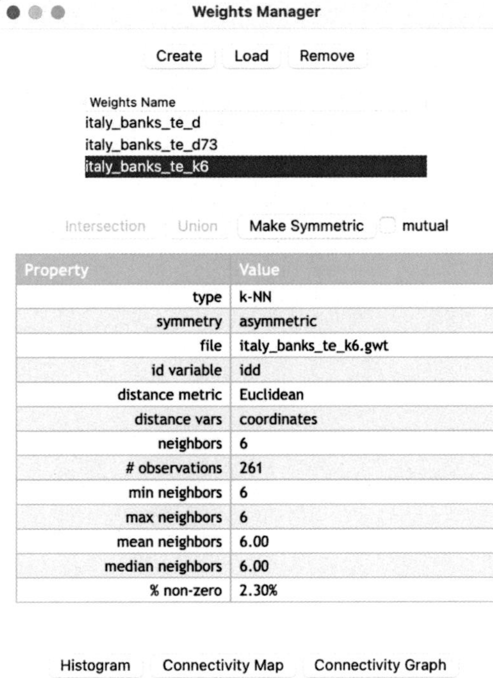

Figure 11.10: Weights summary properties for 6 k-nearest neighbors

In contrast, the connectivity graph, shown in Figure 11.11, clearly demonstrates how each point is connected to six other points. In the example, this yields a fully connected graph.

11.4.3.2 Caveat

One drawback of the k-nearest neighbor approach is that it ignores the distances involved. The first k neighbors are selected, irrespective of how near or how far they may be. This suggests a notion of distance decay that is not absolute, but relative, in the sense of *intervening opportunities* (e.g., one considers the two closest grocery stores, irrespective of how far they may be).

In addition, the requirement to have exactly k neighbors may create artificial connections, spanning large distances or bridging across natural and other barriers. For example, in the graph in Figure 11.11, the blue lines highlight some instances where this is the case. In the case of the northern isolate, the Aosta location becomes connected to observations located near Milan. For an observation on the north coast of Sicily, two long connections are included across the straight of Messina to the boot of Italy. A similar distant neighbor is identified for a location in Calabria to a far away one in the region of Basilicata. It is unlikely that these connections correspond with meaningful interactions among the community banks.

The relative distance effect should be kept in mind before mechanically applying a k-nearest neighbor criterion.

11.4.4 General distance matrix

The **Variables** tab in the distance weights dialog (Figure 11.3) also allows for the computation of a general weights matrix. This is based on the distance between observations in

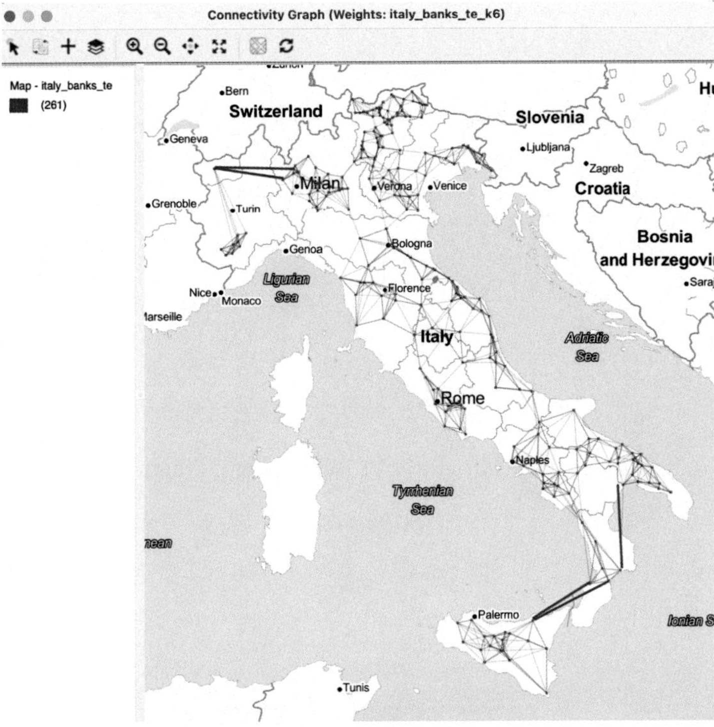

Figure 11.11: Connectivity graph for 6 k-nearest neighbors

multi-attribute space. Instead of selections for the X and Y coordinates, a drop-down list with all the variables is available. From this, any number of variables can be selected, not just two, as in the standard distance case.

In addition, there is a **Transformation** option that allows the usual standardizations (see Section 2.4.2.3). The **distance metric** can be either **Euclidean Distance** or **Manhattan Distance** (great circle distance is not meaningful in this context).

Typically, the variables are first standardized, such that their mean is zero and variance one, which is the default setting.

All the standard options are available, such as specifying a distance band (in multi-attribute distance units) or k-nearest neighbors. Such a general distance matrix or *dissimilarity* matrix is a required input in the multivariate clustering methods considered in Volume 2.

11.5 Broadening the Concept of Contiguity

The concept of contiguity can be generalized to point layers by converting the latter to a tessellation, such as Thiessen polygons (see Section 3.3). Queen or rook contiguity weights can then be created for these polygons, in the usual way.

Similarly, the concepts of distance-band weights and k-nearest neighbor weights can be generalized to polygon layers by considering their centroids (see Section 3.3.1). The polygons are represented by their central points and the standard distance computations are applied.

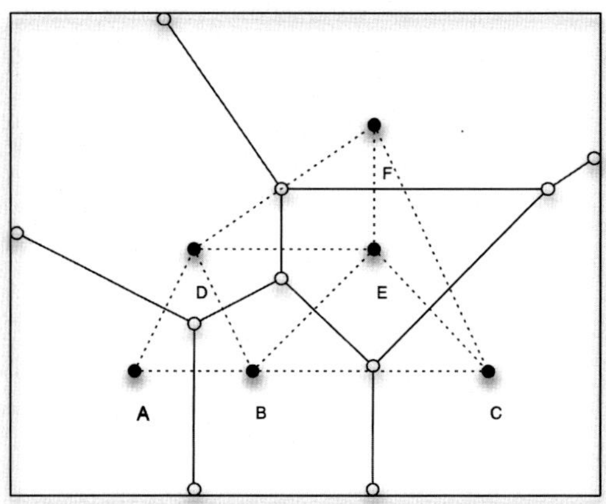

Figure 11.12: Contiguity for points from Thiessen polygons

Weights Manager

Create Load Remove

Weights Name
italy_banks_te_d
italy_banks_te_d73
italy_banks_te_k6
italy_banks_te_q

Intersection Union Make Symmetric mutual

Property	Value
type	queen
symmetry	symmetric
file	italy_banks_te_q.gal
id variable	idd
order	1
# observations	261
min neighbors	3
max neighbors	11
mean neighbors	5.78
median neighbors	6.00
% non-zero	2.21%

Histogram Connectivity Map Connectivity Graph

Figure 11.13: Weights summary properties for point contiguity (Thiessen polygons)

These operations can be carried out explicitly, by actually creating a separate Thiessen polygon layer or centroid point layer, and subsequently applying the weights operations. However, in `GeoDa`, this is not necessary, since the computations happen *under the hood*. In this way, it is possible to create contiguity weights for points or distance weights for polygons directly in the **Weights File Creation** dialog.

11.5.1 Contiguity-based weights for points

To illustrate the concept of contiguity weights for a point layer, consider the layout of six points depicted in Figure 11.12. The Thiessen polygons are drawn as solid lines. This clearly

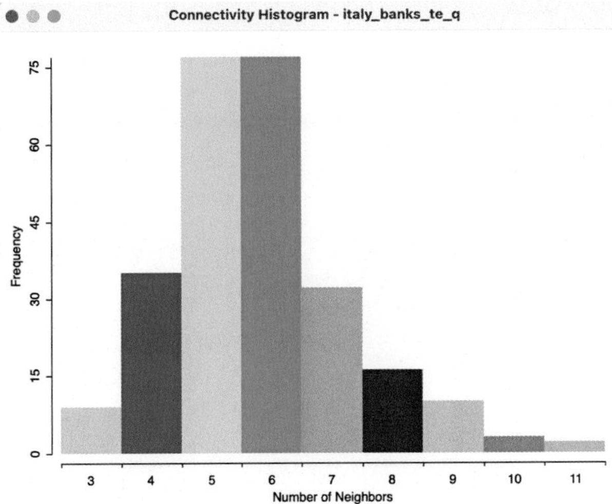

Figure 11.14: Connectivity histogram for point contiguity (Thiessen polygons)

demonstrates how they are perpendicular to the dashed lines that connect the points. The latter correspond to a queen contiguity relation.

Contiguity weights for a point layer are created in the same way as such weights for polygon layers. In the **Weights File Creation** dialog, one selects the **Contiguity Weight** option and specifies the type of contiguity. For the Italian community bank example, the summary properties of a queen contiguity (*italy_banks_te_q.gal*) are shown in Figure 11.13.[7]

Compared to the other weights for this point layer, the mean neighbors is 5.78 and the median neighbors is 6, similar to the k-nearest neighbor results in Figure 11.10. In addition, the resulting weights are still a bit less dense (or, sparser), with a % non-zero of 2.21% (compared to 2.30% for k-nearest neighbors).

11.5.1.1 Weights characteristics

These attractive characteristics are also reflected in the connectivity histogram, shown in Figure 11.14. The graph is nicely balanced with a mode of 5-6, near equal to mean and median. There is only a slight skewness to the right.

While, at first sight, the queen contiguity weights derived from the Thiessen polygons seem to have attractive properties, there is also a serious drawback. As depicted in Figure 11.15, the Thiessen polygons create several artificial contiguities, making connections between point locations that are totally unrealistic. For example, the connectivity graph shows several links between locations on the island of Sicily and the mainland, crossing the Tyrrhenian Sea. Similarly, a location in Trieste (bordering Slovenia) is connected to several points on the other side of the Adriatic Sea. It is unlikely that these links represent actual interaction between the community banks.

A potential solution to this problem is to combine the contiguity weights with other criteria, such as a meaningful distance band, by means of a set operation like intersection (see Section 11.6.1).

[7]Note that the file extension is **GAL**, as for other contiguity weights.

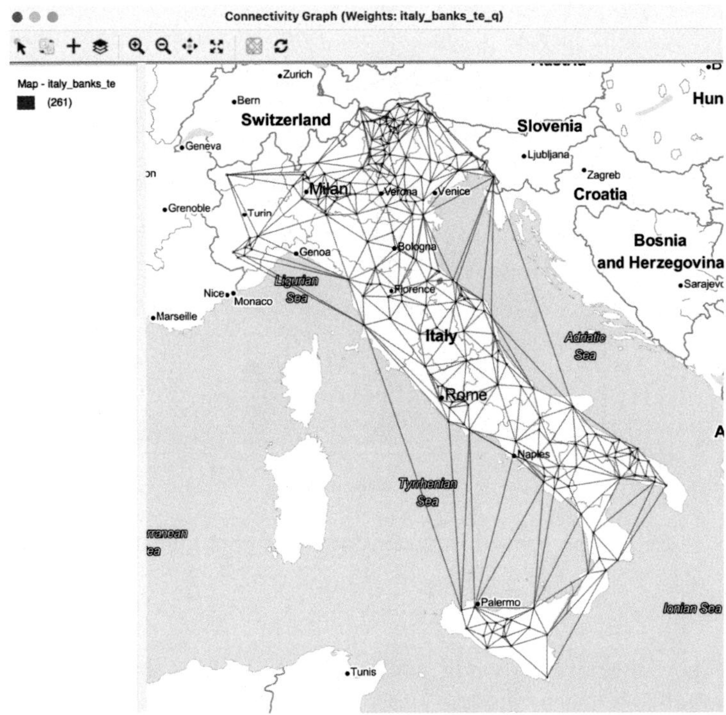

Figure 11.15: Connectivity graph for point contiguity (Thiessen polygons)

11.5.2 Distance-based weights for polygons

As discussed in Section 3.3.1, a polygon layer has a series of **Shape Center** options to add the centroid or mean center information to the data table, display those points on the map, or save them as a separate point layer. Such points can then be used to construct distance-based weights in the usual way.

However, as is the case for contiguity weights for points, in GeoDa this operation is carried out invisibly, so that it is not necessary to create a separate layer. Distance-band weights and k-nearest neighbor weights are obtained in the same way as for any other point layer (see Section 11.4).

11.5.3 Block weights

A slightly different concept of neighbor follows when a block structure is imposed, in which all observations in the same block are considered to be neighbors. This is a form of hierarchical spatial model, in which all units that share a common higher order level are *contiguous*.

Familiar examples are counties as part of a state or province, census tracts belonging to a larger neighborhood, etc. In some contexts, the block structure corresponds to the notion of *local interaction*, in the sense of reflecting the average behavior of a reference group.[8]

The adoption of a block structure became more common in spatial analysis after its application in a study of innovation diffusion by Case (1991).[9] In her study, individual observations

[8]See, for example, Brock and Durlauf (2001), for a formal discussion.
[9]In some literature, block weights are referred to as *Case weights*.

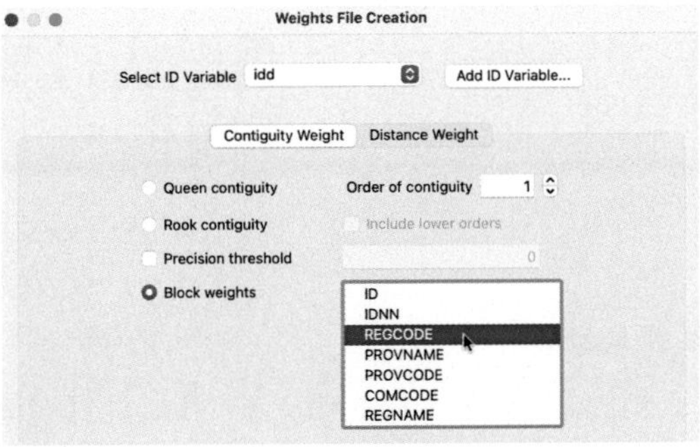

Figure 11.16: Block weights in the Weights File Creation interface

Figure 11.17: Connectivity graph for block weights by region

on farmers were grouped into *districts*, and all the farms within the same district were considered to be neighbors. However, the neighbor relation did not extend across districts. In other words, farms that may be physically adjoining but in different districts would not be considered neighbors, in effect turning each district into an island.

The result of this approach is a block-diagonal spatial weights structure. Within each region or district, all observations are neighbors (row elements of 1, except for the diagonal), but there is no neighbor relation between the regions, hence the block-diagonal structure.

```
0 1710 oaxaca_st STID
1 6
362 298 264 218 178 177
2 3
405 281 25
3 3
409 389 85
4 5
539 457 438 400 205
5 5
479 403 184 153 38
6 11
509 475 406 306 268 126 58 39 30 29 19
```

```
571 6
932 868 834 788 748 747
572 3
975 851 595
573 3
979 959 655
574 5
1109 1027 1008 970 775
575 5
1049 973 754 723 608
576 11
1079 1045 976 876 838 696 628 609 600 599 589
```

Figure 11.18: Space-time weights GAL file

The block weights structure has important consequences for the properties of estimators in spatial regression models, but is less useful in an exploratory context, unless used in combination with other criteria.[10]

11.5.3.1 Weights characteristics

Block weights are invoked as an option of the **Contiguity Weight** tab in the **Weights File Creation** dialog. A drop-down list shows potential variables that could be used to define the blocks, as illustrated in Figure 11.16 for the Italian community bank example. In this instance, the variable **REGCODE** is selected, which provides an integer value for each of the 20 regions in Italy.

The resulting spatial weights have quite different characteristics. The matrix is not sparse (10.28% non-zero), but not as dense as some distance-band weights. The number of neighbors ranges from 0 (a single observation in a region) to 62. The corresponding graph structure, shown together with the regional boundaries in Figure 11.17, reveals several separate *components*, with a strong internal connection, but no links between them.[11] The sole community bank in the Aosta region is an isolate.

11.5.4 Space-time weights

A special type of block weights is created as an option of the **Time Editor**. As mentioned in Section 9.2.1.2, the **Save Space-Time Table/Weights** button in the interface of Figure 9.2 can generate both a pooled data set as well as a weights file. The latter requires that there is an active weights file in the **Weights Manager**.

The resulting file follows a GAL format and repeats the cross-sectional contiguity structure for each time period, by using a new space-time index, **STID**. This index matches the values contained in the corresponding space-time data table. It is a sequential value going from 1 to $n \times T$, where n is the cross-sectional dimension and T is the time dimension.

To illustrate this feature, consider the Oaxaca time-enabled data set from Section 9.2.1 with a first-order queen contiguity weights active (*oaxaca_q.gal*). The contents of the space-time GAL file are illustrated in Figure 11.18.

On the left side, the header line shows how the space-time ID variable, **STID** takes on values from 1 to 1710 (570 cross-sectional units × 3 time periods). The same contiguity structure repeats itself every 570 observations. In the figure, observation 1 has 6 neighbors, with IDs

[10]For a technical discussion of the spatial econometric aspects, see, e.g., Anselin and Arribas-Bel (2013).

[11]The figure was created by combining the point layer with a polygon layer for the 20 regions, *regions.shp*, using the multilayer functionality (see Section 3.6).

362, 298, 264, 218, 178 and 177. On the right hand side, observation 571, which corresponds to the first cross-sectional unit in the second time period, has the same six neighbors, but all ID values are incremented by 570. For example, the first neighbor, with STID 932, is obtained as 362 + 570.

The resulting space-time weights take on a block-diagonal structure in the sense that the same cross-sectional block repeats itself for each time period. Formally, this corresponds to the matrix expression $\mathbf{I} \otimes \mathbf{W}$, where $\mathbf{I}$ is a T-dimensional identity matrix and $\mathbf{W}$ are the cross-sectional spatial weights.[12]

The specialized weights structure can be used in a pooled analysis, where the space-time aspect of the data is *hidden*. The observations are represented as one large cross-section, with variables stacked by time period, and spatial weights block-diagonal by time period. Consequently, all time periods are assumed to have the same parameters, such as the same spatial autocorrelation coefficient.

11.6 Manipulating Weights

The information in spatial weights can be manipulated in a limited number of ways. Set operations allow the creation of a new set of weights by union or intersection operations on two existing weights files. In addition, asymmetric weights, typically k-nearest neighbor weights, can be made symmetric.

11.6.1 Set operations

Once multiple weights are included in the **Weights Manager**, it becomes possible to carry out set operations on any pair. With two weights selected, say *italy_banks_te_d73* (for a distance band of 73 km) and *italy_banks_te_k6* (for 6 k-nearest neighbors), the buttons **Intersection** and **Union** become active (placed directly above **Property** in the dialog).

Figure 11.19 shows the graph structure that results when these two weights are combined through an **Intersection** operation.[13] The only links that remain are those where some of the six nearest neighbors fall within 73 km. The new structure illustrates how some of the drawbacks of the individual weights were remedied.

The very dense connections in the north, shown in Figure 11.9, become much sparser, since only six of the distance-band neighbors are retained. On the other hand, the artificial connections in the k-nearest neighbor graph in Figure 11.11 are removed by imposing a 73 km distance constraint. This yields the same two isolates as for the original distance-band, highlighted in the figure.

Nevertheless, the resulting compromise provides a better reflection of potential interaction pairs for the community banks.

The **Union** function operates in a similar way (again, two weights need to be selected to activate this functionality). An obvious application would be to combine *pure* first- and second-order contiguity weights, although this functionality is built-in through the **Include lower orders** option for contiguity weights.

[12]The symbol $\otimes$ stands for a Kronecker product. This means that each element of the first matrix is multiplied by the second matrix. In other words, since the first matrix is a $T \times T$ identity matrix, the result is a block diagonal matrix with a block of $\mathbf{W}$ for each time period.

[13]This is the operational implementation used by Algeri et al. (2022), with the exception that the two isolates were removed in the analysis.

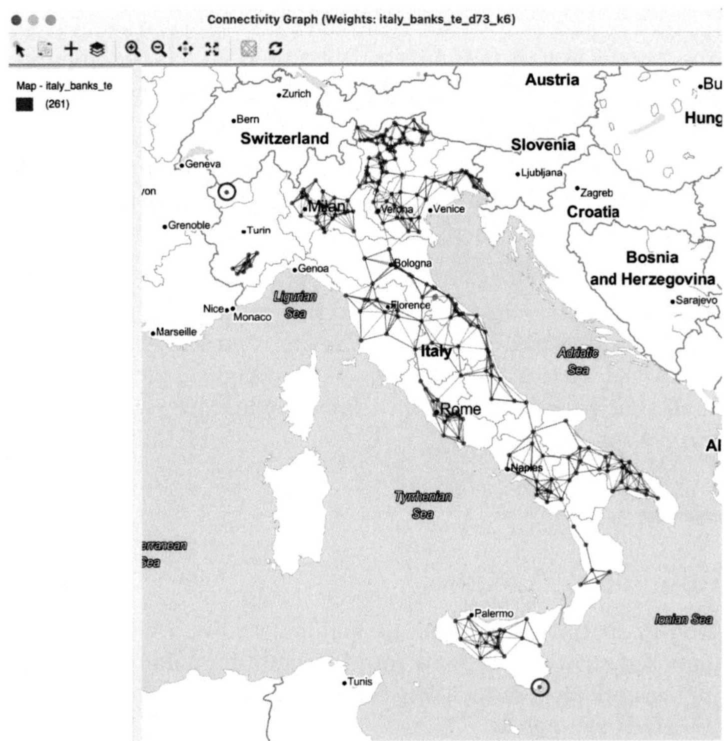

Figure 11.19: Connectivity graph for intersection distance 73 km and knn 6

11.6.2 Make symmetric

As discussed in Section 11.3.2, there are two straightforward ways to turn an asymmetric k-nearest neighbor matrix into a symmetric one. One approach is *inclusive* and identifies a link as long as either w_{ij} or w_{ji} is non-zero. This is the default solution activated when the **Make Symmetric** button is selected in the **Weights Manager**. As a result, the new symmetric weights will be denser than the original k-nearest neighbor weights.

The second approach is *exclusive* in that a link is only counted when it exists in *both* directions: $w_{ij} = w_{ji} = 1$. This is activated when the **mutual** box is checked, next to the **Make Symmetric** button. The resulting symmetric weights will be sparser than the original k-nearest neighbor weights.

For example, when applied to the 6 k-nearest neighbor weights for the Italian community banks, the inclusive symmetric weights have neighbors ranging from 6 to 13 (clearly, the minimum will be 6, as in the original weights, but the maximum can be more than 6). The median number of neighbors has increased to 7, with an overall density of 2.8 /%.

In contrast, the exclusive criterion yields weights with 0 to 6 neighbors (in this case, 6 is the maximum), and thus creates some isolates. The median number of neighbors has decreased to 5, with an overall density of 1.73 /%.

Which of these transformation is appropriate, will depend on each specific empirical circumstance. As a rule, a careful consideration of the weights characteristics is strongly recommended, especially by means of the connectivity graph.

Finally, it is important to save the created weights in a **Project File**, so that metadata can be preserved in later analyses.

12

Special Weights Operations

In this third chapter devoted to spatial weights, I consider some more specialized topics, in the sense that they are less frequently encountered in empirical practice. There are two main subjects.

First, I discuss two situations where the actual values for the spatial weights take on a special meaning. So far, only the presence or absence of a neighbor relation has been taken into account. In this chapter, this is generalized to weights that are transformations of the pairwise distances between neighbors, i.e., inverse distance functions and kernel weights.

The resulting weights files primarily provide the basis for creating new *spatially explicit variables* for use in further analyses, such as in spatial regression specifications.[1] The actual value of the weights themselves is *not* used in measures of spatial autocorrelation or other exploratory analyses in `GeoDa`. As mentioned before, only the existence of a neighbor relation is taken into account.

There are two important applications for these *spatial transformations*. One pertains to the construction of so-called spatially lagged variables. Such variables are used in the exploration of spatial autocorrelation and in spatial regression analysis. In a second set of applications, the spatial weights are included as part of rate smoothing operations, as an extension of the methods covered in Section 6.4.

To illustrate these techniques, I continue to use the point layer from the *Italy Community Banks* sample data set and the *Oaxaca Development* polygon layer.

12.1 Topics Covered

- Compute inverse distance functions
- Compute kernel weights functions
- Assess the characteristics of weights based on distance functions
- Understand the contents of KWT format weights files
- Create a spatially lagged variable as an average or sum of the neighbors
- Create a spatially lagged variable as a window sum or average
- Create a spatially lagged variable based on inverse distance weights
- Create a spatially lagged variable based on kernel weights
- Compute and map spatially smoothed rates
- Compute and map spatial Empirical Bayes smoothed rates

[1] The distance functions in `GeoDa` provide an alternative and more user-friendly way to calculate the weights included in PySAL and GeoDaSpace (see Anselin and Rey, 2014, for details).

DOI: 10.1201/9781003274919-12

GeoDa Functions

- Weight File Creation dialog
  - inverse distance weights
  - kernel weights
  - bandwidth options
  - diagonal element options
- Table > Calculator > Spatial Lag
  - select spatial weights
  - row-standardized weights or not
  - include diagonal or not
- Map > Rates-Calculated Map
  - Spatial Rate
  - Spatial Empirical Bayes
- Table > Calculator > Rates
  - Spatial Rate
  - Spatial Empirical Bayes

12.2 Spatial Weights as Distance Functions

In the creation of distance band or k-nearest neighbor weights, the pairwise distance between neighbors is included in the GWT file (see Figure 11.5). However, these values are not used in any actual operations.

The distances themselves are not that useful, but transformations of distance lead to weights that enable *spatial transformations.* Two types of such functions are considered here: inverse distance weights and kernel weights.

12.2.1 Inverse distance weights

Spatial weights based on a distance cut-off can be viewed as representing a step function, with a value of 1 for neighbors with $d_{ij} < \delta$, and a value of 0 for others. As before, d_{ij} stands for the distance between observations i and j, and δ is the bandwidth.

A straightforward extension of this principle is to specify the spatial weights as a continuous parameterized function of distance itself:

$$w_{ij} = f(d_{ij}, \theta), \tag{12.1}$$

with f as a functional form and θ a vector of parameters.

In order to conform to *Tobler's first law of geography*, a distance *decay* effect must be respected. In other words, the value of the function of distance needs to *decrease* with a growing distance. More formally, this is expressed by the requirement that the partial derivative of the distance function with respect to distance should be negative:

$$\partial w_{ij} / \partial d_{ij} < 0.$$

Commonly used distance functions are the inverse, with $w_{ij} = 1/d_{ij}^{\alpha}$ (and α as a parameter), and the negative exponential, with $w_{ij} = e^{-\beta d_{ij}}$ (and β as a parameter). The functions are typically combined with a distance cut-off criterion, such that for $d_{ij} > \delta$, it follows that $w_{ij} = 0$.

In practice, the parameters are often not estimated, but instead set to a fixed value, such as $\alpha = 1$ for *inverse distance* weights ($1/d_{ij}$), and $\alpha = 2$ for *gravity* weights ($1/d_{ij}^2$). By convention, the diagonal elements of the spatial weights are set to zero and not computed. In fact, if the actual value of $d_{ii} = 0$ were used in the inverse distance function, this would result in a division by zero.

An often overlooked feature in the computation of distance functions is that the resulting values depend not only on the parameter value and functional form but also on the metric and *scale* used for the distance measure. Since the weights are inversely related to distance, large values for the latter will yield small values for the former, and vice versa. This may be a problem in practice when the distances are so large (i.e., measured in small units) that the corresponding inverse distance weights become close to zero, possibly resulting in an all zero spatial weights matrix. This issue is directly related to the scale of the coordinates. It is a common concern in practice when a projection yields distances expressed in feet or meters.

In addition, a potential problem may occur when the distance metric is such that distances take on values less than one. As a consequence, some inverse distance values may be larger than one, which is typically not a desired result.

A simple rescaling of the coordinates on which the distance computation is based will fix both problems.

12.2.1.1 Implementation

Inverse distance weights are specified through the **Weights File Creation** interface, with the **Distance Weight** tab selected, as in Figure 11.3. The check box next to **Use inverse distance?** enables this functionality. This also activates the **Power** drop-down list, where a coefficient can be specified, the default being **1**.

The remainder of the interface works in the same fashion as before, with a query for a file name for the new weights.

Using the *Italy Community Banks* sample data with the default settings yields a spatial weights file like *italy_banks_te_id1*, the summary properties of which are shown in Figure 12.1. The descriptive statistics are identical to those for the distance band weights in Figure 11.4. The only differences are that **inverse distance** takes on the value of **true**, and the **power** is listed as **1**.

By design, `GeoDa` only takes into account the existence of a neighbor relation. The actual values of the weights are ignored. Consequently, the properties of any distance function (irrespective of the power) will be identical to those of the corresponding simple weights for the same distance band. Also, the connectivity histogram, connectivity map and connectivity graph will be the same as before. They do not provide any new information.

Inverse distance weights can be calculated for any bandwidth specified. While the default is the max-min bandwidth, in some applications this is not useful. For example, the bandwidth can be set to the maximum inter-point distance. As a result, the calculations will be for a *full* $n \times n$ distance matrix. This is not recommended for larger data sets, but it can provide a useful point of departure to compute accessibility indices formulated as spatially lagged variables (see Section 12.3.1).

In addition, the coordinate option is perfectly general, and any two variables contained in the data set can be specified as X and Y coordinates in the distance band setup. For

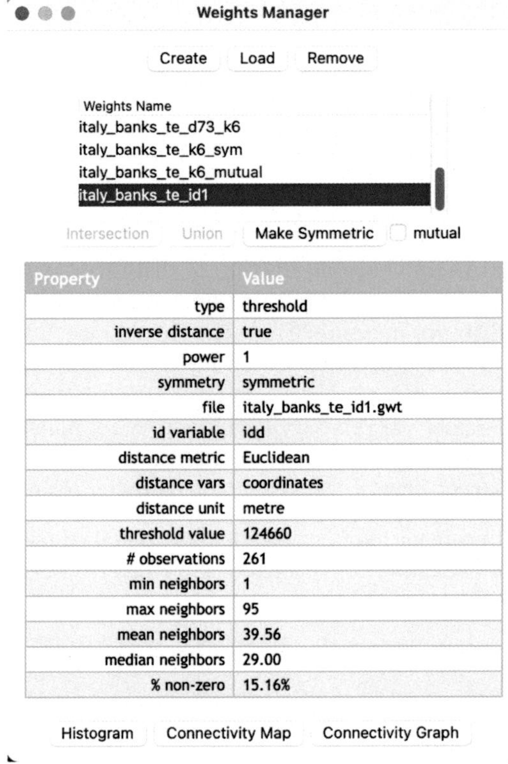

Figure 12.1: Summary properties of inverse distance weights

example, this allows for the computation of so-called *socio-economic* weights, where the inverse operation is applied to the Euclidean distance computed in the attribute domain between two locations, using any two (but only two) variables.

In contrast, with the **Variables** tab checked in the **Weights File Creation** interface, fully multivariate inverse distance weights can be computed as well, for as many variables as required. However, for such distances to be meaningful, one has to be mindful of the scale in which the variables are expressed. Standardization is highly recommended.

12.2.1.2 K-nearest neighbor inverse distance weights

The computation of inverse distance weights extends to the k-nearest neighbor design in a straightforward manner. The only difference is that the **K-Nearest neighbors** tab must be checked. All the operations are the same as for distance band weights.

For example, using the original coordinates with k=6, this yields a GWT file, e.g., *italy_banks_te_idk6.gwt*, and with the coordinates in kilometers, a file *italy_banks_te_idk6km.gwt*.

12.2.1.3 GWT file

The results of an inverse distance calculation are written to a **GWT** file, which has the same structure as before. However, in contrast to the standard case, the third column now contains the value for the inverse distance rather than the distance itself.

```
0 261 italy_banks_te idd        0 261 italy_banks_te idd
1 95     2.01549423e-05         1 95       0.0201549423
1 96     2.01694085e-05         1 96       0.0201694085
1 267     2.09357517e-05        1 267       0.0209357517
1 309    2.94645422e-05         1 309       0.0294645422
1 272     4.95814043e-05        1 272       0.0495814043
1 64     2.57304036e-05         1 64       0.0257304036
8 304     1.34577532e-05        8 304       0.0134577532
```

Figure 12.2: GWT files with distance for original and rescaled coordinates

To illustrate this, consider the two **GWT** files just created, using 6 nearest neighbors as the bandwidth. The left-hand panel of Figure 12.2 shows the outcome when using the default coordinates (*italy_banks_te_idk6.gwt*). Note how the value of 2.01549423e-05 for the pair 1-95 is exactly the inverse of the distance of 49,615.6221 (meters), listed on the matching line of Figure 11.5.

This illustrates the problem just mentioned. Since the projected coordinates for the Italian bank point file are expressed in meters, the distances are fairly large (e.g., almost 50 km in the example). As a consequence, the values obtained for inverse distance are very small. This has undesirable effects for the computation of spatial transformations like a spatially lagged variable.

The right-hand panel in Figure 12.2 shows the result after the coordinates were rescaled to kilometers.[2] The outcome is simply multiplied by 1,000.

12.2.2 Kernel weights

Kernel smoothing is a core element in nonparametric statistics (Silverman, 1986; Racine, 2019). It is essentially a weighted average of observations, centered on a focal point, with decreasing weights as the distance to the focal point increases. The magnitude of the weights is determined by a *kernel function*, i.e., a decreasing function of the distance to the focal point (in attribute space), and a *bandwidth*, i.e., the range within which observations are weighted.

In spatial analysis, kernel weights are probably best known for their inclusion in geographically weighted regression, or GWR (Fotheringham et al., 2002). However, they are also an essential part of non-parametric approaches to model spatial covariance (Hall and Patil, 1994), considered more closely in Chapter 13.[3]

More precisely, kernel weights are defined as a function $K(z)$ of the ratio between the distance d_{ij} from i to j, and the bandwidth h_i, with $z = d_{ij}/h_i$. This ensures that z is always less than 1. For distances greater than the bandwidth, $K(z) = 0$.

Two important considerations in the use of kernel weights are the selection of the kernel function and the determination of the bandwidth.

12.2.2.1 Specification of the kernel weights

Kernel functions determine how quickly the weights get smaller as the distance to the focal point increases. In GeoDa, there currently is support for five commonly used kernel weights

[2]In GeoDa, this is implemented in the **Calculator**. Assuming that **COORD_X** and **COORD_Y** are the coordinates in meters, a new set of variables **XKM** and **YKM** are computed by dividing these values by 1,000. The new **XKM** and **YKM** then need to be entered as respectively **X_coordinate variable** and **Y_coordinate variable** in the **Weights File Creation** interface.

[3]An application in spatial econometrics is the heteroskedasticity and autocorrelation consistent (HAC) estimation of the regression error variance, due to Kelejian and Prucha (2007). See Anselin and Rey (2014), for implementation details in GeoDaSpace and PySal.

functions:

- Uniform, $K(z) = 1/2$ for $|z| < 1$,
- Triangular, $K(z) = (1 - |z|)$ for $|z| < 1$,
- Quadratic or Epanechnikov, $K(z) = (3/4)(1 - z^2)$ for $|z| < 1$,[4]
- Quartic, $K(z) = (15/16)(1 - z^2)^2$ for $|z| < 1$ and
- Gaussian. $K(z) = (2\pi)^{(-1/2)} \exp(-z^2/2)$.[5]

Typically, the value for the diagonal elements of the weights is set to 1, although sometimes the actual kernel value is used. For example, for a quadratic of Epanechnikov kernel function, the weight for $d_{ij} = 0$ would in principle be 3/4, although in practice a value of 1 is often used as well.

Many careful decisions must be made in selecting a kernel function. Apart from the choice of a functional form for $K(\ )$, a crucial aspect is the selection of the *bandwidth*. In the literature, the latter is found to be more important than the functional form. Several *rules* have been developed to determine an *optimal* bandwidth (e.g., see Silverman, 1986), although in practice, trial and error may be more appropriate.

An important choice is whether the bandwidth should be the same for all observations, a so-called *fixed bandwidth*, or instead allowed by vary, as a *variable* bandwidth.

A drawback of fixed bandwidth kernel weights is that the number of non-zero weights can differ considerably from one observation to the next, especially when the density of the point locations is not uniform throughout space. This is the same problem encountered for the distance band spatial weights (Section 11.3.1).

`GeoDa` includes two different implementations of the concept of fixed bandwidths. One is the max-min distance used earlier (the largest of the nearest-neighbor distances). The other is the maximum distance for a given specification of k-nearest neighbors. For example, for a specified number of nearest neighbors, this is the distance between the k-nearest neighbors pairs that are the farthest apart. In other words, this is an extension of the max-min concept to the specific context of k neighbors, rather than nearest neighbor pairs.

To correct for the issues associated with a fixed bandwidth, a *variable bandwidth* approach can be taken. It adjusts the bandwidth for each location to ensure equal or near-equal coverage. One common approach is to take the k-nearest neighbors, and to adjust the bandwidth for each location such that exactly k neighbors are included in the kernel function. The bandwidth specific to each location is then any distance larger than its k nearest neighbor distance, but less than the k+1 nearest neighbor distance.

When the kernel weights are implemented for the nearest neighbor concept, a final decision pertains to the value of k. In `GeoDa`, the default value for k equals the cube root of the number of observations.[6] In general, a wider bandwidth gives smoother and more robust results, so the bandwidth should always be set at least as large as the recommended default. Some experimentation may be needed to find a suitable combination of bandwidth and/or number of nearest neighbors in each particular application.

---

[4] Note that the Epanechnikov kernel is sometimes referred to without the (3/4) scaling factor. `GeoDa` implements the scaling factor.

[5] While the Gaussian kernel is in principle without a bandwidth constraint, in `GeoDa` it is implemented with the same bandwidth option as the other kernel functions.

[6] This is based on the recommendation by Kelejian and Prucha (2007) in the context of HAC estimation and typically forms a good starting point

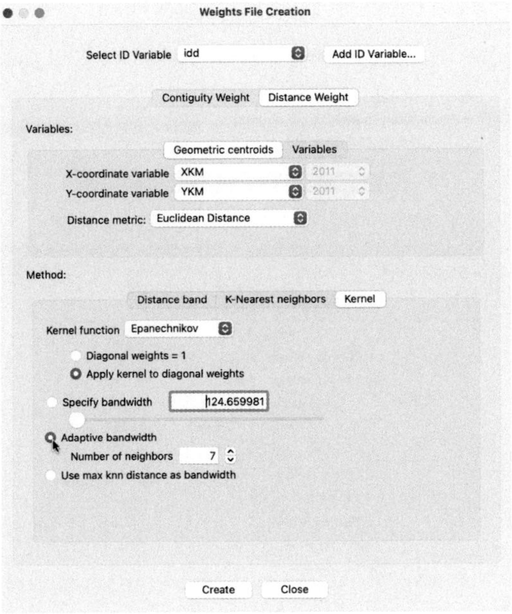

Figure 12.3: Kernel weights in Weights File Creation interface

12.2.2.2 Creating kernel weights

Kernel weights are created with the **Distance Weight** tab in the **Weights File Creation** interface. The **Method** must be specified as **Kernel**, as shown in Figure 12.3. In this illustration for the Italian community bank example, the coordinates have been converted to kilometers, specified as **XKM** and **YKM**.

The **Kernel** functionality brings up several options, as illustrated in the figure. The **Kernel function** drop-down list contains the five supported functions. In the example, the **Epanechnikov** kernel is selected. Furthermore, the actual weights are used for the diagonal elements, with the radio button checked next to **Apply kernel to diagonal weights**. The default is actually to have **Diagonal weights = 1**.

The next set of options pertains to the bandwidth. The default max-min distance is listed in the **Specify bandwidth** box as 124.659981 km. In the example, the **Adaptive bandwidth** is checked, with **7** neighbors. A final option is to instead specify a fixed bandwidth with **max knn distance as bandwidth**.

The suggested value of **k** in the interface is based on the cube root criterion. In the example, this is the cube root of 261, or 6.39, rounded up to the next integer (7). Note that the convention used in `GeoDa` is to count the observation itself as one of k, so that in fact only k-1 true nearest *neighbors* are taken into account.

The result is saved in a file with a **KWT** extension, e.g., *italy_banks_te_epa.kwt*, for consistency with the approach taken in the PySAL library. Except for the inclusion of the diagonal element, its structure is the same as a **GWT** format file.

12.2.2.3 Characteristics of kernel weights

As soon as the weights are created, their properties appear in the weights manager. As illustrated in Figure 12.4, the descriptive statistics are again the same as for standard knn

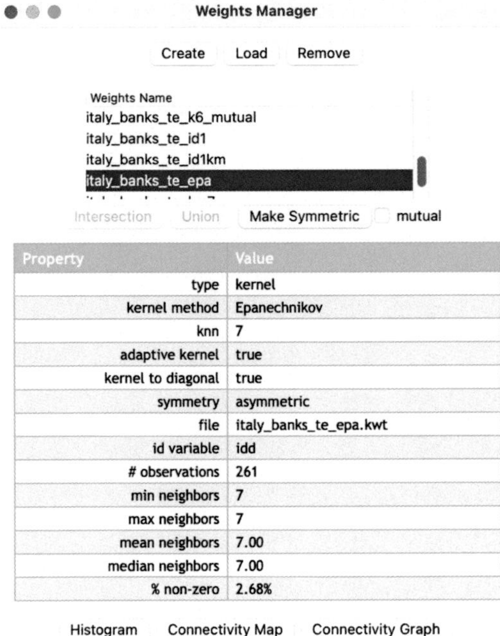

Figure 12.4: Properties of kernel weights

weights. The differences are in the first six items. The **type** of weights is given as **kernel**, the **kernel method** is identified as **Epanechnikov**, with the bandwidth definition (**knn 7**) and **adaptive kernel** set to **true**. It is also indicated that the kernel is applied to the diagonal elements, since **kernel to diagonal** is **true**. Also, as for the k-nearest neighbor weights, the resulting weights are **asymmetric**.

Since the connectivity histogram, map and graph ignore the actual weights values and are solely based on the implied connectivity structure, they are identical to those obtained for the corresponding k-nearest neighbor weights. Consequently, they are not informative with respect to the properties of the weights themselves.

12.2.2.4 KWT file

The contents of the **KWT** file in our example are shown in the right-hand panel of Figure 12.5, compared to the knn distances in the corresponding GWT file on the left (based on coordinates expressed as kilometers).

A few characteristics of the results should be noted. First, the bandwidth is determined by the largest distance among the seven neighbors. In the current example, for observation 1, the panel on the left shows the distance to the seventh neighbor (249) as 54.251 km. This is the bandwidth. As a result, the kernel weight value for the link 1 to 249 will be 0 (right hand panel). The value for z, to be used in the kernel function, is therefor obtained by dividing the pairwise distances by 54.251. For example, for the link between 1 and 95, who are 49.615 km apart, this yields 0.915. The value for the Epanechnikov kernel is $0.75(1 - z^2) = 0.75(1 - 0.836) = 075 \times 0.164 = 0.123$, the result listed in the third column of the matching row on the right-hand side.

There are eight entries for observation 1: the seven neighbors as well as the diagonal. For the latter, listed as the pair 1 to 1, the value is 0.75.

```
0 261 italy_banks_te idd
1 249          54.2507586
1 95           49.6156221
1 64           38.8645283
1 96            49.580036
1 309          33.9390986
1 272          20.1688519
1 267          47.7651824
8 417          77.6220085
```

```
0 261 italy_banks_te idd
1 249                   0
1 95          0.122683784
1 64          0.365092447
1 96          0.123583329
1 309          0.456471525
1 272          0.646339813
1 267          0.168603365
1 1                  0.75
```

Figure 12.5: KWT file

12.3 Spatial Transformations

The main form of spatial transformation considered here is the so-called *spatial lag* operation. This summarizes the observations at neighboring locations into a single new variable. The weights matrix is central in this operation, since it identifies which observations need to be selected and if any weights need to be applied.

12.3.1 Spatially lagged variables

With a neighbor structure defined by the non-zero elements of the spatial weights matrix **W**, a spatially lagged variable is a weighted sum or a weighted average of the neighboring values for that variable (Anselin, 1988). In the typical notation, the spatial lag of y is then expressed as Wy.

Formally, for observation i, the spatial lag of y_i, referred to as $[Wy]_i$ (the variable Wy observed for location i) is:

$$[Wy]_i = w_{i1}y_1 + w_{i2}y_2 + \cdots + w_{in}y_n,$$

or,

$$[Wy]_i = \sum_{j=1}^{n} w_{ij}y_j,$$

where the weights w_{ij} consist of the elements of the i-th row of the matrix **W**, matched up with the corresponding elements of the vector **y**.

In other words, the spatial lag is a weighted sum of the values observed at neighboring locations, since the non-neighbors are not included (those j for which $w_{ij} = 0$). Typically, the weights matrix is very sparse, so that only a small number of neighbors contribute to the weighted sum. For row-standardized weights, with $\sum_j w_{ij} = 1$, the spatially lagged variable becomes a weighted average of the values at neighboring observations.

In matrix notation, the spatial lag expression corresponds to the matrix product of the $n \times n$ spatial weights matrix **W** with the $n \times 1$ vector of observations **y**, or $\mathbf{W} \times \mathbf{y}$. The matrix **W** can therefore be considered to be the spatial lag *operator* on the vector **y**.

In a number of applied situations, it may be useful to include the observation at location i itself in the weights computation. This implies that the diagonal elements of the weights matrix must be non-zero, i.e., $w_{ii} \neq 0$. Depending on the context, the diagonal elements may take on the value of one or equal a specific value, such as for kernel weights where the kernel function is applied to the diagonal.

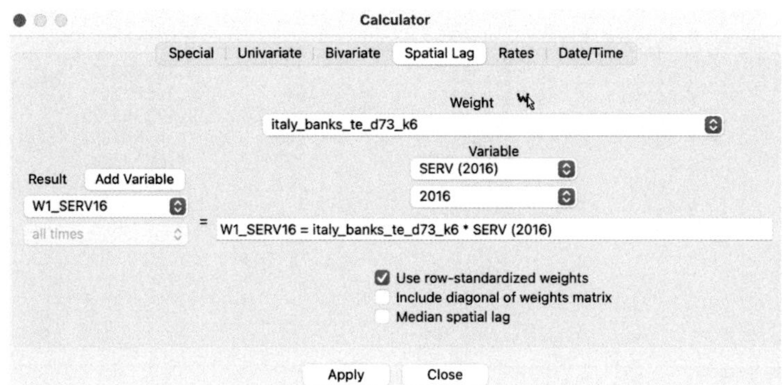

Figure 12.6: Spatial lag in calculator

An alternative concept of spatial lag is the so-called *median spatial lag*. Instead of computing a (weighted) average of the neighboring observations, the median is selected.

12.3.1.1 Creating a spatially lagged variable

The spatial lag computation is part of the **Calculator** functionality associated with a data table (Section 2.4.2). It is started from the menu as **Table > Calculator**, or by right clicking in a table to bring up the list of options.

The lag operation requires the **Spatial Lag** tab to be active. The **Weight** drop-down list contains all the spatial weights available to the project, with the currently active weights listed. In the example illustrated in Figure 12.6, a time-enabled version of the Italian bank data is used, with the 6 k-nearest neighbor weights after intersection with the 73 km distance band. The spatial weights are contained in the file *italy_banks_te_d73_k6.gal* (see Section 11.6.1). The spatial lag operator will be applied to the variable **SERV (2016)**, i.e., the time enabled ratio of net interest income over total operating revenues in 2016.

To compute the spatially lagged variable, first **Add Variable** is selected to create a name for the new variable, **W1_SERV16** in the example. At this point, the expression for the calculation of the lag is filled out in the box, listing both the weights and the original variable.

The default option is to **Use row-standardized weights** without including the diagonal elements. The result is added to the data table, as shown in Figure 12.7, in the column **W1_SERV16**.

The file *italy_banks_te_d73_k6.gal* contains the neighbor list for the first observation, with **idd = 1**. These are the observations with **idd** equal to, respectively, **64**, **95**, **96**, **267**, **272** and **309** (corresponding to table rows 15, 41, 42, 167, 172 and 209). The value of **SERV (2016)** for the first observation is listed as **0.640208**. A look up in the data table reveals the values for the neighboring observations as **0.560816**, **0.748250**, **0.659608**, **0.615197**, **0.349620** and **0.701551**. The average of these six values is **0.605840**, which is exactly the entry on the first row of the **W1_SERV16** column.

As illustrated in Section 11.6.1, the spatial weights intersection yields two isolates, observations with **idd 269** (Valley of Aosta, row 169) and **287** (Sicily, row 187). Since a spatially lagged variable is undefined for an isolate (an actual computation would yield a value of zero), the corresponding entries in the table are empty. This avoids complications from mistakenly associating a value of zero in further calculations, such as in the computation

Table - italy_banks_te

	SERV (2016)	W1_SERV16	W2_SERV16	W3_SERV16	W4_SERV16	W5_SERV16
1	0.640208	0.605840	3.635042	0.637403	0.610750	4.275251
2	0.502599	0.592221	2.961105	0.614457	0.577284	3.463704
3	0.567946	0.597560	3.585360	0.576170	0.593329	4.153305
4	0.426885	0.542616	3.255695	0.551646	0.526083	3.682580
5	0.576827	0.635728	3.814368	0.645612	0.627314	4.391195
6	0.667058	0.705937	4.235621	0.718108	0.700383	4.902678
7	0.490465	0.594648	2.973238	0.614457	0.577284	3.463704
8	0.617039	0.597322	3.583931	0.596177	0.600139	4.200970
9	0.636955	0.660822	3.964933	0.685617	0.657413	4.601888
10	0.615309	0.643255	3.859531	0.624045	0.639263	4.474840
11	0.516654	0.700213	4.201279	0.634439	0.673990	4.717933
12	0.624910	0.576627	2.306507	0.576271	0.586283	2.931417
13	0.424762	0.727681	1.455362	0.727681	0.626708	1.880124
14	0.613671	0.586066	3.516398	0.569395	0.590010	4.130068
15	0.560816	0.714160	4.284958	0.709145	0.692254	4.845775

Figure 12.7: Spatially lagged variables added to table

of a spatial autocorrelation coefficient. The outcome of the empty entry is that the isolate observations are excluded from such computations.

The effect of the spatial lag operation is a form of *smoothing* of the original data. The mean is roughly the same (the theoretic expected values are the same), e.g., 0.6363 for the original variable vs. 0.6378 for the spatial lag, but the standard deviation is much smaller, i.e., 0.0943 for the original vs. 0.0562 for the spatial lag. Similarly, a form of shrinking has occurred: the original variable has a minimum of 0.3496 and a maximum of 0.9143, whereas for the spatial lag this has narrowed to 0.5124 to 0.7478.[7]

A measure of spatial correlation can be computed as the correlation between the original and the spatially lagged variable. Note that this is NOT the accepted notion of spatial autocorrelation considered in the literature (see Chapter 13). Nevertheless, as long as it is not interpreted in the context of a spatial autoregressive model, it remains a useful measure of the linear association between a variable and the average of its neighbors. In the example, this value is 0.350.[8]

12.3.1.2 Spatial lag as a sum of neighboring values

Combining different settings for the **row-standardized** and **Include diagonal** options allows for other notions of spatially lagged variables to be implemented. For example, a spatial lag as a *sum* of the neighboring values is obtained by unsetting all options.

The result is listed in column **W2_SERV16** of the table (Figure 12.7). For the first observation, the corresponding value is **3.635042**, which is the sum of the values for the six neighbors.

12.3.1.3 Median spatial lag

The **Median spatial lag** is obtained by checking the corresponding box (the two other options become unavailable). In the example, there are six neighbors, so the median is computed as the midpoint between the third (0.615197) and fourth (0.659608) values in the ranked set, yielding **0.637403**. This is the entry in the first row of the **W3_SERV16** column in Figure 12.7.

[7]The statistics can be readily obtained from a box plot of each variable.

[8]The correlation can be calculated wit h the **Data > View Standardized Data** in a scatter plot of **SERV (2016)** on **W1_SERV16**.

12.3.1.4 Spatial window transformations

So far, the spatial lag operations did not include the observations itself. However, by checking the **Include diagonal of weights matrix** option, an average or sum over all observations in the *window* can be computed. The window consists of the observation and its neighbors, a notion commonly used in spatial smoothing operations (see also Section 12.4).

The *spatial window average* is obtained by checking the **Use row-standardized weights** option. The result is an average of the seven values (observation and six neighbors), or **0.610750**, the entry in the first row of the **W4_SERV16** column in Figure 12.7.

Unchecking the **row-standardized** box yields the *spatial window sum.* In the example, this is **4.275251**, listed under column **W5_SERV16**. Clearly, this is equivalent to the addition of **SERV (2016)** and **W2_SERV16**.

12.3.2 Using inverse distance weights – *potential* measure

The spatial lag operation can also be applied using spatial weights calculated from the inverse distance between observations. This and kernel weights (Section 12.3.3) are the only two instances in `GeoDa` where the actual values of the weights are used in an operation.

As mentioned in Section 12.2.1, the magnitude of the inverse distance weights depends on the units in which the coordinates are expressed. Therefore, care is needed before using inverse distance weights in spatial transformations. When the coordinates are expressed in feet or meters, as is often the case, the corresponding distances will tend to be very large. Consequently, the inverse distances may take on very small values, and any spatial lag could end up being essentially zero.

Formally, the spatial lag operation uses the same principle as before. It amounts to a weighted average of the neighboring values, with the inverse distance function as the weights:

$$[Wy]_i = \sum_j \frac{y_j}{d_{ij}^{\alpha}},$$

In practice, this calculation is typically combined with the application of a bandwidth. Such a bandwidth can be expressed in the form of a distance-band weight, with $w_{ij} = 0$ for $d_{ij} > \delta$, with δ as the bandwidth. Formally:

$$[Wy]_i = \sum_j \frac{w_{ij} y_j}{d_{ij}^{\alpha}},$$

As mentioned, α is either 1 or 2. This expression for the spatial lag corresponds to the concept of *potential*, familiar in geo-marketing analyses.[9] The potential is a measure of how accessible location i is to opportunities located in the neighboring locations (as defined by the weights).

12.3.2.1 Default inverse distance weights setting

Spatial lags constructed from inverse distance weights operate in the same way as outlined in Section 12.3.1.1. The default setting is that both **Use row-standardized weights** and **Include diagonal of weights matrix** are **unchecked**. This yields the potential measure as the sum of the neighboring values weighted by the inverse distance (or its square).

[9] The concept of population potential goes back to the early literature in *social physics*, e.g., Stewart (1947). See also Harris (1954) and Isard (1960) for extensive early discussions.

Table - italy_banks_te

	SERV (2016)	ID_SERV16	IDK_SERV16	IDR_SERV16	IDS_SERV16	EPA_SERV16
1	0.640208	0.000094	0.093700	0.564334	0.733909	1.508157
2	0.502599	0.003625	3.625202	0.492436	4.127801	1.957092
3	0.567946	0.000105	0.105363	0.583162	0.673308	1.968662
4	0.426885	0.000232	0.232036	0.549579	0.658921	1.791276
5	0.576827	0.000125	0.124682	0.641227	0.701508	1.917262
6	0.667058	0.000346	0.346269	0.717616	1.013327	2.608324
7	0.490465	0.003713	3.712901	0.504361	4.203366	1.953950
8	0.617039	0.000238	0.237659	0.603408	0.854698	1.668986
9	0.636955	0.000139	0.138704	0.632328	0.775659	1.769977
10	0.615309	0.000247	0.246890	0.656369	0.862199	2.248963
11	0.516654	0.000120	0.119754	0.675322	0.636408	1.738610
12	0.624910	0.000072	0.071550	0.590604	0.696459	2.474655
13	0.424762	0.000055	0.055300	0.636649	0.480063	1.168037
14	0.613671	0.000125	0.125436	0.598754	0.739106	2.194710
15	0.560816	0.000168	0.167833	0.711382	0.728649	2.086366

Figure 12.8: Variables computed with inverse distance weights added to table

Using the Italian bank example, with the inverse distance weights computed from the original coordinates (in meters) with a 6 nearest neighbor bandwidth requires the file *italy_banks_te_idk6.gwt*. The result is shown in the column **ID_SERV16** of Figure 12.8. The small values illustrate the problem associated with the original weights.

Instead, with the inverse distances computed from coordinates in kilometers, the result in column **IDK_SERV16** follow (using the weights file *italy_banks_te_idk6km.gwt*), essentially a rescaled version of **ID_SERV16**.

For example, for observation with **idd=1**, the inverse distance weights are given in the right-hand panel of Figure 12.2. Combining these weights with the observation values for the neighbors listed in Section 12.3.1.1 yields the value **0.093700** given in the first row of the **IDK_SERV16** column.

12.3.2.2 Row-standardized inverse distance weights

The original inverse distance weights are highly scale dependent. This can be remedied by expressing them in row-standardized form. The spatial lag then takes on the standard meaning of a weighted average of the values at neighboring observations. The main difference with lags computed for connectivity weights is that now the neighbors are weighted differently. In contrast, for the straightforward spatial lag calculation of Section 12.3.1.1, all the neighboring values get the same weight.

The result of applying the row-standardized inverse distance weights from the right-hand panel in Figure 12.2 yields **0.564334**, listed on the first row of the **IDR_SERV16** column. In contrast to the calculation for the unstandardized inverse weights, the result is now similar in scale to the original variable (and similar to the spatial lags based on connectivity weights).

12.3.2.3 Spatial window inverse distance sum

Similar to the operations in Section 12.3.1.4, the inverse distance spatial lag can be computed for all observations in a *spatial window*. This is accomplished by checking the **Include diagonal of weights matrix** option.

However, unlike what is the case for the simple spatial lag, there is a complication on how to deal with the diagonal value. In the simple spatial lag calculation this value is either added to the neighboring observations, or averaged with them. In the inverse distance case, the neighbors are weighted, but it is not obvious what the weight should be for the diagonal, since a distance of zero does not result in a valid weight.

The convention taken in GeoDa is to keep the diagonal value unweighted. For example, without row-standardization, this amounts to:

$$[Wy]_i = y_i + \sum_j \frac{y_j}{d_{ij}^\alpha}.$$

Again, this can be implemented in combination with a specific bandwidth for the distance.

In some contexts, this may be the desired result, but it is by no means the most intuitive concept. It should therefore be used sparingly and only when there is a strong substantive motivation. The corresponding result is listed in the first row of the **IDS_SERV16** column. It is simply the sum of the value for **SERV (2016)** and **IDK_SERV16**.

The same problem occurs when using row-standardized weights. Again, in GeoDa the diagonal is unweighted, which yields:

$$[Wy]_i = y_i + \sum_j w_{ij} y_j.$$

where w_{ij} are the row-standardized inverse distance weights.

12.3.3 Using kernel weights

Spatially lagged variables can also be computed from kernel weights. However, in this instance, only one of the options with respect to row-standardization and diagonal weights makes sense. Since the kernel weights are the result of a specific kernel function, they should not be altered. Also, each kernel function results in a specific value for the diagonal element, which should not be changed. As a result, the only viable option to create spatially lagged variables based on kernel weights is to have *no* row-standardization and to have the diagonal elements included. Other options are not available in the interface.

Kernel-based spatially lagged variables correspond to a form of local smoothing. They can be used in specialized regression specifications, such as geographically weighted regression. The resulting spatially lagged variable is

$$[Wy]_i = \sum_j K_{ij} y_j,$$

When kernel weights are used, the concept of a bandwidth is already implemented in the calculation of the kernel.

For the example using Epanechnikov weights with 6 nearest neighbors (*italy_banks_te_epa.kwt*), the result is listed on the first row of the column **EPA_SERV16**. The value of **1.508157** can be obtained by combining the weights in the right-hand panel of Figure 12.5 with the neighboring observations from Section 12.3.1.1).

Creating spatially lagged variables and their use in spatial rate smoothing are the only operations in GeoDa where kernel weights can be applied. They are not available for the analysis of spatial autocorrelation.

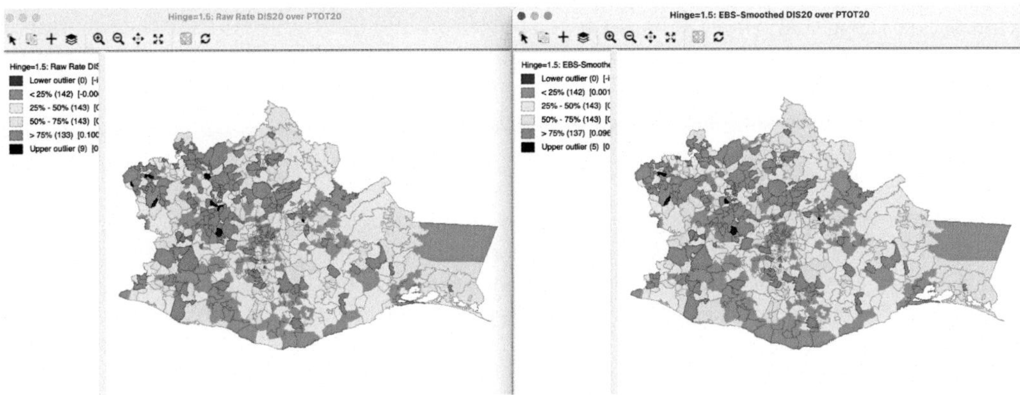

Figure 12.9: Disability rate: raw rate and EB smoothed rate

12.4 Spatial Weights in Rate Smoothing

In the discussion of rate smoothing in Chapter 6, two approaches were left until after spatial weights were introduced. **Spatial rate smoothing** and **spatial Empirical Bayes smoothing** are extensions of, respectively, the raw rate map and Empirical Bayes smoothing. The spatial versions are distinct in that they incorporate values for neighboring observations into the numerator and denominator of the rate calculation. This is made operational through the use of spatially lagged variables.

These methods will be illustrated with the municipal disability rate in 2020 from the *Oaxaca Development* sample data set. In addition to variables, five different spatial weights matrices will be employed: first-order queen contiguity (*oaxaca_q.gal*); inclusive second-order queen contiguity (*oaxaca_q2.gal*); nearest neighbor weights, with k=6 (*oaxaca_k6.gwt*); inverse distance weights with a 6 nearest neighbor bandwidth (*oaxaca_idk6.gwt*); and block weights based on the region classification (*oaxaca_reg.gal*).

12.4.0.1 Oaxaca disability rate

The disability rate is obtained as the ratio of the number of disabled persons (**DIS20**) over the population (**PTOT20**) in 2020.

This rate provides a prime example of variance instability (see Section 6.4.1), due to the wide range of population totals for municipalities, including several with very small values. In 2020, the municipal population for the 570 municipalities ranged from 81 to 270,955, and 47 locations had fewer than 500 inhabitants. The average population was 7,249, but the median was much smaller, at 2,881, reflecting the effect of upper outliers (there are 62 such outliers, with populations ranging from 14,198 to 270,955). This will yield substantial variation in the precision of the computed raw rates.

To provide some reference for the later illustrations, Figure 12.9 includes the raw rate map (Section 6.2.2) in the left-hand panel and the Empirical Bayes (EB) smoothed rate map (Section 6.4.2.2) in the right-hand panel.

The effect of the smoothing is to eliminate a few upper outliers in the map. Otherwise, the relative ranking of observations in quartiles is not much affected. However, there is a real impact on the actual values. The range is reduced from 0.0149 to 0.3161 for the raw rate, to a much narrower 0.0174 to 0.2164 for the smoothed rate. The correlation between the two rates is 0.984.

12.4.1 Spatial rate smoothing

A spatial rate smoother is a special case of a nonparametric rate estimator, based on the principle of locally weighted estimation (see, e.g., Waller and Gotway, 2004, pp. 89–90). Rather than applying a local average to the rate itself, as would be the case in an application of a spatial window average, the weighted average is applied separately to the numerator and denominator.

The spatially smoothed rate for a given location i is then given as:

$$\pi_i = \frac{\sum_{j=1}^{n} w_{ij} O_j}{\sum_{j=1}^{n} w_{ij} P_j},$$

where O_j is the event count in location j, P_j is the population at risk and w_{ij} are the spatial weights. In contrast to most procedures covered so far, here the weights typically include the diagonal element as well, i.e., $w_{ii} \neq 0$.

Different smoothing functions are obtained for different spatial definitions of neighbors and/or different weights applied to those neighbors, such as contiguity weights, inverse distance weights, or kernel weights.

An early application of this principle was the spatial rate smoother outlined in Kafadar (1996), based on the notion of a spatial moving average or *window average* (see also Kafadar, 1997). The window average is not applied to the rate itself, but it is computed separately for the numerator and denominator.

The simplest case boils down to applying the idea of a *spatial window sum* (Section 12.3.1.4) to the numerator and denominator (i.e., with binary spatial weights in both, and including the diagonal term):

$$\pi_i = \frac{O_i + \sum_{j=1}^{J_i} w_{ij} O_j}{P_i + \sum_{j=1}^{J_i} w_{ij} P_j},$$

A map of spatially smoothed rates tends to emphasize broad spatial trends and is useful for identifying general features of the data. However, it is not suitable for the analysis of spatial autocorrelation, since the smoothed rates are autocorrelated by construction. It is also not very insightful for identifying outlying observations, since the values portrayed are *regional* averages and not specific to an individual location. By construction, the values shown for individual locations are determined by both the events and the population sizes of adjoining spatial units, which can lead to misleading impressions. More specifically, the counts and populations of small areas are dwarfed by those or larger surrounding areas. Therefore, inverse distance weights are sometimes applied to both the numerator and denominator, e.g., as in the early discussion by Kafadar (1996). In `GeoDa` this is not implemented in the rates calculation, but can be carried out *manually* in the data table (see Section 12.4.1.2).

12.4.1.1 Spatially smoothed rate map

A spatially smoothed rate map can be created from the list of map options by selecting **Rates > Spatial Rate**. Alternatively, from the menu, it is obtained as **Map > Rates-Calculated Map > Spatial Rate**.

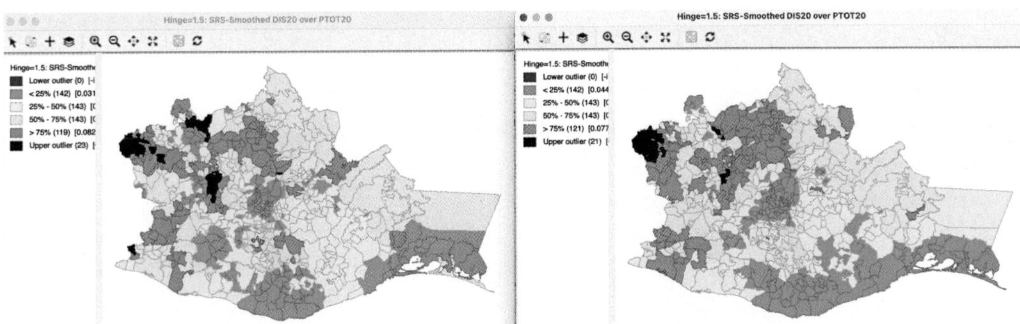

Figure 12.10: Spatial rate map: first- and second-order queen contiguity

The map construction follows the same approach as for other rate maps (see Section 6.2.2). The numerator in the ratio is set as the **Event Variable**, here **DIS20**, and the denominator as the **Base Variable**, here **PTOT20**. At the bottom of the variable selection dialog are two drop-down lists. One contains different **Map Themes** and the other the spatial **Weights**. In the example, two **Box map** are created, using first- and second-order queen contiguity, **oaxaca_q** and **oaxaca_q2**. The left panel of Figure 12.10 illustrates the former, the right panel the latter.

Compared to the maps in Figure 12.9, a lot of detail is lost, and the emphasis is on broader regional patterns, which tend to be dominated by the data for the larger municipalities. Increasing the range of the contiguity to second-order yields somewhat larger *regions* of similar values (in the sense of falling in the same quartile).

The spatial rate smoothing yields an even greater reduction in the variable range than the non-spatial EB. For first-order contiguity, the range becomes 0.0390 to 0.1370, and for second-order contiguity 0.0468 to 0.1164. In contrast to EB rates, which remain highly correlated with the original raw rate, this is no longer the case for spatially smoothed rates. The averaging over spatial units breaks the connection with the original values somewhat. In the current example, the correlation between raw rate and the two spatial rates is, respectively, 0.504 for first-order and 0.468 for second-order contiguity. This should be kept in mind when interpreting the results.

The spatially smoothed rate map retains all the standard options of a map in `GeoDa`, as well as the special options of other rate maps (see Section 6.2.2). The default variable name for the **Save Rates** option is **R_SPAT_RT**. Currently, there are no metadata that keep track of the spatial weights used in the smoothing operation.

12.4.1.2 Spatially smoothed rate in table

As is the case for other rate calculations (see, e.g., Section 6.2.2.1), the rates can also be computed by means of **Table > Calculator**. The only difference is that now there is also a drop-down list to specify the spatial weights. All other operations are identical to the non-spatial case.

As mentioned, sometimes the use of inverse distance or kernel weights is advocated to lessen the effect of neighboring values on the computed spatial smoothed rate. While it may be tempting to carry out the rate calculation using such spatial weights, this does not work correctly in `GeoDa`. Figure 12.11 illustrates this issue. It portrays spatially smoothed rates using two different spatial weights. The map on the left is based on k-nearest neighbor

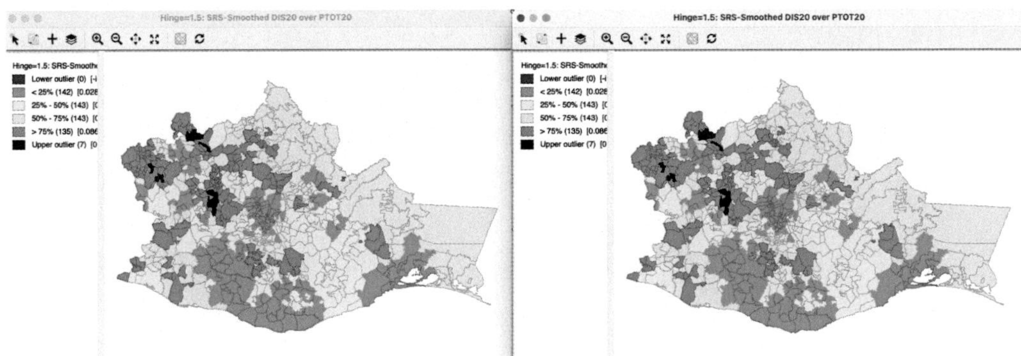

Figure 12.11: Spatial rate map: k-nearest neighbor and inverse distance weights

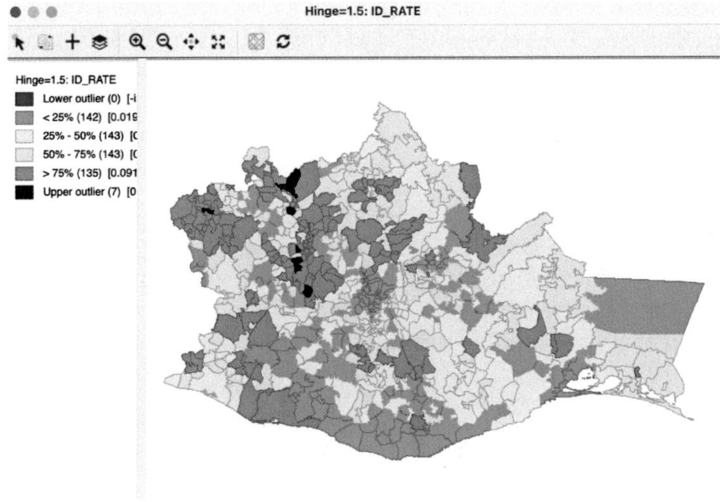

Figure 12.12: Spatial rate map: inverse distance weights from Calculator

weights, with k = 6, **oaxaca_k6**,[10] the map on the right on inverse distance weights using the 6 nearest neighbors as a variable bandwidth, **oaxaca_idk6**. The maps are identical because only the contiguity information is used in the weights operations.

The proper way to compute spatially smoothed rates for inverse distance or kernel weights is through the **Calculator**. A spatially lagged variables must be computed separately for the numerator and denominator, using either inverse distance or kernel weights. These new variables are then used to compute a ratio, as a division operation in the calculator. Figure 12.12 shows a box map with the result (the computed variable is **ID_RATE**). The pattern clearly differs from that in the maps in Figure 12.11. The effect of large neighbors is lessened because of the distance decay effect embedded in the inverse distance weights.

12.4.2 Spatial Empirical Bayes smoothing

The *spatial* Empirical Bayes rate smoothing method is based on the same principle as the standard Empirical Bayes smoother, covered in Chapter 6. The main difference is that the

[10]Since the coordinates for the Oaxaca data set are in decimal degrees latitude-longitude, the weights were constructed using **Arc Distance (km)** option in the **Weights File Creation** dialog.

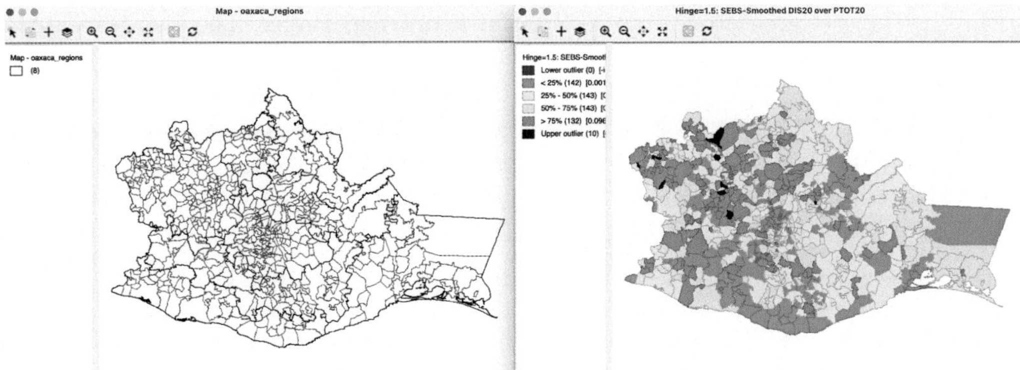

Figure 12.13: Spatial Empirical Bayes smoothed rate map using region block weights

reference rate is computed for a spatial window, specific to each individual observation, rather than taking the same overall reference rate for all observations. This approach only works when the number of observations in the reference window, as defined by the spatial weights, is large enough to allow for effective smoothing. This is a context where *block weights* become very useful (Section 11.5.3), since they avoid the irregularity in the number of neighbors on which the reference rate is based. All the observations in the same *block* have an identical number of reference points from which the reference rate is computed.[11]

Similar to the standard EB principle, a reference rate (or prior) is computed from the data. It is estimated from the *spatial window* surrounding a given observation.

Formally, the reference mean for location i is then:

$$\mu_i = \frac{\sum_j w_{ij} O_j}{\sum_j w_{ij} P_j},$$

with w_{ij} as binary spatial weights, and $w_{ii} = 1$.

The local estimate of the prior variance follows the same logic as for EB, but replacing the population and rates by their local counterparts:

$$\sigma_i^2 = \frac{\sum_j w_{ij}[P_j(r_i - \mu_i)^2]}{\sum_j w_{ij} P_j} - \frac{\mu_i}{\sum_j w_{ij} P_i/(k_i + 1)}.$$

Note that the average population in the second term pertains to all locations within the window, therefore, this is divided by $k_i + 1$ (with k_i as the number of neighbors of i). As in the case of the standard EB rate, it is quite possible (and quite common) to obtain a negative estimate for the local variance, in which case it is set to zero.

The spatial EB smoothed rate is computed as a weighted average of the crude rate and the prior, in the same manner as for the standard EB rate (Section 6.4.2.2).

12.4.2.1 Spatial Empirical Bayes smoothed rate map

A spatial Empirical Bayes smoothed rate map is selected from the list of map options as **Rates > Spatial Empirical Bayes**. Alternatively, from the menu, it is obtained from

[11]Since the observation itself is included in the reference window, all observations in the same block will have the same mean and variance for the prior.

Map > Rates-Calculated Map > Spatial Empirical Bayes. The map is created in exactly the same way as for the spatial rate map.

In the example, shown in Figure 12.13, block weights based on the region to which a municipality belongs (**oaxaca_reg**) are used to define the reference window for each observation. These weights have on average 95 neighbors, ranging from a minimum of 19 to 154. The left panel of the figure illustrates the regional division, whereas the smoothed rates are mapped in the right panel.

While there are many similarities with the pattern for the standard EB rate, shown in Figure 12.9, there are also a few differences, such as an additional upper outlier and a number of observations that shift from the first quartile to the second quartile, as well as from the fourth quartile to the third. The correlation with the original EB rate is very high, at 0.996, suggesting overall a great similarity between the two results.

All the usual options are in effect. The rate can be saved with a default variable name of **R_SPAT_EBS**.

12.4.2.2 Spatial Empirical Bayes smoothed rate in table

Identical to the treatment of the other rates, a spatial Empirical Bayes smoothed rate can be computed through the table **Calculator** option, as an item under the **Rates** tab.

Part IV

Global Spatial Autocorrelation

13

Spatial Autocorrelation

In this chapter, I begin the discussion of spatial autocorrelation, both in general as well as focused on the special case of the Moran's I statistic. I start off with some definitions, more specifically the notion of spatial randomness as the null hypothesis, and positive and negative spatial autocorrelation as the alternative hypotheses.

This is followed by an overview of the general concept of a spatial autocorrelation statistic, with a brief discussion of several specific implementations. Then attention shifts to arguably the best known such statistic, Moran's I (Moran, 1948). The formal structure, inference and interpretation of the statistic are covered, followed by its visualization through the so-called Moran scatter plot (Anselin, 1996). The chapter closes with an outline of the implementation of the Moran scatter plot in `GeoDa`.

To illustrate these concepts, I use the *Chicago Community Areas* sample data set with socio-economic variables for the 77 community areas in the city of Chicago (from the American Community Survey).

13.1 Topics Covered

- Understand spatial randomness
- Appreciate the difference between positive and negative spatial autocorrelation
- Understand the structure of a spatial autocorrelation statistic
- Moran's I
- Visualize Moran's I by means of the Moran scatter plot
- Carry out inference using the permutation approach
- Make analyses reproducible using the random seed setting
- Nonlinear LOWESS smooth of the Moran scatter plot
- Brush Moran scatter plot to assess regional Moran's I
- Appreciate the difference between dynamic weights and static weights in Moran scatter plot regime regression

GeoDa Functions

- Table > Calculator
  - Univariate > Shuffle
- Space > Univariate Moran's I
  - permutation inference
  - setting the random seed
  - LOWESS smoother of the Moran scatter plot
  - brushing the Moran scatter plot
  - save results (standardized value and spatial lag)

DOI: 10.1201/9781003274919-13

Toolbar Icons

Figure 13.1: Moran Scatter Plot | Spatial Correlogram | Cluster Maps

13.2 Spatial Randomness

Spatial randomness is the point of departure in any statistical analysis of spatial pattern. It is the *null hypothesis* against which the data are compared.

Spatial randomness is the *absence* of any systematic spatial structure in the data, i.e., there are no distinct patterns. The location of a given observation is of no value as a piece of information, such that the *where* does not matter. In other words, the notion of spatial randomness is not very interesting other than as a statistical concept. The main goal of an analysis is to *reject* the null hypothesis of spatial randomness in favor of the presence of some type of pattern. This is accomplished by means of a test statistic.

Spatial randomness can be interpreted in two main ways, which correspond to different approaches toward spatial analysis. In one, representing a *simultaneous* view, the observed spatial pattern is viewed as equally likely as any other spatial pattern. More specifically, it is the pattern as a whole that is considered. Another view represents a *conditional* perspective. It implies that the value observed at a given location does not depend on the values observed at neighboring locations. Again, this requires that location does not matter. Fundamentally, the two notions are equivalent, but they result in different statistical approaches.

Before considering specific statistics (see Section 13.4), two perspectives toward making spatial randomness operational are briefly outlined. One is based on a parametric assumption, specifying observations as independent and identically distributed (*i.i.d.*). The other perspective does not assume a particular distribution, although it still requires identical distributions, especially a constant variance. In such an approach, the location of the observations may be altered without affecting the information content of the data. This is referred to as *randomization*, in the sense that each observed value has the same probability of being associated with any of the locations, i.e., the map is irrelevant.

13.2.1 Parametric approach

The parametric approach requires the assumption of a specific distribution. Typically, this is the normal (or Gaussian) distribution, since it has the important property that lack of correlation also implies independence. For other distributions, this is not the case.

The simplest implementation of spatial randomness is to take a collection of uncorrelated (independent) standard normal variates. Since nothing *spatial* is assumed for this distribution, the result is spatially random.

For example, Figure 13.2 illustrates the layout of 100 standard normal variates on a 10×10 square grid, using a quantile map with six categories. Even though one may be tempted to *see* structure in the data, this would be purely coincidental, since the map is devoid of pattern by construction.

Figure 13.2: Spatially random observations – i.i.d

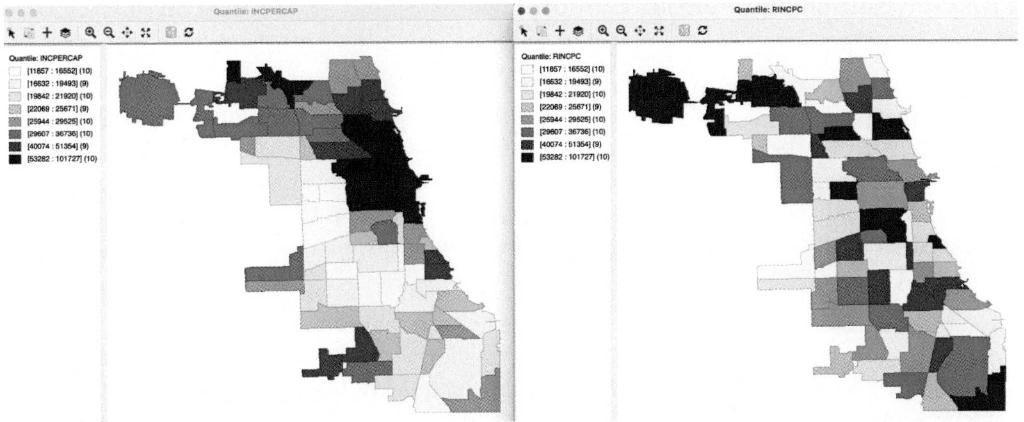

Figure 13.3: Spatially random observations – true and randomized per capita income

13.2.2 Randomization

The randomization approach does not require the selection of a specific parametric distribution for the data. Instead, it takes advantage of the property that under the null hypothesis, each observation is equally likely to be at any location. By randomly *permuting* or reshuffling the data across locations, the concept of spatial randomness can be simulated.

For example, the left hand panel of Figure 13.3 shows the spatial distribution of the income per capita for the 77 Chicago community areas (from the 2015–2019 ACS) as a quantile map with eight categories. On the right, the same 77 values are randomly allocated to community areas, resulting in a totally different map. The true map illustrates the well-known pattern of higher incomes in the northern part of the city, contrasting with low income in the south and west. By contrast, the randomized map shows no such pattern. High income areas (dark brown) are found all over the city, without any apparent structure to their locations.

Parenthetically, the contrast between the two maps illustrates the value of a spatial perspective. Taken as such, the statistical distribution of the original and randomized variables are identical. In Figure 13.4, an eight category histogram highlights this property. In other words, the a-spatial perspective on the data distribution offered by the histogram for the two variables cannot account for the important distinction between their spatial patterns.

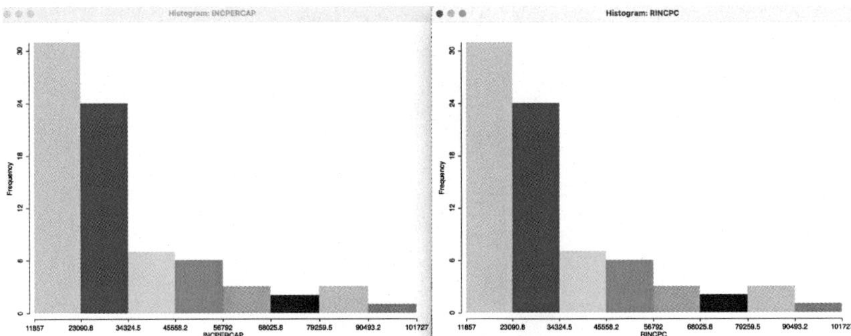

Figure 13.4: True and randomized per capita income – histograms

13.2.3 Simulating spatial randomness

The spatially randomized map in Figure 13.3 is obtained by creating a new variable, taking advantage of the **Calculator** functionality in the table. Specifically, as listed in Section 2.4.2.2, the **Univariate** operations include a **Shuffle** function, which implements randomization. Operationally, the list of index numbers of the observations is randomly reshuffled and the original observations are assigned to the new locations. For example, if the original list was 1, 2, 3, 4, 5, and the randomized list yields 3, 2, 4, 5, 1, then the value at location 1 will be assigned to location 3, 2 will remain in place, 3 will move to spot 4, 4 to spot 5 and 5 to location 1.

The randomization (or permutation) approach is an essential tool to create many versions of the data under the null hypothesis. This is particularly useful, since, except in highly stylized situations (e.g., standard normal distribution), it quickly becomes very difficult to obtain the properties of any spatial autocorrelation statistic by means of an analytical derivation.

13.3 Positive and negative spatial autocorrelation

Rejecting the null hypothesis of spatial randomness suggests that there is some spatial structure in the data. When the focus in on dependence (as opposed to heterogeneity), the two *alternative hypotheses* are positive and negative spatial autocorrelation. Unlike what holds for the familiar notion of correlation (without space), negative spatial autocorrelation is not simply the negative of positive spatial autocorrelation, i.e., a negatively sloped linear relation contrasted with a positively sloped on. Negative spatial autocorrelation yields a totally different type of spatial pattern.

13.3.1 Positive spatial autocorrelation

Positive spatial autocorrelation implies that like values in neighboring locations occur more frequently than they would under spatial randomness. This results in an impression of *clustering*, the clumping in space of like values. These like values can be either high (*hot spots*) or low (*cold spots*). The key aspect is *attribute similarity* among neighboring locations.

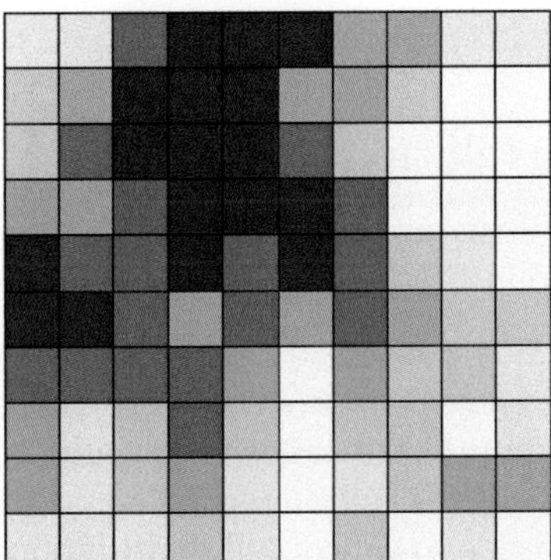

Figure 13.5: Positive spatial autocorrelation

Figure 13.6: Negative spatial autocorrelation

Figure 13.5 illustrates strong positive spatial autocorrelation on a 10 × 10 regular grid.[1] While the pattern may suggest some clustering, this is very difficult to assess by eye, especially relative to the true spatially random pattern of Figure 13.2.

13.3.2 Negative Spatial Autocorrelation

Negative spatial autocorrelation is suggested when neighboring values are more *dissimilar* than they would be under spatial randomness. This typically yields a checkerboard-like pattern. The dissimilarity is often associated with spatial heterogeneity, although the latter is not limited to neighboring units. Negative spatial autocorrelation implies alternating high and low values, whereas spatial heterogeneity can pertain to larger regions that include many neighbors.

Figure 13.6 illustrates how difficult it is to distinguish negative spatial autocorrelation from spatial randomness (Figure 13.2). It portrays a strong degree of negative spatial autocorrelation on a 10 × 10 regular grid.[2] Whereas there are quite a few dark values surrounded by lighter ones, this is by no means a perfect checkerboard.

[1]The pattern is obtained by imposing a spatial autoregressive structure with a coefficient of 0.9 on 100 standard normal variates.

[2]The pattern is obtained by imposing a negative spatial autoregressive structure with a coefficient of −0.9 on 100 standard normal variates.

13.4 Spatial Autocorrelation Statistic

Relying on identifying spatial structure in the data by eye is bound to lead to spurious results, especially since the human brain is wired to find patterns. Instead, a spatial autocorrelation statistic is used.

A *statistic* is a summary measure computed from the data that allows one to assess the extent to which a given null hypothesis holds. Typically, this is accomplished by deriving the distribution of the statistic if the null hypothesis were true. If the observed value of the statistic is *extreme* relative to this so-called null distribution (i.e., in the extreme tails of the distribution), the null is rejected.

In classical statistics (as opposed to a Bayesian perspective), a fundamental notion in this regard is the *Type I error*, or the probability that the null is rejected, when in fact it is true. In other words, this is the chance one takes to make the wrong decision. A critical value of the Type I error is the so-called significance or *p-value*, a target cut-off point. If the probability that the value of the statistic (as computed from the data) follows the null distribution is *less* than the p-value, one typically rejects the null hypothesis. In this context, the choice of the p-value becomes crucial. Customary, this has often been set to 0.05, but recently there has been a lot of discussion around this choice in the context of replicability of scientific experiments. The current consensus is that a p-value should be much smaller than 0.05, or even be replaced by alternative concepts (Efron and Hastie, 2016; Benjamin and 72 others, 2018). The choice of a suitable p-value will be revisited in the discussion of inference for local indicators of spatial association (LISA, Chapter 16).

Spatial autocorrelation is about the coincidence of *attribute similarity* (how attribute values are alike) with *locational similarity* (how close observations are to each other). One aspect of the spatial case that is distinct from a standard correlation coefficient is that the attribute similarity pertains to the *same* variable at different locations, hence *auto* correlation. More precisely, it is a univariate measure, in contrast to the bivariate correlation coefficient.

A spatial autocorrelation statistic is then a summary computed from the data that combines these two notions. The objective is to assess the observed value of the statistic relative to the distribution of values it would take on under the null hypothesis of spatial randomness.

Formally, such a statistic can be expressed in cross-product form as:

$$SA = \sum_i \sum_j f(x_i, x_j).l(i, j),$$

where the sum is over all *pairs* of observations, x_i and x_j are observations on the variable x at locations i and j, $f(x_i, x_j)$ is a measure of attribute similarity and $l(i, j)$ is a measure of locational similarity between i and j.

Different choices for the functions f and l lead to specific spatial autocorrelation statistics. A few common options are briefly reviewed next.

13.4.1 Attribute similarity

The measure of attribute similarity $f(x_i, x_j)$ is a summary of the similarity or dissimilarity over all pairs of observations. Typical choices are the cross product, $x_i.x_j$, the squared difference $(x_i - x_j)^2$, and the absolute difference, $|x_i - x_j|$. The cross product focuses on similarity, with extreme large or extreme small values indicating greater similarity (i.e., the product of adjoining large values, or the product of adjoining small values). In contrast, the squared and absolute differences are measures of dissimilarity, with smaller values indicating more alike neighbors.

13.4.2 Locational similarity

There are two main ways to incorporate locational similarity in spatial autocorrelation statistics. One approach uses the spatial weights matrix, discussed at length in Chapters 10 through 12. The corresponding spatial autocorrelation statistics take the generic form:

$$SA = \sum_i \sum_j f(x_i, x_j).w_{ij},$$

i.e., the sum over all neighbors (i.e., pairs where $w_{ij} \neq 0$) of a measure of attribute similarity between them. This approach is most appropriate when the observations pertain to a set of discrete locations, a so-called lattice data structure.

A different perspective is taken when the observations are viewed as a sample from a continuous surface, although this approach can also be applied to discrete locations. Instead of subsuming the neighbor relations in the spatial autocorrelation statistic, a measure of attribute similarity is expressed as a function of the distance that separates pairs of observations:

$$f(x_i, x_j) = g(d_{ij}),$$

where g is a function of distance. The function can take on a specific form, as in a variogram or semi-variogram, or be left unspecified, in a non-parametric approach. This form of modeling is representative of geostatistical analysis, which concerns itself with modeling and interpolation on spatial surfaces (Isaaks and Srivastava, 1989; Cressie, 1993; Chilès and Delfiner, 1999; Stein, 1999). The discussion of this approach is deferred to Chapter 15.

13.4.3 Examples of spatial autocorrelation statistics

Each combination of a measure of attribute similarity with a spatial weights matrix yields a different spatial autocorrelation statistic.

An early example of a very generic spatial autocorrelation statistic is the Mantel test (Mantel, 1967), popularized in spatial analysis through the work of Hubert and Golledge (Hubert et al., 1981, 1985). It combines the elements of a pairwise (dis)similarity matrix **A** with a spatial weights matrix **W**, yielding the Gamma statistic:

$$\Gamma = \sum_i \sum_j \mathbf{A}_{ij}.\mathbf{W}_{ij}.$$

Some commonly used statistics can be viewed as special cases of this statistic, using a particular expression for $\mathbf{A}_{ij}$.

13.4.3.1 Cross-product

Arguably the most commonly used spatial autocorrelation statistic is Moran's I (Moran, 1948), based on a pairwise cross-product as the measure of similarity. Moran's I is typically expressed in deviations from the mean, using $z_i = x_i - \bar{x}$, with $\bar{x}$ as the mean. The full expression is:

$$I = \frac{\sum_i \sum_j z_i.z_j.w_{ij}/S_0}{\sum_i z_i^2/n}, \tag{13.1}$$

with $S_0 = \sum_i \sum_j w_{ij}$ as the sum of the weights, and the other notation as before.

This statistic is examined more closely in Section 13.5.1.

13.4.3.2 Squared difference

Whereas squared difference is most commonly used as an attribute dissimilarity measure in geostatistics, it is combined with spatial weights in Geary's c statistic (Geary, 1954):

$$c = \frac{\sum_i \sum_j (x_i - x_j)^2 w_{ij} / 2S_0}{\sum z_i^2 / (n-1)}.$$

Since the numerator consists of a squared difference, it is not necessary to take the deviations from the mean (the two means would cancel each other out in the difference operation).

In contrast to Moran's I, which takes values between roughly (not exactly) −1 and +1, Geary's c is always positive, with values between 0 and 2. The theoretical mean for Geary's c is 1. Because Geary's c is based on a dissimilarity measure, positive spatial autocorrelation corresponds with small values for the statistic (less than 1), and negative spatial autocorrelation with large values (larger than 1).

13.4.3.3 Absolute difference

Although used less frequently in practice, attribute dissimilarity can also be based on absolute difference. This yields a slightly more robust measure in the sense that the influence of outliers is lessened. An early discussion of this criterion was contained in Sokal (1979). It will not be considered further.

13.5 Moran's I

Moran's I is arguably the most used spatial autocorrelation statistic in practice, especially after its discussion in the classic works by Cliff and Ord (Cliff and Ord, 1973, 1981). In this section, I briefly review its statistical properties and introduce an approach toward its visualization by means of the Moran scatter plot (Anselin, 1996). The implementation of the Moran scatter plot in `GeoDa` is further detailed in Section 13.6.

13.5.1 Statistic

Moran's I statistic, given in Equation (13.1), is a cross-product statistic similar to a Pearson correlation coefficient. However, the latter is a bivariate statistic, whereas Moran's I is a univariate statistic. Also, the magnitude for Moran's I critically depends on the characteristics of the spatial weights matrix **W**. This has as the unfamiliar consequence that, as such, without standardization, the values of Moran's I statistics are *not directly comparable*, unless they are constructed with the same spatial weights. This contrasts with the interpretation of the Pearson correlation coefficient, where a larger value implies stronger correlation. For Moran's I, this is not necessarily the case, depending on the structure of the spatial weights involved.

The scaling factor for the denominator of the statistic is the familiar n, i.e., the total number of observations (sample size), but the scaling factor for the numerator is unusual. The latter is S_0, the sum of all the (non-zero) spatial weights. However, since the cross-product in the numerator is only *counted* when $w_{ij} \neq 0$, this turns out to be the proper adjustment. For example, dividing by n^2 would also account for all the zero cross-products, which would not be appropriate.

For row-standardized weights, in the absence of unconnected observations (isolates), the sum of weights equals the number of observations, $S_0 = n$. As a result, the scaling factors in numerator and denominator cancel out. Therefore, the statistic can be written concisely in matrix form as:

$$I = \frac{\mathbf{z}'\mathbf{W}\mathbf{z}}{\mathbf{z}'\mathbf{z}},$$

with $\mathbf{z}$ as a $n \times 1$ vector of observations as deviations from the mean.[3]

13.5.2 Inference

Inference is carried out by assessing whether the observed value for Moran's I is compatible with its distribution under the null hypothesis of spatial randomness. This can be approached from either an analytical or a computational perspective.

13.5.2.1 Analytical inference

Analytical inference starts with assuming an uncorrelated distribution, which then allows for asymptotic properties of the statistic to be derived. An exception is the result for an exact (finite sample) distribution obtained by Tiefelsdorf and Boots (1995), under fairly strict assumptions of normality. The statistic is treated as a special case of the ratio of two quadratic forms of random variables. The resulting exact distribution depends on the eigenvalues of the weights matrix.

In Cliff and Ord (1973) and Cliff and Ord (1981), the point of departure is either a Gaussian distribution or an equal probability (randomization) assumption. In each framework, the mean and variance of the statistic under the null distribution can be derived, which can then be used to construct an approximation by a normal distribution. The derivation of moments leads to some interesting and rather surprising results.

In both approaches, the theoretical mean of the statistic is:

$$E[I] = -\frac{1}{n-1}.$$

In other words, under the null hypothesis of spatial randomness, the mean of Moran's I is *not zero*, but slightly negative relative to zero. Clearly, as $n \rightarrow \infty$, the mean will approach zero.

In the case of a Gaussian assumption, the second-order moment is obtained as:

$$E[I^2] = \frac{n^2 S_1 - n S_2 + 3 S_0^2}{S_0^2 (n^2 - 1)},$$

with $S_1 = (1/2) \sum_i \sum_j (w_{ij} + w_{ji})^2$, $S_2 = \sum_i (\sum_j w_{ij} + \sum_j w_{ji})^2$ and S_0 as before.

Interestingly, the variance depends only on the elements of the spatial weights, irrespective of the characteristics of the distribution of the variable under consideration, except that the latter needs to be Gaussian. As a consequence, as long as the same spatial weights are used, the variance computed for different variables will always be identical.

[3]Expressed in this manner, the statistic takes on the same form as the familiar Durbin-Watson statistic for serial correlation in regression residuals.

This property no longer holds under the randomization assumption. Now, the second-order moment is obtained as:

$$E[I^2] = \frac{n((n^2 - 3n + 3)S_1 - nS_2 + 3S_0^2) - b_2((n^2 - n)S_1 - 2nS_2 + 6S_0^2)}{(n-1)(n-2)(n-3)S_0^2},$$

with $b_2 = m_4/m_2^2$, the ratio of the fourth moment of the variable over the second moment squared.

In other words, under randomization, the variance of Moran's I does depend on the moments of the variable under consideration, which is a more standard result.

Actual inference is then based on a normal approximation of the distribution under the null. Operationally, this is accomplished by transforming Moran's I to a *standardized* z-variable, by subtracting the theoretical mean and dividing by the theoretical standard deviation:

$$z(I) = \frac{I - E[I]}{SD[I]},$$

with $SD[I]$ as the theoretical standard deviation of Moran's I. The probability of the result is then obtained from a standard normal distribution.

Note that the approximation is asymptotic, i.e., in the limit, for an imaginary infinitely large number of observations. It may therefore not work well in small samples.[4]

13.5.2.2 Permutation inference

An alternative to the analytical approximation of the distribution is to *simulate* spatial randomness by permuting the observed values over the locations. The statistic is calculated for each of such artificial and spatially random data sets. The total collection of such statistics then forms a *reference* distribution, mimicking the null distribution.

This procedure maintains all the properties of the attribute distribution, but alters the spatial layout of the observations.

The comparison of the Moran's I statistic computed for the actual observations to the reference distribution allows for the calculation of a *pseudo p-value*. This provides a summary of the position of the statistic relative to the reference distribution. With m as the number of times the computed Moran's I from the simulated data equals or exceeds the actual value out of R permutations, the pseudo p-value is obtained as:

$$p = \frac{m+1}{R+1},$$

with the 1 accounting for the observed Moran's I. In order to obtain nicely rounded figures, R is typically taken to be a value such as 99, 999, etc.

The pseudo p-value is only a *summary* of the results from the reference distribution and should *not* be interpreted as an analytical p-value. Most importantly, it should be kept in mind that the extent of *significance* is determined in part by the number of random permutations. More precisely, a result that has a p-value of 0.01 with 99 permutations is not necessarily less significant than a result with a p-value of 0.001 with 999 permutations. In each instance, the observed Moran's I is so extreme that none of the results computed for simulated data sets exceeds its value, hence $m = 0$.

[4] An alternative alternative approximation based on the saddlepoint principle is given in Tiefelsdorf (2002), but it is seldom used in practice.

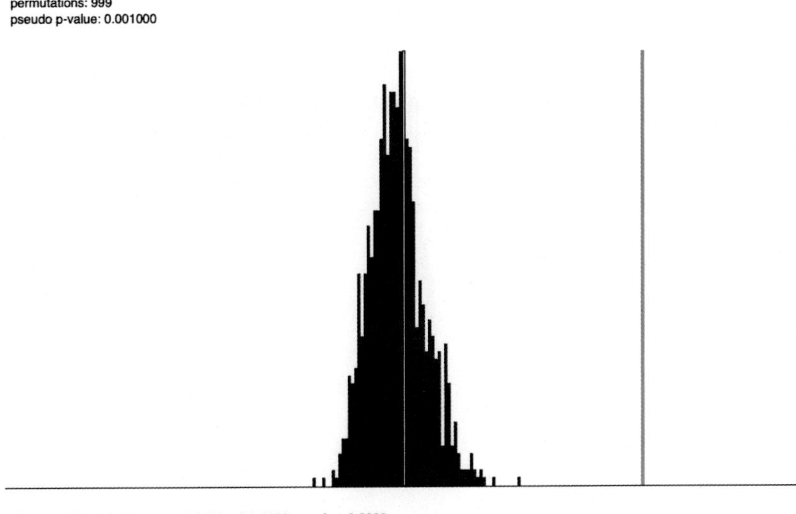

Figure 13.7: Reference distribution, Moran's I for per capita income

Figure 13.7 illustrates a reference distribution based on 999 random permutations for the Chicago community area income per capita, using queen contiguity for the spatial weights. The green bar corresponds with the Moran's I statistic of 0.6011. The dark bars form a histogram for the statistics computed from the simulated data sets. The theoretical expected value for Moran's I depends only on the number of observations. In this example, $n = 77$, such that $E[I] = -1/76 = -0.0132$. This compares to the mean of the reference distribution of -0.0125. The standard deviation from the reference distribution is 0.0696, which yields the z-value as:

$$z(I) = \frac{0.6011 + 0.0125}{0.0696} = 8.82,$$

as listed at the bottom of the graph. Since none of the simulated values exceed 0.6011, the pseudo p-value is 0.001.

In contrast, Figure 13.8 shows the reference distribution for Moran's I computed from the randomized per capita incomes from the right-hand panel in Figure 13.3. Moran's I is -0.0470, with a corresponding z-value of -0.4361, and a pseudo p-value of 0.357. In the figure, the green bar for the observed Moran's I is slightly to the left of the center of the graph. In other words, it is indistinguishable from a statistic that was computed for a simulated spatially random data set with the same attribute values.

13.5.3 Interpretation

The interpretation of Moran's I is based on a combination of its sign and its significance. More precisely, the statistic must be compared to its theoretical mean of $-1/(n-1)$. Values larger than E[I] indicate positive spatial autocorrelation, values smaller suggest negative spatial autocorrelation.

However, this interpretation is only meaningful when the statistic is also significant. Otherwise, one cannot really distinguish the result from that of a spatially random pattern, and thus the sign is meaningless.

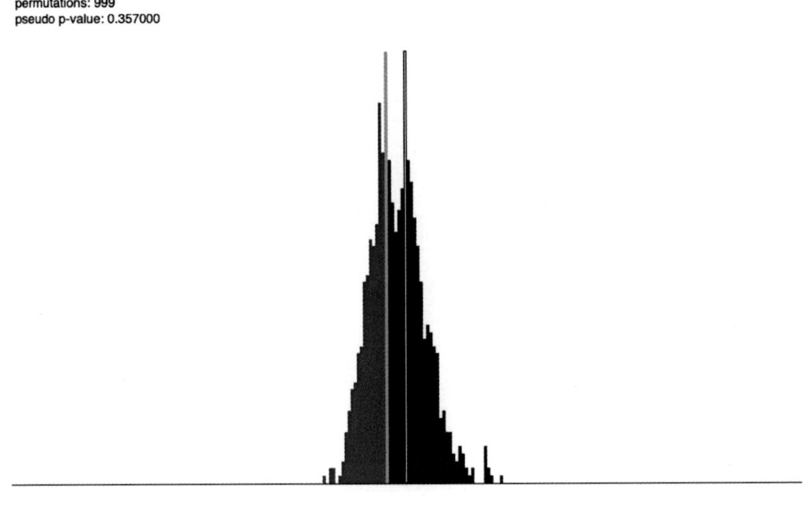

Figure 13.8: Reference distribution, Moran's I for randomized per capita income

It should be noted that Moran's I is a statistic for *global* spatial autocorrelation. More precisely, it is a single value that characterizes the complete spatial pattern. Therefore, Moran's I may indicate *clustering*, but it does *not* provide the location of the clusters. This requires a local statistic.

Also, the indication of clustering is the property of the *pattern*, but it does not suggest what *process* may have generated the pattern. In fact, different processes may result in the same pattern. Without further information, it is impossible to infer what process may have generated the pattern.

This is an example of the *inverse problem* (also called the identification problem in econometrics). In the spatial context, it boils down to the failure to distinguish between *true contagion* and *apparent contagion*. True contagion yields a clustered pattern as a result of spatial interaction, such as peer effects and diffusion. In contrast, apparent contagion provides clustering as the result of spatial heterogeneity, in the sense that different spatial structures create local similarity.

This is an important limitation of cross-sectional spatial analysis. It can only be remedied by exploiting external information, or sometimes by using both temporal and spatial dimensions.

13.5.3.1 Comparison of Moran's I results

In contrast to a Pearson correlation coefficient, the value of Moran's I is not meaningful in and of itself. Its interpretation and the comparison of results among variables depend critically on the spatial weights used. Only when the same weights are used are results among variables directly comparable. To avoid this problem, a reliable comparison should be based on the z-values.

For example, using queen contiguity, Moran's I is 0.601 for the per capita income, but 0.472 using k-nearest neighbor weights, with k=8. This would suggest that the spatial correlation is stronger for queen weights than for knn weights. However, when considering the corresponding z-values, this turns out not to be the case. For queen weights, the z-value

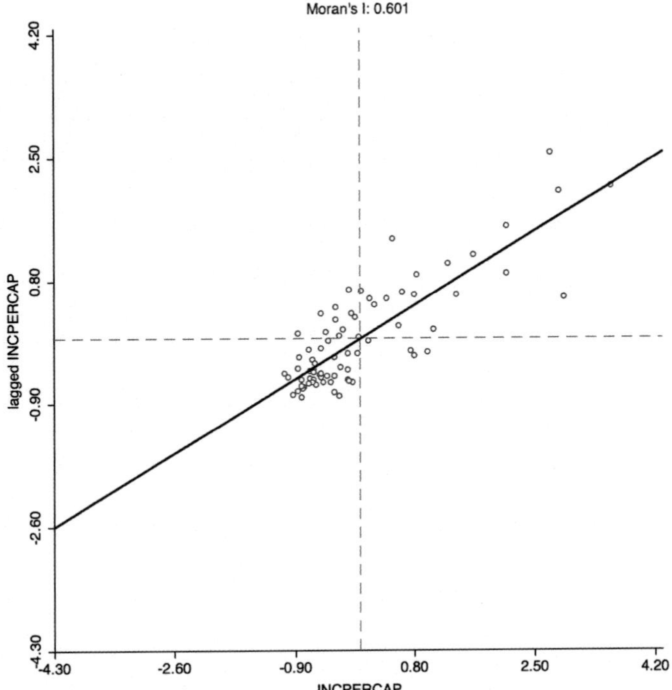

Figure 13.9: Moran scatter plot, community area per capita income

is 8.82, whereas for k-nearest neighbor weights, it is 9.46, instead suggesting the latter shows stronger spatial correlation.

Comparison of Moran's I among different variables is only valid when the same weights are used. For example, percent population change between 2020 and 2010 yields a Moran's I of 0.410 using queen contiguity, with a corresponding z-value of 6.30. In this instance, there is consistent evidence that income per capita shows a stronger pattern of spatial clustering.

13.5.4 Moran scatter plot

The Moran scatter plot, first outlined in Anselin (1996), consists of a plot with the spatially lagged variable on the y-axis and the original variable on the x-axis. The slope of the linear fit to the scatter plot equals Moran's I.

As pointed out, with row-standardized weights, the sum of all the weights (S_0) equals the number of observations (n). As a result, the expression for Moran's I simplifies to:

$$I = \frac{\sum_i \sum_j w_{ij} z_i . z_j}{\sum_i z_i^2} = \frac{\sum_i (z_i \times \sum_j w_{ij} z_j)}{\sum_i z_i^2},$$

with z in deviations from the mean.

Upon closer examination, this expression turns out to be the slope of a regression of $\sum_j w_{ij} z_j$ on z_i.[5] This becomes the principle underlying the Moran scatter plot. While the plot provides

[5]In a bivariate linear regression $y = \alpha + \beta x$, the least squares estimate for β is $\sum_i (x_i \times y_i) / \sum_i x_i^2$. In the Moran scatter plot, the role of y is taken by the spatial lag $\sum_j w_{ij} z_j$.

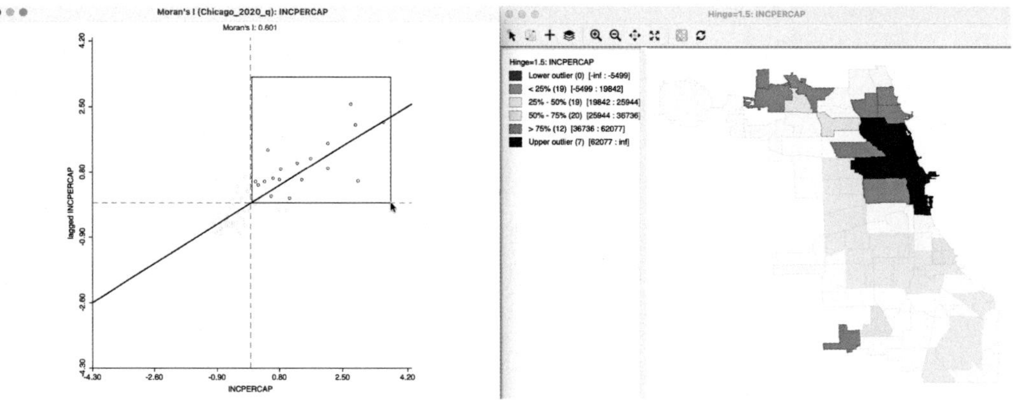

Figure 13.10: Positive spatial autocorrelation, high-high

a way to visualize the magnitude of Moran's I, inference should still be based on a permutation approach, and *not* on a traditional t-test to assess the significance of the regression line.

Figure 13.9 illustrates the Moran scatter plot for the income per capita data for Chicago community areas (based on queen contiguity). The value for income per capita is shown in standard deviational units on the horizontal axis, with mean at zero. The vertical axis corresponds to the spatial lag of these standard deviational units. While its mean is close to zero, it is not actually zero. Hence the horizontal dashed line is typically not exactly at zero. The slope of the linear fit, in this case 0.611, is listed at the top of the graph.

13.5.4.1 Categories of spatial autocorrelation

An important aspect of the visualization in the Moran scatter plot is the classification of the *nature* of spatial autocorrelation into four categories. Since the plot is centered on the mean (of zero), all points to the right of the mean have $z_i > 0$ and all points to the left have $z_i < 0$. These values can be referred to respectively as *high* and *low*, in the limited sense of higher or lower than average. Similarly, the values for the spatial lag can be classified above and below the mean as *high* and *low*.

The scatter plot is then easily decomposed into four quadrants. The upper-right quadrant and the lower-left quadrant correspond to *positive* spatial autocorrelation (similar values at neighboring locations). They are referred to as respectively *high-high* and *low-low* spatial autocorrelation. In contrast, the lower-right and upper-left quadrant correspond to *negative* spatial autocorrelation (dissimilar values at neighboring locations). Those are referred to as respectively *high-low* and *low-high* spatial autocorrelation.

The classification of the spatial autocorrelation into four types begins to make the connection between *global* and *local* spatial autocorrelation. However, it is important to keep in mind that the classification as such does not imply significance. This is further explored in the discussion of local indicators of spatial association in Chapter 16.

Figure 13.10 illustrates the high-high quadrant for the income per capita Moran scatter plot. The points selected in the left-hand graph are linked to their locations in the map on the right. Clearly, these tend to be locations in the upper quartile (as well as upper outliers) surrounded by locations with a similar classification.

Negative spatial autocorrelation of the low-high type is illustrated in Figure 13.11. The community area selected in the map on the right is in the second quartile, surrounded by

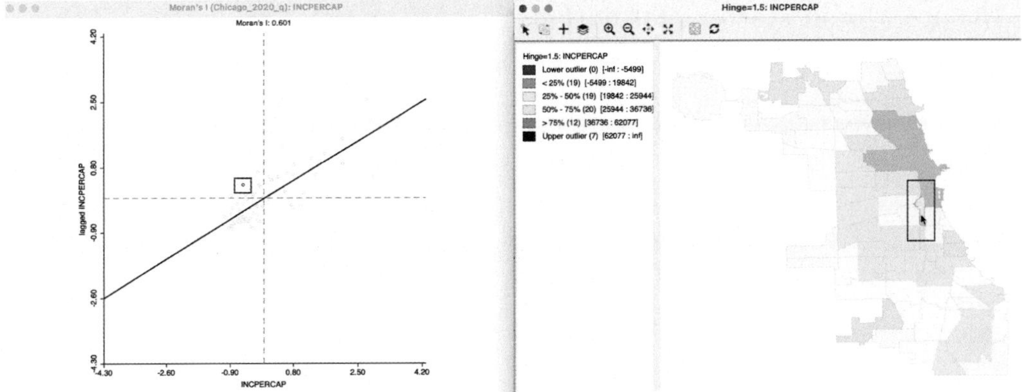

Figure 13.11: Negative spatial autocorrelation, low-high

neighbors that are mostly of a higher quartile. One neighbor in particular belongs to the upper outliers, which will pull up the value for the spatial lag. The corresponding selection in the scatter plot on the left is indeed in the low-high quadrant.

One issue to keep in mind when assessing the relative position of the value at a location relative to its neighbors is that a box map such as Figure 13.3 reduces the distribution to six discrete categories. These are based on quartiles, but the range between the lowest and highest observation in a given category can be quite large. In other words, it is not always intuitive to assess the high-low or low-high relationship from the categories in the map.

13.6 Visualizing spatial autocorrelation with the Moran scatter plot

In `GeoDa`, a Moran scatter plot is created from the menu as **Space > Univariate Moran's I**. Alternatively, it can be invoked through the left most icon in the spatial subset of the toolbar, highlighted in Figure 13.1, followed by selecting **Univariate Moran's I** from the drop-down list.

13.6.1 Creating a Moran scatter plot

The first requirement for the Moran scatter plot is to select a variable, through the familiar **Variable Settings** interface. The community area income per capita variable is **INCPERCAP**. The **Weights** need to be chosen from the matching drop-down list.

If the **Weights Manager** does not contain at least one spatial weight, then one must be created or loaded. The example uses first-order queen contiguity, as the file *Chicago_2020_q.gal*. Selecting **OK** brings up the Moran scatter plot shown in Figure 13.9. This graph has the same features as the standard scatter plot (Section 7.3), except that no statistics are provided at the bottom. Also, in contrast to what holds for the standard scatter plot, the **Regimes Regression** option is *not* active.

13.6.2 Moran scatter plot options

In the usual way, a right click on the graph brings up a list of available options. These are largely identical to those for a standard scatter plot (Section 7.3.1), with a few exceptions. Since the variables in the Moran scatter plot are standardized by design, there is no **Data** option.

One additional feature is to **Save Results**. This saves two new variables to the data table, one with the standardized variable and one with its spatial lag. The default variable names for these are, respectively, **MORAN_STD** and **MORAN_LAG**.

13.6.2.1 Randomization

As such, the slope of the linear fit only provides an estimate of Moran's I, but it does not reveal any information about the significance of the test statistic. The latter is obtained by means of the **Randomization** option, which implements a permutation approach. Four hard-coded choices are provided for the number of permutations: **99**, **199**, **499** and **999**. In addition, there is an **Other** option, which allows for any value up to **99,999**. As a consequence, the *smallest* result that can be obtained for the pseudo p-value is **0.00001**. This implies that out of 99,999 randomly generated data sets, none yielded a computed Moran's I that equals or exceeds the one obtained from the actual data.

The result for a choice of **999** is shown in Figure 13.7, with an associated pseudo p-value of **0.001** for the spatial distribution of the Chicago community area per capita income data. The graph provides strong evidence of spatial clustering of similar values.

As detailed in Section 13.5.2.2, the reference distribution is accompanied by several descriptive statistics, such as the theoretical (E[I]) and empirical means, the standard deviation and the associated z-value.

The **Run** button allows for repeated simulations in order to assess the sensitivity of the result to the particular run of random numbers that is used. For a large number of permutations such as 999, one run is typically sufficient.

13.6.2.2 Reproducing results - the random seed

In order to facilitate replication, the default setting in `GeoDa` is to use a specific seed for the random number generator. This is evidenced by the check mark next to **Use Specified Seed** in the **Randomization** options. The default seed value used is shown when selecting the **Specify Seed ...** box, where it can also be changed.

The random seed can also be set globally in the **Preferences**, under the **System** tab, selected under the **GeoDa** item on the menu. At the bottom of the dialog, under the item **Method:**, the box for using the specified seed is checked by default, and the value **123456789** is used as the default seed. This can be changed by typing in a different value.

The same random seed is used in all operations in `GeoDa` that rely on some type of random permutations, such as all flavors of the Moran scatter plot, as well as all the local spatial autocorrelation statistics. This ensures that the results will be identical for each analysis that uses the same sequence of random numbers.

However, due to the way the random number generator is implemented in each operating system, there may be slight differences in the results depending on the hardware, especially

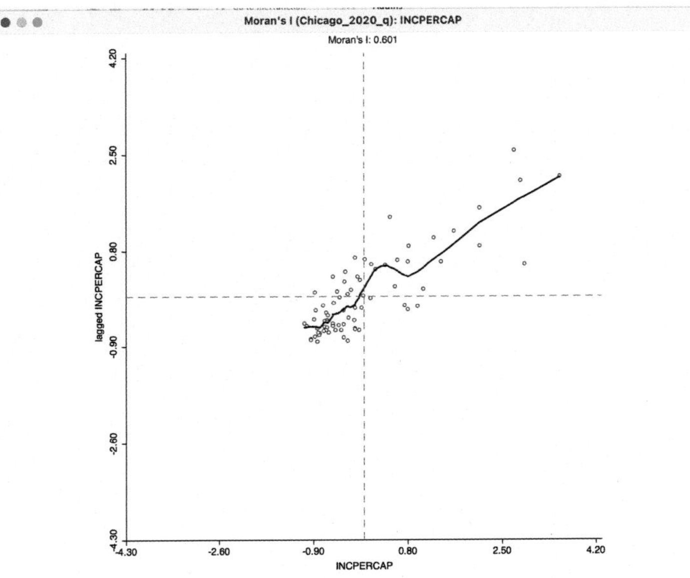

Figure 13.12: Moran scatter plot LOWESS smoother

when a different number of CPU cores is used.[6] In order to control for any such discrepancies, it is possible to **Set the number of CPU cores manually** in the **Preferences**.

13.6.2.3 LOWESS smoother

For the Moran scatter plot, the default is a **Linear Smoother**, but just as for the standard scatter plot, it is also possible to use a nonlinear smoother. The **View > LOWESS Smoother** works in exactly the same way as for the standard scatter plot.

With the **View > Linear Smoother** option turned off, only the nonlinear fit from the local regression is shown on the Moran scatter plot, as in Figure 13.12. The LOWESS smoother can be explored to identify potential structural breaks in the pattern of spatial autocorrelation. For example, in some parts of the data set, the curve may be very steep and positive, suggesting strong positive spatial autocorrelation, whereas in other parts, it could be flat, indicating no autocorrelation.

The local smoother is fit through the points in the Moran scatter plot by utilizing only those observations within a given bandwidth. As for the standard scatter plot, the default is to use 0.20 of the range of values. This can be specified by selecting the **View > Edit LOWESS Parameters** item in the options menu, in the usual way.

Using the default bandwidth, the nonlinear smoother in Figure 13.12 suggests some instability in the global spatial autocorrelation for values of z between 0 and 0.80, where the generally positive slope turns into a negative one for a brief interval. Through linking (and brushing), the corresponding observations can be identified on the associated map.

[6]The implementation of the random numbers in `GeoDa` takes advantage of multi-threading, a built-in form of parallel computing.

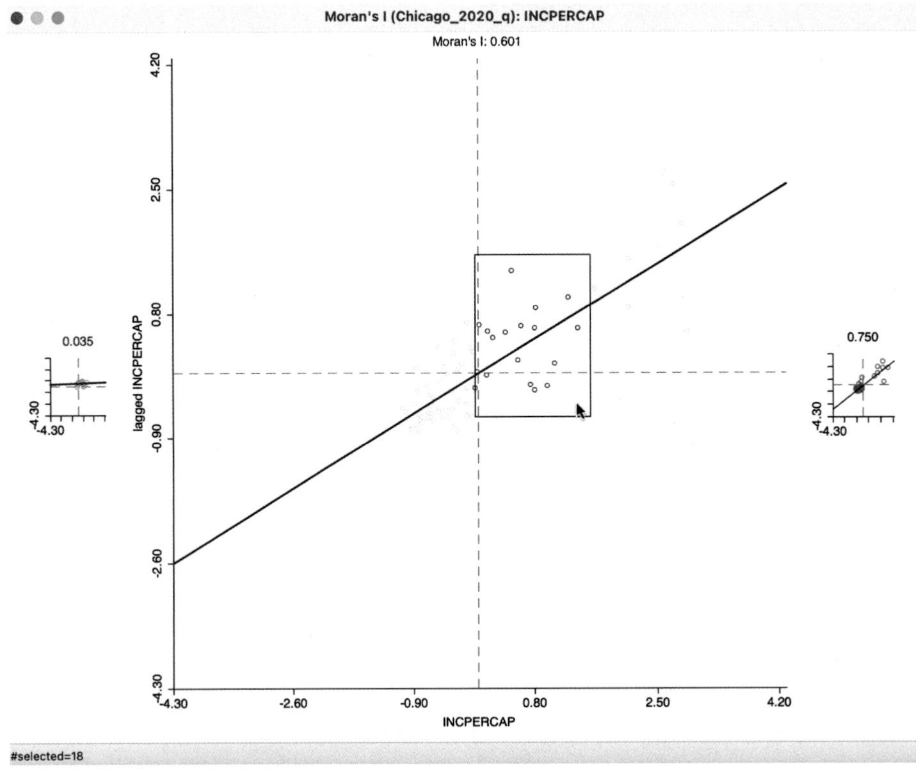

Figure 13.13: Brushing the Moran scatter plot

13.6.3 Brushing the Moran scatter plot

An important option to assess the stability of the spatial autocorrelation throughout the data set, or, alternatively, to search for potential spatial heterogeneity, is the exploration of the scatter plot through dynamic brushing.

This is invoked by selecting **Regimes Regression** in the **View** option. As in the standard scatter plot, with this option selected, statistics for the slope and intercept are recomputed as observations are selected and unselected. However, this is implemented somewhat differently for the Moran scatter plot.

As soon as this option is activated, three Moran scatter plots are shown, as in Figure 13.13. The one to the left (in red) is for the **selected** observations. The one to the right is for the complement, referred to as **unselected** (in black).

In the figure, 18 observations from the high-high and high-low quadrants have been selected, as indicated in the status bar. These yield a Moran's I of 0.035, illustrated by the plot on the left. The Moran's I for the complement is 0.750. While there is no significance associated with these measures, they suggest the presence of spatial heterogeneity.

The distinguishing characteristics of the selection in the Moran scatter plot (and any associated brushing) is that the spatial weights are *dynamically updated* for both the selection and the unselected observations. More precisely, for each of the subset of observations, selected and unselected, the spatial weights are adjusted to remove links to observations belonging to the other subset. In other words, there are no edge effects and each subset is treated as a self-contained entity.

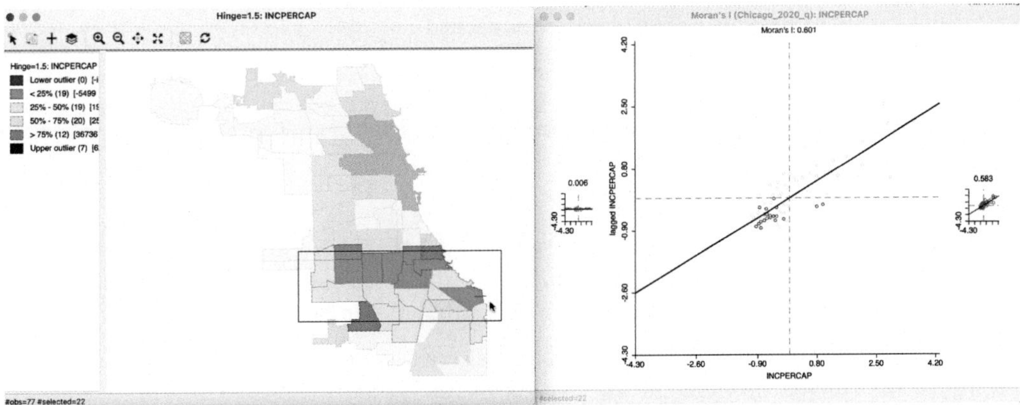

Figure 13.14: Brushing and linking the Moran scatter plot

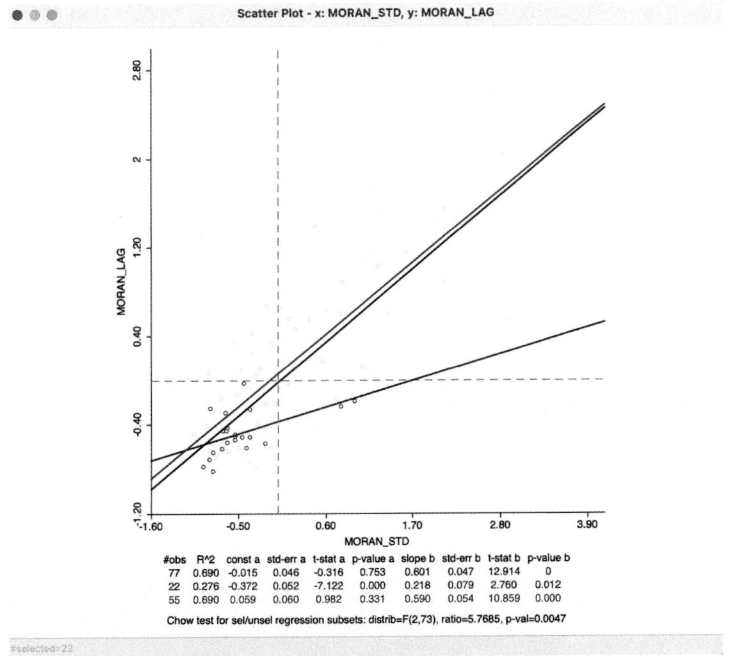

Figure 13.15: Static spatial weights in Moran scatter plot

As a result, the slope of the Moran scatter plots on the left and the right are as if the data set consisted only of the selected/unselected points. This is a visual implementation of a *regionalized* Moran's I, where the indicator of spatial autocorrelation is calculated for a subset of the observations (Munasinghe and Morris, 1996). As the *brush* is moved over the graph, the selected and unselected scatter plots are updated in real time, showing the corresponding regional Moran's I statistics in the panels.

An alternative perspective is to approach the brushing starting with the map, as in Figure 13.14. In the left panel, 22 community areas are selected, with the associated observations highlighted in the Moran scatter plot to the right. The Moran's I corresponding to the selection is 0.006, with 0.583 for the complement, again suggesting somewhat of a structural break.

13.6.3.1 Static and dynamic spatial weights

The behavior of the Moran scatter plot with **Regimes Regression** turned on is to have the spatial weights updated dynamically, in order to remove any edge effects. This contrasts with the treatment in a conventional scatter plot constructed with the standardized variable and its spatial lag.

In Figure 13.15, the two variables are used that were formed by the **Save Results** option, i.e., **MORAN_STD** on the horizontal axis and **MORAN_LAG** on the vertical axis. Since this is a standard scatter plot, the **Regimes Regression** option is on by default, resulting in the statistics for selected and unselected at the bottom.

The graph is for the same 22 observations as selected in Figure 13.14. However, in contrast to that case, there is no dynamic adjustment of the spatial weights.

More precisely, this means that the spatial lag for a selected/unselected observation is computed from all its original neighbors, including the ones that belong to the other subset. In the example, the corresponding slopes are, respectively, 0.218 and 0.590, compared to 0.006 and 0.583 for the dynamic weights. The associated Chow test shows strong evidence of structural instability.

However, this result needs to be interpreted with caution, since the weights structure has not been adjusted to reflect the actual *breaks* within the data set. Specifically, the areas at the edge of the selection are still connected to their neighbors, whereas this is not the case in the dynamic weights approach. The standard scatter plot *ignores* boundary effects, in the sense that it includes neighbors *outside* the regimes. In some contexts, this may be the interpretation sought, whereas in others the dynamic weights adjustment is more appropriate.

14

Advanced Global Spatial Autocorrelation

In this chapter, I continue with the treatment of global spatial autocorrelation. First, I consider some special applications of the Moran scatter plot principle to time differenced variables and to proportions. These cases provide some short cuts to the computations needed to construct the graph, but otherwise are identical to the standard plot covered in Chapter 13.

Next, I cover the topic of bivariate spatial correlation, a common source of confusion in practice. After presenting the principle behind a bivariate Moran scatter plot, attention focuses on the special bivariate situation where the observation on a variable at one point in time is related to its neighbors in a different time period, i.e., space-time correlation. A critical review is provided of different perspectives and their relative merits and drawbacks.

To illustrate these methods, I go back to the *Oaxaca Development* sample data set.

14.1 Topics Covered

- Visualize spatial correlation among time-differenced values
- Correct Moran's I for the variance instability in rates
- Visualize bivariate spatial correlation
- Assess different aspects of bivariate spatial correlation
- Alternative computations of space-time autocorrelation

GeoDa Functions

- Space > Differential Moran's I
  - saving the first differences and spatial lag
- Space > Moran's I with EB Rate
  - saving the rate and spatial lag
- Space > Bivariate Moran's I
  - variable selection
  - permutation inference
  - saving the spatial lag

14.2 Specialized Moran Scatter Plots

In this section, two special applications of the Moran scatter plot principle are considered. Each pertains to a different type of situation, both commonly encountered in practice.

DOI: 10.1201/9781003274919-14

In the first example, the data for the same variable are available at different points in time. Rather than having to explicitly compute change values between two points in time, the *differential Moran scatter plot* provides a short cut, in the sense that the differences are computed under the hood, as part of the scatter plot construction.

The second situation is the familiar case where the variable of interest is a rate or proportion. As discussed at length in Chapter 6, the inherent variance instability of such data is something that needs to be accounted for. This is especially important in the context of spatial autocorrelation, since an important requirement for such statistics to be valid, like Moran's I, is to have a constant variance. The *Morans' I with EB Rate* implements the necessary standardization.

14.2.1 Differential Moran scatter plot

The differential Moran scatter plot is a short cut to compute Moran's I for the *first difference* of a variable measured at two points in time. Specifically, for a variable y observed for each location i at times t and $t-1$, the spatial autocorrelation coefficient is computed for the difference $y_{i,t} - y_{i,t-1}$.

The difference operation takes care of any temporal correlation that may be the result of a *spatial fixed effect*, i.e., one or more variables determining the value of y_i that remain unchanged over time. Examples of such fixed effects are locational advantages (e.g., a port city), or regulatory differences (state income tax vs. no income tax).

More formally, with μ_i as the spatial fixed effect associated with location i, the value at each location for times t and $t-1$ can be expressed as the sum of some intrinsic value u and the fixed effect:

$$y_{i,t} = u_{i,t} + \mu_i,$$

and

$$y_{i,t-1} = u_{i,t-1} + \mu_i,$$

where μ_i is unchanged over time (hence, fixed).

The presence of μ_i in both time periods induces a *temporal* correlation between $y_{i,t}$ and $y_{i,t-1}$, above and beyond the intrinsic correlation between $u_{i,t}$ and $u_{i,t-1}$. Even when $u_{i,t}$ and $u_{i,t-1}$ are uncorrelated, the correlation between $y_{i,t}$ and $y_{i,t-1}$ would be $\mathrm{E}[\mu_i^2]$, or the variance of the fixed effect.

Taking the first difference eliminates the fixed effect and ensures that any remaining correlation is solely due to u:

$$y_{i,t} - y_{i,t-1} = u_{i,t} + \mu_i - u_{i,t-1} - \mu_i = u_{i,t} - u_{i,t-1}.$$

A differential Moran's I is then the slope in a regression of the spatial lag of the difference, i.e., $\sum_j w_{ij}(y_{j,t} - y_{j,t-1})$ on the difference $(y_{i,t} - y_{i,t-1})$.

In `GeoDa`, the slope computation is applied to the standardized value of the difference (i.e., the standardization of $y_{i,t} - y_{i,t-1}$), and not to the difference between the standardized values.

14.2.1.1 Implementation

The differential Moran scatter plot functionality is the third item in the drop-down list activated by the Moran scatter plot toolbar icon (Figure 13.1). Alternatively, it can be started from the main menu as **Space > Differential Moran's I**.

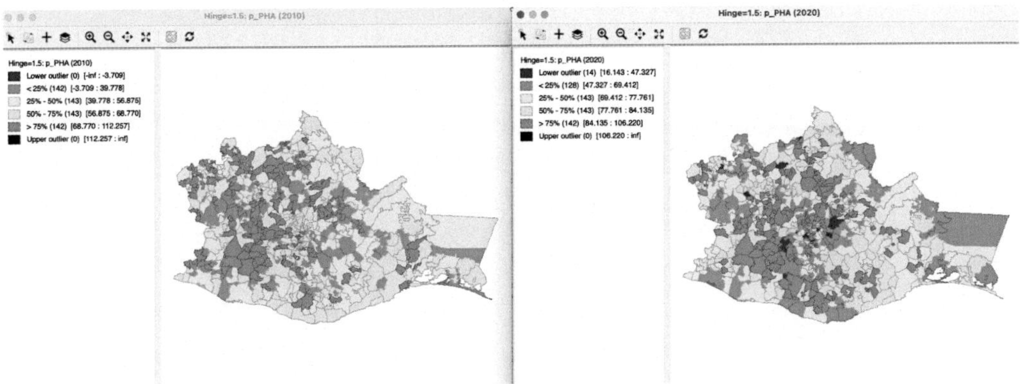

Figure 14.1: Access to health care in 2010 and 2020

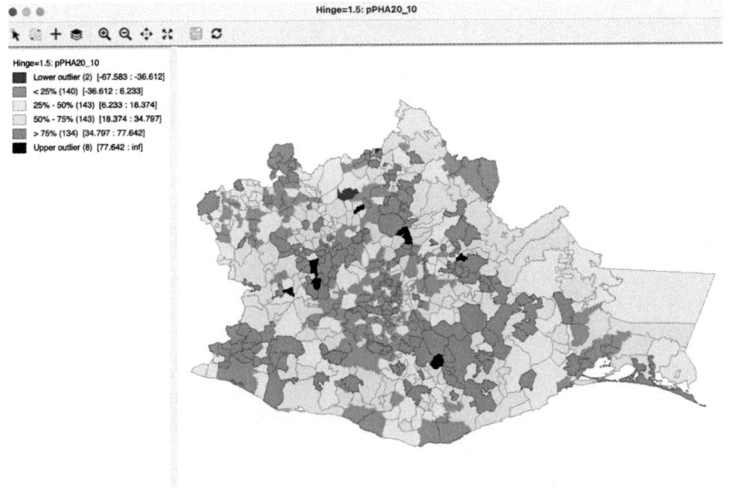

Figure 14.2: Change in access to health care between 2010 and 2020

Note that this option is only available for data sets with *grouped* variables (see Section 9.2.1). Without grouped variables, a warning message is generated.

The functionality is illustrated with a time enabled version of the *Oaxaca Development* sample data set (see Chapter 9). The relevant variables include **p_P614NS**, **p_JOB** and **p_PHA**, among others.

Before proceeding with the actual differential Moran graph, Figure 14.1 illustrates the spatial pattern of access to health care in 2010 and 2020, **p_PHA**, as box maps.[1] The classification is relative and therefore does not reflect that the values for access to health care have improved rather dramatically between these two years: for example, the mean changed from 53.4% to 75.8%. The spatial distribution shows some differences, especially in the center of the map, where several municipalities moved from a higher quartile (brown) to a lower quarter (blue), including 14 lower outliers in 2020.

[1]Note that the item in the variable selection list is initially given as **p_PHA (all times)**. After selection, this becomes, respectively, **p_PHA (2010)** and **p_PHA (2020)**.

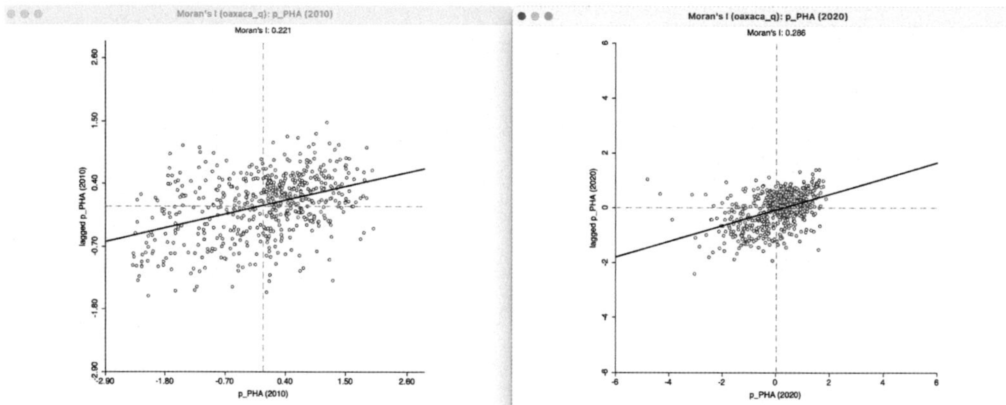

Figure 14.3: Moran scatter plot for access to health care in 2010 and 2020

A box map for the first difference between 2020 and 2010 (computed in the table) is shown as Figure 14.2. In contrast to the map for the individual years, the first differences include both lower (2) and upper outliers (8), i.e., areas that improved much less or much more than others, relatively speaking. It is for this spatial pattern that the differential Moran's I is computed.

To provide some context, the individual Moran scatter plots for **p_PHA** in 2010 and 2020 are shown in Figure 14.3, computed using queen contiguity weights (*Oaxaca_q.gal*). Both Moran's I statistics are positive and highly significant, suggesting strong clustering. For 2010, the statistic is 0.221, with an associated z-value of 8.371, for 2020, it is 0.286, with an associated z-value of 11.058. Using 999 permutations, both cases yield a pseudo p-value of 0.001.

After starting the differential Moran's I functionality, a **Differential Moran Variable Settings** interface provides the means to select the variable and time interval. The drop-down list by **Select variable** contains only the *grouped* variables in the data set, such as **p_PHA** for the example. Next, **two time periods** need to be selected. The values for the time periods are determined by their definition in the **Time Editor** (see Section 9.2.1). For the illustration, the years **2020** and **2010** are selected (in this order). Finally, the **Weights** drop-down list contains the available spatial weights, here using **oaxaca_q**.

The resulting differential Moran scatter plot is shown in Figure 14.4. Moran's I is 0.279, with an associated z-value of 10.611 (from a randomization), again suggesting a strong pattern of clustering. In other words, not only is there clustering in the values of the percent health access, there is also clustering in their change over the ten year period. However, the *global* Moran's I provides no information as to *where* the clustering may happen, nor whether it is driven by high or by low values. This is a common misinterpretation of the global Moran's I statistic.

The differential Moran scatter plot has the same options as the regular Moran scatter plot, with one exception. The **Save Results** option provides a way to not only save the **Standardized Data** (as **MORAN_STD**) and the **Spatial Lag** (as **MORAN_LAG**), as in the regular case, but also **Diff Values** (as **DIFF_VAL**), the value of the difference before standardization. Clearly, applying a regular Moran scatter plot to this new variable provides the same result as the differential Moran scatter plot. The latter simply saves the effort of first having to construct the differences.

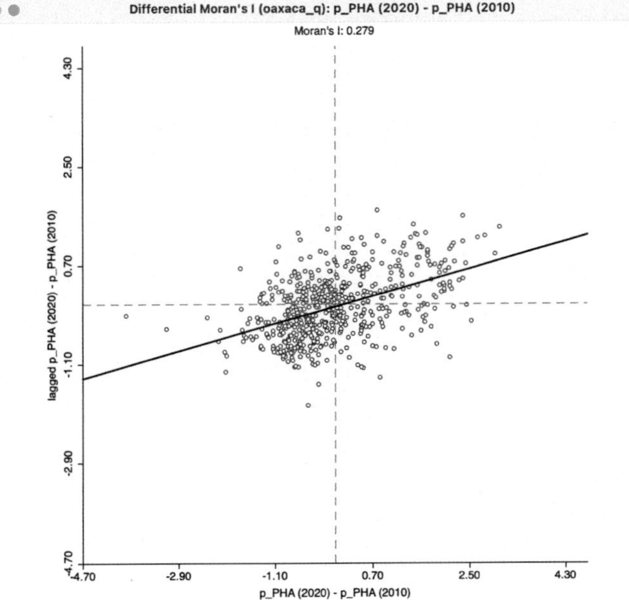

Figure 14.4: Differential Moran scatter plot for access to health care 2020–2010

14.2.2 Moran scatter plot for EB rates

In Chapter 6, the issue of the intrinsic variance instability associated with rates or proportions was introduced. The proposed solution was to utilize the Bayesian idea of *borrowing strength* through various smoothing operations, including an Empirical Bayes (EB) procedure (Section 6.4.2.2).

One may be tempted to carry out spatial autocorrelation tests on EB smoothed rates, but this is not appropriate, since their construction already induces a degree of correlation. Instead, Assuncao and Reis (1999) suggested a slightly different procedure to correct Moran's I spatial autocorrelation test statistic for varying population densities across observational units, when the variable of interest is a rate or proportion. In this approach, the spatial autocorrelation is not computed for a smoothed version of the original rate, but for a *transformed standardized* random variable. In other words, the crude rate is turned into a new variable that has a constant variance. The mean and variance used in the transformation are computed for each individual observation.

The principle is similar to the *shrinkage estimator* presented in Section 6.4.2.2, in that it adjusts the crude rate with a mean and variance obtained from a prior distribution. As in the case of EB smoothing, these moments are computed from the data, hence the designation as *empirical* Bayes.

As before, the point of departure is the crude rate, $r_i = O_i/P_i$, where O_i is the count of events at location i and P_i is the corresponding population at risk.

The rationale behind the Assunção-Reis approach is to standardize each r_i as

$$z_i = \frac{r_i - \beta}{\sqrt{\alpha + (\beta/P_i)}},$$

using an estimated mean β and standard deviation $\sqrt{\alpha + \beta/P_i}$. The parameters α and β are related to the prior distribution for the risk, similar to the treatment in Section 6.4.2.2.

In practice, the prior parameters are estimated by means of the so-called method of moments (e.g., Marshall, 1991), yielding the following expressions:

$$\beta = O/P,$$

with O as the total event count ($\sum_i O_i$), and P as the total population count ($\sum_i P_i$), and

$$\alpha = [\sum_i P_i(r_i - \beta)^2]/P - \beta/(P/n),$$

with n as the total number of observations (in other words, P/n is the average population).

One problem with the method of moments estimator is that the expression for α could yield a negative result. In that case, its value is typically set to zero, i.e., $\alpha = 0$. However, in Assuncao and Reis (1999), the value for α is only set to zero when the resulting estimate for the variance is negative, that is, when $\alpha + \beta/P_i < 0$. Slight differences in the standardized variates may result depending on the convention used. In `GeoDa`, when the variance estimate is negative, the original crude rate is used.

Also, after the EB adjustment, the rates are further standardized to have a mean of zero and a variance of one, in the usual fashion for a Moran scatter plot.

14.2.2.1 Implementation

The Moran scatter plot for standardized rates is invoked from the main menu as **Space > Moran's I with EB Rate**, or as the fourth item in the drop-down list from the Moran scatter plot toolbar icon, Figure 13.1.

To illustrate this procedure, the same disability rate example for Oaxaca municipalities is used as in Section 12.4.0.1, i.e., the ratio of the number of disabled persons (**DIS20**) over the municipal population (**PTOT20**) in 2020. The crude rate and EB smoothed rate are mapped in Figure 12.9.

The left panel of Figure 14.5 shows the Moran scatter plot obtained for the crude rate, again using queen contiguity for the spatial weights. The Moran statistic is 0.2649, with an associated z-value (from 999 permutations) of 10.354.

The right hand panel is for the EB Moran scatter plot. This is obtained through the same variable selection interface as for spatially smoothed rates in Section 12.4.1.1, with a variable for the numerator (**DIS20**) and for the denominator (**PTOT20**), as well as the spatial weights (**oaxaca_q**).

The result for the EB Moran's I is only slightly different from the crude rate, yielding a statistic of 0.2680, with associated z-value of 10.452. However, the values along the horizontal axis show the effect of the standardization, with a much smaller range than for the crude rate.

In practice, the differences between Moran's I for the crude rate and EB standardized rate tend to be minor.

The EB standardized Moran scatter plot has all the same options as the conventional Moran scatter plot, except again for the **Save Results** feature. In addition to the **Standardized Data** (the values used in the calculation of the Moran scatter plot), and their **Spatial Lag**, the **EB Rates** (i.e., before the standardization used in the Moran computation) can be saved, with default variable name **MORAN_EB**.

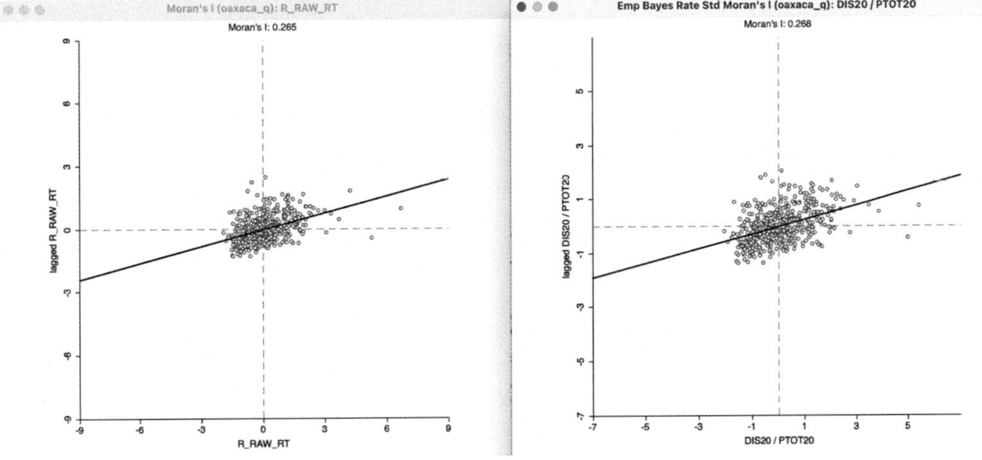

Figure 14.5: Moran scatter plot for raw rate and EB Moran scatter plot

14.3 Bivariate Spatial Correlation

The concept of bivariate spatial correlation is complex and often misinterpreted. It is typically considered to be the correlation between one variable and the spatial lag of another variable, as originally implemented in the precursor of `GeoDa` (described in Anselin et al., 2002). However, this formulation does not necessarily take into account the inherent correlation between the two variables. More precisely, in a bivariate Moran's I, the correlation is between x_i at location i and its neighbors $\sum_j w_{ij} y_j$, where x and y are different variables. As formulated, this concept does not take into account the correlation between x_i and y_i, i.e., between the two variables at the same location.

As a result, a bivariate Moran's I statistic is often interpreted incorrectly, as it may overestimate the spatial aspect of the correlation that instead may be due mostly to the in-place correlation.[2]

14.3.1 Bivariate Moran scatter plot

In its initial conceptualization in Anselin et al. (2002), a bivariate Moran scatter plot extends the idea of a Moran scatter plot with a variable on the horizontal axis and its spatial lag on the vertical axis to a bivariate context. The fundamental difference is that in the bivariate case the spatial lag pertains to a different variable. In essence, this notion of bivariate spatial correlation measures the degree to which the value for a given variable at a location is correlated with its neighbors for a *different variable.*

As in the univariate Moran scatter plot, the interest is in the slope of the linear fit to the scatter plot. This yields a Moran's I-like statistic as:

$$I_B = \frac{\sum_i (\sum_j w_{ij} y_j \times x_i)}{\sum_i x_i^2},$$

or, the slope of a regression of Wy on x.

[2]Based on some strong assumptions, Lee (2001) provides an alternative that considers a separation between the correlative aspect and the spatial aspect. This is not pursued further.

As before, all variables are expressed in standardized form, such that their means are zero and their variance one. In addition, the spatial weights are row-standardized.

Note that, unlike in the univariate autocorrelation case, the regression of x on Wy also yields an unbiased estimate of the slope, providing an alternative perspective on bivariate spatial correlation (see Section 14.4). In the case of the regression of x on Wx, the explanatory variable Wx is *endogenous*, so that the ordinary least squares estimation of the linear fit is *biased*. However, with Wy referring to a different variable, the so-called simultaneous equation bias becomes a non-issue, and OLS in a regression of x on Wy has all the standard properties (such as unbiasedness).

A special case of bivariate spatial autocorrelation is when the variable is measured at two points in time, say $z_{i,t}$ and $z_{i,t-1}$, as in Section 14.2.1. The statistic then pertains to the extent to which the value observed at a location at a given time is correlated with its value at neighboring locations at a different point in time.

The natural interpretation of this concept is to relate $z_{i,t}$ to $\sum_j w_{ij} z_{j,t-1}$, i.e., the correlation between a value at t and its neighbors in a previous time period:

$$I_T = \frac{\sum_i (\sum_j w_{ij} z_{j,t-1} \times z_{i,t})}{\sum_i z_{i,t}^2},$$

which expresses the effect of neighbors in $t-1$ on the present value.

Alternatively, and maybe less intuitively, one can relate the value at a previous time period z_{t-1} to its neighbors in the future, $\sum_j w_{ij} z_t$, as:

$$I_T = \frac{\sum_i (\sum_j w_{ij} z_{j,t} \times z_{i,t-1})}{\sum_i z_{i,t-1}^2},$$

or the effect of a location at $t-1$ on its neighbors in the future.

While formally correct, this may not be a proper interpretation of the dynamics involved. In fact, the notion of spatial correlation pertains to the effect of neighbors on a central location, not the other way around. While the Moran scatter plot seems to reverse this logic, this is purely a formalism, without any consequences in the univariate case. However, when relating the slope in the scatter plot to the dynamics of a process, this representation should be interpreted with caution.

A possible source of confusion is that the proper *regression* specification for a dynamic process would be as:

$$z_{i,t} = \beta_1 \sum_j w_{ij} z_{j,t-1} + u_i,$$

with u_i as the usual error term, and *not* as:

$$z_{i,t-1} = \beta_2 \sum_j w_{ij} z_{j,t} + u_i,$$

which would have the future *predicting the past*. This contrasts with the linear regression specification used (purely formally) to estimate the bivariate Moran's I, for example:

$$\sum_j w_{ij} z_{j,t-1} = \beta_3 z_{i,t} + u_i.$$

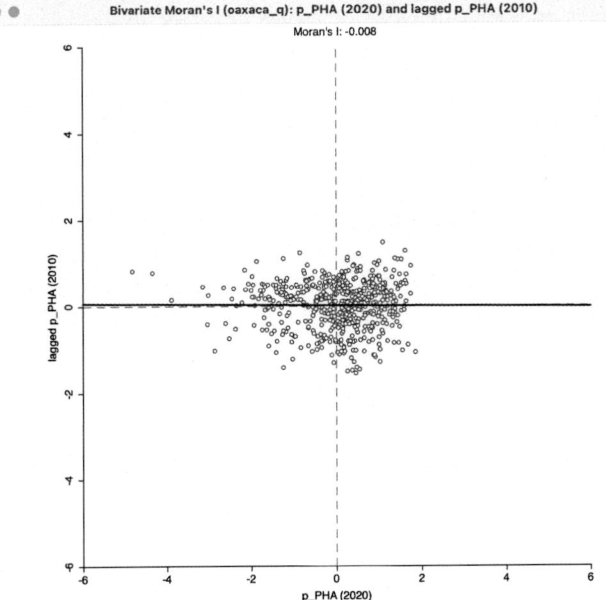

Figure 14.6: Bivariate Moran scatter plot for access to health care 2020 and its spatial lag in 2010

In terms of the interpretation of a dynamic process, only β_1 has intuitive appeal. However, in terms of measuring the degree of *spatial* correlation between past neighbors and a current value, as measured by the linear fit in a Moran scatter plot, β_3 is the correct interpretation.

In the univariate case, only the specification with the spatially lagged variable on the left hand side yields a valid estimate. As a result, for the univariate Moran's I, there is no ambiguity about which variables should be on the x-axis and y-axis. In contrast, in the bivariate case, both options are valid, although with a different interpretation.

Inference is again based on a permutation approach, but with an important difference. Since the interest focuses on the bivariate spatial association, the values for x and y are fixed at their locations, and only the remaining values for y are randomly permuted. In the usual manner, this yields a reference distribution for the statistic under the null hypothesis that the spatial arrangement of the remaining y values is random. It is important to keep in mind that since the focus is on the correlation between the x value at i and the y values at the neighboring locations, the correlation between x and y at location i is ignored.

14.3.1.1 Creating a bivariate Moran scatter plot

A bivariate Moran scatter plot is created as the second item in the drop-down list activated by the Moran scatter plot toolbar icon (Figure 13.1). Alternatively, it can be started from the main menu as **Space > Bivariate Moran's I**.

The variables and spatial weights are selected through the **Bivariate Moran Variable Settings** dialog. This allows for the selection of the **First Variable (X)**, the **Second Variable (Y)** and the **Weights** from corresponding drop-down lists. Since the version of the *Oaxaca Development* data set used here is time enabled, there are also drop-down lists for the **Time** period.

To illustrate this functionality, the same variables are used as in Section 14.2.1, again with queen contiguity spatial weights (*oaxaca_q*). For the bivariate Moran's I, the x-variable is then **p_PHA (2020)** and the y-variable its spatial lag in 2010, **W_pPHA (2010)**.

The resulting graph is shown in Figure 14.6. The linear fit is almost horizontal, reflected by a Moran's I of −0.008. Randomization, using 999 permutations, yields a pseudo p-value of 0.363, clearly not significant. This result may seem surprising, given the strong spatial autocorrelation found for **p_PHA** in each of the individual years (Figure 14.3). One possible explanation for this finding is that whereas there is strong evidence of clustering in each year, the *location* of the clusters may be different. Therefore, the relationship between the value at a given location and its neighbors in a different year may be weak, as is the case in this example.

All the same options as for the univariate Moran scatter plot apply here as well. Also, the standardized value of the x variable and the spatial lag of the y-variable (applied to its standardized value) can be saved to the table, in the same way as for the univariate Moran scatter plot.

14.4 The Many Faces of Space-Time Correlation

In contrast to the univariate Moran scatter plot, where the interpretation of the linear fit is unequivocally Moran's I, there is no such clarity in the bivariate case. In addition to the interpretation offered above, which is a *traditional* Moran's I-like coefficient, there are at least four additional perspectives that are relevant. Each is briefly considered in turn.

14.4.1 Serial (temporal) correlation

A first interpretation is the regression/correlation between the variable under consideration at two points in time, i.e., the linear relationship between $z_{i,t}$ and its time lag $z_{i,t-1}$. Since all variables are expressed as standardized entities, regression and correlation are equivalent.[3]

The result for the relationship between **p_PHA (2020)** and **p_PHA (2010)** is shown in Figure 14.7. The slope of the regression line is slightly positive and highly significant (p-value of 0.000) at 0.157. This suggests an overall positive correlation between the values at the two points in time. However, the overall fit of the regression is very poor, with an R^2 of only 0.025.

14.4.2 Serial (temporal) correlation between spatial lags

A second perspective is the linear relation between the spatial lag of the variable of interest between the two time periods, i.e., a regression of the variable $\sum_j w_{ij} y_{j,t}$ on $\sum_j w_{ij} y_{j,t-1}$, or **W_pPHA (2020)** on **W_pPHA (2010)**. In some sense, if there is a strong temporal correlation between the variable *in-situ*, as well as spatial correlation in each of the time periods, one may expect a temporal correlation between their neighbors in the form of the associated spatial lags. However, this assumes that the *location* of the clusters in unchanging over time, a property that the global Moran's I is unable to assess.

[3]The regression can be carried out by selecting **Data > View Standardized Data** in the scatter plot, or by using the saved standardized values and their spatial lags from the Moran scatter plot for each individual year (by means of the **Save Results** option).

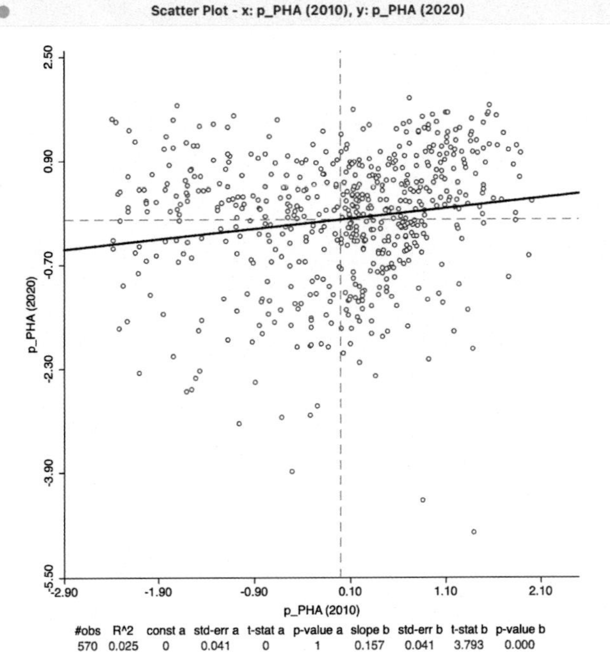

Figure 14.7: Correlation between access to health care in 2020 and 2010

The result is shown in Figure 14.8. The regression slope of 0.073 is not significant, with a p-value of 0.132. Also, the overall fit is very poor, given the R^2 of 0.004. This is no surprise, given the ball-like shape of the scatter plot. In other words, while there is a relationship between the health access for two different time periods, this does not hold for the neighbors (spatial lags). This finding provides additional support for the notion that the clusters in the two years may be in different locations. If they were in the same locations, then the spatial lags for the two time periods would tend to be correlated.

14.4.3 Space-time regression

A third perspective is offered by a regression of the value of a variable at time t on the value of its neighbors at time $t-1$, formally, a regression of $y_{i,t}$ on $\sum_j w_{ij} y_{j,t-1}$. This is the most natural formulation of a *space-time regression*, expressing how the values at neighboring locations at the previous time period *diffuse* to the location at the next time period.

In Figure 14.9, this is illustrated for a regression of **p_PHA (2020)** on **W_pPHA (2010)**. Again, the scatter plot is highly circular, yielding an almost horizontal slope for the linear fit. The regression coefficient of -0.028 is not significant, with a p-value of 0.718. The R^2 is actually zero!

In other words, here we find no direct relationship between the value at a location and its neighbors in a previous location, consistent with the lack of significance of the bivariate Moran's I in Figure 14.6.

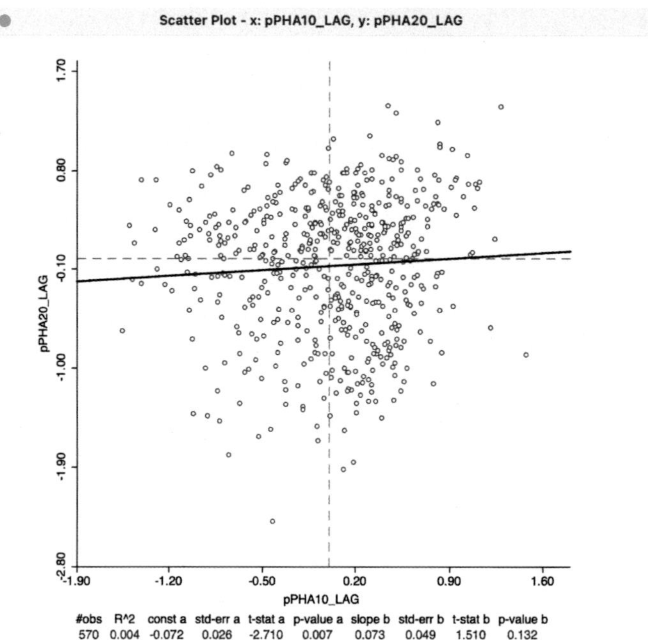

Figure 14.8: Correlation between the spatial lag of access to health care in 2020 and the spatial lag in 2010

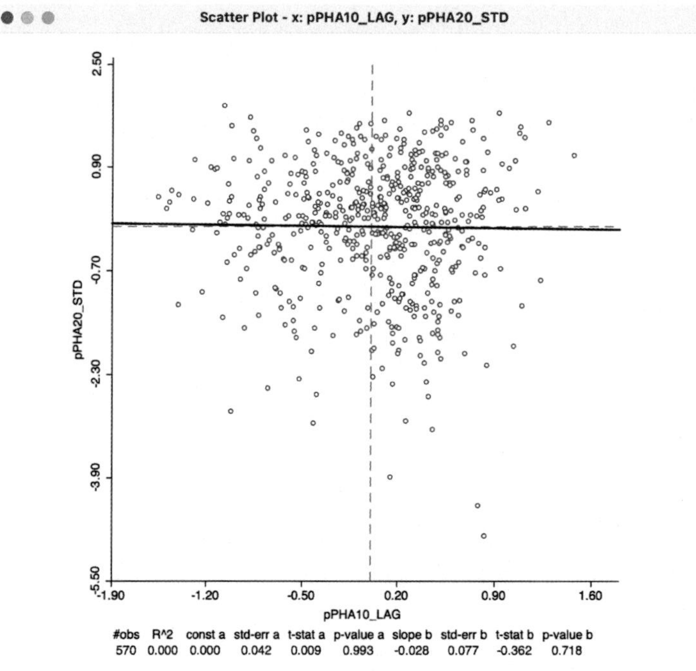

Figure 14.9: Correlation between access to health care in 2020 and its spatial lag in 2010

```
SUMMARY OF OUTPUT: ORDINARY LEAST SQUARES ESTIMATION
Data set             :  oaxaca_master
Dependent Variable   :  pPHA20_STD  Number of Observations:  570
Mean dependent var   : 2.92943e-16  Number of Variables   :    3
S.D. dependent var   :    0.999122  Degrees of Freedom    :  567

R-squared            :    0.032167  F-statistic           :     9.42242
Adjusted R-squared   :    0.028753  Prob(F-statistic)     : 9.42815e-05
Sum squared residual:      550.697  Log likelihood        :    -798.976
Sigma-square         :    0.971247  Akaike info criterion :     1603.95
S.E. of regression   :    0.985519  Schwarz criterion     :     1616.99
Sigma-square ML      :    0.966135
S.E of regression ML:     0.982922

-----------------------------------------------------------------------------
       Variable       Coefficient       Std.Error     t-Statistic   Probability
-----------------------------------------------------------------------------
          CONSTANT     0.00243308       0.0412952       0.0589193       0.95297
       pPHA10_STD        0.195549       0.0452086         4.32549       0.00002
       pPHA10_LAG       -0.173956       0.0831597        -2.09183       0.03690
-----------------------------------------------------------------------------
```

Figure 14.10: OLS regression of pPHA 2020 on pPHA 2010 and W PHA 2010

14.4.4 Serial and space-time regression

A final perspective is offered by including both in-situ correlation and space-time correlation in a regression specification. Specifically, this is expressed as a regression of $y_{i,t}$ on both $x_{i,t-1}$ and $\sum_j w_{ij}x_{j,t-1}$. In the absence of temporal error autocorrelation, standard ordinary least squares regression will yield unbiased estimates of the coefficients.

The results for a regression of **p_PHA (2020)** on **p_PHA (2010)** *and* **W_pPHA (2010)** are shown in Figure 14.10. All variables are in standardized form. Purely as an illustration, this uses the regression functionality of `GeoDa`, a discussion of which is beyond the scope of this book.[4]

The specification offers yet a different perspective on the space-time relationships. Both regression coefficients are significant, but the coefficient of the space-time lag is negative and only weakly significant, with a p-value of 0.04. The overall adjusted R^2 is only 0.03. This would suggest that after correcting for in-situ correlation, the effect of neighbors in the past was negative, rather than positive.

Needless to say, focusing the analysis solely on the bivariate Moran scatter plot provides only a limited perspective on the complexity of the dynamics of space-time patterns. In practice, the full space-time regression may provide the most reliable insights.

[4]See Anselin and Rey (2014) for specifics on both the methods and their implementation in `GeoDa`.

15

Nonparametric Spatial Autocorrelation

In this chapter, the spatial weights matrix is abandoned in the formulation of a spatial autocorrelation statistic. Instead, a measure of attribute (dis)similarity is related directly to the distance that separates pairs of observations, in a non-parametric fashion. Two approaches covered are the *spatial correlogram* and the *smoothed distance scatter plot.*

To illustrate these methods, I will use the *Italy Community Banks* point layer sample data set.

15.1 Topics Covered

- Analyze the range of spatial autocorrelation by means of a spatial correlogram
- Assess the sensitivity of the spatial correlogram to its settings (maximum distance, number of bins, smoother)
- Analyze the range of spatial autocorrelation by means of the smoothed distance scatter plot
- Assess the sensitivity of the smoothed distance scatter plot to its settings (maximum distance, number of bins, smoother)
- Address the computation of the correlogram and distance plots for large(r) data sets by relying on random sampling

GeoDa Functions

- Space > Spatial Correlogram
  - variable selection
  - selecting the number of bins
  - selecting the maximum distance
  - using a random sample of locations
  - changing the smoothing parameters
- Space > Distance Scatter Plot
  - variable selection
  - selecting the number of bins
  - selecting the maximum distance
  - using a random sample of locations
  - changing the smoothing parameters

DOI: 10.1201/9781003274919-15

Toolbar Icons

Figure 15.1: Moran Scatter Plot | Spatial Correlogram | Cluster Maps

15.2 Non-Parametric Approaches

An alternative to the spatial weights and cross-product formulation of a spatial autocorrelation statistic is to establish the connection between attribute (dis)similarity and locational similarity in a different way (Section 13.4.2). This is implemented by expressing a measure of attribute (dis)similarity directly as a function of the distance that separates pairs of observations:

$$f(x_i, x_j) = g(d_{ij}),$$

where g is a function that expresses the *distance decay* in the strength of association between the values observed at a pair of locations i, j.

In this chapter, two approaches are outlined that let the function g be specified in a non-parametric way, directly from the data. This contrasts with the standard practice in geo-statistics, where a specific function (semi-variogram) is fit. The *spatial correlogram* approach uses a cross-product measure of attribute similarity, whereas the *smoothed distance scatter plot* is based on a squared difference dissimilarity.

15.3 Spatial Correlogram

A spatial correlogram expresses the covariances or correlations between all pairs of observations as a function of the distance that separates them. The estimate is obtained from a non-parametric local regression fit, such as LOWESS (Section 7.3.2) or kernel regression. Its properties are based on the theory of nonparametric estimators of the autocovariance for a stationary random field, first outlined in Hall and Patil (1994; see also Bjornstad and Falck, 2001).

With the variables expressed in standardized form as z, this approach boils down to a local regression:

$$z_i.z_j = g(d_{ij}) + \epsilon,$$

where d_{ij} is the distance between a pair of locations i, j, ϵ is an error term and g is the non-parametric function to be determined from the data. Since the variables are standardized, the result is a *spatial correlogram* which shows how the strength of correlation varies with the distance separating the observations.

In `GeoDa`, this is implemented by first arranging pairs of observations by distance *bin*, similar to the approach taken in the estimation of a semi-variogram in geostatistics. The correlogram is then the nonlinear curve fit to the average correlation for all pairs of observations in a distance bin by means of a LOWESS regression .

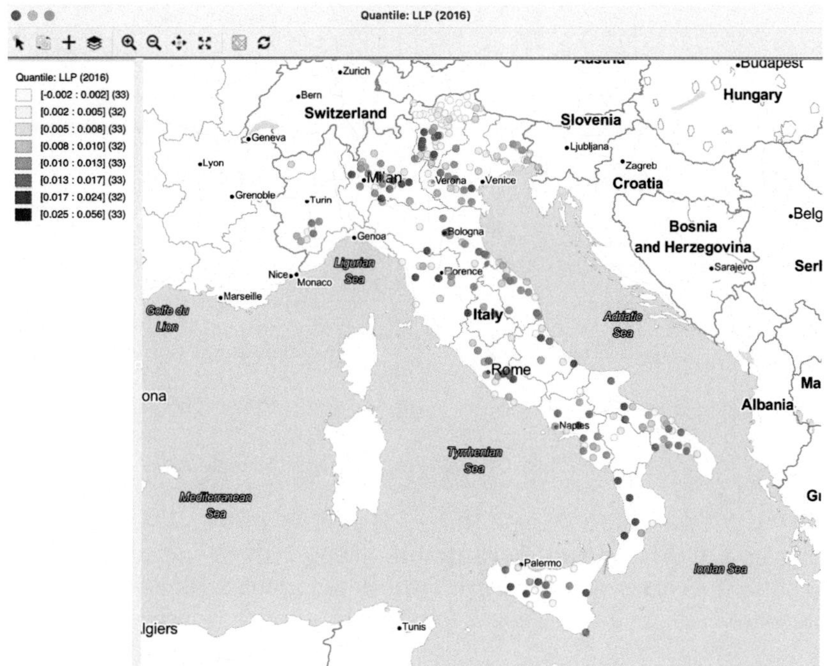

Figure 15.2: Spatial distribution of loan loss provisions – Italian banks 2016

The main interest in the spatial correlogram is to determine the *range* of interaction, i.e., the distance within which the spatial correlation is positive. In addition, following Tobler's law (Tobler, 1970), the strength of the correlation should *decrease* with distance.

Beyond the range of interaction, the correlation should become negligible, although in practice it can still show fluctuations.[1]

15.3.1 Creating a spatial correlogram

To illustrate the spatial correlogram, I use the *Italy Community Banks* sample data set from Chapter 11 in time-enabled form. Specifically, the spatial pattern of the loan loss provisions over customer loans is investigated for 2016 (**LLP (2016)**). The presumption is that this variable addresses similar local risk environments, which would suggest spatial autocorrelation.

A quantile map with eight categories is shown in Figure 15.2. The suggestion of patterning is confirmed by a Moran's I of 0.179, with associated z-value of 5.41 (based on 999 permutations) for the six neighbor k-nearest neighbor weights truncated at 73 km (*italy_banks_te_d73_k6.gal*).[2]

The spatial correlogram functionality is invoked by selecting the middle icon in the spatial analysis group, as in Figure 15.1, and choosing **Spatial Correlogram**. Alternatively, from the menu, it can be started by selecting **Space > Spatial Correlogram** (the item near the very bottom of the list of options).

[1]Some autoregressive spatial processes induce a pattern of alternating positive and negative autocorrelations, but a further consideration of this is beyond the current scope.

[2]The Moran scatter plot calculation in `GeoDa` removes the two isolates from the computations.

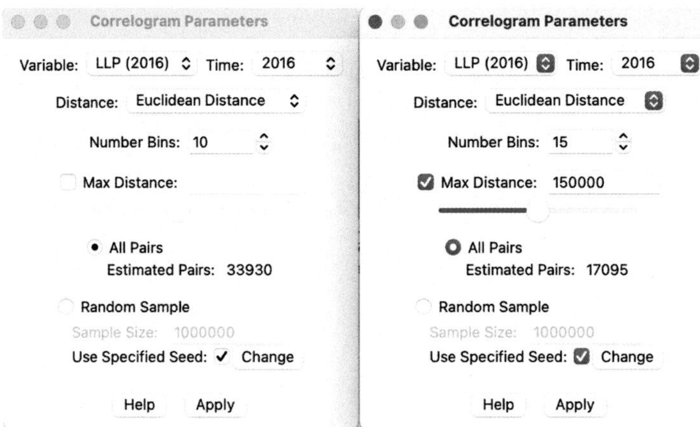

Figure 15.3: Correlogram parameter selection

This brings up a **Correlogram Parameters** dialog with the default parameter settings and a graph in the background. However, this initial graph is almost never informative at this point, since it shows a correlogram for the first variable, which is usually an ID value.

The initial layout of the interface is shown in the left-hand panel of Figure 15.3. The items at the top of the dialog are the **Variable** (available from a drop-down list), the **Distance** metric (default is **Euclidean Distance**, but **Arc Distance** is also available, expressed either as miles or as kilometers). In the example, since the data set is time enabled, the **Variable** also has an associated **Time** drop-down list. For the initial default setting, **LLP (2016)** and **2016** are used with **Euclidean Distance** (given the projection used, this distance is in meters).

Next follows the **Number of Bins**. The non-parametric correlogram is computed by means of a local regression on the spatial correlation estimated from the pairs that fall within each distance bin. The number of bins determines the distance range of each bin. This range is the maximum distance divided by the number of bins. The more bins are chosen, the more fine-grained the correlogram will be. However, this also potentially can lead to too few pairs in some bins (the rule of thumb is to have at least 30).

The number of pairs contained in each bin is a result of the interplay between the number of bins and the maximum distance. As the default, **All Pairs** are used, with **Max Distance** unchecked. In the example, with 261 observations, this yields: $[261^2 - 261]/2 = \mathbf{33930}$ pairs, as listed in the **Estimated Pairs** box.

In many instances in practice, using all pairs may not be a good choice. For example, when there are many observations, the number of pairs quickly becomes unwieldy and the program will run out of memory. Also, the correlations computed for pairs that are far apart are not that meaningful, since they should be zero due to Tobler's law. Therefore it is often practical to truncate the observations to only those pairs that are within a reasonable distance range.

The bottom half of the dialog provides options for fine-tuning these choices. These are considered in Sections 15.3.2.1 and 15.3.2.2.

With the default set as in the left-hand panel of Figure 15.3, the spatial correlogram shown in Figure 15.4 is created.

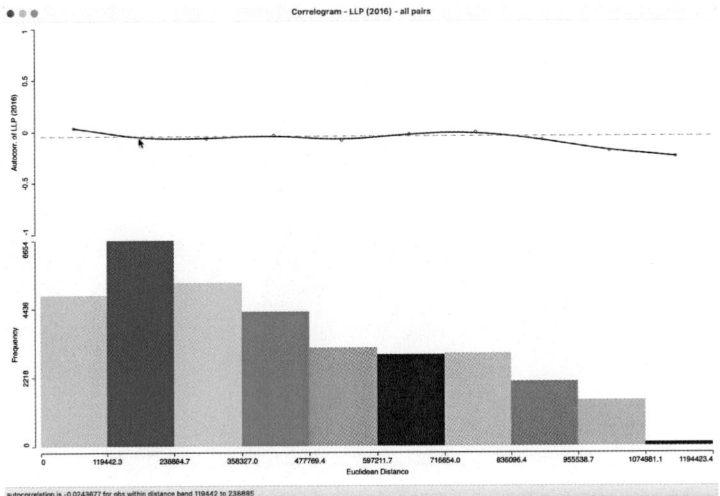

Figure 15.4: Default spatial correlogram – LLP (2016)

15.3.1.1 Interpretation

The top half of the graph in Figure 15.4 is the actual correlogram that depicts how the spatial autocorrelation changes with distance. Hovering the pointer over each blue dot gives the spatial autocorrelation associated with that distance band in the status bar. For example, the second dot corresponds with an autocorrelation of -0.024 for observations within the distance band from 119.4 km to 238.9 km, as shown in the status bar (the distances in the graph are expressed in meters). The first dot corresponds with an autocorrelation of 0.081.

The intersection between the correlogram and the dashed zero axis, which determines the range of spatial autocorrelation, happens in the midpoint of the second range, or roughly around 179 km. Beyond that range, the autocorrelation fluctuates around the zero line.

The bottom half of the graph consists of a histogram that shows the number of pairs of observations in each bin. Hovering the pointer over a given bin shows in the status bar how many pairs are contained in the bin. In the example, each bin has more than sufficient observation pairs. Even the last bin, which seems small (a function of the vertical scale), contains 102 pairs to compute the average autocorrelation.

15.3.2 Spatial correlogram options

The spatial correlogram has four main options, invoked in the usual way by right clicking on the graph.

The **Change Parameters** item brings back the **Correlogram Parameters** dialog and allows for fine tuning of the number of bins, maximum distance and randomization options (Sections 15.3.2.1 and 15.3.2.2).

The **Edit LOWESS Parameters** option manipulates the bandwidth and other technical parameters that can be specified for any LOWESS smoother. The default is to use a bandwidth of 0.20, which works well in most situations. As usual, a larger bandwidth will yield a (slightly) smoother curve, and a smaller bandwidth will work in the opposite way. This option works in exactly the same manner as for the standard LOWESS nonlinear smoother (Section 7.3.2.1).

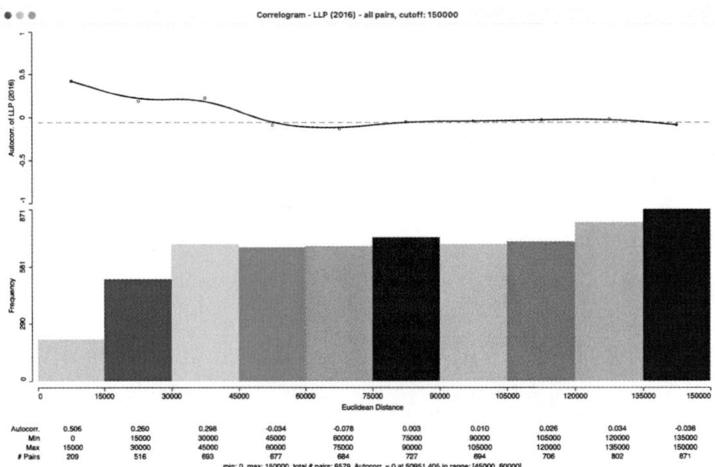

Figure 15.5: Spatial correlogram – LLP, 150 km cut off, 10 bins

The **View > Display Statistics** option generates a list of descriptive statistics for each bin. The computed autocorrelation is provided, as well as the distance range for the bin (lower and upper bound), and the number of pairs used to compute the statistic. In addition, there is a summary with the minimum and maximum distance, the total number of pairs and an estimate for the range, i.e., the distance at which the estimated autocorrelation first becomes zero (see the discussion of Figures 15.5 and 15.6 for an illustration). The other **View** options are the familiar **Set Display Precision**, **Set Display Precision on Axes** and **Show Status Bar**. The latter is active by default.

Finally, **Save Results** provides a record of the descriptive statistics of the correlogram in a text file (with file extension csv). The file contains the information listed at the bottom of the graph when **View > Display Statistics** has been selected, with a column matching each bin. The file includes the estimates of the spatial autocorrelation, bin ranges, number of observations in each bin, as well as the summary.

15.3.2.1 Maximum distance and number of bins

In most situations in practice, the default use of all the distance pairs is not informative. In addition, there is a good reason to limit the maximum distance considered in the selection of pairs, since correlations at large(r) distances are both sparser (fewer pairs in a bin, which leads to less precise estimates) and supposed to be near zero (Tobler's law).

The **Correlogram Parameters** dialog contains options to set the **Max Distance**. When this option is checked, half of the largest inter-observation distance is used.[3] In the example, that would be about 597 km, resulting in **17095 Estimated Pairs**, about half of the number when all pairs are used (not shown).[4]

Instead of using a rule of thumb such as half the total distance, the maximum distance can be specified explicitly, as in the right-hand panel of Figure 15.3. Here, a value of **150000**

[3]This follows one of the most common rules of thumb from geostatistics, e.g., as in Journel and Huijbregts (1978) and Deutsch and Journel (1998).

[4]Since the estimated number of pairs is computed on the fly, as the maximum distance is adjusted by means of the slider, it is only an approximation. The actual number of pairs used in the calculation of the pairwise correlations is given in the status bar of the correlogram.

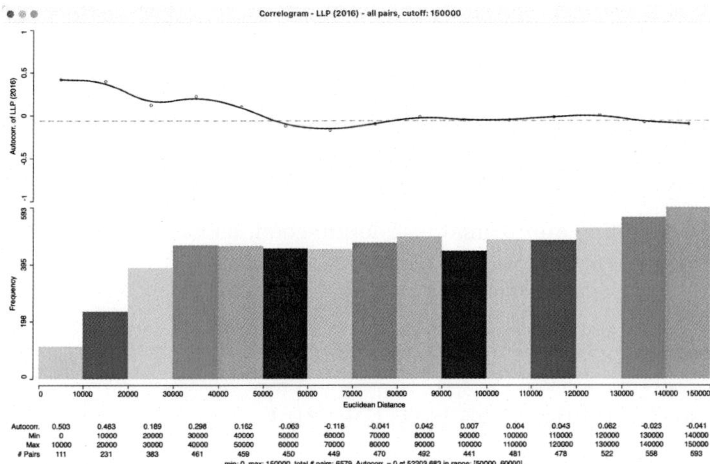

Figure 15.6: Spatial correlogram – LLP, 150 km cut off, 15 bins

(150 km) is entered. Combined with the default **Number Bins** of 10, and with **View > Display Statistics** activated, this yields the correlogram in Figure 15.5.

The graph header lists the **cutoff** but also gives that **all pairs** have been used. This should be interpreted in the sense that all pairs that meet the distance cutoff are used in the computations, as opposed to a random sample (Section 15.3.2.2).

The distance range for each bin is now 15 km (15,000 in the graph) and, as a result, the range is estimated to be between 45 km and 60 km, contrasted with more than double the figure obtained in Figure 15.4. Also, the initial spatial autocorrelation, for pairs within 15 km of each other is 0.506, compared to 0.081 for the first range in the default setting. The descriptive statistics depict the change of the estimated autocorrelation coefficient with distance at a much more fine grained level than before. The correlogram shows the desired shape, decreasing with distance until the range is reached and more or less flat afterward. The computations are based on 6,579 pairs, less than a fifth of the original total.

A further refinement can be obtained by increasing the number of bins for the same maximum distance. In Figure 15.6, the correlogram is depicted for 15 bins with the 150 km cut-off (the settings shown in the right-hand panel of Figure 15.3), resulting in a range of 10 km for each bin. The zero autocorrelation is estimated to occur between 50 and 60 km, an even narrower band than for Figure 15.5. The autocorrelation in the first 10 km range is 0.503, basically the same as in the first 15 km range shown before. The overall shape of the correlogram also remains the same.

Overall, this suggests quite a strong pattern of positive spatial autocorrelation at short distances, but negligible association beyond distances of 60 km. This highlights a different aspect of spatial autocorrelation than can be provided by a statistic like Moran's I, which considers all pairs that meet the spatial weights criterion.

15.3.2.2 Random sampling

A final feature of the correlogram is the ability to compute the spatial correlations for a sample of locations, which reduces the number of pairs used in the calculations. This is especially useful when the data set is large, in which case the number of pairs could quickly become prohibitive.

In the example used here, this is not really necessary, since the default sample size of 1,000,000 that is used to generate the random sample exceeds the current total number of pairs in the data set. To implement this option, the **Random Sample** radio button needs to be checked and a sample size specified if the default of 1 million is not desired. Also, to allow exact replication, the **Use Specified Seed** option should be checked (the seed can be adjusted by means of the **Change** button).

In practice, the sampling approximation is quite good, as long as the selected sample size is not too small relative to the original data size.

15.4 Smoothed Distance Scatter Plot

The *smoothed distance scatter plot* (Anselin and Li, 2020) offers an alternative approach to relating a measure of attribute dissimilarity to distance. It is similar in spirit to the familiar semi-variogram from geostatistics (Isaaks and Srivastava, 1989; Cressie, 1993; Chilès and Delfiner, 1999; Stein, 1999) but also different in some important respects.

In geostatistics, the point of departure for the semi-variogram is to express the magnitude of the *variance of the difference* between observations at two points in space as a function of the distance that separates them:

$$\gamma(h) = \frac{1}{2}\text{Var}[z_s - z_{s+h}],$$

where γ is the semi-variogram function and h is the separation between two observations.

Exploiting assumptions of spatial stationarity (such as constant mean and constant variance), this results in a simpler expression that relates the expected value of the squared difference to distance:

$$\gamma(h) = \frac{1}{2}\text{E}[z_s - z_{s_h}]^2.$$

An *empirical* variogram is estimated from the data by computing the average squared difference for several distance bins, similar to the computation of the spatial correlogram (Section 15.3.2.1). A number theoretical variograms have been proposed, such as the spherical, exponential and Matérn (see, e.g., Banerjee et al., 2015, for an overview). These differ in terms of characteristics, such as the *range*, at which point the semi-variogram becomes flat, and the shape of the function between the origin and the range.

The smoothed distance scatter plot is similar in spirit in that it relates a measure of dissimilarity to the distance separating the observations, but it is not based on the typical assumptions from geostatistics. It is a simple scatter plot of the distance in attribute space (on the vertical axis) against the distance in geographic space (on the horizontal axis). The distance in attribute space for a single variable z can be expressed as:

$$v_{ij} = \sqrt{(z_i - z_j)^2},$$

with i and j as the observation locations. Note that this is not the squared difference, as in the semi-variogram, but its square root, as a direct measure of distance.

The geographic Euclidean distance between the observation point is then the familiar:

$$d_{ij} = \sqrt{(x_i - x_j)^2 + (y_i - y_j)^2},$$

with $(x_{i,j}, y_{i,j})$ as the observation coordinates.

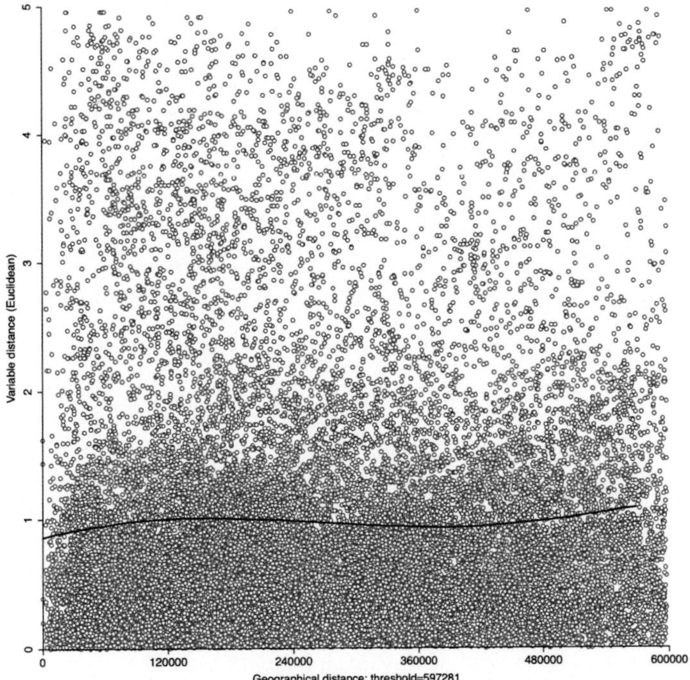

Figure 15.7: Smoothed distance scatter plot

Whereas the geographic distance has an immediate interpretation in terms of distance units (e.g., kilometers), this is not the case for the attribute distance matrix. In practice, it is therefore advisable to first standardize the variable, so that attribute distance is expressed in standard deviational units.

An example of the resulting scatter plot is shown in Figure 15.7 for the variable **LLP (2016)** from the *Italy Community Banks* data set. The graph shows a slow increase up to about 120 km, after which it is mostly flat (but see Section 15.4.3 for more detailed options and interpretation).

The idea of combining a measure of distance in attribute space with geographic distance goes back at least to an example in Oden and Sokal (1986). They implemented a Mantel test (Section 13.4.3) to assess the similarity between the elements of a geographic distance matrix and a variable dissimilarity matrix. However, their approach is less flexible than the smoothed distance scatter plot and does not lend itself readily to an extension to multiple variables (Section 15.4.1).

One may be tempted to apply a linear fit to the scatter plot as an intuitive measure of the association between the two variables. However, this would imply a *linear* relationship between the two, whereas Tobler's law suggests a non-linear distance decay, with a *range* beyond which there is no association, in the same fashion as shown for the spatial correlogram. However, in contrast to the correlogram, the smoothed distance scatter plot *increases* with distance until the range is reached, since it involves a measure of dissimilarity.

A non-parametric local regression fit is applied to the smoothed distance scatter plot to extract the overall pattern. Specifically, the implementation in `GeoDa` is based on the *loess* approach (Cleveland et al., 1992).

As for the spatial correlogram, the main interest in this graph is to obtain an estimate of the range of interaction, i.e., the point where the curve ceases to increase with distance and begins to flatten out.

15.4.1 Multivariate extension

As explored in more depth in Chapter 18, a spatial autocorrelation measure based on a cross-product (like Moran's I) is difficult to extend to multiple variables. In contrast, as demonstrated in Anselin (2019a), this is not the case for statistics that use a squared difference (or its square root) as an indicator of attribute dissimilarity.

More specifically, the univariate attribute distance is readily extended to k variables as:

$$v_{ij} = \sqrt{\sum_k (z_{ki} - z_{kj})^2},$$

for k variables z_k observed at locations i and j. This Euclidean distance in multi-attribute space will play an important role in the treatment of clustering considered in Volume 2. Instead of a Euclidean distance, a Manhattan metric could be used. Neither of these takes into account the inherent correlation among the variables, which could be addressed by means of a Mahalanobis distance, although this requires a reliable estimate of the variance matrix (more precisely, the inverse of the variance matrix).[5] This approach is not considered further.

The multivariate smoothed distance plot replaces the univariate attribute distance by its multivariate counterpart on the vertical axis. In all other respects, the method is the same as the univariate case.

15.4.2 Creating a smoothed distance scatter plot

The smoothed distance plot is invoked as the second option from the spatial correlogram icon shown in Figure 15.1, or as **Space > Distance Scatter Plot** from the menu. The dialog that follows allows for the selection of the variable(s) and the settings to be used in the analysis. In contrast to the spatial correlogram, multiple variables can be selected. To illustrate the univariate case, the same variable as before is employed, **LLP (2016)**. The extension to multiple variables is straightforward, the only difference being that more than over variable must be selected in the dialog. All other options and interpretations are the same as for the univariate case.

Other important initial settings in the dialog are the **Variable Distance**, either **Euclidean Distance** (the default) or **Manhattan Distance**, and the **Geographic Distance**, as **Euclidean Distance** (the default, for projected coordinates), or two versions of **Arc Distance** (in miles or kilometers, for coordinates as latitude-longitude).

In addition, a few options need to be set to make the plot operational. As argued for the spatial correlogram, it is usually not advised to compute the attribute distance for all the observation pairs in the data set. By default, the **Max Distance** box is checked, which results in **1/2 maximum pairwise distance** to be used as the largest pairwise distance, following standard practice in empirical variography (Journel and Huijbregts, 1978; Deutsch and Journel, 1998). In the example, this is about 597 km (compared to a maximum distance

[5]The Mahalanobis distance between vectors $\mathbf{x}_i$ and $\mathbf{x}_j$ is $d_{ij} = \sqrt{(\mathbf{x}_i - \mathbf{x}_j)'\Sigma^{-1}(\mathbf{x}_i - \mathbf{x}_j)}$, with Σ as the covariance matrix.

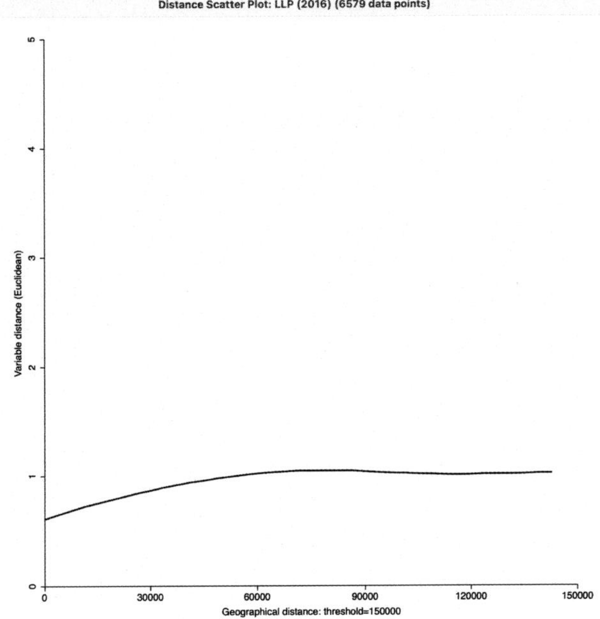

Figure 15.8: Smoothed distance scatter plot with 150 km distance cut-off

of some 1,200 km). As in the spatial correlogram, this yields **17095** pairs to be considered, compared to the 33,930 pairs in the default case. A different option is to use **1/2 of the diagonal of the bounding box**, another rule of thumb often used for empirical variograms.

The **All Pairs** button is checked by default. The alternative is to use **Random Sample**, which works in the same way as for the spatial correlogram.

These default settings yield the plot in Figure 15.7, but without the actual points, which are not shown by default. Using a more reasonable cut-off distance of 150 km yields the graph in Figure 15.8, based on the computation for 6579 observation pairs. In this plot, the range seems to be between 70 and 80 km, slightly higher than for the spatial correlogram in Figure 15.6.

Further options to refine the visualization are described in Sections 15.4.3 and 15.4.3.1.

15.4.3 Smoothed distance scatter plot options

The plot has several options that are invoked in the usual fashion, by right clicking on the graph. There are five items:

- **LOESS Setup**
- **View**
- **Color**
- **Axis Option**
- **Save Results**.

The **LOESS Setup** determines the settings for the nonparametric local regression and is considered in Section 15.4.3.1.

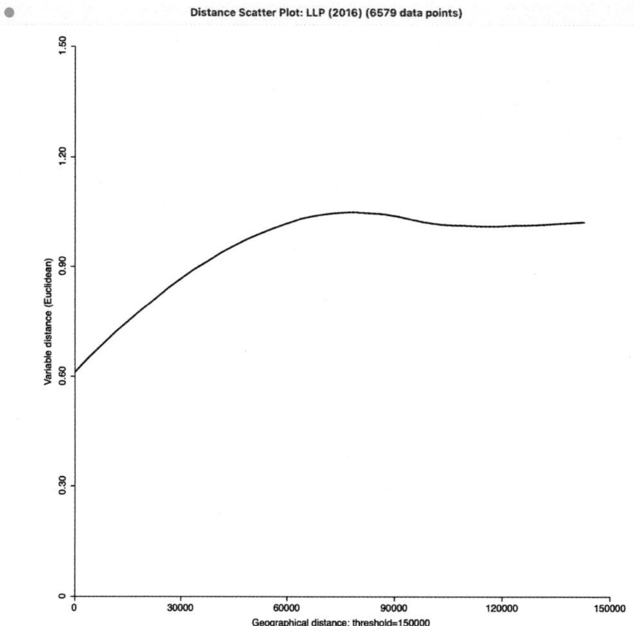

Figure 15.9: Smoothed distance scatter plot with 150 km distance cut-off and customized Y-Axis

The default **View** setting is to show the status bar. **View > Show Data Points** adds all the points to the scatter plot. As mentioned, this is turned off by default. The **Color** options allow the color to be specified for the regression line, the scatter plot points and the background.

An important option for effective visualization is to adjust the range on the Y-axis. This is implemented through **Axis Option > Adjust Value Range of Y-Axis**. It allows the minimum and maximum values for the Y-Axis to be customized, resulting in a clearer view of the shape of the graph. For example, setting the maximum at 1.5 in Figure 15.8, yields the plot shown in Figure 15.9.

In addition, the **Axis Option** can be used to specify the display precision, in the usual way.

Finally, **Save Results** will create a comma-separated value (csv) file that contains all the scatter plot points with their X and Y coordinates. This file can then be used as input to more advanced statistical analyses.

15.4.3.1 Loess settings

The *loess* approach implements a local polynomial regression method as outlined in Cleveland et al. (1992). The implementation in `GeoDa` is based on the code contained in the *netlib* library. The *local* regression uses a subset of the observations within a close attribute distance from each observation point to compute the slopes in a polynomial regression, which are then combined to produce a smooth function of predicted values. Since the smoothed distance scatter plot only has one explanatory variable (geographic distance), the loess implementation is fairly straightforward.

The **LOESS Setup** option allows for three important parameters to be set. The first is the **Degree of smoothing (span)**, i.e., the share of total observations that is used. The

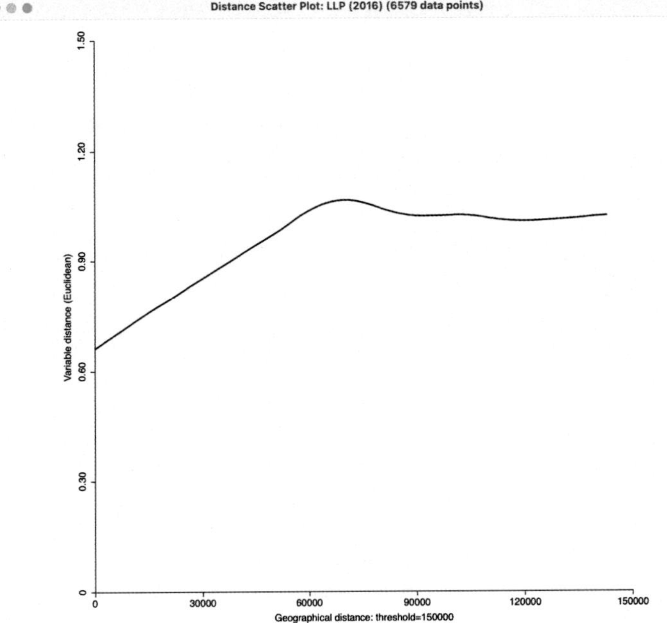

Figure 15.10: Smoothed distance scatter plot with 150 km distance cut-off and span 0.5

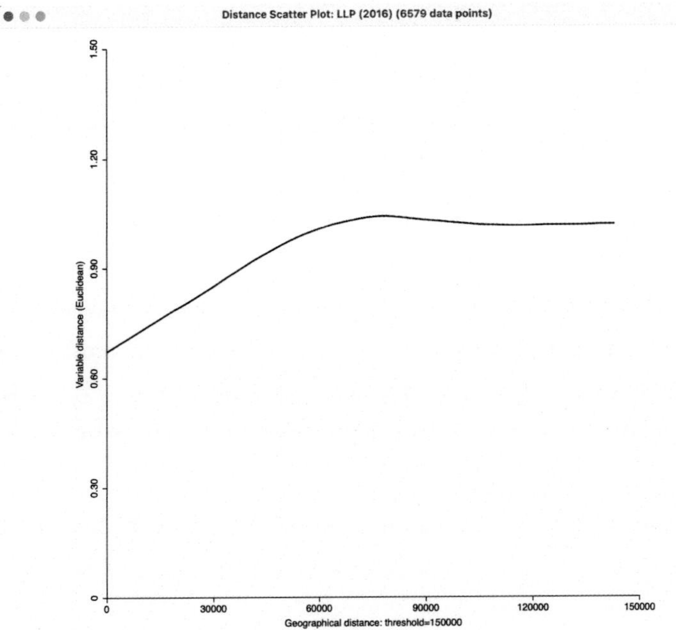

Figure 15.11: Smoothed distance scatter plot with 150 km distance cut-off, span 0.5, linear fit

netlib default is 0.75, which is also the default value in `GeoDa`. This corresponds to the formal smoothing parameter α which is used in some other software implementations. The

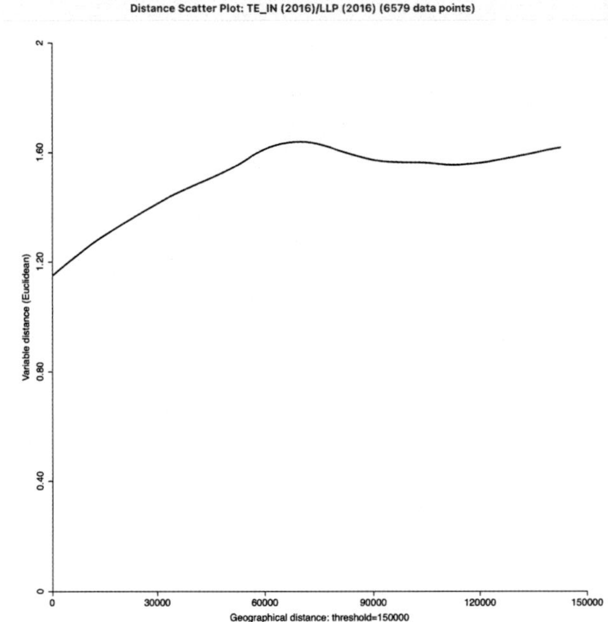

Figure 15.12: Bivariate smoothed distance scatter plot

observations within the maximum range from the fit point are weighted as in a kernel regression.[6]

The **Degree of the polynomials** is **2** by default, for a quadratic model, but **1** (linear) is another (less often used) option. Finally, the **Family** setting selects the type of fitting procedure. The default is **gaussian**, which boils down to a least squares fit. The alternative is a sometimes more robust **symmetric** option.

To illustrate the sensitivity of the graph to some of these settings, Figure 15.10 shows the effect of setting the span to 0.50, for a more *local* fit.

Figure 15.11 illustrates the effect of a linear fit vs a quadratic fit (with the same span of 0.5).

In practice, some experimentation may be necessary to yield the most effective graph.

Finally, Figure 15.12 illustrates a smoothed distance scatter plot for two variables, technical input efficiency in 2016, **TE_IN (2016)** and loan loss provision **LLP (2016)**. The span is set to 0.5, with a quadratic function and the vertical axis truncated at 2. The graph shows a gradually increasing function up to about 70 km, after which it is more or less horizontal. This suggests the presence of a spatial relationship of the tuples of the two variables observed within this range.

[6]The *netlib* implementation uses a tricubic weighting scheme, i.e., $(1 - (\text{distance}/\text{maximum distance})^3)^3$.

Part V

Local Spatial Autocorrelation

16

LISA and Local Moran

This chapter is the first in a series where attention shifts to the detection of the *location* of clusters and spatial outliers. These are collections of observations that are surrounded by neighbors that are more similar to them or more different from them than would be expected under spatial randomness. The general principle by which this is approached is based on the concept of *Local Indicators of Spatial Association*, or *LISA* (Anselin, 1995).

I begin by outlining the basic idea underlying the concept of LISA. This is followed by an in-depth coverage of its most common application, in the form of the *Local Moran* statistic. This statistic becomes a powerful tool to detect hot spots, cold spots, as well as spatial outliers when combined with the classification of spatial autocorrelation in the Moran scatter plot. The ultimate result is a local cluster map. An extensive discussion of its properties and interpretation is provided, with special attention to the notion of significance.

To illustrate these methods, I will employ the *Oaxaca Development* sample data set.

16.1 Topics Covered

- Understand the concept of a Local Indicator of Spatial Autocorrelation
- Identify clusters with the Local Moran cluster map and significance map
- Interpret the spatial footprint of spatial clusters
- Assess the significance by means of a randomization approach
- Assess the sensitivity of different significance cut-off values
- Interpret significance by means of Bonferroni bounds and the False Discovery Rate (FDR)
- Assess potential interaction effects by means of conditional cluster maps

GeoDa Functions

- Space > Univariate Local Moran's I
  - significance map and cluster map
  - permutation inference
  - setting the random seed
  - selecting the significance filter
  - saving LISA statistics
  - select all cores and neighbors
  - local conditional map

DOI: 10.1201/9781003274919-16

Toolbar Icons

Figure 16.1: Moran Scatter Plot | Spatial Correlogram | Cluster Maps

16.2 LISA Principle

As pointed out in the discussion of global spatial autocorrelation, statistics designed to detect the presence of such correlation, such as Moran's I and Geary's c, are formulated to reject the null hypothesis of spatial randomness, typically in favor of an alternative of *clustering* (occasionally in favor of negative spatial autocorrelation). Such *clustering* is a characteristic of the complete spatial pattern, but it does *not* provide an indication of the *location* of the clusters.

An early suggestion of a test for local spatial autocorrelation was formulated by Getis and Ord (1992) (see Chapter 17). This was generalized in Anselin (1995), where the concept of a local indicator of spatial association, or *LISA* was introduced. A LISA is viewed as having two important characteristics. First, it provides a statistic for each location with an assessment of significance. Second, and less importantly, it establishes a proportional relationship between the sum of the local statistics and a corresponding global statistic.

As shown in section 13.4, most global spatial autocorrelation statistics can be expressed as a double sum over the i and j indices, for all pairs of observations, as $\sum_i \sum_j g_{ij}$. The local form of such a statistic would then be, for each observation (location) i, the sum of the relevant expression over the j index, $\sum_j g_{ij}$.

More precisely, and as covered previously, a spatial autocorrelation statistic consists of a combination of a measure of attribute similarity between a pair of observations, $f(x_i, x_j)$, with an indicator for geographical or locational similarity, in the form of spatial weights, w_{ij}. For a global statistic, this takes on the form:

$$\sum_i \sum_j w_{ij} f(x_i, x_j)$$

. A generic expression for a local indicator of spatial association is then:

$$\sum_j w_{ij} f(x_i, x_j).$$

For each different measure of attribute similarity f, a different statistic for global spatial autocorrelation results. Consequently, there will be a corresponding LISA for each such global statistic. First, the local counterpart of Moran's I is considered.

16.3 Local Moran

The Local Moran (Anselin, 1995) is by far the most commonly used LISA statistic. It follows from a straightforward application to the global Moran statistic of the principles just outlined.

16.3.1 Formulation

As discussed in Chapter 13, the Moran's I statistic is obtained as:

$$I = \frac{\sum_i \sum_j z_i.z_j.w_{ij}/S_0}{\sum_i z_i^2/n}, \tag{16.1}$$

with the variable of interest (z) expressed as deviations from the mean, and S_0 as the sum of all the weights. In the row-standardized case, the latter equals the number of observations, n. As a result, as shown in the discussion of the Moran scatter plot, the Moran's I statistic simplifies to:

$$I = \frac{\sum_i \sum_j w_{ij} z_i z_j}{\sum_i z_i^2}.$$

Using the logic just outlined, a corresponding *Local* Moran statistic would consist of the component in the double sum that corresponds to each observation i, or:

$$I_i = \frac{\sum_j w_{ij} z_i z_j}{\sum_i z_i^2}.$$

In this expression, the denominator is fixed and can thus further be ignored. To keep the notation simple, it can be replaced by a, so that the Local Moran expression becomes $a.\sum_j w_{ij} z_i z_j$. After some re-arranging, a simple expression is:

$$I_i = a \times z_i \sum_j w_{ij} z_j,$$

or, the (scaled) product of the value at location i with its spatial lag, the weighted sum or the average of the values at neighboring locations.

A special case occurs when the variable z is fully standardized. Then its variance $\sum_i z_i^2/n = 1$, so that $\sum_i z_i^2 = n$. Consequently, the global Moran's I can be written as:

$$I = \sum_i (a.z_i \sum_j w_{ij} z_j) = (1/n) \sum_i (z_i \sum_j w_{ij} z_j),$$

or the average of the local Moran's I. This case illustrates a direct connection between the local and the global Moran's I.

Significance can be based on an analytical approximation, but, as argued in Anselin (1995), this is not very reliable in practice.[1] A preferred approach consists of a *conditional* permutation method. This is similar to the permutation approach considered in the Moran scatter plot, except that the value of each z_i is held fixed at its location i. The remaining n-1 z-values are then randomly permuted to yield a *reference distribution* for the local statistic, one for each location.

The randomization operates in the same fashion as for the global Moran's I (see Section 13.5.2.2), except that the permutation is carried out for each observation in turn. The result is a pseudo p-value for each location, which can then be used to assess significance. Note that this notion of significance is not the standard one, and should not be interpreted that way (see the discussion in section 16.5).

[1] Further discussion of analytical results can be found in Sokal et al. (1998a), for an asymptotic approach, and in Tiefelsdorf (2002), where a saddlepoint approximation is suggested.

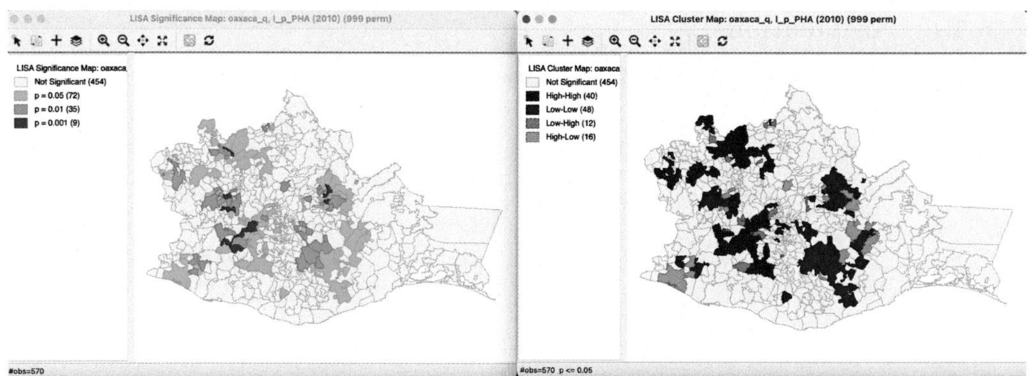

Figure 16.2: Significance map and cluster map

Assessing significance in and of itself is not that useful for the Local Moran. However, when an indication of significance is combined with the quadrant location of each observation in the Moran Scatter plot, a very powerful interpretation becomes possible. The combined information allows for a classification of the significant locations as High-High and Low-Low *spatial clusters*, and High-Low and Low-High *spatial outliers*. It is important to keep in mind that the reference to high and low is *relative* to the mean of the variable, and should not be interpreted in an absolute sense. The notions of clusters and outliers are considered more in-depth in section 16.4.

16.3.2 Implementation

The **Univariate Local Moran's I** is started from the **Cluster Maps** toolbar icon, the right-most icon in the spatial correlation group, highlighted in Figure 16.1. This brings up a drop-down list, with the univariate Local Moran as the top-most item. Alternatively, this option can be selected from the main menu, as **Space > Univariate Local Moran's I**.

Either approach brings up the familiar **Variable Settings** dialog which lists the available variables as well as the default weights file, at the bottom (e.g., **oaxaca_q** in the example). As in Chapter 14, the variable is **p_PHA(2010)**, the 2010 percentage of the municipal population with access to health care. In the example, this variable is time enabled, which results in the selected year (2010) being listed in the dialog as well.

As a reference, a box map reflecting the spatial distribution of this variable is shown in the left-hand panel of Figure 14.1 (see also the left-hand panels in Figures 16.6 and 16.8 in this Chapter).

16.3.2.1 Significance map and cluster map

Clicking **OK** at this point brings up a dialog to select the number and types of graphs and maps to be created. The default is to provide just the **Cluster Map**, which is typically the most informative graphic. In addition, a **Significance Map** and a **Moran Scatter Plot** option is included as well.

To illustrate the Local Moran, selecting both cluster map and significance map brings up Figure 16.2. This visualizes the result of the default randomization test, using 999 permutations and a p-value cut-off of 0.05.

Figure 16.3: Significance map and cluster map – 99999 permutations

The significance map on the left shows the locations with a significant local statistic, with the degree of significance reflected in increasingly darker shades of green. The map starts with $p < 0.05$ and shows all the categories of significance that are meaningful for the given number of permutations. In the example, since there were 999 permutations, the smallest pseudo p-value is 0.001, with nine such locations (the darkest shade of green).

The cluster map augments the significant locations with an indication of the type of spatial association, based on the location of the value and its spatial lag in the Moran scatter plot (see section 16.4). In this example, all four categories are represented, with dark red for the High-High clusters (40 in the example), dark blue for the Low-Low clusters (48 locations), light blue for the Low-High spatial outliers (12 locations) and light red for the High-Low spatial outliers (16 locations).

16.3.2.2 Randomization options

The **Randomization** option is the first item in the options menu for both the significance map and the cluster map. It operates in the same fashion as in the Moran scatter plot. As before, up to 99,999 permutations are possible (for each observation in turn), preferably using a specified random seed to allow replication.

The effect of the number of permutations is typically marginal relative to the default of 999. In the example, selecting 99,999 results in one more significant location (117 vs 116). As shown in the left-hand panel of Figure 16.3, there are now 70 locations significant at $p < 0.05$ (compared to 72), 34 locations at $p < 0.01$ (compared to 35), 11 locations at $p < 0.001$ (compared to 9) and one each at $p < 0.0001$ and $p < 0.00001$ (the smallest possible p-value). They are illustrated by different shades of green in the significance map.

The cluster map, in the right-hand panel, is similarly only marginally affected. There are two new observations in the High-High category, and one of the High-Low outliers disappears. Otherwise, the map is the same as for 999 permutations.

16.3.2.3 Saving the Local Moran statistic

The third item in the options menu for the Local Moran is to **Save Results**.

This brings up a dialog with three potential variables to save to the table: **Lisa Indices**, **Clusters** and **Significance**. The default variable names are **LISA_I**, **LISA_CL** and **LISA_P**. The first are the actual values for the local statistics, which are typically not that useful. The clusters are identified by an integer that designates the type of spatial association: 0 for non-significant (for the current selection of the cut-off p-value, e.g., 0.05 in the example), 1 for High-High, 2 for Low-Low, 3 for Low-High and 4 for High-Low.

The default variable names would typically be changed, especially when more than one variable is considered in an analysis (or different spatial weights for the same variable).

16.3.2.4 Other options

In addition to the three options considered so far, there are several specialized operations for the cluster and significance map. These are covered in section 16.4 on clusters and spatial outliers (**Select All**) and 16.6 on conditional cluster maps (**Show As Conditional Map**).

The other options are customary, and include **Selection Shape**, **Color** and **Show Status Bar**, **Save Selection**, **Copy Image to Clipboard** and **Save Image As**. In addition, when the data set is time-enabled, the **Time Variable Options** are available. Finally, the various options associated with the **Connectivity** item were covered in Section 10.4.5.1.

These other options operate in the same way as discussed before.

16.4 Clusters and Spatial Outliers

A critical link in the interpretation of significant Local Moran statistics is the connection between the Moran scatter plot and the cluster map. As shown in Chapter 14, the Moran scatter plot provides a classification of spatial association into four categories, corresponding to the location of the observations in the four quadrants of the plot. These categories are referred to as High-High, Low-Low, Low-High and High-Low, relative to the mean, which is the center of the graph. It is important to keep in mind that there is a difference between a location being in a given quadrant of the plot, and that location being a *significant* local cluster or spatial outlier.

The cluster map provides a way to interpret the results and classify observations. However, this is only valid for those locations where the local statistic is significant.

16.4.1 Clusters

Clusters are centered around locations with a significant *positive* Local Moran statistic, i.e., locations with *similar* neighbors. In the terminology of the Moran scatter plot, these are either High-High or Low-Low locations.

Figure 16.4 illustrates the connection between the Moran scatter plot and the cluster map. On the left, the 230 observations in the High-High quadrant of the Moran scatter plot for **p_PHA (2010)** (using queen contiguity) are selected. Through the process of *linking*, the corresponding observations are highlighted in the cluster map on the right. It is clear that only a fraction of the locations in the High-High quadrant are actually significant (42), shown as the bright red municipalities in the map. The non-significant ones are identified in a grey color, as illustrated by the locations within the red rectangle on the map.

The reverse logic works as well, as shown in Figure 16.5. The High-High locations are selected in the cluster map on the right (by clicking on the red rectangle in the legend) and the corresponding 42 observations are highlighted in the Moran scatter plot on the left. Again,

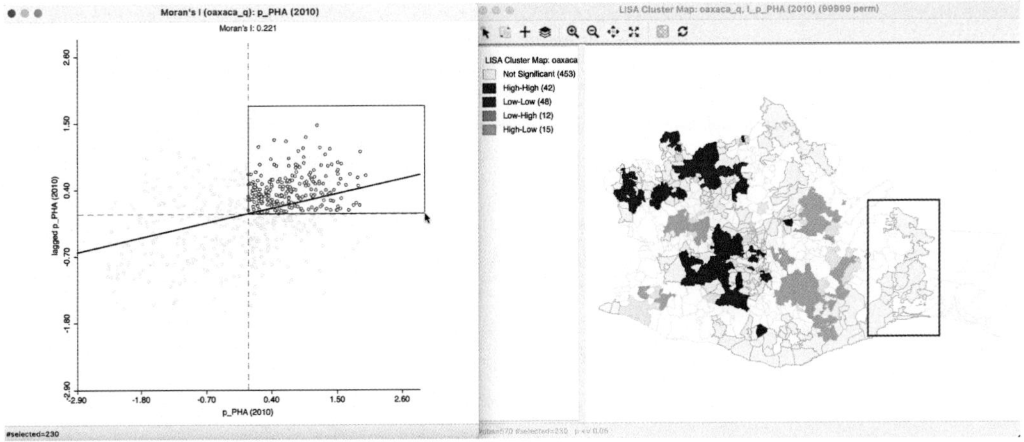

Figure 16.4: High-High observations in Moran scatter plot

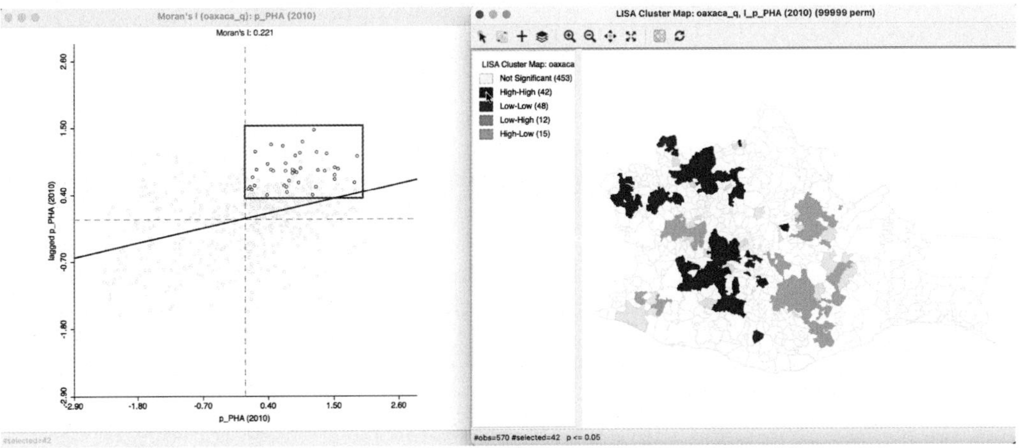

Figure 16.5: Significant High-High observations in Moran scatter plot

this confirms that just belonging to the High-High quadrant in the Moran scatter plot does not imply significance.

A similar exercise can be carried out for observations in the Low-Low quadrant, which also correspond with positive local spatial autocorrelation.

The interpretation of clusters is not always that straightforward (see also Section 16.5.3), especially after inspecting a choropleth map of the variable of interest. In some instances, it is very clear that an observations is selected as High-High when it is in the top category and surrounded by neighbors that are also in the top category. For example, in Figure 16.6, the location selected within the black rectangle on the cluster map (with the arrow pointed at it) belongs to the top quartile in the box map on the left. In addition, its neighbors all also belong to the top quartile, hence a natural interpretation as a cluster.

However, in other instances, the identification as a High-High (or similarly, Low-Low) cluster may seem counter-intuitive. Consider the High-High location identified with the blue arrow on the cluster map. In the box map to the left, that location is shown as belonging to the second quadrant, i.e., *below* the median, which would seem to contradict the characterization

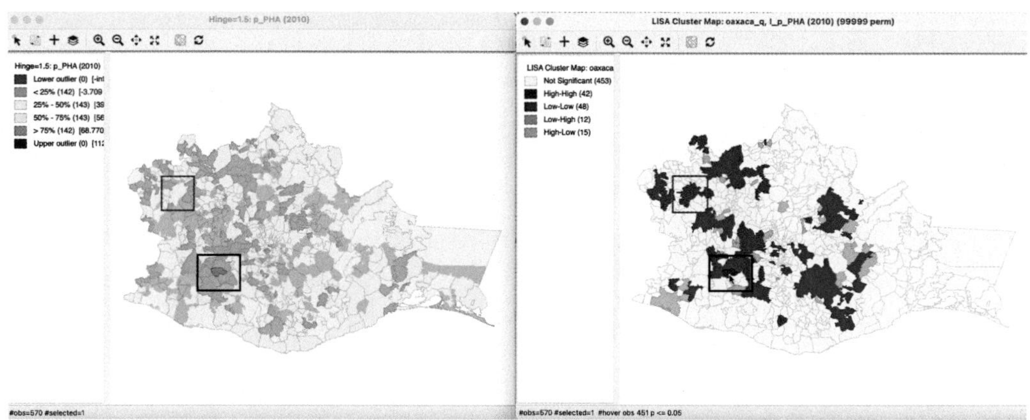

Figure 16.6: High-High cluster location

as *high*. However, as it turns out, for 2010 the median value of access to health care (56.9) is above the mean (53.4), so that it is possible for an observation to be below the median, but still above the mean. Moreover, the simplification of the continuous distribution into a small number of discrete map categories may be confusing. In this particular example, the spatial lag is well above the mean, even though the neighbors of the location in question consist of a mix of municipalities from all four categories on the map. It is important to keep in mind that the spatial lag is the *average* of the neighbors, and its value can be easily influenced by a few extreme values. In addition, the categories on the map may contain observations with very disparate values (depending on the range in the given category), which may yield less than intuitive insights into the similarity of neighbors.

16.4.2 Spatial outliers

Spatial outliers are locations with a significant *negative* Local Moran statistic, i.e., locations with *dissimilar* neighbors. In the terminology of the Moran scatter plot, these are observations in the off-diagonal quadrants, i.e., High-Low or Low-High locations.

Again, the significant spatial outliers are only a subset of the observations in the respective off-diagonal quadrant of the Moran scatter plot. For example, in Figure 16.7, the selected High-Low spatial outliers in the cluster map on the right correspond to only 15 of the 106 observations in the lower left quadrant, as shown in the left-hand panel.

Here also, there may be some counter intuitive results. On the one hand, in the case of the selected High-Low outlier in the purple rectangle in the right-hand panel of Figure 16.8, the interpretation is straightforward. As seen in the left-hand panel, it consists of an observation in the top quartile (indicated by the dark brown color), surrounded by neighbors in the first or second quartile (blue colors), corresponding with much lower values. However, the interpretation of the spatial outlier in the blue rectangle is less straightforward. As shown in the left-hand panel, this corresponds with an observation in the second quadrant, i.e., below the median, but, as before, above the mean, hence classified as high. This observation is surrounded by neighbors in the first quadrant (i.e., with lower values), but one neighbor belongs to the top quadrant. Again, what matters here is the average of neighboring values, not the classifications within which they fall.

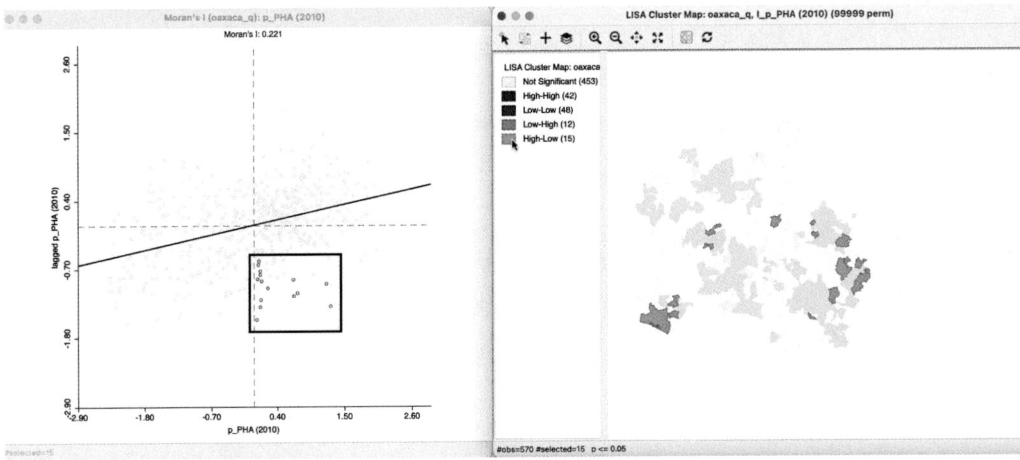

Figure 16.7: Significant High-Low observations in Moran scatter plot

Figure 16.8: High-Low spatial outlier location

16.5 Significance and Interpretation

The interpretation of the results in a cluster map needs to proceed with caution. Because the permutation test is carried out for each location in turn, the resulting p-values no longer satisfy the classical assumptions. More precisely, because the same data are re-used in several of the tests, the problem is one of *multiple comparisons*. Moreover, the p-values of the local statistic may be affected by the presence of global spatial autocorrelation, e.g., as investigated by Sokal et al. (1998b), Ord and Getis (2001), among others. Consequently, the interpretation of the *significance* of the clusters and outliers is fraught with problems.

In addition, the cluster map only provides an indication of significant *centers* of a cluster, but the notion of a cluster involves both the core as well as the neighbors.

These issues are considered more closely in the remainder of the current section.

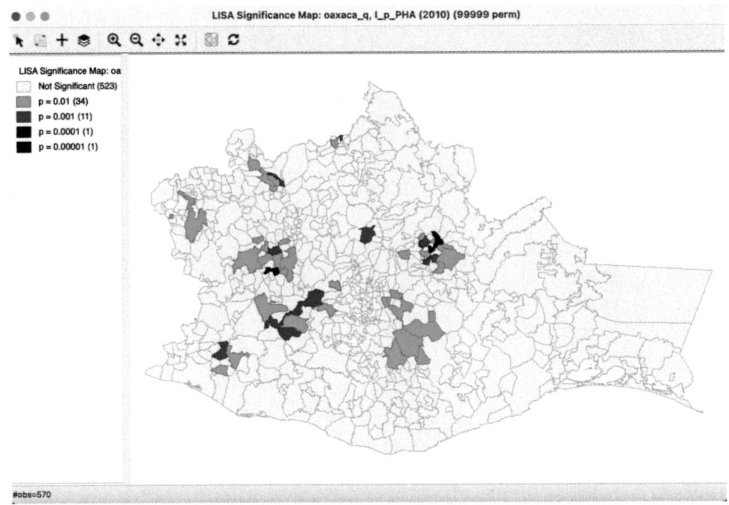

Figure 16.9: Significance map – $p < 0.01$

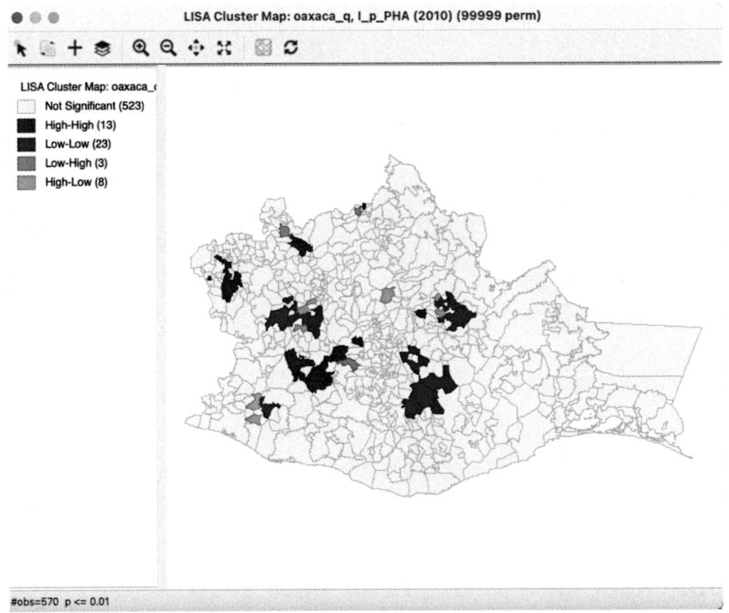

Figure 16.10: Cluster map – $p < 0.01$

16.5.1 Multiple comparisons

An important methodological issue associated with the local spatial autocorrelation statistics is the selection of the p-value cut-off to properly reflect the desired Type I error. Not only are the pseudo p-values not analytical, since they are the result of a computational permutation process, but they also suffer from the problem of *multiple comparisons* (for a detailed discussion, see de Castro and Singer, 2006; Anselin, 2019a). The bottom line is that a traditional choice of 0.05 is likely to lead to many false positives, i.e., rejections of the null when in fact it holds.

There is no completely satisfactory solution to this problem, and no strategy yields an unequivocal *correct* p-value. A number of approximations have been suggested in the literature, of which the best known are the *Bonferroni bounds* and *False Discovery Rate* approaches. However, in general, the preferred strategy is to carry out an extensive sensitivity analysis, to yield insight into what may be termed *interesting* locations, rather than *significant* locations, following the suggestion in Efron and Hastie (2016).

In GeoDa, these strategies are implemented through the **Significance Filter** option (the second item in the options menu).

16.5.1.1 Adjusting p-values

Adjusting the cut-off p-value as an option of the **Significance Filter** is straightforward. One can select one of the pre-selected p-values from the list, i.e., 0.05, 0.01, 0.001, or 0.0001. For example, choosing 0.01 immediately changes the locations that are displayed in the significance and cluster maps, as shown in Figures 16.9 and 16.10. The number of significant locations drops from 117 to 47 (for 99,999 permutations), with the 70 observations initially labeled as significant at $p < 0.05$ removed from the significance map.

Whereas for $p < 0.05$ the High-High and Low-Low clusters were of comparable size, for $p < 0.01$ there are almost twice as many Low-Low clusters as High-High. Similarly, for the spatial outliers, which were roughly equally represented for $p < 0.05$, for $p < 0.01$ there are more than twice as many High-Low outliers than Low-High outliers.

A sensitivity analysis by adjusting the cut-off p-value in this manner provides some insight into which selected clusters and spatial outliers stand up to high scrutiny vs. which are more likely to be spurious.

A more refined approach to select a p-value is available through the **Custom Inference** item of the significance filter. The associated interface provides a number of options to deal with the multiple comparison problem.

The point of departure is to set a target α value for an overall Type I error rate. In a multiple comparison context, this is sometimes referred to as the Family Wide Error Rate (FWER). The target rate is selected in the input box next to **Input significance**. Without any other options, this is the cut-off p-value used to select the observations. For example, this could be set to 0.1, a value suggested by Efron and Hastie (2016) for *big data analysis*. In the example, this value is kept to the conventional 0.05.

16.5.1.2 Bonferroni bound

The first custom option in the inference settings dialog is the **Bonferroni bound** procedure. This constructs a bound on the overall p-value by taking α and dividing it by the number of multiple comparisons. In the LISA context, the latter is typically taken as the number of observations, n, although alternatives have been proposed as well. These include the average or maximum number of neighbors, or some other measure of the range of interaction.

With the number of observations at 570, the Bonferroni bound in the example would be $\alpha/n = 0.00008772$, the cut-off p-value to be used to determine significance.

Checking the Bonferroni bound radio button in the dialog immediately updates the significance and cluster maps. This typically is a very strict criterion. For example, in the cluster map in Figure 16.11, only two observations meet this criterion, in the sense that their pseudo p-value is less than the cut off.

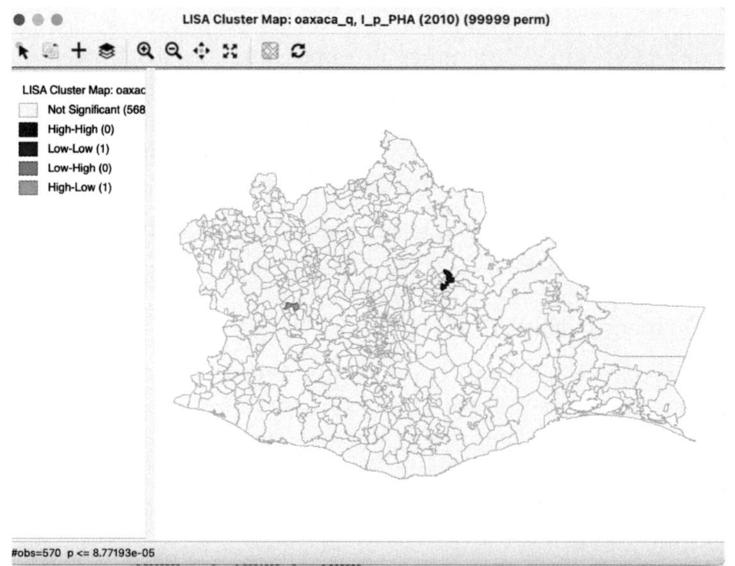

Figure 16.11: Cluster map – Bonferroni bound

16.5.1.3 False Discovery Rate

In practice, the application of a Bonferroni bound is often too strict, in that very few (if any) observations turn out to be significant. In part, this is due to the use of the total number of observations as the adjustment factor, which is almost certainly too large, hence the resulting critical p-value will be too small.

A slightly less conservative option is to use the False Discovery Rate (FDR), first proposed by Benjamini and Hochberg (1995). It is based on a ranking of the pseudo p-values from the smallest to the largest. The FDR approach then consists of selecting a cut-off point as the first observation for which the pseudo p-value is not smaller than the product of the rank with α/n. With the sorted order as i, the critical p-value is the value in the sorted list that corresponds with the sequence number i_{max}, the largest value for which $p_{i_{max}} \leq i \times \alpha/n$.

To illustrate this method, a few additional columns are added to the data table in the example. First, as shown in Figure 16.12, the column **LISA_p** containing the pseudo p-values is sorted in increasing order (signified by the > symbol next to the variable name). Next, a column with the sorted order is added, as **pOrder**.[2] A new column is added with the **FDR** reference value. This is the product of the sequence number in **pOrder** with α/n, which for the first observations is the same as the Bonferroni bound, shown as 0.000087 in the table. The value for the second observation is then 2×0.000087 or 0.000174.[3]

The full result is shown in Figure 16.12, with the observations that meet the FDR criterion highlighted in yellow. For the 17th ranked observation (location 55), the pseudo p-value of 0.00139 still meets the criterion of 0.001479, but for the 18th ranked observation, this is no longer the case ($0.001730 > 0.001556$).

[2]This is accomplished by means of the **Table Calculator**, using **Special > Enumerate**.

[3]Again, this is accomplished in the **Table Calculator**. First, the new column/variable **FDR** is set to the constant value α/n, using **Univariate > Assign**. Next, the **FDR** column is multiplied with the value for **pOrder**, using **Bivariate > Multiply**.

	MORAN_LAG	LISA_I	LISA_CL	LISA_P >	pOrder	FDR
158	-1.415779	1.729314	2	0.000010	1	0.000087
254	-1.509848	-0.083773	4	0.000070	2	0.000174
110	1.047833	0.871688	1	0.000160	3	0.000261
260	1.038013	1.960268	1	0.000170	4	0.000348
261	-1.467932	1.842264	2	0.000190	5	0.000435
116	-1.487060	2.378390	2	0.000240	6	0.000522
451	1.281311	1.236016	1	0.000340	7	0.000609
467	1.117812	0.201827	1	0.000450	8	0.000696
262	-1.130977	-0.757187	4	0.000560	9	0.000783
194	-1.198495	-0.148593	4	0.000720	10	0.000870
232	1.470119	1.703623	1	0.000800	11	0.000957
563	-1.071213	0.409235	2	0.000890	12	0.001044
264	-1.309143	-0.140566	4	0.000920	13	0.001131
435	-1.562965	3.279877	2	0.001060	14	0.001218
399	1.091277	0.994853	1	0.001100	15	0.001305
239	-1.045961	0.480765	2	0.001200	16	0.001392
55	-1.180935	1.560607	2	0.001390	17	0.001479
548	-0.824969	0.373297	2	0.001730	18	0.001566
495	-0.897751	0.702272	2	0.001840	19	0.001653
433	-1.302451	2.997747	2	0.002330	20	0.001740

Figure 16.12: False Discovery Rate calculation

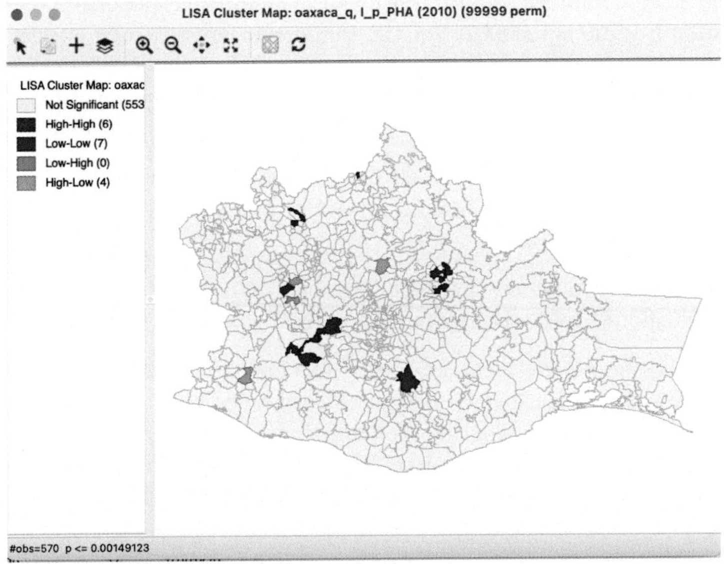

Figure 16.13: Cluster map – FDR

Checking the **False Discovery Rate** radio button will update the significance and cluster maps accordingly, displaying only 17 significant locations, as shown in Figure 16.13. At this point, the High-High and Low-Low clusters are again fairly balanced (respectively, 6 and 7), but there are no longer any Low-High spatial outliers (there are 4 remaining High-Low outliers).

16.5.2 Interpretation of significance

As mentioned, there is no fully satisfactory solution to deal with the multiple comparison problem. Therefore, it is recommended to carry out a sensitivity analysis and to identify the stage where the results become *interesting*. A mechanical use of 0.05 as a cut off value is definitely *not* the proper way to proceed.

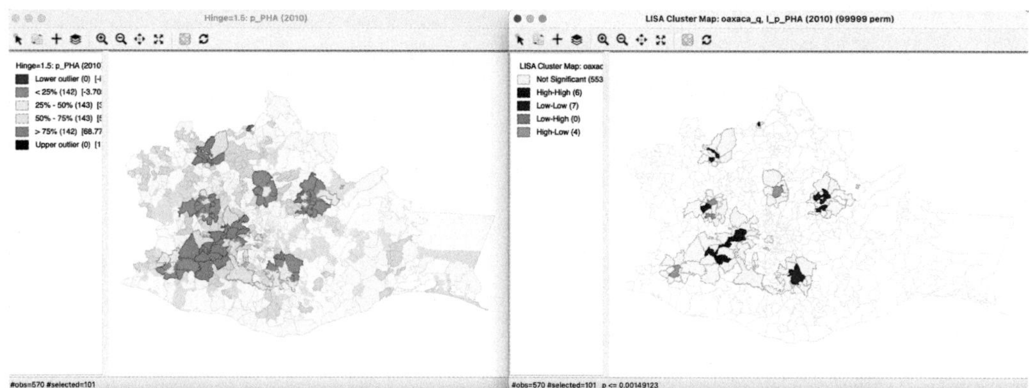

Figure 16.14: Cores and neighbors

Also, for the Bonferroni and FDR procedures to work properly, it is necessary to have a large number of permutations, to ensure that the minimum p-value can be less than α/n. Currently, the largest number of permutations that `GeoDa` supports is 99,999, and thus the *smallest* possible p-value is 0.00001. For the Bonferroni bound to be effective, α/n must be greater than 0.00001, or n should be less than $\alpha/0.00001$. This is not due to a characteristic of the data, but to the lack of sufficient permutations to yield a pseudo p-value that is small enough.

In practice, this means that with $\alpha = 0.01$, data sets with $n > 1000$ cannot have a single significant location using the Bonferroni criterion. With $\alpha = 0.05$, this value increases to 5000, and with $\alpha = 0.1$ to 10,000.

The same limitation applies to the FRD criterion, since the cut-off for the first sorted observation corresponds with the Bonferroni bound $(1 \times \alpha/n)$.

Clearly, this drives home the message that a mechanical application of p-values is to be avoided. Instead, a careful sensitivity analysis should be carried out, comparing the locations of clusters and spatial outliers identified for different p-values, including the Bonferroni and FDR criteria, as well as the associated neighbors to suggest *interesting* locations that may lead to new hypotheses and help to *discover the unexpected.*

16.5.3 Interpretation of clusters

Strictly speaking, the locations shown as significant on the significance and cluster maps are not the actual *clusters*, but the *cores* of a cluster. In contrast, in the case of spatial outliers, the observations identified are the actual locations of interest.

In order to get a better sense of the spatial extent of the cluster, the **Select All...** option offers a number of ways to highlight cores, their neighbors, or both.

The first option selects the cores, i.e., all the locations shown as non-white in the map. This is not so much relevant for the cluster or significance map, but rather for any maps that are linked. The selection of the **Cores** will select the corresponding observations in any other map or graph window.

The next option does not select the cores themselves, but their **Neighbors**. Again, this is most relevant when used in combination with linked maps or graphs.

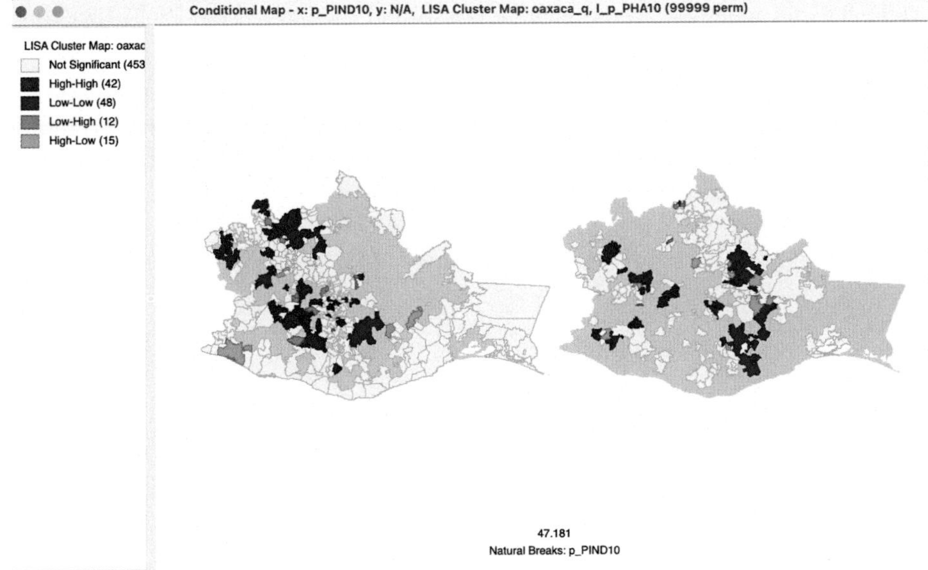

Figure 16.15: Conditional cluster map

The third option selects both the cores and their neighbors (as defined by the spatial weights). This is most useful to assess the spatial range of the areas identified as *clusters*. For example, in Figure 16.14, the **Cores and Neighbors** option is applied to the significant locations obtained using the FDR criterion. The non-significant neighbors are shown in grey on the cluster map on the right, with all the matching cores and neighbors highlighted in the box map on the left.

A careful application of this approach provides insight into the spatial range of interaction that corresponds with the respective cluster cores.

16.6 Conditional Local Cluster Maps

A final option for the Local Moran statistic is that the cluster maps can be incorporated in a conditional map view, similar to the conditional maps considered in Section 8.4.3. This is accomplished by selecting the **Show As Conditional Map** option (the fourth entry in the options menu). The resulting dialog is the same as for the standard conditional map.

In Figure 16.15, the cluster map for access to health care is turned into two micromaps, conditioned by the percentage of the population that speaks an indigenous language on the horizontal axis (**p_IND10**). An additional variable could be taken into account to condition on the vertical axis, but that is not considered here. The number of intervals (**Horizontal Bin Breaks**) was changed to two (from the default three), using **Natural Breaks** as the criterion to determine the cut-off value.

The point of departure is that, in the absence of any interaction effect, the two micromaps should have the same mix of High-High and Low-Low clusters and spatial outliers. In the example, that does not seem to be the case, with a larger share of the Low-Low clusters on the right (i.e., for more indigenous communities), and the bulk of the High-High clusters on the left.

It should be noted that this example is purely illustrative of the functionality available through the conditional cluster map feature, rather than as a substantive interpretation. As always, the main focus is on whether the micromaps suggest different patterns, which would imply an interaction effect with the conditioning variable(s).

17

Other Local Spatial Autocorrelation Statistics

This chapter continues the exploration of local spatial autocorrelation statistics. First, several extensions of the Local Moran are introduced, such as the *Median Local Moran*, the *Differential Local Moran* and a specialized version that deals with the variance instability in rates or proportions, the *EB Local Moran*. In addition, a local version of Geary's c statistic is discussed. The chapter also takes an in-depth look at an important class of local statistics introduced by Getis and Ord. These statistics are not a LISA in a strict sense, but nevertheless are effective tools to discover local hot spots and cold spots.

The chapter closes with a brief discussion of the comparative merits of the local statistics covered so far.

These methods are again illustrated by means of the *Oaxaca Development* sample data set.

17.1 Topics Covered

- Assess the sensitivity of Local Moran to the use of a median spatial lag instead of an average spatial lag
- Identify clusters and outliers in the change of a variable over time
- Correct the local Moran statistic for variance instability in rates
- Identify clusters and outliers by means of the Local Geary statistic
- Identify hot spots and cold spots by means to the Gi and Gi* statistics
- Carry out sensitivity analysis
- Compare the results yielded by different LISA

GeoDa Functions

- Space > Median Local Moran's I
- Space > Differential Local Moran's I
- Space > Local Moran's I with EB rate
- Space > Univariate Local Geary
- Space > Local G
- Space > Local G*

DOI: 10.1201/9781003274919-17

17.2 Extensions of the Local Moran

Three extensions of the Local Moran are considered. Two of these are local extensions of the specialized Moran scatter plots considered in Section 14.2.1 for the differential Moran scatter plot, and in Section 14.2.2 for the Moran scatter plot with EB rates. The third extension uses a different concept of spatial lag. Instead of the average of the values in neighboring locations, the *Median Local Moran* uses the median. This is considered first.

As was the case for their global counterparts, the other two extensions are primarily shortcuts to facilitate computations. They are straightforward applications of the Local Moran statistic to pre-computed variables. In one case, it is the difference between observations at two periods in time, yielding the *Differential Local Moran*, in the other case, rates are standardized to control for variance instability, resulting in the *EB Local Moran.*

17.2.1 Median Local Moran

In the discussion of spatial autocorrelation statistics so far, a spatially lagged variable was defined as $\sum_j w_{ij} z_j$, or the average of the values observed at the neighboring locations. However, as is well known, the average may be sensitive to the presence of outliers. This may pull the average up (or down), even when many of the neighbors do not have high (low) values, creating a potentially misleading impression of a cluster or spatial outlier.

An alternative can be based loosely on the idea of a *median smoother* (e.g., Wall and Devine, 2000). In the latter, the value at a location (typically a rate) is replaced by the median of the neighboring locations (see also Section 12.3.1.3). In the context of the Local Moran, the median of the neighbors is used *in the place of* the average as a *median spatial lag.*

Consequently, the Median Local Moran becomes:

$$I_i^M = z_i \times \text{med}(z_j, j \in N_i),$$

where N_i is the neighbor set of location i (i.e., those locations for which $w_{ij} \neq 0$).

Inference and interpretation are identical to that for the original Local Moran, based on a conditional permutation approach (see Section 16.5).

17.2.1.1 Implementation

The Median Local Moran is invoked as the second item in first group of the **Cluster Maps** drop-down list from the toolbar, or as **Space > Univariate Median Local Moran's I** from the menu.

This brings up the usual variable selection dialog. All the options are the same as for the conventional Local Moran, i.e., the randomization, significance filter and saving of the results (see Chapter 16).

These features are illustrated using the variable **p_PHA(2010)** (or, **p_PHA10**) from the Oaxaca data set, with queen contiguity. The significance map for 99,999 permutations is shown in Figure 17.1, with the result for the Median Local Moran in the right-hand panel, compared to the conventional Local Moran on the left.

Overall, with a $p < 0.05$ cut-off, there are 18 fewer significant locations for the Median Local Moran compared to the conventional version. These are distributed over the different

Figure 17.1: Significance maps conventional and Median Local Moran

categories as 10 fewer for $p < 0.05$, 5 fewer for $p < 0.01$, 2 fewer for $p < 0.001$ and one at $p < 0.00001$, but none at $p < 0.0001$. However, the changes work in both directions, with some locations becoming significant for the Median Local Moran, that were not for the conventional Local Moran, and the other way around. In addition, there are changes in the category of significance in both directions (both less and more significant). For example, the blue highlight in Figure 17.1 points to a municipality that was not significant for the conventional Local Moran, but becomes so at $p < 0.01$ for the Median Local Moran. In contrast, the red highlighted location was significant for the conventional Local Moran and becomes insignificant for the median version.

17.2.1.2 Clusters and spatial outliers

The classification of significant locations for the Median Local Moran uses the same principle as for the conventional Local Moran, but now the reference points are the *median* of the (standardized) variable and the *median* of its median spatial lag. In a scatter plot of the median lag on the original variable, the center point is drawn at the location of the means, not the medians. Consequently, the resulting quadrants are not appropriate to derive the categories of spatial association. For example, for the standardized variable computed from **p_PHA10**, the mean is obviously 0 by construction, but the median is 0.1613. Similarly, the mean and median for the median spatial lag (on the vertical axis) are not the same, with the latter equal to 0.1786.

As a result, the classification into High-High or Low-Low clusters or spatial outliers must be done relative to the median values and *not* relative to the means shown in a conventional scatter plot.

This is further illustrated in Figures 17.2 and 17.3 for spatial clusters, and Figures 17.4 and 17.5 for spatial outliers.[1]

In the left-hand panel of Figure 17.2, the observations are selected that are above the median for the (standardized) variable and above the median for the median spatial lag. Clearly, this selection does not match the conventional High-High criterion based on the means. In the cluster map on the right, the subset of the selected observations that is also significant (see Figure 17.1) is classified as High-High in red, the non-significant locations are colored grey. A similar operation is carried out (under the hood) to find the Low-Low clusters, shown in

[1]The selection is obtained using two steps in the **Selection Tool**: first selecting the observations that meet the criterion for the variable, followed by a **Select From Current Selection** for the median spatial lag.

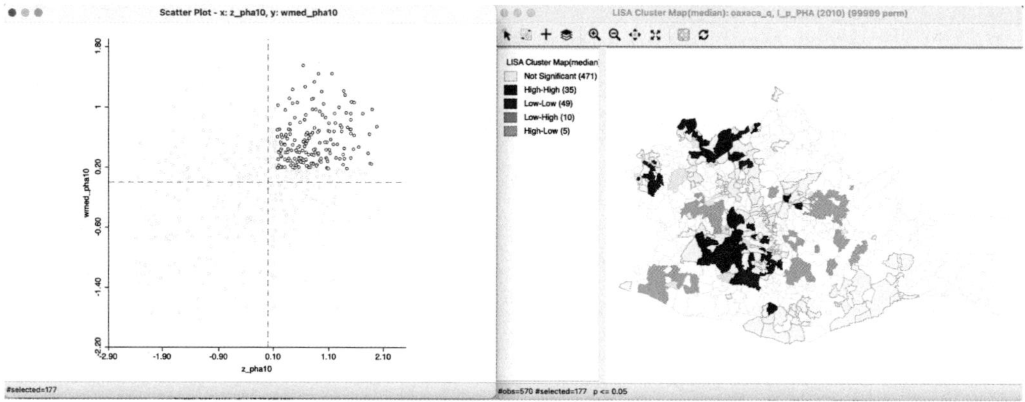

Figure 17.2: High-High clusters in Median Local Moran

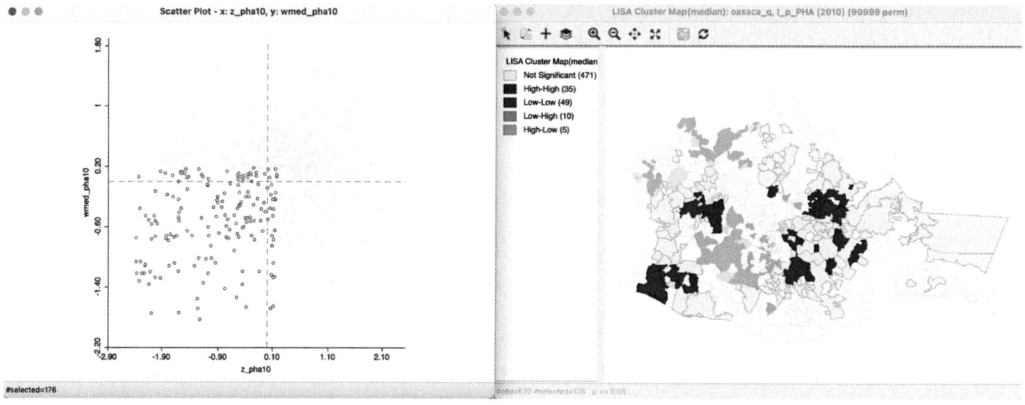

Figure 17.3: Low-Low clusters in Median Local Moran

Figure 17.3. In this case, since the median is above the mean for both axes, the selection includes observations that fall in different quadrants for the conventional Local Moran.

The same principle is applied to the spatial outliers, illustrated in Figures 17.4 and 17.5. Again, the classification into the respective categories differs from what the conventional scatter plot would yield. The significant observations correspond with locations classified as spatial outliers in the cluster map on the right.

17.2.1.3 Median Moran cluster map

A comparison of the cluster map obtained by the conventional Local Moran and the Median Local Moran (for 99,999 permutations, with $p < 0.05$) is illustrated in Figure 17.6. In the current example, the difference in significance results in clusters that are much more limited in range compared to the conventional Local Moran. For example, both the High-High cluster highlighted in the blue rectangle and the Low-Low cluster highlighted in the red rectangle are much smaller for the Median Local Moran. In addition, the classification of the significant locations changes as well. For example, in the green rectangle, two High-Low outliers for the conventional Local Moran become part of a Low-Low cluster.

Overall, a comparison of the results for the Median Local Moran to the conventional Local Moran provides insight into the sensitivity of the results to potential outliers. It should

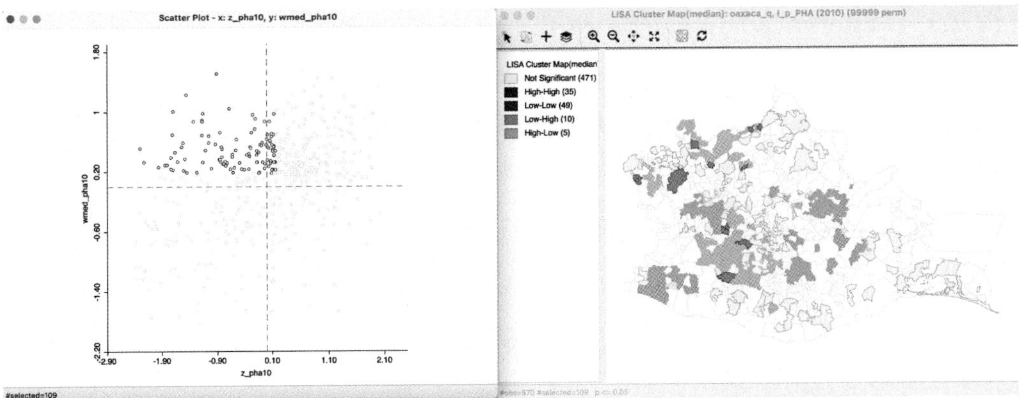

Figure 17.4: Low-High spatial outliers in Median Local Moran

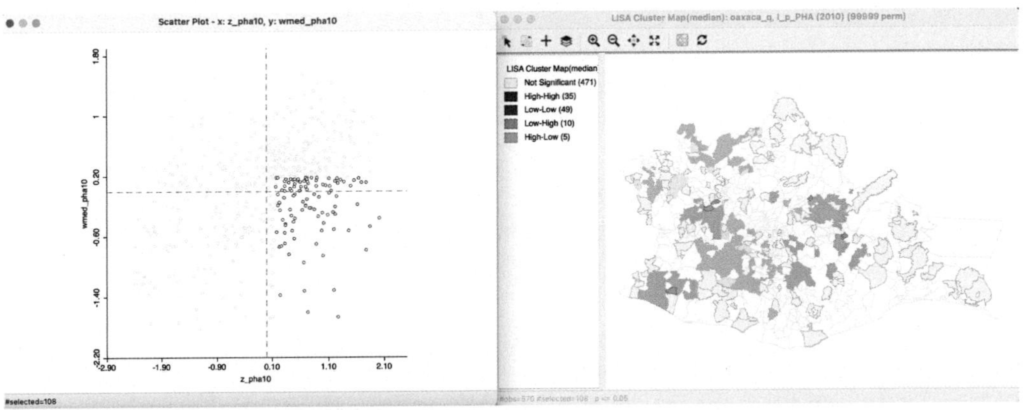

Figure 17.5: High-Low spatial outliers in Median Local Moran

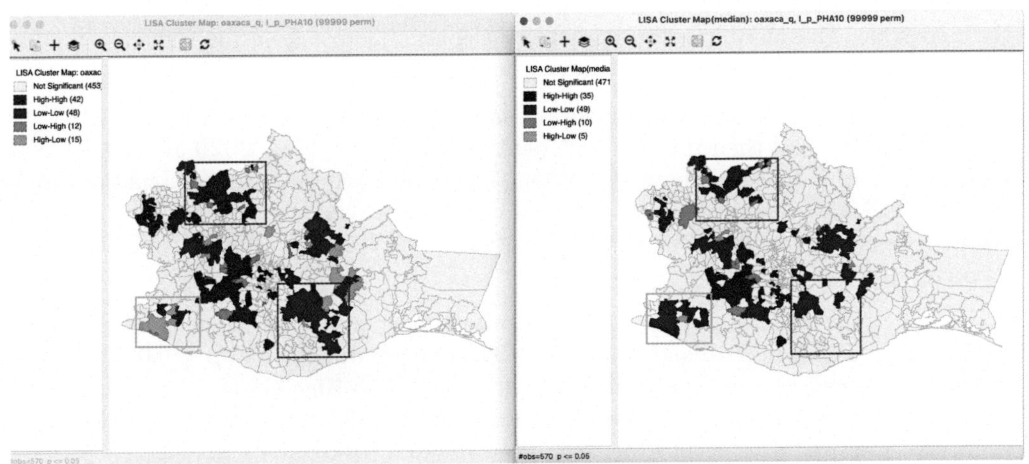

Figure 17.6: Cluster maps conventional and Median Local Moran

be part of a standard sensitivity analysis, together with an assessment of different p-value cut-offs.

17.2.2 Differential Local Moran

The Differential Local Moran statistic is the local counterpart to the Differential Moran scatter plot, discussed in Section 14.2.1. Instead of using the observations on a variable at two different time periods separately, this statistic is based on the change over time, i.e., the difference between y_t and y_{t-1}. Note that this is the actual difference and not the absolute difference, so that a positive change will be viewed as *high*, and a negative change as *low*. The differences are used in standardized form, i.e., they are not the differences between the standardized variable at two points in time, but the standardized differences between the original values for the variable.

The formal expression for this statistic follows the same logic as before, and consists of the cross product of the difference between y_t and y_{t-1} at i with the associated spatial lag:

$$I_i^D = a(y_{i,t} - y_{i,t-1}) \sum_j w_{ij}(y_{j,t} - y_{j,t-1}).$$

The scaling constant a can be ignored. In essence, this is the same as the conventional Local Moran applied to the difference, but the implementation is based on selecting the two variables and computing the difference under the hood, rather than computing the difference separately.

As before, inference is based on conditional permutation. All the usual caveats hold about multiple comparisons and the choice of a p-value. In all respects, the interpretation is the same as for the conventional Local Moran statistic.

17.2.2.1 Implementation

The Differential Local Moran is invoked as the fourth item in first group of the **Cluster Maps** drop-down list from the toolbar, or as **Space > Differential Local Moran's I** from the menu.

As in the global case, the variables under consideration must be time-enabled (grouped) in the data table. The variable selection dialog is slightly different from the standard interface, but the same as for the differential Moran scatter plot.

Continuing to use the (time-enabled) Oaxaca data set, first, the variable of interest is selected (here, **p_PHA**), and then the two time periods are chosen (here, **2020** and **2010**). Note that the system is agnostic about the actual time periods, so that any combination can be selected. The statistic is computed for the difference between the time period specified as the first item and that given as the second item. In the example, the spatial weights are **oaxaca_q**.

To provide some context, Figure 17.7 shows the cluster maps for the conventional Local Moran for **p_PHA** in 2010 and 2020 (using 99,999 permutations and $p < 0.05$). The local patterns in the two years are very different, with many High-High clusters from 2010 labeled as Low-Low in 2020, and vice versa. Note that the classification is relative to the mean, which has improved considerably between the two years (from 53.4% to 75.8%).

The significance map and cluster map for the Differential Local Moran (99,999 permutations with $p < 0.05$) are shown in Figure 17.8. Note how the High-High clusters correspond to locations that were Low-Low in 2010, but High-High in 2020, suggesting a grouping of large increases. Reversely, the Low-Low clusters for the difference correspond to locations that were High-High in 2010, but Low-Low in 2020, suggesting a grouping of small increases, or even decreases.

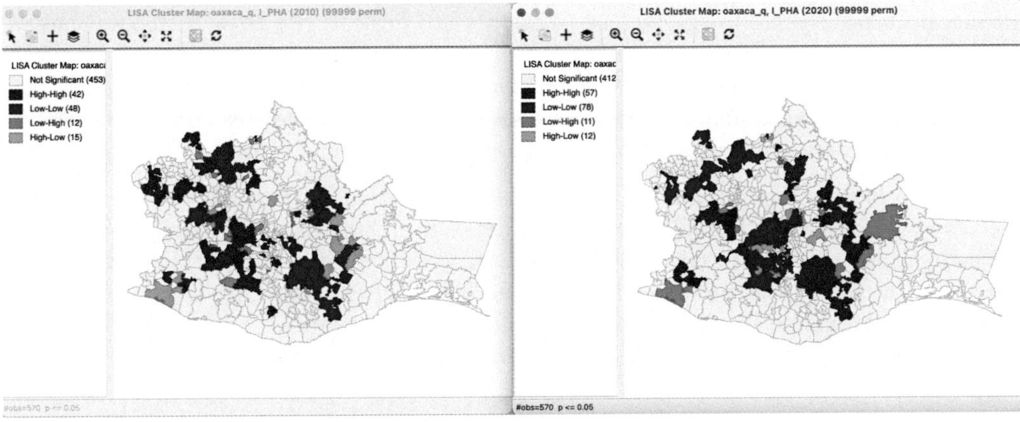

Figure 17.7: Cluster maps for Local Moran in 2010 and 2020

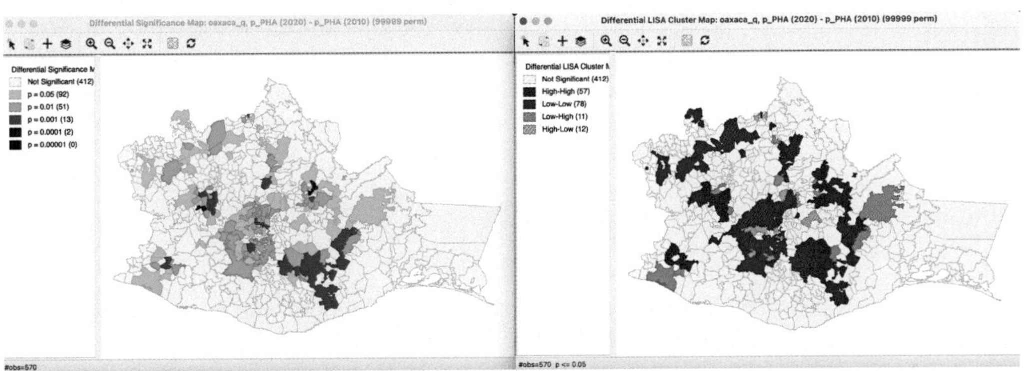

Figure 17.8: Differential Local Moran significance and cluster maps

All the options, such as randomization and significance filter are the same as for the conventional local Moran and will not be further discussed here. The only slight difference is in how the results are saved. Similar to the functionality for the differential Moran scatter plot, the **Save Results** option includes an item to save the actual change variable (in raw form, not in standardized form). In the dialog, this corresponds to the **Diff Values** item with default variable name **DIFF_VAL2**. The other options are the same as for all local spatial autocorrelation statistics, i.e., the value of the statistic (**LISA_I**), cluster type (**LISA_CL**) and p-value (**LISA_P**).

Once the difference is saved as a separate variable, it can be used in a conventional univariate Local Moran operation.

17.2.2.2 Interpretation

The significance and cluster maps for a Differential Local Moran identify the locations where the change in the variable over time is matched by similar/dissimilar changes in the surrounding locations. It is important to keep in mind that the focus is on change, and there is no direct connection to whether this changes is from high or from low values.

Two situations can be distinguished, depending on whether the change variable takes on both positive and negative values, or when all the changes are of the same sign (i.e., all observations either increase or decrease over time).

When both positive and negative change values are present, the *High-High* locations will tend to be locations with a large increase (positive change), surrounded by locations with similar large increases. The *Low-Low* locations will be observations with a large decrease (negative change), surrounded by locations with similar large decreases. Spatial outliers will be locations where an increase is surrounded by a decrease and vice versa.

When all changes are of the same sign, the interpretation of High-High and Low-Low is relative to the mean. When all the changes are positive, large increases surrounded by other large increases will be labeled High-High, whereas small(er) increases surrounding by other small(er) increases will be labeled Low-Low. When all the signs are negative, small(er) decreases surrounded by other small(er) decreases will be labeled as High-High, whereas large decreases surrounded by other large decreases will be labeled Low-Low.

17.2.3 Local Moran with EB Rate

The third extension of the Local Moran pertains to the special case where the variable of interest is a rate or proportion. As discussed for the Moran scatter plot in Section 14.2.2, the resulting variance instability can cause problems for the Moran statistic. The EB standardization suggested by Assuncao and Reis (1999) for the global case can be extended to the local statistic in a straightforward manner. The statistic has the usual form, but is computed for the standardized rates, z:

$$I_i^{EB} = az_i \sum_j w_{ij} z_j,$$

The standardization of the raw rate r_i is the same as before and is repeated here for completeness (for a more detailed discussion, see Section 14.2.2):

$$z_i = \frac{r_i - \beta}{\sqrt{\alpha + (\beta / P_i)}}$$

with β as an estimate of the mean and the denominator as an estimate of the standard error.[2]

All inference and interpretation is the same as for the conventional case and is not further pursued here.

17.2.3.1 Implementation

The local Moran functionality for standardized rates is invoked as the last item in the Moran group on the **Cluster Maps** toolbar icon. Alternatively, it can be selected from the menu as **Space > Local Moran's I with EB Rate**.

Since the rate standardization is computed as part of the operation, the variable selection interface is similar to that used for rate maps. To illustrate this feature, the rate is computed with **DIS20** (number of people with disabilities in 2020) as the **Event Variable**, and **PTOT20** (total population in 2020) as the **Base Variable**, as done earlier in Section 14.2.2.1. A box map of the raw rate is shown in the left hand panel of Figure 12.9. The spatial weights are queen contiguity, **oaxaca_q**.

[2]To recap, $\beta = \sum_i O_i / \sum_i P_i$, where O_i is the number of events at i and P_i is the population at risk. The estimate of $\alpha = [\sum_i P_i (r_i - \beta)^2]/P - \beta/(P/n)$,, with n as the total number of observations, such that P/n is the average population. Note that the estimate of α can be negative, in which case it is set to zero.

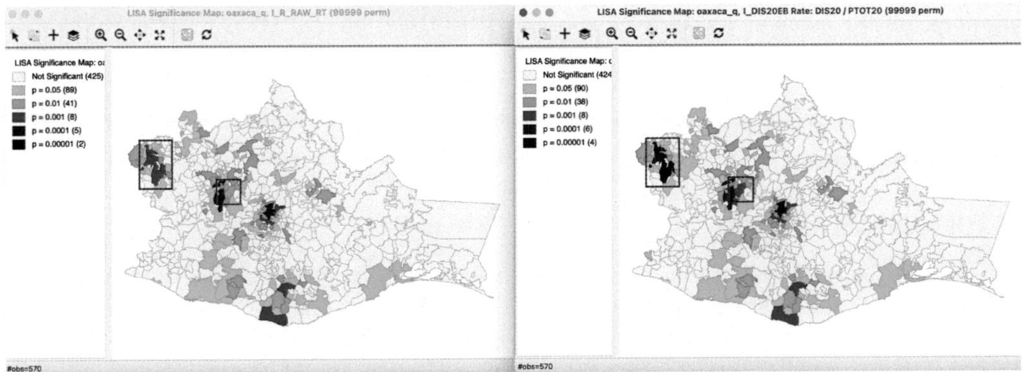

Figure 17.9: Significance maps – raw rate and EB rate

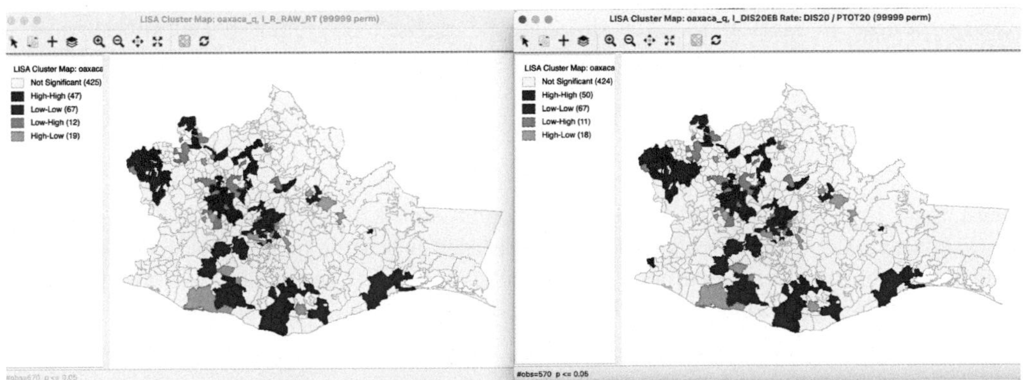

Figure 17.10: Cluster maps – raw rate and EB rate

The resulting significance map is shown in the right-hand panel of Figure 17.9, next to the corresponding map for the crude rate (for 99,999 permutations with $p < 0.05$). Compared to the map for the crude rate, there is one more significant location. Interestingly, the main difference is at the high end of the significance, where the EB Local Moran has slightly more significant locations. For example, the location highlighted in the blue rectangle is significant at $p < 0.00001$ for EB Local Moran, but at $p < 0.001$ for the crude rate. Similarly, the location highlighted in the red rectangle is significant at $p < 0.001$ for the EB Local Moran, but at $p < 0.01$ for the crude rate. Since the same number of permutations is used in both cases, the p-values are comparable. Overall, however, the differences are minimal and rather subtle, confirming the findings for the Moran scatter plot.

The differences between the cluster map for the crude rate and for the EB Local Moran, shown in Figure 17.10, are equally subtle. The EB Local Moran cluster map has three more locations in the High-High group, and one less in each of the Low-High and High-Low spatial outliers. Otherwise, the identified locations and groups match exactly.

17.2.3.2 Saving the results

As for the EB Moran scatter plot, the **Save Results** option includes an item to save the actual EB rate. This operation is similar to the EB Rate standardization that can be computed with the **Table Calculator** option (see Section 6.4.3.2). In the dialog, it corresponds to the **EB Rates** item, with default variable name **LISA_EB**. The other options are the same as for all local spatial autocorrelation statistics, i.e., the value of the statistic, cluster type and p-value, with the same default variable names.

17.3 Local Geary

The Local Geary statistic, first outlined in Anselin (1995), and further elaborated upon in Anselin (2019a), is a Local Indicator of Spatial Association (LISA) based on a measure of attribute similarity that differs from the cross-product. As in its global counterpart, the focus is on squared differences, or, rather, *dissimilarity*. In other words, small values of the statistics suggest positive spatial autocorrelation, whereas large values suggest negative spatial autocorrelation.

As introduced in Section 13.4.3.2, the Geary c statistic of global spatial autocorrelation (Geary, 1954) takes on the following form:

$$c = \frac{\sum_i \sum_j w_{ij}(x_i - x_j)^2 / 2S_0}{\sum_i (x_i - \bar{x})^2/(n-1)},$$

with $S_0 = \sum_i \sum_j w_{ij}$, and where the x in the numerator do not need to be in standardized form, due to the squared difference. The statistic has a mean value of 1 under the null hypothesis of spatial randomness. Significant values less than 1 indicate positive spatial autocorrelation and values larger than 1 negative spatial autocorrelation.

After controlling for the parts in the expression that do not change with i, a local version of the statistic can be found as (for technical details, see Anselin, 1995):

$$c_i = \sum_j w_{ij}(x_i - x_j)^2,$$

or as:

$$c_i = (1/m_2) \sum_j w_{ij}(x_i - x_j)^2,$$

with $m_2 = \sum z_i^2/n$. Again, because of the squared difference, there is no need to standardize x.

The sum over all the local statistics is:

$$\sum_i c_i = n[\sum_i \sum_j w_{ij}(x_i - x_j)^2 / \sum_i z_i^2].$$

In comparison, Geary's global c statistic can be reformulated as:

$$c = [(n-1)/2S_0][\sum_i \sum_j w_{ij}(x_i - x_j)^2 / \sum_i z_i^2].$$

This establishes the connection between the local and the global as:

$$\sum_i c_i = 2nS_0/(n-1)c.$$

Hence, the Local Geary is a LISA statistic in the sense established in Section 16.2.

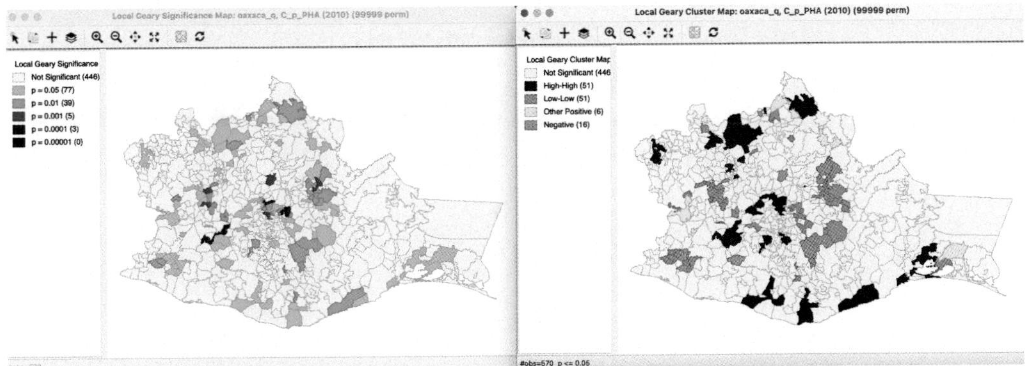

Figure 17.11: Local Geary significance and cluster maps

Closer examination reveals that the Local Geary statistic consists of a weighted sum of the squared distance in attribute space for the geographical neighbors of observation i. Since there is no cross-product involved, there is no direct relation to *linear* similarity. In other words, since the Local Geary uses a different criterion of attribute similarity, it may detect patterns that escape the Local Moran, and vice versa.

As for the Local Moran, analytical inference is based on an approximation and generally not very reliable. Instead, the same conditional permutation procedure as for the Local Moran is implemented. The results are interpreted in the same way, with the caveat regarding the p-values and the notion of significance.

17.3.1 Implementation

the Local Geary can be invoked from the **Cluster Maps** toolbar icon, as the first item in the fourth block in the drop-down list. Alternatively, it can be started from the main menu, as **Space > Univariate Local Geary**.

The subsequent step is the same as before, bringing up the **Variable Settings** dialog that contains the names of the available variables as well as the spatial weights. Everything operates in the same way for all local statistics.

The final dialog offers window options. In the case of the local Geary, there is no Moran scatter plot option, but only the **Significance Map** and the **Cluster Map**. The default is that only the latter is checked, as before.

With the variable **p_PHA(2010)** (or **p_PHA10**), queen contiguity (**oaxaca_q**) and 99,999 permutations, the significance map and cluster map for the Local Geary statistic, using $p < 0.05$ are shown in Figure 17.11.

The significance map uses the same conventions as before, but the cluster map is different. There are four different types of associations, three for clusters and one for spatial outliers. The classification is further elaborated upon in Section 17.3.2. Comparison with the results for a Local Moran statistic is considered in Section 17.3.3.

Overall, there are 108 locations indicating positive spatial autocorrelation, contrasted with only 16 locations for negative spatial autocorrelation.

All the options operate the same for all local statistics, including the randomization setting, the selection of significance levels, the selection of cores and neighbors, the conditional map

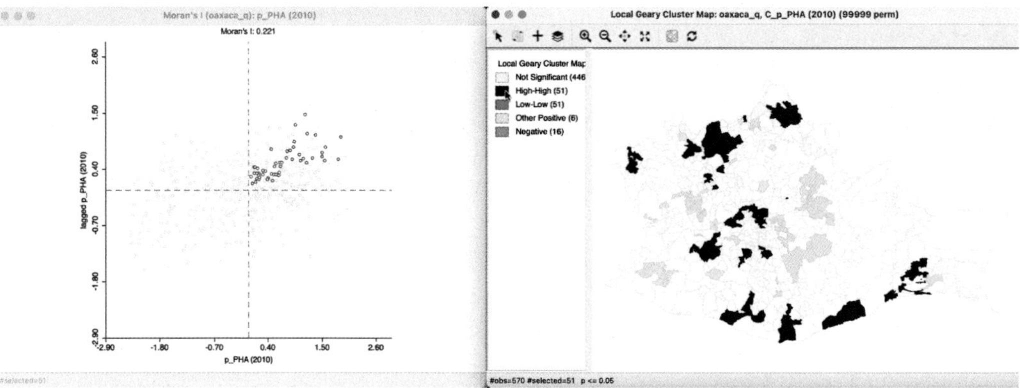

Figure 17.12: Local Geary – High-High clusters

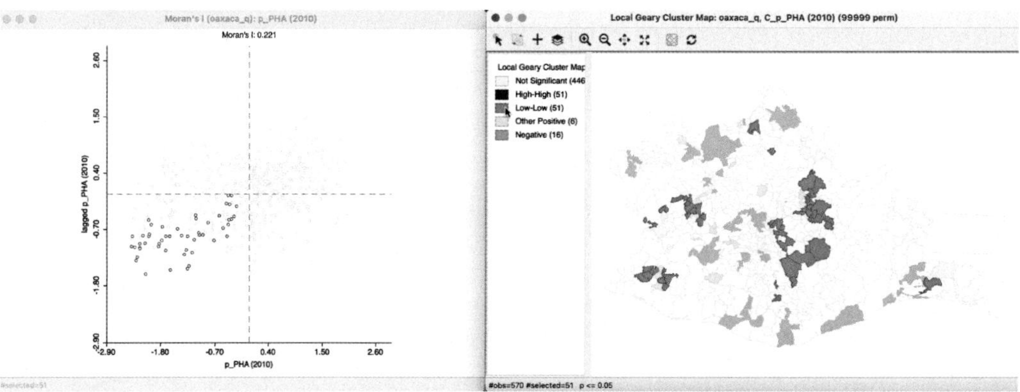

Figure 17.13: Local Geary – Low-Low clusters

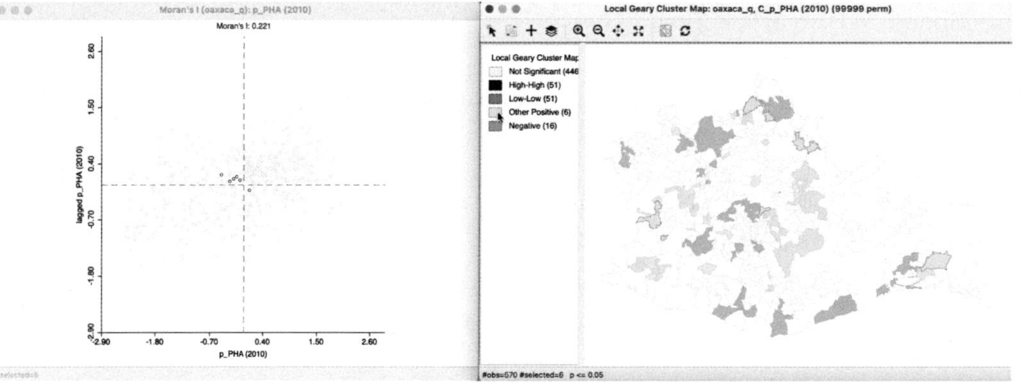

Figure 17.14: Local Geary – Other clusters

option, as well as the standard operations of setting the selection shape and saving the image.

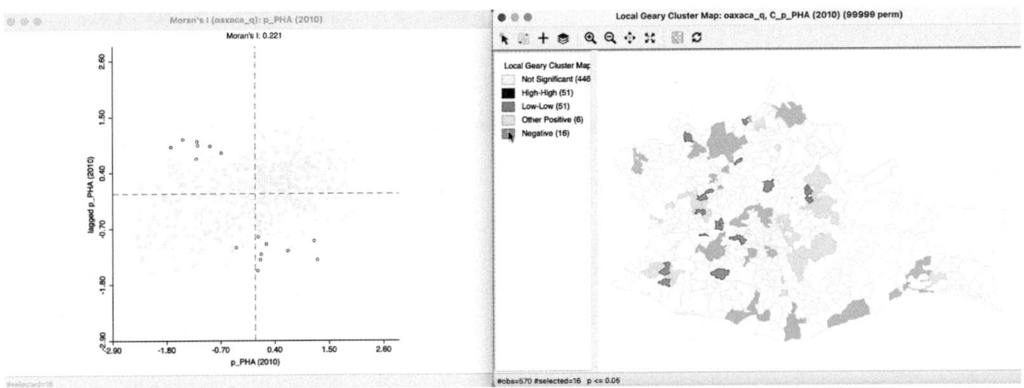

Figure 17.15: Local Geary – Spatial outliers

17.3.1.1 Saving the results

The **Save Results** option again includes the possibility to save the statistic itself (**Geary_I**), the cluster indication (**Geary_CL**) and the significance (**Geary_P**). The code for the cluster classification for the Local Geary is 0 for not significant, 1 for a High-High cluster core, 2 for a Low-Low cluster core, 3 for other positive spatial autocorrelation and 4 for negative spatial autocorrelation.

17.3.2 Clusters and spatial outliers

The interpretation of significant locations in terms of the type of association is not as straightforward for the Local Geary as it was for the Local Moran. In essence, this is because the attribute similarity is a squared difference and not a cross-product, and thus has no direct correspondence with the slope in a scatter plot. Nevertheless, the linking capability within `GeoDa` can be exploited to accomplish classification, albeit an incomplete one.

The locations identified as significant and with the Local Geary statistic smaller than its mean, suggest positive spatial autocorrelation (small differences imply similarity). Those observations that can be classified in the upper-right or lower-left quadrants of a matching Moran scatter plot, are identified as High-High or Low-Low. However, given that the squared difference can cross the mean, there may be observations for which such a classification is not possible. Those are referred to as *other positive* spatial autocorrelation.

For negative spatial autocorrelation (large values imply dissimilarity), it is not possible to assess whether the association is between High-Low or Low-High outliers, since the squaring of the differences removes the sign.

This is illustrated in Figures 17.12 through 17.15.

Figure 17.12 shows the case where significant local positive spatial autocorrelation (small Local Geary) corresponds with points in the upper right quadrant of the Moran scatter plot. This matches the concept of High-High clusters introduced before. The corresponding locations are labeled as such in the Local Geary cluster map.

The case where significant local positive spatial autocorrelation corresponds with points in the lower right quadrant of the Moran scatter plot is illustrated in Figure 17.13. This matches the concept of Low-Low clusters, with the matching observations classified as such in the cluster map.

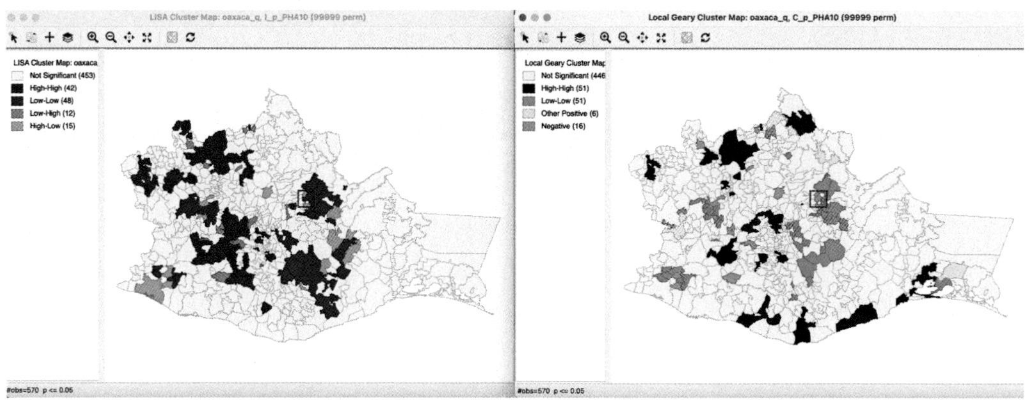

Figure 17.16: Local Geary and Local Moran cluster maps

The third case of significant positive spatial autocorrelation is illustrated in Figure 17.14. There are six observations for which the Moran scatter plot points appear in the lower right and upper left quadrants, suggesting a mismatch between the cross-product and squared difference attribute measures. Those locations are classified as **Other Positive** in the cluster map.

Finally, observations with significant negative spatial autocorrelation (large Local Geary) cannot be classified by type of spatial outlier. They are labeled as **Negative** in the cluster map, as shown in Figure 17.15. All but one of these can also be found in the negative spatial autocorrelation quadrants of the Moran scatter plot, but one is in the lower left quadrant, again suggesting a mismatch between the cross-product and squared difference.

17.3.3 Comparing Local Geary and Local Moran

To gain further insight into the (subtle) difference between the measures of attribute (dis)similarity for the Local Moran and Local Geary statistics, the indications of clusters and spatial outliers obtained by each are considered more closely in what follows. Of course, the findings are not general, and pertain only to the empirical illustration considered here. However, it is useful to carry out such a comparison, especially when interest lies in potential nonlinear associations, which the Local Moran is not able to capture.

Overall, the Local Geary cluster map identifies somewhat more significant locations, yielding 124 such observations relative to 117 for the Local Moran (at $p < 0.05$), as shown in Figure 17.16.

17.3.3.1 Clusters

Figure 17.17 identifies the matches for the 51 High-High cluster locations in the Local Geary cluster map (on the right) in the Local Moran cluster map (on the left). Only 16 of the 42 Local Moran High-High cluster locations belong to the set identified for the Local Geary, but there are no mismatches in the sense of Low-Low clusters identified. Instead, the non-matches are all not significant for the Local Moran, indicated by their grey color in the map. Similarly, the significant High-High locations in the Local Moran cluster map that are not identified with Local Geary are not significant in the latter (not shown).

As shown in Figure 17.18, only a subset of the 48 Low-Low clusters for the Local Moran correspond to the 51 such clusters identified by the Local Geary. The other locations are not

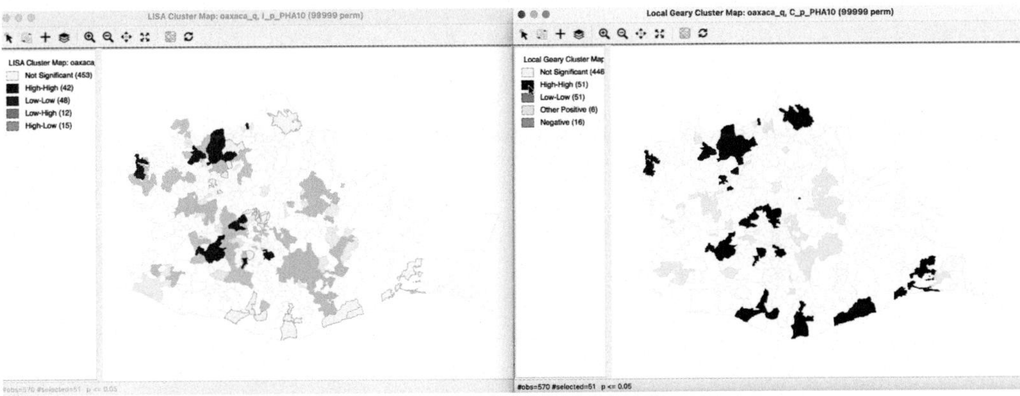

Figure 17.17: Local Geary and Local Moran – High-High clusters

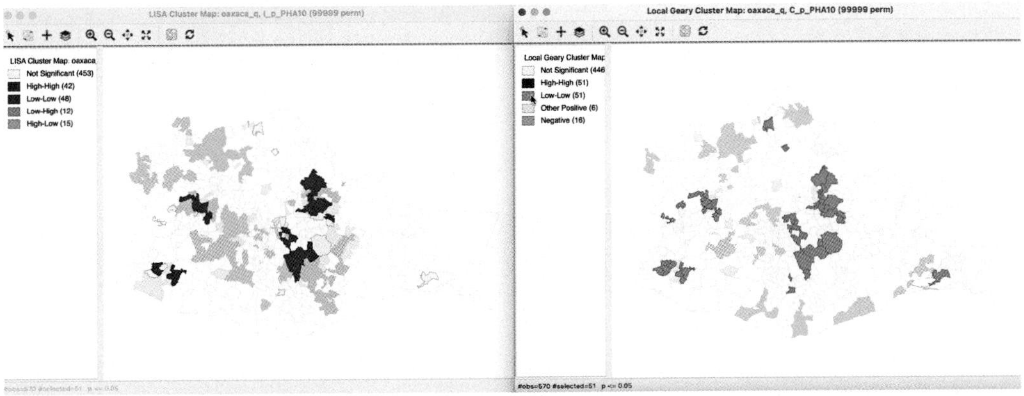

Figure 17.18: Local Geary and Local Moran – Low-Low clusters

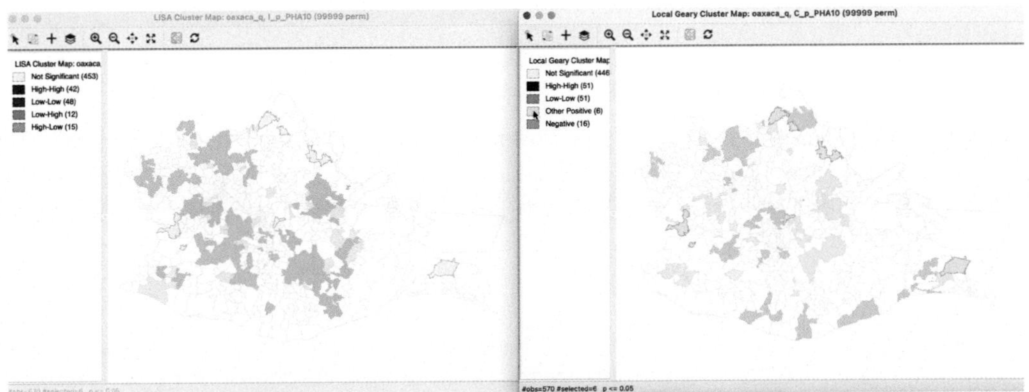

Figure 17.19: Local Geary and Local Moran – Other positive clusters

significant. However, in the reverse order, there is one exception. One of the locations given as a Low-Low cluster by the Local Moran is identified as a spatial outlier (negative) by the Local Geary. The location in question is highlighted in a red rectangle in Figure 17.16. The others are again not significant for Local Geary.

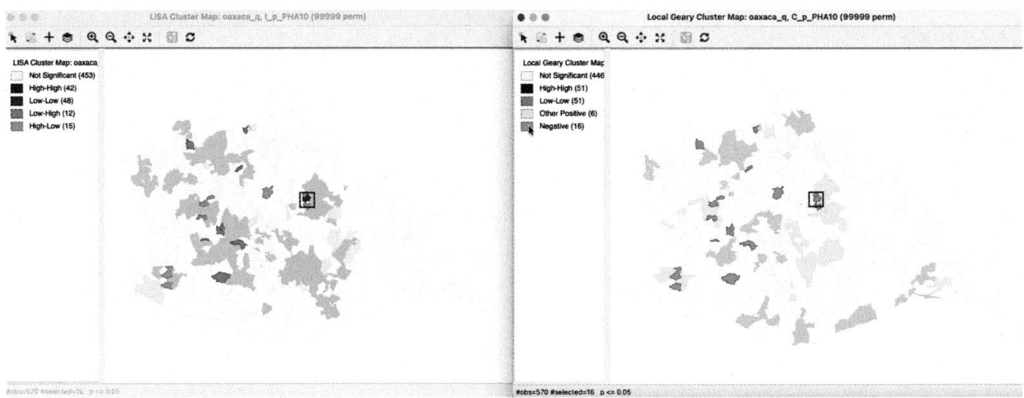

Figure 17.20: Local Geary and Local Moran – Spatial outliers

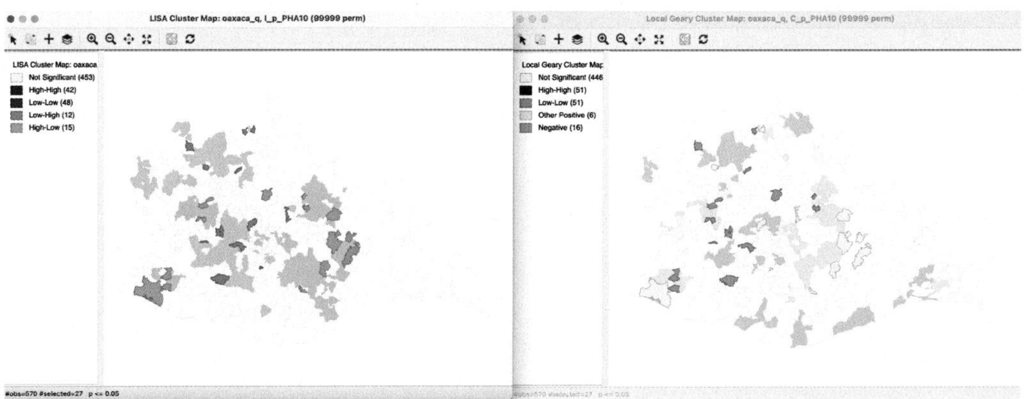

Figure 17.21: Local Moran and Local Geary – Spatial outliers

Finally, the six *other* spatial clusters identified in the Local Geary cluster map correspond to non-significant locations for the Local Moran, as shown in Figure 17.19.

A careful consideration of the differences and similarities between the identified cluster locations for the two local statistics sheds light on the extent to which the association may be nonlinear. This should be followed by an in-depth inspection of the actual statistics and associated attribute values.

17.3.3.2 Spatial outliers

The Local Geary cluster map indicates 16 locations with significant negative spatial autocorrelation, compared to 27 such outliers for Local Moran. As shown in Figure 17.20, all but one of the spatial outliers for Local Geary match outliers for Local Moran. The same location mentioned earlier is part of a Low-Low cluster for the Local Moran, highlighted by the red rectangle in the maps.

As Figure 17.21 illustrates, the spatial outliers in the Local Moran cluster map all match such outliers for Local Geary or are not significant in the Local Geary cluster map, indicated by the 11 grey polygons.

17.4 Getis-Ord Statistics

An early class of statistics for local spatial autocorrelation was proposed by Getis and Ord (1992), and further elaborated upon in Ord and Getis (1995). It is derived from a point pattern analysis logic. In its earliest formulation, the statistic consisted of a ratio of the number of observations within a given range of a point to the total count of points. In a more general form, the statistic is applied to the values at neighboring locations (as defined by the spatial weights). There are two versions of the statistic. They differ in that one takes the value at the given location into account, and the other does not.

The G_i statistic consist of a ratio of the weighted average of the values in the neighboring locations, to the sum of all values, **not including the value at the location** (x_i).

$$G_i = \frac{\sum_{j \neq i} w_{ij} x_j}{\sum_{j \neq i} x_j}$$

In contrast, the G_i^* statistic includes the value x_i **in both numerator and denominator**:

$$G_i^* = \frac{\sum_j w_{ij} x_j}{\sum_j x_j}.$$

Note that in this case, the denominator is constant across all observations and simply consists of the total sum of all values in the data set. The statistic is the ratio of the average values in a window centered on an observation to the total sum of observations.

The interpretation of the Getis-Ord statistics is very straightforward: a value larger than the mean (or, a positive value for a standardized z-value) suggests a High-High cluster or hot spot, a value smaller than the mean (or, negative for a z-value) indicates a Low-Low cluster or cold spot. In contrast to the Local Moran and Local Geary statistics, the Getis-Ord approach does not consider negative spatial autocorrelation (spatial outliers).[3]

Inference can be derived from an analytical approximation, as given in Getis and Ord (1992) and Ord and Getis (1995). However, similar to what holds for the Local Moran and Local Geary, such approximation may not be reliable in practice. Instead, conditional random permutation can be employed, using an identical procedure as for the other statistics.

17.4.1 Implementation

The implementation of the Getis-Ord statistics is largely identical to that of the other local statistics. The **Local G** and **Local G*** options can be selected from the second group in the drop-down menu generated by the **Cluster Maps** toolbar icon. Alternatively, they can be invoked from the menu as **Space > Local G** or **Space > Local G***.

The next step brings up the **Variable Settings** dialog, followed by a choice of windows to be opened. The latter is again slightly different from the previous cases. The default is to **use row-standardized weights** and to generate only the **Cluster Map**. The **Significance Map** option needs to be invoked explicitly by checking the corresponding box. In contrast to previous cases, it is also possible to compute the Getis-Ord statistics using binary (not row-standardized) spatial weights, yielding a simple count of the neighboring values.

[3]When all observations for a variable are positive, as is the case in our examples, the G statistics are positive ratios less than one. Large ratios (more precisely, less small values since all ratios are small) correspond with High-High hot spots, small ratios with Low-Low cold spots.

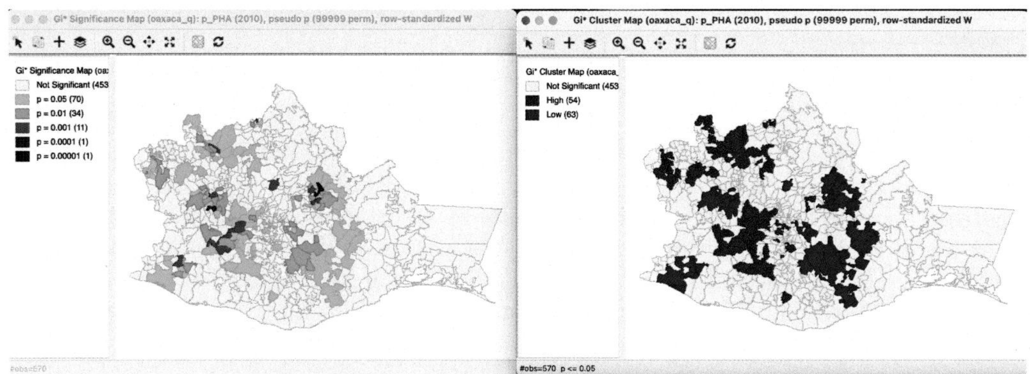

Figure 17.22: Gi* signficance map and cluster map

Continuing with the same example, using **p_PHA10** (or **p_PHA(2010)**) from the Oaxaca data set, with queen contiguity weights, 99,999 permutations and $p < 0.05$ yields the significance and cluster maps for the G_i^* statistic shown in Figure 17.22. In this example, the results are identical to those for the G_i statistic, which is not shown separately.

Overall, 117 locations are identified as significant, the same total as for the Local Moran. In contrast to the latter, the cluster map for the Getis-Ord statistics only takes on two colors, red for hot spots (High-High), and blue for cold spots (Low-Low).

The options menu contains the same items as for the other local statistics, such as the randomization setting, the selection of significance levels, the selection of cores and neighbors, the conditional map option, as well as the standard operations of setting the selection shape and saving the image.

17.4.1.1 Saving the results

The **Save Results** option includes the possibility to save the statistic itself (**G** or **G_STR**), the cluster indication (**C_ID**) and the significance (**PP_VAL**). For the Getis-Ord statistics, there are only three cluster categories, with observations taking the value of 0 for not significant, 1 for a High-High cluster and 2 for a Low-Low cluster.

17.4.2 Clusters and outliers

The interpretation of High-High and Low-Low clusters for the Getis-Ord statistics is somewhat different from that of the Local Moran or Local Geary. As illustrated in Figures 17.23 and 17.24, the emphasis in the Getis-Ord statistics is on the magnitude of the *neighbors.*

The cluster maps on the right have, respectively, the observations in the High-High and Low-Low clusters selected. The corresponding observations are highlighted in the Moran scatter plot on the left.

In Figure 17.23, the High-High cluster observations all have spatial lags (average of neighbors) above the mean, whereas the variable itself takes values both above and below its mean. For the Low-Low cluster observations in Figure 17.24, all the spatial lags are below the mean, with again the variable itself taking values both above and below the mean.

Consequently, the Getis-Ord statistics include significant locations as High-High clusters that would be classified as Low-High spatial outliers according to the Local Moran. Similarly, the Getis-Ord statistics include significant locations as Low-Low clusters that would be

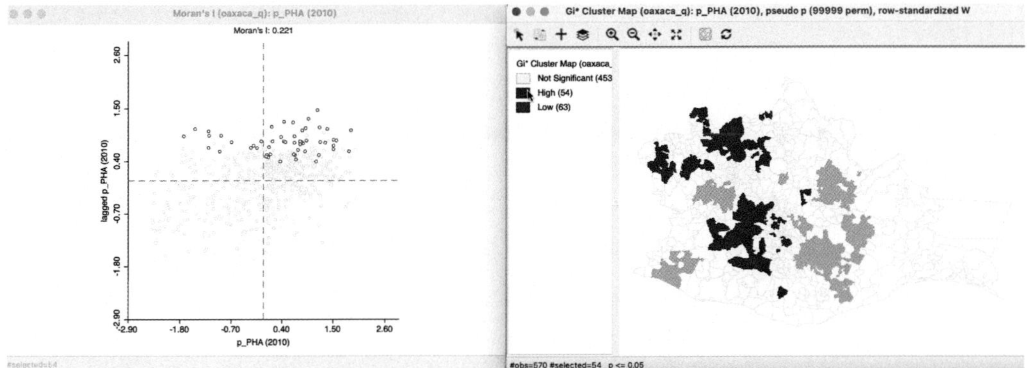

Figure 17.23: Gi* High-High clusters

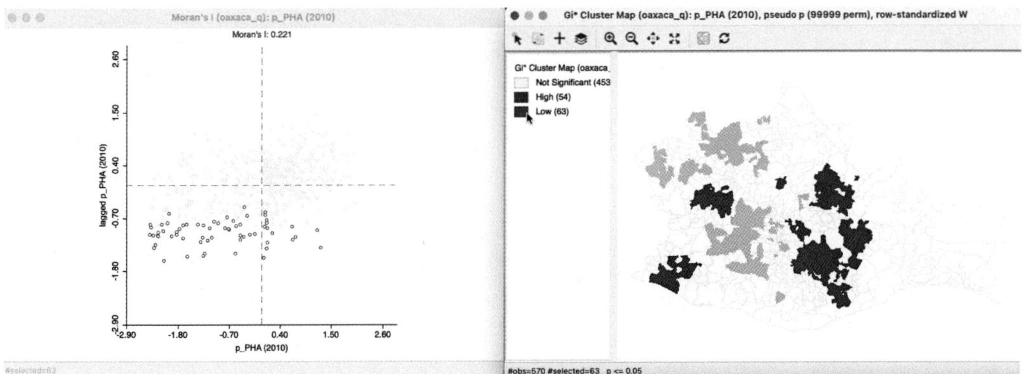

Figure 17.24: Gi* Low-Low clusters

classified as High-Low spatial outliers according to the Local Moran. The extent to which this affects the interpretation of the spatial extent of hot spots or cold spots depends on the relative importance of spatial outliers in the sample, but it can lead to quite different conclusions between the two types of statistics.

17.4.3 Comparing G statistics and Local Moran

The Getis-Ord statistics typically identify the same locations as significant as the Local Moran, but their interpretation is different. This is primarily due to the fact that the Getis-Ord statistics do not account for spatial outliers. How much this affects results in practice depends on each particular instance, but it is something one should be aware of.

17.4.3.1 Clusters

Figures 17.25 and 17.26 show the hot spots and cold spots selected in the Getis-Ord cluster map on the right and identify the corresponding locations in the Local Moran cluster map on the left. As mentioned, all the significant locations match exactly, but their classification does not.

In Figure 17.25, the 54 hot spots in the Getis-Ord cluster map match the 42 High-High clusters in the Local Moran cluster map as well as the 12 Low-High spatial outliers, similar to what the Moran scatter plot in Figure 17.23 indicated.

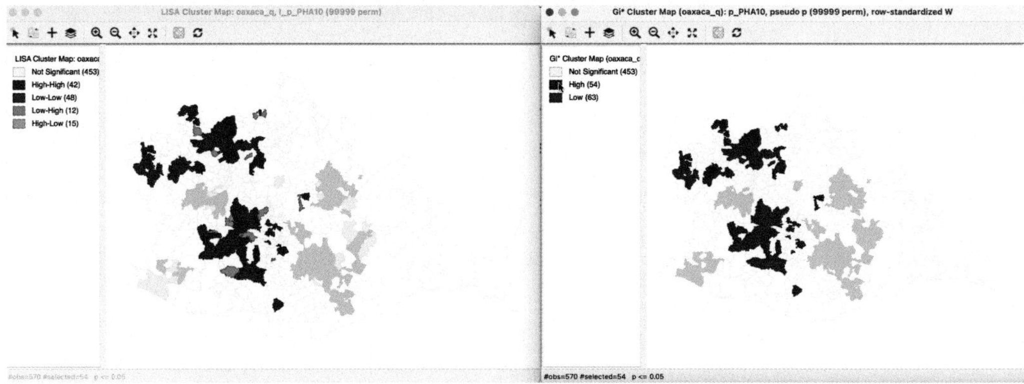

Figure 17.25: Gi* and Local Moran High-High clusters

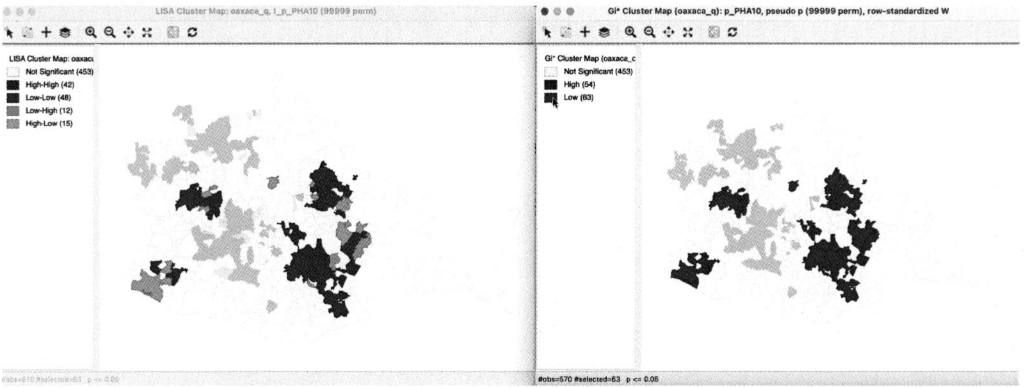

Figure 17.26: Gi* and Local Moran Low-Low clusters

On the other hand, in Figure 17.26, the 63 cold spots in the Getis-Ord cluster map correspond with the 48 Low-Low clusters in the Local Moran cluster map, in addition to the 15 High-Low spatial outliers. Again, this confirms the indication from the Moran scatter plot in Figure 17.24.

17.4.3.2 Spatial outliers

Figures 17.27 and 17.28 illustrate the same properties, but approached from the significant spatial outliers in the Local Moran cluster map on the left.

In Figure 17.27, the 12 locations of significant Low-High spatial outliers are selected in the map on the left, and their corresponding observations identified in the Getis-Ord cluster map on the right. All locations are classified as High-High hot spots.

In Figure 17.28, the reverse is illustrated. The 15 significant High-Low spatial outliers are selected in the Local Moran cluster map on the left. Their matching locations in the Getis-Ord cluster map on the right are all classified as Low-Low cold spots.

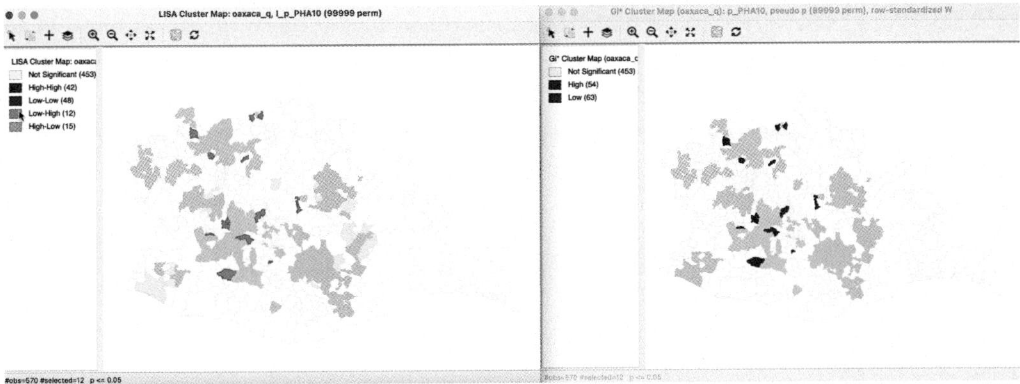

Figure 17.27: Gi* and Local Moran Low-High spatial outliers

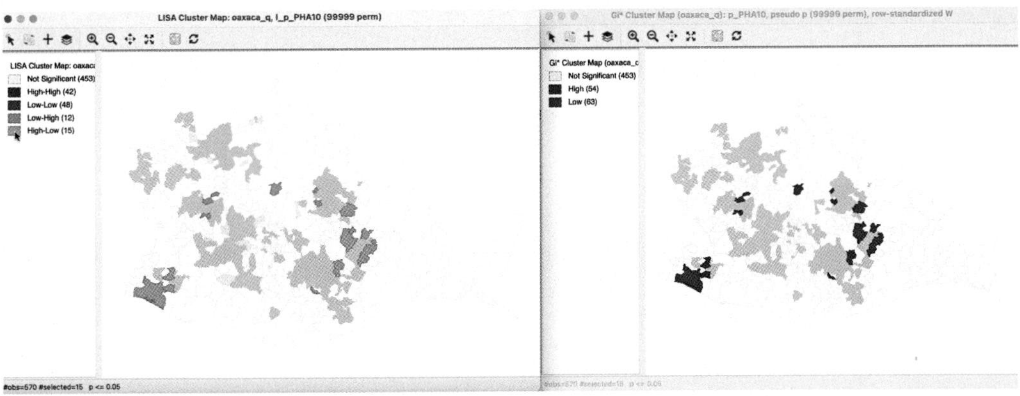

Figure 17.28: Gi* and Local Moran High-Low spatial outliers

17.5 Which Local Statistic to Use?

In practice, the choice between the three main classes of local spatial autocorrelation statistics covered so far may seem a bit bewildering. They each provide a slightly different perspective on the notion of clusters and spatial outliers. The main difference is between the Local Moran and the Local Geary, with the latter potentially picking up nonlinear relationships which the Local Moran is not able to. However, the extent to which this matters in practice depends on the particular application.

When the results between the two approaches differ much, a closer examination is in order. For example, this could consist of inspecting the values of identified clusters and their neighbors to look for the potential impact of outliers. Such an investigation may also be in order when the results for the conventional and Median Local Moran differ much.

The Getis-Ord statistics are different in that they do not account for spatial outliers. Whether this matters depends on the particular context. In general, they tend to identify the same locations as significant as the Local Moran.

For all the local statistics, a careful consideration of *significance* is in order. Locations that remain identified as clusters or outliers under different criteria (as well as using different statistics) are likely *interesting* locations. On the other hand, observations that move in and out of significance as the criteria change are likely spurious. The only way to address this problem confidently is though a careful sensitivity analysis.

18

Multivariate Local Spatial Autocorrelation

In this chapter, the concept of local spatial autocorrelation is extended to the multivariate domain. This turns out to be particularly challenging, due to the difficulty in separating spatial effects from the pure attribute correlation among multiple variables.

Three methods are considered. First, a bivariate version of the Local Moran is introduced, which, similar to what is the case for its global counterpart, needs to be interpreted with great caution. Next, an extension of the Local Geary statistic to the multivariate Local Geary is considered, proposed in Anselin (2019a). The final approach is not based on an extension of univariate statistics, but uses the concept of distance in attribute space, in the form of a *Local Neighbor Match Test* (Anselin and Li, 2020).

To illustrate these methods, the *Chicago SDOH* sample data set is employed. It contains observations on socio-economic determinants of health in 2014 for 791 census tracts in Chicago (for a detailed discussion, see Kolak et al., 2020).

18.1 Topics Covered

- Understand the issues related to extending a spatial autocorrelation measure to a multivariate setting
- Identify clusters and outliers with the bivariate extensions of the Local Moran statistic
- Identify clusters and outliers with the Multivariate Local Geary
- Understand the issues pertaining to interpreting the results of the multivariate Local Geary
- Identify clusters by means of the local neighbor match test

GeoDa Functions

- Space > Bivariate Local Moran's I
- Space > Multivariate Local Geary
- Space > Local Neighbor Match Test

18.2 The Multivariate Spatial Autocorrelation Problem

Designing a spatial autocorrelation statistic in a multivariate setting is fraught with difficulty. The most common statistic, Moran's I, is based on a cross-product association, which is in

DOI: 10.1201/9781003274919-18

the same spirit as a bivariate correlation statistic. As a result, it is difficult to disentangle whether the correlation between multiple variables at adjoining locations is due to the correlation among the variables in-situ, or a similarity due to being neighbors in space.

Early attempts at extending Moran's I to multiple variables focused on principal components, as in the suggestion by Wartenberg (1985), and later work by Dray et al. (2008). However, these proposals only dealt with a global statistic. A more local perspective along the same lines is presented in Lin (2020), although it is primarily a special case of a geographically weighted regression, or GWR (Fotheringham et al., 2002).

Lee (2001) outlined a way to separate a bivariate Moran-like spatial correlation coefficient into a spatial part and a Pearson correlation coefficient. However, this approach relies on some fairly strong simplifying assumptions that may not be realistic in practice.

An alternative perspective is offered in Anselin (2019a). The central idea underlying this approach is to focus on the distance between observations in both attribute and geographical space. A multivariate local spatial autocorrelation statistic then assesses the match between those two concepts of distance.

More formally, the squared multi-attribute Euclidean distance between a pair of observations i, j on k variables is given as:

$$d_{ij}^2 = ||x_i - x_j|| = \sum_{h=1}^{k} (x_{ih} - x_{jh})^2,$$

with x_i and x_j as vectors of observations. In some expressions, the squared distance will be preferred, in others, the actual distance (d_{ij}, its square root) will be used.

In this approach, the overarching objective is to identify observations that are *close* in both multi-attribute space and in geographical space, i.e., those pairs of observations where the two types of distances *match.*

18.3 Bivariate Local Moran

The treatment of the Bivariate Local Moran's I closely follows that of its global counterpart (see Section 14.3.1). In essence, as outlined in more detail in Anselin et al. (2002), it is intended to capture the relationship between the value for one variable at location i, x_i, and the average of the neighboring values for *another variable*, i.e., its spatial lag $\sum_j w_{ij} y_j$. Apart from a constant scaling factor (that can be ignored), the statistic is the product of x_i with the spatial lag of y_i (i.e., $\sum_j w_{ij} y_j$). In order to make this operational and easier to interpret, both variables should be standardized, such that their means are zero and variances equal one. The Bivariate Local Moran is then:

$$I_i^B = x_i \sum_j w_{ij} y_j,$$

in the usual notation.

As is the case for its global counterpart, this statistic needs to be interpreted with caution, since it ignores the correlation between x_i and y_i at location i (see also Section 14.3.1).

A special case of the Bivariate Local Moran statistic is when the same variable is compared in neighboring locations at different points in time. The most meaningful application is where one variable is for time period t, z_t, and the other variable is for the neighbors in the previous time period, $\sum_j w_{ij} z_{t-1}$. This formulation measures the extent to which the value at a location in a given time period is correlated with the values at neighboring locations in a previous time period, or an inward influence. An alternative view is to consider z_{t-1} and the neighbors at the current time, $\sum_j w_{ij} z_t$. This would measure the correlation between a location and its neighbors at a later time, or an outward influence. The first specification is more accepted, as it fits within a standard space-time regression framework.

Inference proceeds similar to the global case, but is now conditional upon the tuple x_i, y_i observed at location i. This somewhat controls for the correlation between x and y at i. The remaining values for y are randomized and the statistic is recomputed to create a reference distribution. The usual caveats regarding the interpretation of significance apply here as well.

18.3.1 Implementation

The Bivariate Local Moran I is invoked as the third item in the drop-down list associated with the **Cluster Maps** toolbar icon, or, from the menu, as **Space** > **Bivariate Local Moran's I**. The next dialog is the customary **Variable Settings**, which now has two columns, one for **First Variable (X)**, and one for **Second Variable (Y)**. Since the Bivariate Local Moran is *not symmetric*, the order in which the variables are specified matters. At the bottom of the dialog, the **Weights** need to be selected.

To illustrate this feature, two variables are used from the *Chicago SDOH* sample data set: the percentage children in poverty in 2014 (**ChldPvt14**), and a crowded housing index (**EP_CROWD**). The spatial weights are nearest neighbor, with $k = 6$ (**Chi-SDOH_k6**).

Before proceeding with the actual bivariate analysis, the univariate characteristics of the spatial distribution of each variable are considered more closely.

18.3.1.1 Univariate description of two variables

Figure 18.1 shows a box map for each of the variables. On the left, for **ChldPvt14**, there are no outliers, so the box map corresponds with a simple quartile map. In contrast, the map for **EP_CROWD** has 27 upper outliers, somewhat scattered around the city.

The correspondence between the two spatial distributions is highlighted in the co-location map in Figure 18.2.[1] This map shows the census tracts where the observations for both variables fall in the same quartile category. In this example, this is appropriate, since both variables work in the same direction (higher values are worse conditions). The classifications match for 284 out of 791 observations, or 36%. The bulk of the matches is for the lowest quartile (i.e., better conditions), where 109 out of 198 observations match. About half that many locations also match for the three other categories (clearly, there are no matches for the outliers, since those are only present in one of the maps). The co-location map provides an initial sense of the degree of spatial correspondence between the two variables. More formally, the (non-spatial) correlation coefficient between the two is 0.392.

[1]The map is constructed after saving the map categories for both box maps, followed by **Map** > **Co-location Map**, using the classification codes and a **Box Map** classification for the co-location map.

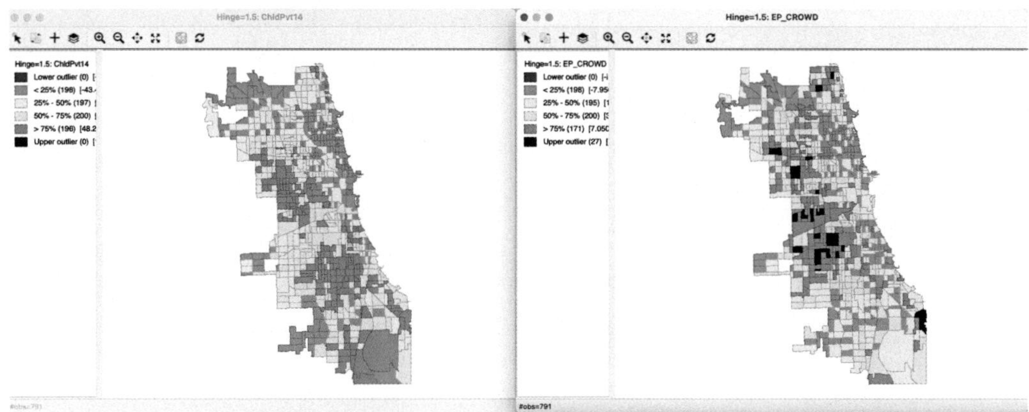

Figure 18.1: Box maps – Child Poverty and Crowded Housing

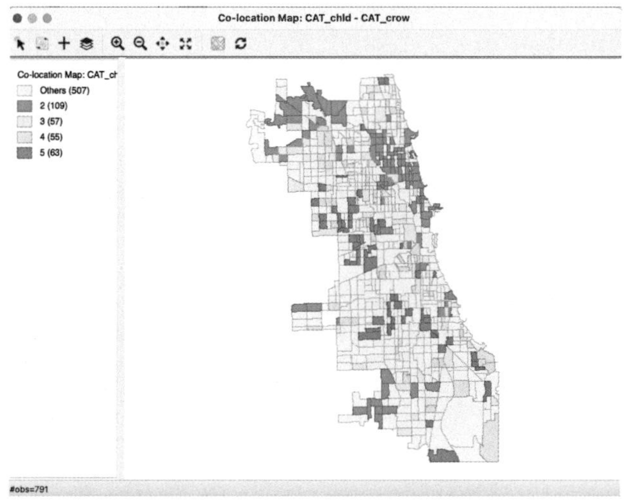

Figure 18.2: Co-location of Child Poverty and Crowded Housing

18.3.1.2 Univariate Local Moran for each variable

To provide further context, the cluster maps for the univariate Local Moran for each variable (using 99,999 permutations with a cut-off p-value of 0.01) are shown in Figure 18.3. In both maps, the largest share consists of Low-Low cluster cores (118 for child poverty and 102 for crowded housing), but for child poverty this is almost equally balanced by High-High cores (101). For crowded housing, there are clearly much fewer High-High cores (67). In both instances, the cluster cores constitute a small number of larger *regions*, with very few spatial outliers.

The correspondence between the two cluster maps is further highlighted in Figure 18.4, which shows a co-location map for the cluster classifications.[2] Almost 3/4 of the Low-Low cluster cores match between the two variables, but this is only the case for 15 High-High cores. Since a *cluster* consists of the core and its neighbors, the actual match has a somewhat wider reach than just the cluster cores. None of the spatial outliers occur in the same location.

[2]This is accomplished after saving the cluster classifications for both cluster maps, followed by **Map > Co-location Map**, using the cluster codes and a **LISA Map** classification for the co-location map.

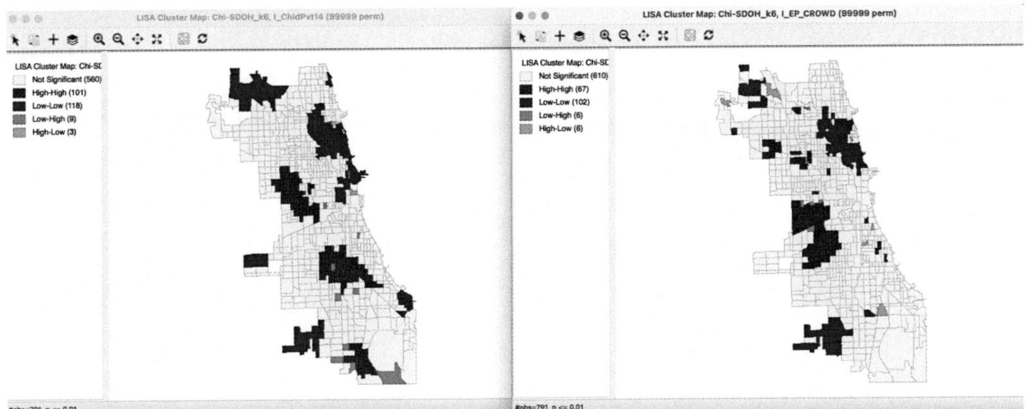

Figure 18.3: Local Moran cluster map – Child Poverty and Crowded Housing

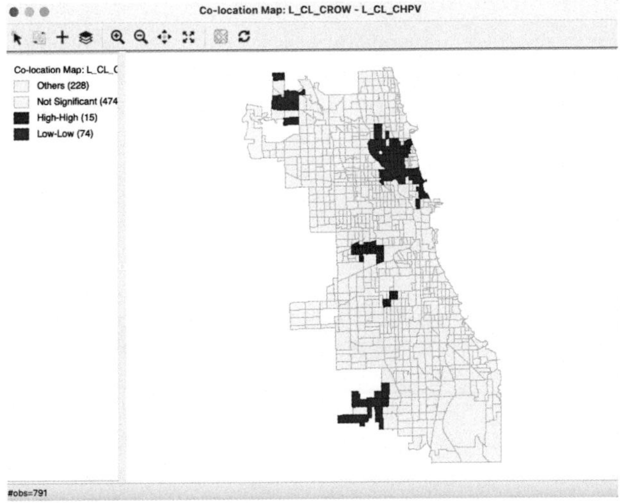

Figure 18.4: Co-location of Local Moran for Child Poverty and Crowded Housing

18.3.1.3 Bivariate analysis

The cluster maps for the Bivariate Local Moran are shown in Figure 18.5, using 99,999 permutations and a p-value cut-off of 0.01. The map on the left is for child poverty surrounded by crowded housing, the map on the right for crowded housing surrounded by child poverty. Clearly, the cluster maps are different, in accordance with the lack of symmetry for this statistic. Both maps are characterized by many more spatial outliers than in the univariate cluster maps, highlighting locations where a high/low value for one variable is surrounded by neighbors with the opposite, i.e., low/high values.

The common cluster cores from the univariate maps (Figure 18.4) are present in both bivariate maps. In fact, a co-location map based on the two bivariate maps yields the exact same result as Figure 18.4. This makes intuitive sense, since a bivariate High-High cluster requires a High value for one variable (cluster core) to be surrounded by High values for the other variable (neighbors part of the cluster surrounding the matching cluster core). The same holds for the Low-Low clusters.

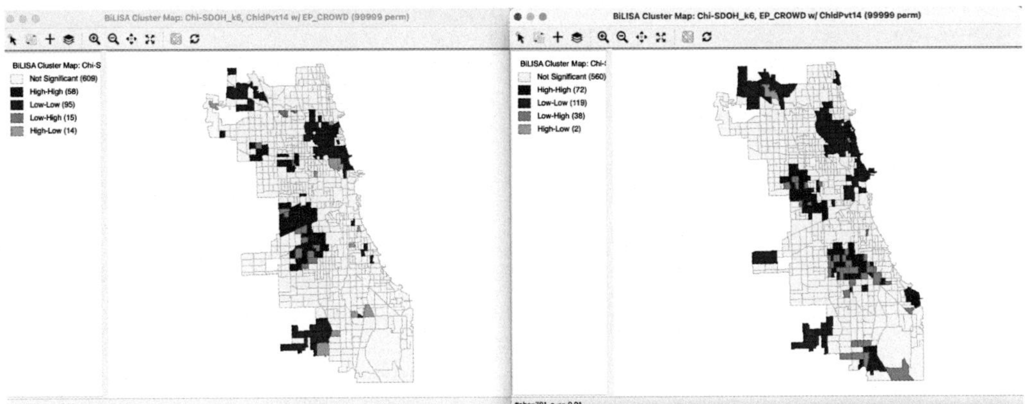

Figure 18.5: Bivariate Local Moran cluster map – Child Poverty and Crowded Housing

Arguably, the more interesting information follows from the location of the spatial outliers, especially the Low-High outliers, of which there are 15 for child poverty-crowded housing and 38 for crowded housing-child poverty. These are census tracts where a Low (good) value for one indicator is surrounded by a High (bad) value for the other, or vice versa, more so than expected under spatial randomness. This is information the univariate cluster maps cannot provide.

18.3.1.4 Options

The Bivariate Local Moran has all the same options as the conventional Local Moran, as detailed in Chapter 16. The cluster codes associated with saved results are the same as well.

18.3.2 Interpretation

The interpretation of the Bivariate Local Moran needs to be carried out very carefully. Aside from the usual caveats about multiple comparisons and p-values, the association between one variable and a different variable at neighboring locations needs to consider the in-situ correlation between the two variables as well. As the discussion in the previous sections illustrates, it is best to combine the bivariate analysis with a univariate analysis for each variable. In addition, it is important to consider both directions of association. This will tend to reveal strong local clustering among the two variables as well as instances where their spatial patterns do not coincide in the form of *bivariate spatial outliers*.

18.4 Multivariate Local Geary

In Anselin (2019a), a multivariate extension of the Local Geary statistic is proposed. This statistic measures the extent to which neighbors in multi-attribute space (i.e., data points that are close together in the multidimensional variable space) are also neighbors in geographical space. While the mathematical formalism is easily extended to many variables, in practice one quickly runs into the *curse of dimensionality*.

In essence, the Multivariate Local Geary statistic measures the extent to which the average distance in attribute space between the values at a location and the values at its geographic neighboring locations are smaller or larger than what they would be under spatial randomness. The former case corresponds to positive spatial autocorrelation, the latter to negative spatial autocorrelation.

An important aspect of the multivariate statistic is that it is not simply the superposition of univariate statistics. In other words, even though a location may be identified as a cluster using the univariate Local Geary for each of the variables separately, this does not mean that it is also a multivariate cluster, and vice versa. The univariate statistics deal with distances in attribute space projected onto a single dimension, whereas the multivariate statistics are based on distances in a higher dimensional space. The multivariate statistic thus provides an additional perspective to measuring the tension between attribute similarity and locational similarity.

The Multivariate Local Geary statistic is formally the sum of individual Local Geary statistics for each of the variables under consideration. For example, with k variables, indexed by h, the corresponding expression is:

$$c_i^M = \sum_{h=1}^{k} \sum_j w_{ij} (x_{hi} - x_{hj})^2.$$

This measure corresponds to a weighted average of the squared distances in multidimensional attribute space between the values observed at a given geographic location i and those at its geographic neighbors $j \in N_i$ (with N_i as the neighborhood set of i).

To achieve comparable values for the statistics, one can correct for the number of variables involved, as c_i^M / k. This is the approach taken in `GeoDa`.

Inference can again be based on a conditional permutation approach. This consists of holding the *tuple* of values observed at i fixed, and computing the statistic for a number of permutations of the remaining tuples over the other locations. Note that because the full tuple is being used in the permutation, the internal correlation of the variables is controlled for, and only the *spatial* component gets altered.

The permutation results in an empirical reference distribution that represents a computational approach at obtaining the distribution of the statistic under the null. The associated pseudo p-value corresponds to the fraction of statistics in the empirical reference distribution that are equal to or more extreme than the observed statistic.

Such an approach suffers from the same problem of multiple comparisons mentioned for all the other local statistics. In addition, there is a further complication. When comparing the results for k univariate Local Geary statistics, these multiple comparisons need to be accounted for as well.

For example, for each univariate test, the target p-value of α would typically be adjusted to α / k (with k variables, each with a univariate test), as a Bonferroni bound. Since the multivariate statistic is in essence a sum of the statistics for the univariate cases, this would suggest a similar approach by dividing the target p-value by the number of variables (k). Alternatively, and preferable, a FDR strategy can be pursued. The extent to which this actually compensates for the two dimensions of multiple comparison (multiple variables and multiple observations) remains to be further investigated.[3]

Consequently, the interpretation of *significance* is difficult. In practice, extensive sensitivity analysis is recommended, using small p-values or an FDR strategy.

[3]See Anselin (2019a) for further technical discussion.

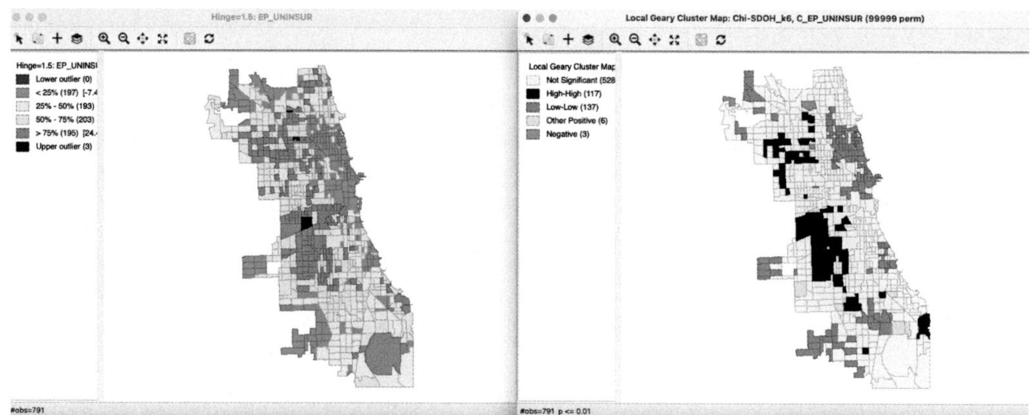

Figure 18.6: Box map and Local Geary cluster map – Uninsured

18.4.1 Implementation

The Multivariate Local Geary is invoked as the second item in third block in the drop-down list associated with the **Cluster Maps** toolbar icon, or, from the menu, as **Space > Multivariate Local Geary**. The next dialog is a slightly different **Multi-Variable Settings**, which now allows for the selection of several variables (not limited to two). As usual, the spatial **Weights** need to be specified at the bottom of the dialog.

To illustrate this functionality, three variables are used from the *Chicago SDOH* sample data set. In addition to the same two as for the Bivariate Local Moran, the percentage children in poverty in 2014 (**ChldPvt14**), and a crowded housing index (**EP_CROWD**), the percentage without health insurance (**EP_UNINSUR**) is included as well. The latter has a correlation of 0.486 with child poverty and 0.632 with crowded housing. The spatial weights are again nearest neighbor, with $k = 6$ (**Chi-SDOH_k6**).

As before, the univariate properties of these variables are considered first, but now from the perspective of the Local Geary.

18.4.1.1 Univariate analysis of multiple variables

Figure 18.6 includes a box map for the percent uninsured on the left and a univariate Local Geary cluster map on the right (for 99,999 permutations with a 0.01 p-value cut-off). The box map contains three upper outliers, but otherwise boils down to the classification of a quartile map. The Local Geary cluster map reveals larger regions of High-High and Low-Low clusters, with very little evidence of negative spatial autocorrelation (only three spatial outliers).

Box maps for the other two variables were included in Figure 18.1. The corresponding cluster maps for the Local Geary (with 99,999 permutations and a p-value cut-off of 0.01) are shown in Figure 18.7. The results are similar to the broad clusters revealed for the Local Moran (Figure 18.3), but with several local differences.

The commonality in local clusters between the three variables is highlighted in the co-location map in Figure 18.8. Overall, the three variables have 39 Low-Low and 10 High-High cluster cores in common, reflecting the overlap in regions of low and high values found in the univariate cluster maps.

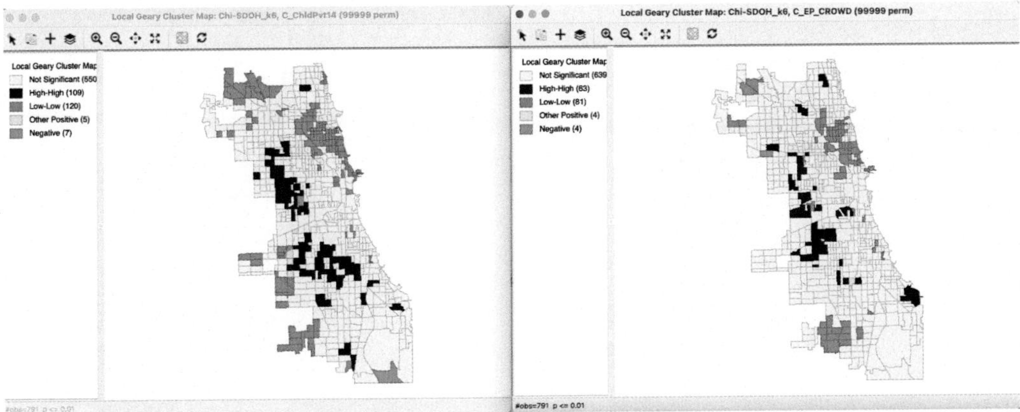

Figure 18.7: Local Geary cluster maps – Child Poverty, Crowded Housing

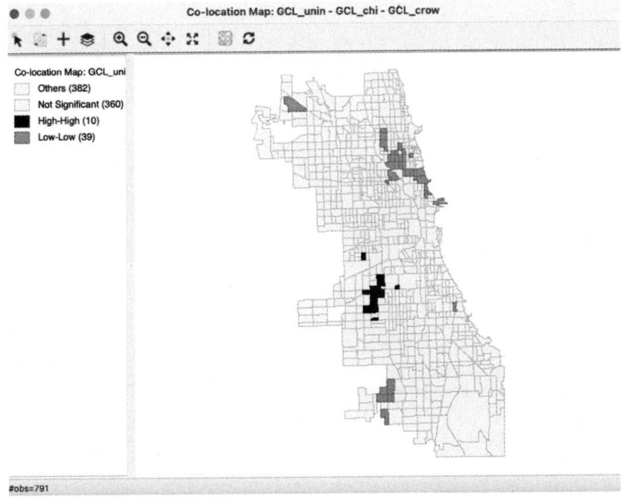

Figure 18.8: Co-location of Local Geary

18.4.1.2 Multivariate Local Geary analysis

The significance map and cluster map for the Multivariate Local Geary, with 99,999 permutations and a p-value cut-off of 0.01 are shown in Figure 18.9. The main item that stands out is the very large number of *significant* observations, something surely due to the use of an inappropriate p-value cut-off. In all, 342 positive cluster cores are identified, but only two locations that suggest negative spatial autocorrelation. Even for a p-value cut-off of 0.0001, there are still 84 positive cluster cores. A careful sensitivity analysis is in order.

18.4.1.3 Options

All the options of the significance and cluster maps considered before remain the same. Specifically, the **Save Results** option offers the same choices as for the univariate Local Geary (Section 17.3.1.1).

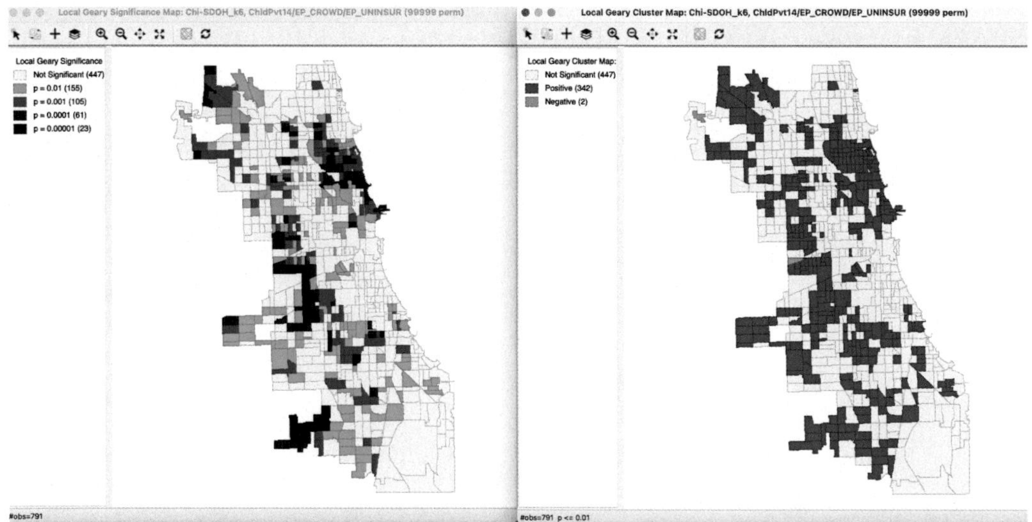

Figure 18.9: Multivariate Local Geary cluster map – Child Poverty, Crowded Housing, Uninsured

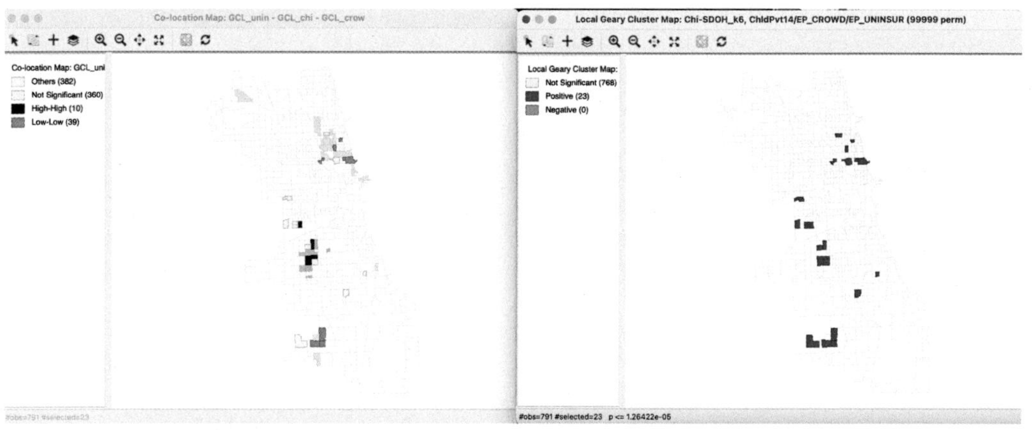

Figure 18.10: Multivariate Local Geary cluster map and Univariate Local Geary co-location

18.4.2 Interpretation

To shed further light on where *interesting* locations can be found, the results for a range of different p-value cut-offs can be investigated. In Figure 18.10, the 23 observations that remain significant under the Bonferroni bound (p = 0.0000126) are linked to the univariate co-location map from Figure 18.8. Only about half (12) of these locations match between the two, pointing to other dimensions of association beyond the univariate overlap.

Further insight can be gained by focusing closer on those observations that are identified by the Multivariate Local Geary, but not by the univariate overlap. This then becomes much more of an exploratory exercise than a clean p-value interpretation. It should therefore be carried out with caution.

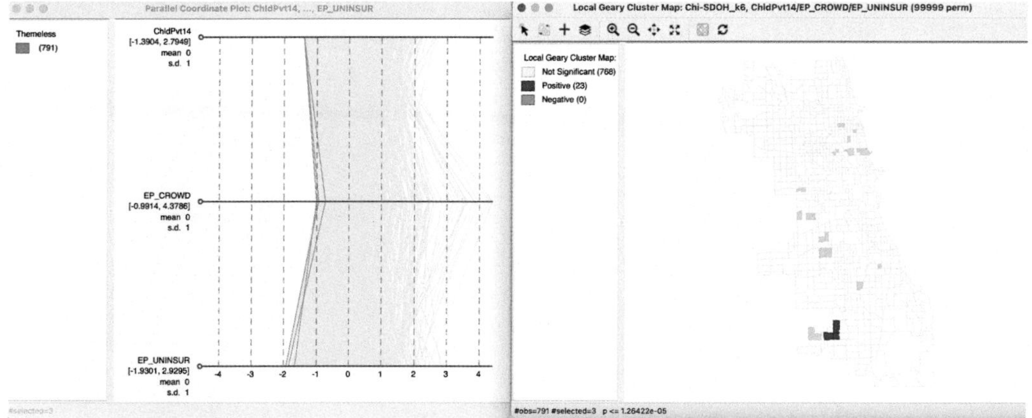

Figure 18.11: Multivariate Local Geary cluster in PCP

18.4.2.1 Multivariate clusters and PCP

Further insight into what the Multivariate Local Geary identifies as interesting locations can be gained from the parallel coordinate plot in the left-hand panel of Figure 18.11. The three observations in the multivariate local cluster identified using the Bonferroni bound in the right hand map are shown as three very close lines in the parallel coordinate plot. This suggests that neighbors in geographic space are also close neighbors in multi-attribute space.

The connection between neighbors in geographic space and in multi-attribute space is further leveraged by the concept of a Local Neighbor Match Test, considered next.

18.5 Local Neighbor Match Test

An alternative approach to visualize and quantify the trade off between geographical and attribute similarity was suggested by Anselin and Li (2020) in the form of what is called a *Local Neighbor Match Test*. The basic idea is to assess the extent of overlap between k-nearest neighbors in geographical space and k-nearest neighbors in multi-attribute space.

In `GeoDa`, this becomes a simple intersection operation between two k-nearest neighbor weights matrices (see Section 11.6.1). One matrix is derived from the distances in multi-attribute space, the other using geographical distance. With the intersection in hand, the probability that an overlap occurs between the two neighbor sets can be quantified. This corresponds to the probability of drawing v common neighbors from the k out of $n-1-k$ possible choices as neighbors, a straightforward combinatorial calculation.

More formally, the probability of v shared neighbors out of k is:

$$p = C(k, v).C(N-k, k-v)/C(N, k),$$

where $N = n - 1$ (one less than the number of observations), k is the number of nearest neighbors considered in the connectivity graphs, v is the number of neighbors in common and C is the combinatorial operator. Alternatively, a pseudo p-value can be computed based on a permutation approach, although that is not pursued in `GeoDa`.

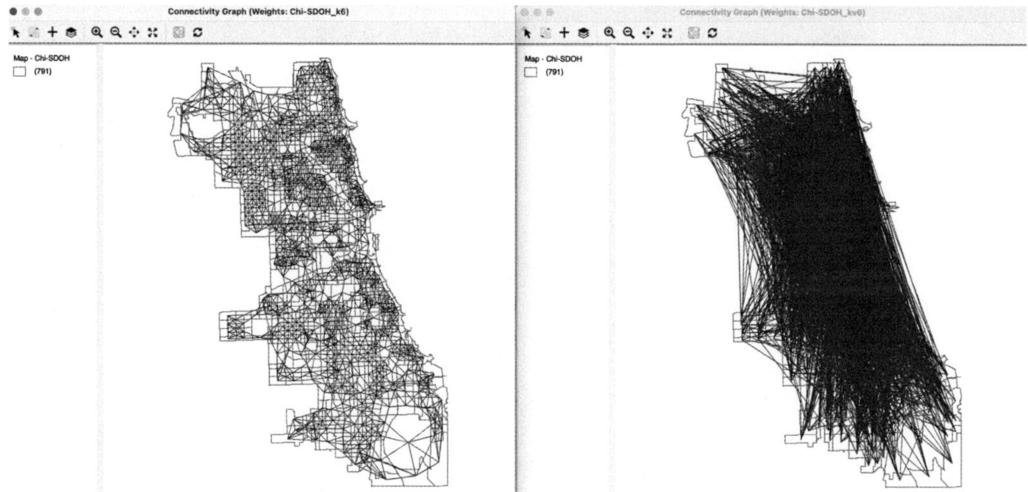

Figure 18.12: K nearest neighbor connectivity graphs

The degree of overlap can be visualized by means of a cardinality map, a special case of a unique values map. In this map, each location indicates how many neighbors the two weights matrices have in common. In addition, different p-value cut-offs can be employed to select the *significant* locations, i.e., where the probability of a given number of common neighbors falls below the chosen p.

In practice, the value of k may need to be adjusted (increased) in order to find meaningful results. In addition, the k-nearest neighbor calculation becomes increasingly difficult to implement in very high attribute dimensions, due to the empty space problem (Section 8.2.1).

The idea of matching neighbors can be extended to distances among variables obtained from dimension reduction techniques, such as multidimensional scaling, covered in Volume 2 (see also the more detailed discussion in Anselin and Li, 2020).

In contrast to the approach taken for the Multivariate Local Geary, the local neighbor match test focuses on the pairwise distances directly, instead of converting these into a weighted average. Both measures have in common that they focus on squared distances in attribute space, rather than a cross-product as in the Moran statistics.

18.5.1 Implementation

The Nearest Neighbor Match test is invoked as the last item in the drop-down list associated with the **Cluster Maps** toolbar icon, or, from the menu, as **Space > Local Neighbor Match Test**. This is followed by a dialog to select the variables.

In addition to the variable names, the dialog includes options for four important parameters. The **Number of Neighbors** specifies the range for which the match is explored, i.e., the value of k in the k-nearest neighbor weights that are computed under the hood. In the example, **6** is used, to allow comparison with the other approaches considered in this chapter. Next is the **Variable Distance Function**, with **Euclidean** as the default, but **Manhattan** distance as the other option. A third option pertains to the **Geographic Distance Metric**. Here again, **Euclidean Distance** is the default, but **Arc Distance** is available as well, for the case where the map layer is unprojected. Finally, **Transformation**

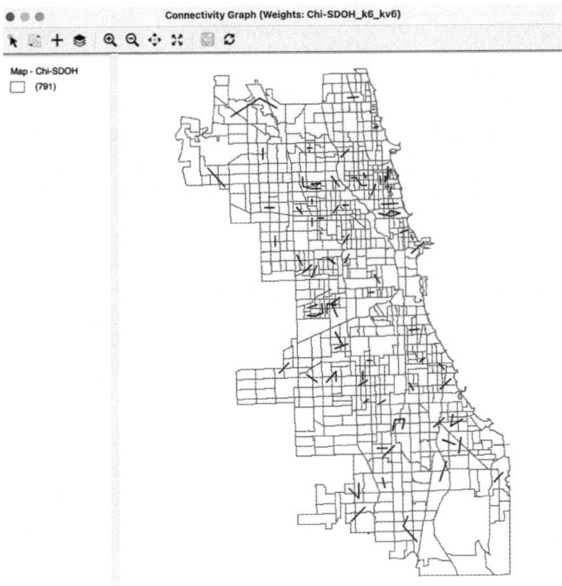

Figure 18.13: Intersection of K nearest neighbor connectivity graphs

offers six options to adjust the variables: **Raw**, **Demean**, **Standardize (Z)**, **Standardize (MAD)**, **Range Adjust** and **Range Standardize**. The default of **Standardize (Z)** is the recommended approach. With this information, the two k-nearest neighbor weights are constructed and the cardinality of the intersection computed.

In the illustration, the same three variables are employed as for the Multivariate Local Geary. Before moving to the actual test, the logic of the approach is detailed.

18.5.1.1 Intersection between connectivity graphs

Figure 18.12 shows the connectivity graphs corresponding to geographic 6 nearest neighbors and the multi-attribute 6 nearest neighbors (see Section 11.4.4). The graph on the left takes on a familiar form, but the graph on the right suggests that a lot of multivariate connections span large geographic distances.

The intersection between the two weights yields the connectivity graph illustrated in Figure 18.13.[4] The result is very sparse, with most intersections consisting of a single link. This graph structure forms the point of departure for the Local Neighbor Match Test.

18.5.1.2 Testing the local neighbor match

The degree of local neighbor match is visualized by means of a cardinality map, as shown in Figure 18.14. This is a special case of a unique values map, where the categories correspond to the number of matching neighbors. In this example, there are 101 observations with a single match, and 10 observations with two matches.

The map is accompanied with a prompt to save the cardinality (default variable **card**) and the associated probability (default variable **cpval**). Note that no probabilities are computed for the cases where no matches occur.

[4]This is accomplished by means of the **Intersection** command in the **Weights Manager**.

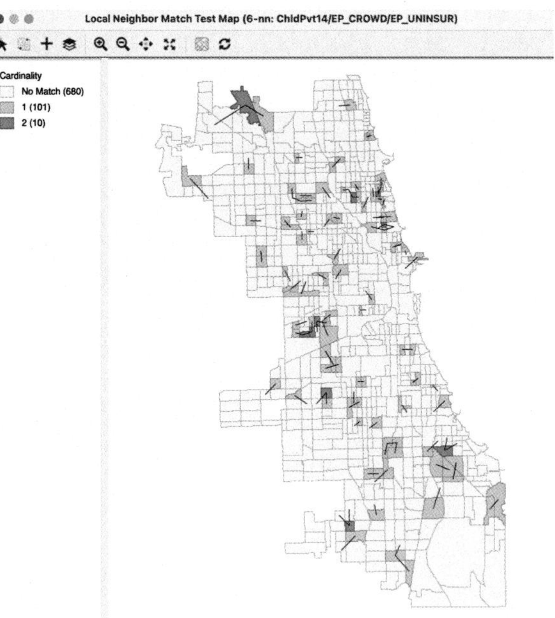

Figure 18.14: Local neighbor match cardinality map

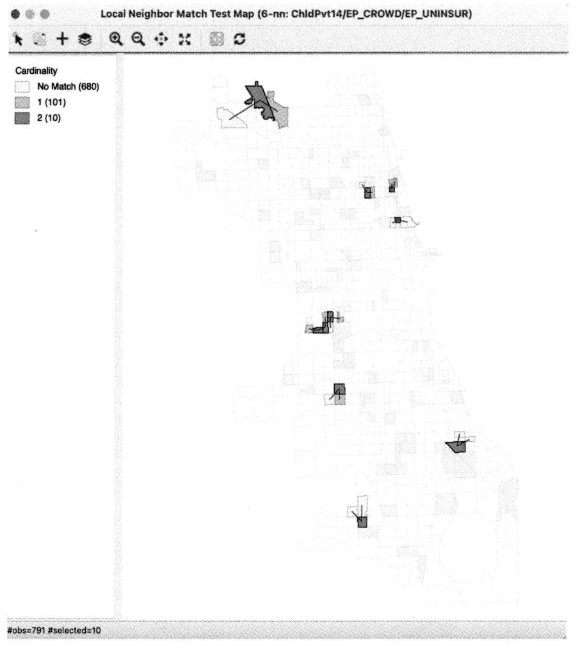

Figure 18.15: Local neighbor match significance map

Even though the largest number of matches is only two, such an occurrence is very rare under spatial randomness, corresponding with a p-value of 0.0007. The 10 such identified observations are highlighted in Figure 18.15, with the matching links shown in red.

Figure 18.16 provides a better sense of the rarity of the coincidence between geographic and multi-attribute nearest neighbors. The two sets of neighbors are shown for the 10 highlighted

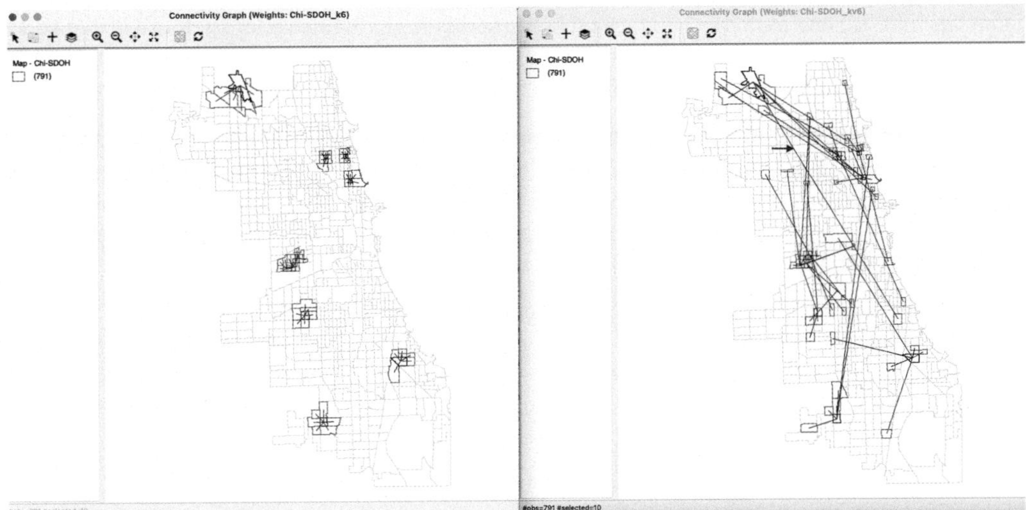

Figure 18.16: Local neighbor significant matches

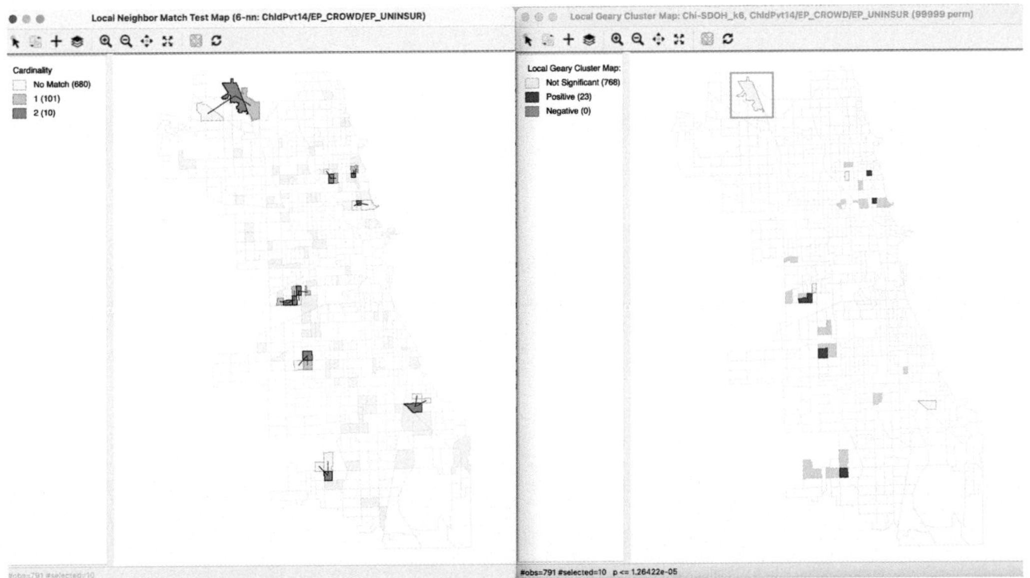

Figure 18.17: Local neighbor match and Multivariate Local Geary

locations. It is clear that several nearest neighbors in attribute space are not so close in geographic space, confirming the tension between the two notions of similarity.

18.5.2 Local neighbor match and Multivariate Local Geary

A final perspective on what insights the local neighbor match provides follows from a comparison with the *interesting* locations identified by means of the Multivariate Local Geary. In Figure 18.17, the 10 locations in the Local Neighbor Match Test map on the left are linked to their corresponding observations in the Multivariate Local Geary cluster map on the right. Of the 10, six are also identified by the Multivariate Local Geary, but four are not. Those four are instances where the average distances in attribute space to the geographic neighbors are much larger than the individual distances for the nearest neighbors in attribute space (as opposed to all the neighbors).

This can be assessed by investigating the linkage structure for the affected observations in the right hand panel of Figure 18.16. For example, for the census tract identified by the green rectangle in Figure 18.17, several of the nearest attribute neighbors are far away in geographic space, as illustrated by the link highlighted by the blue arrow in Figure 18.16, as well as its other links beyond the first two neighbors.

By focusing on a different aspect of the locational-attribute similarity trade-off, the Local Neighbor Match Test provides yet another avenue to explore multivariate local clusters. However, as pointed out, this approach does not scale well to a large number of variables, since the empty space problem creates a large computational burden for the calculation of k-nearest neighbors in high-dimensional attribute space.

19

LISA for Discrete Variables

So far, the application of local spatial autocorrelation statistics has been to continuous variables. In this chapter, discrete variables are considered, and, more specifically, binary variables. To address this context, a univariate *Local Join Count* statistic (Anselin and Li, 2019) and its extension to a multivariate setting are introduced. The latter allows for a distinction between situations where the two discrete variables can co-occur (i.e., take the value of 1 for the same location), and where they cannot (no co-location).

The principle behind the Local Join Count statistic is broadened by applying it to a subset of observations on a continuous variable that satisfy a given constraint. The most common application of this idea is to a specific quantile of the observations, leading to the concept of a *Quantile LISA* (Anselin, 2019b).

The *Chicago SDOH* sample data set is again used to illustrate these methods.

19.1 Topics Covered

- Identify clusters in binary variables by means of the local join count statistic
- Distinguish between co-location and no co-location in a bivariate binary variable case
- Identify multivariate co-location clusters in binary variables with the local join count statistic
- Apply the Quantile LISA concept to simplify the assessment of univariate and multivariate clusters

GeoDa Functions

- Space > Univariate Local Join Count
- Space > Bivariate Local Join Count
- Space > Co-location Join Count
- Space > Univariate Quantile LISA
- Space > Multivariate Quantile LISA

19.2 Univariate Local Join Count Statistic

For binary variables, coded as 0 and 1, the global spatial autocorrelation statistic of choice is the *join-count* statistic (see Cliff and Ord, 1973). This statistic consists of counting the *joins* that correspond to occurrences of value pairs at neighboring locations. The three cases are joins of $1-1$ (so-called BB joins, for *black-black*), $0-0$ (so-called WW joins, for

DOI: 10.1201/9781003274919-19

white-white) and $0 - 1$ (so-called BW joins). The former two are indicators of positive spatial autocorrelation, the latter of negative spatial autocorrelation.

The primary interest lies in identifying co-occurrences of uncommon events, i.e., situations where observations that take on the value of 1 constitute much less than half of the sample. The definition of what is 1 or 0 can easily be reversed to make sure this condition is met. Therefore, the focus is on the BB join counts. While this is not an absolute requirement, the way the inference is obtained requires that the probability of obtaining a large number of like neighbors is small, and thus can form the basis for rejecting the null hypothesis. When the proportion of observations with 1 is larger than half, then the probability of a *small* number of neighbors with a value of 1 will be small, which is counter the overall logic.[1]

With the variable x_i at location i taking either the value of 1 or 0, a global BB join count statistic can be written as:

$$BB = \sum_i \sum_j w_{ij} x_i x_j,$$

where w_{ij} are the elements of a *binary* spatial weights matrix. In other words, a *join* is *counted* when $w_{ij} = x_i = x_j = 1$. In all other instances, either when there is a mismatch in x between i and j, or a lack of a neighbor relation ($w_{ij} = 0$), the term on the right-hand side does not contribute to the double sum.

Following the logic in Anselin (1995), Anselin and Li (2019) recently introduced a local version of the BB join count statistic as:

$$BB_i = x_i \sum_j w_{ij} x_j,$$

where $x_{i,j}$ can only take on the values of 1 and 0, and, again, w_{ij} are the elements of a *binary* spatial weights matrix (i.e., *not* row-standardized).

The statistic is only meaningful for those observations where $x_i = 1$, since for $x_i = 0$ the result will always equal zero. When $x_i = 1$, it corresponds to the sum of neighbors for which the value $x_j = 1$. In this sense, it is similar in spirit to to the local second-order analysis for point patterns outlined in Getis (1984) and Getis and Franklin (1987), where the number of points are counted within a given distance d of an observed point. The distance cut-off d could readily form the basis for the construction of the spatial weights w_{ij}, which yields the join count statistic as a count of events (points) within the critical distance from a given point ($x_i = 1$).

The main difference between the two concepts is the underlying data structure: in the point pattern perspective, the locations themselves are considered to be random, whereas the local join count statistic is based on a lattice perspective. The latter considers a finite set of known locations, for which both events ($x_i = 1$) and non-events ($x_i = 0$) are observed. In point patterns analysis, one does not know the locations where events *might* have happened, but did not.

In addition, the local join count statistic also has the same structure as the numerator in the local G_i statistic of Getis and Ord (1992), when applied to binary observations (and with a binary weights matrix). The numerator in this statistic is $\sum_j w_{ij} x_j$, which is identical to the multiplier in the local join count statistic. However, the difference between the two statistics is that the local G_i includes the neighbors with $x_j = 1$ for all locations, including the ones where $x_i = 0$. Such observations are ignored in the computation of the local join

[1] In practice, this can easily be detected when the results show locations with 1 like neighbor to be significant, and locations with more like neighbors not to be significant.

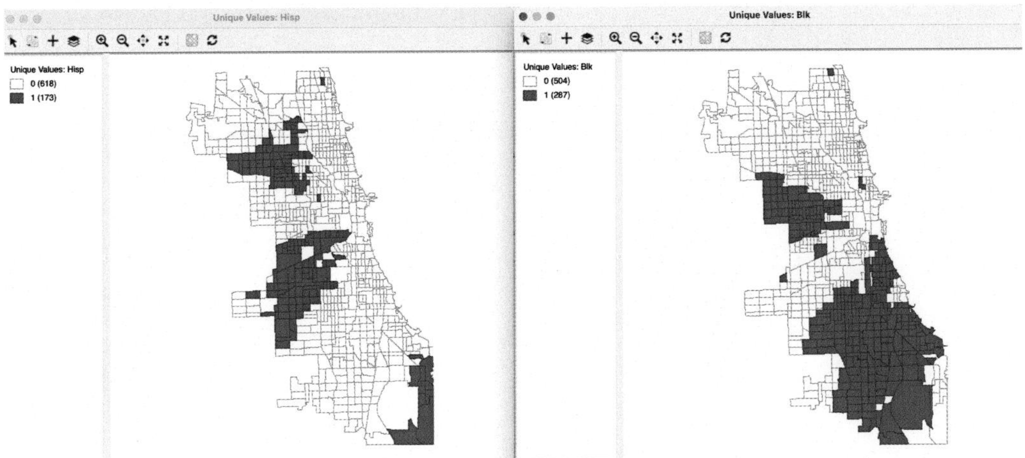

Figure 19.1: Unique values Black-Hispanic tracts

count statistic as outlined above. In a sense, the local join count statistic could thus be considered a *constrained* form of the local G_i statistic, limited to observations where $x_i = 1$.

Inference can be based on a hypergeometric distribution, or, as before, on a permutation approach. Given a total number of events in the sample of n observations as P, the magnitude of interest is the number of neighbors of location i for which $x_i = 1$, i.e., conditional upon observing 1 at this location. The number of neighbors with $x_j = 1$ is represented by p_i. The probability of observing exactly $p_i = p$, conditional upon $x_i = 1$ follows the hypergeometric distribution for $n - 1$ data points and $P - 1$ events:

$$\text{Prob}[p_i = p | x_i = 1] = \frac{\binom{P-1}{p}\binom{N-P-2}{k_i-p}}{\binom{N-1}{k_i}},$$

where k_i is the number of neighbors for observation i.

Instead, a conditional permutation procedure can be followed to compute a pseudo p-value, in the usual fashion. This is the preferred approach since the probability given by the previous expression underestimates uncertainty, because it ignores the uncertainty associated with observing $x_i = 1$.

In practice, the permutation approach does not require any parametric assumptions. It is formulated as a classical one-sided hypothesis test against the null hypothesis of spatial randomness. In what follows, only the conditional permutation approach is considered. Further technical details are provided in Anselin and Li (2019).

19.2.1 Implementation

To illustrate the univariate *Local Join Count* statistic, two variables are considered that correspond to census tracts in Chicago with a predominant ethnic make-up. More precisely, these are the tracts where the Hispanic population makes up more than 50% (**Hisp**), and the tracts where the majority population is Black (**Blk**).[2] A unique values map for these two

[2] Note that in the original data set, the binary variables **HISP50PCT** and **BLCK50PCT** are not computed correctly. In fact, it turns out these variables are computed in the data set using a 49% cut-off percentage, which does not preclude co-location. The variables **Blk** and **Hisp** use the correct cut-off, so that co-location (majority population is more than one ethnic group) is precluded by construction.

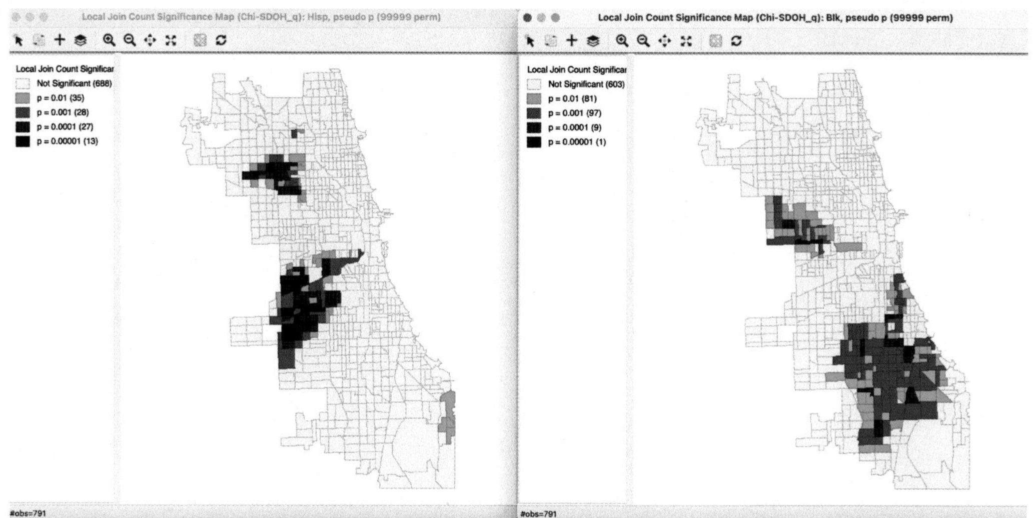

Figure 19.2: Local join count significance map

variables is given in Figure 19.1. The pattern reveals the strong degree of racial segregation, a well-known characteristics of the population distribution in Chicago.

The univariate local join count statistic is invoked in the third group in the **Cluster Maps** drop-down list from toolbar icon, or from the menu, as **Space > Univariate Local Join Count**. This is followed by a **Binary Variable Settings** dialog, where the variable is selected and the spatial weights matrix is specified. In this illustration, the spatial weights are queen contiguity (**Chi_SDOH_q**). This is one of the few instances in `GeoDa` where the spatial weights are *not* row-standardized and instead are used as binary weights.

19.2.1.1 Local significance map

The local significance maps for the Univariate Local Join count statistic for respectively **Hisp** and **Blk** are shown in the panels of Figure 19.2. The result is for 99,999 permutations and a 0.01 p-value cut-off. The strong impression of local clustering given by the unique values map is confirmed by these significance maps. Unlike what is the case for other local statistics, there is no separate cluster map, since only positive spatial autocorrelation is considered, and there is no notion of High-High or Low-Low. The map shows those cores of clusters with a value of 1 for the variable of interest, where the probability of finding the observed number of neighbors with a value of 1 is highly unlikely. As before, the actual notion of a cluster would include those neighbors.

All the usual options for a significance map are available (see Chapter 16).

19.2.1.2 Saving the results

The **Save Results** option saves the statistic, i.e., the number of BB joins (default variable name **JC**), the total number of neighbors (**NN**) and the pseudo p-value (**PP_VAL**). These values are only included for those observations where $x_i = 1$. They can be used to more closely assess the structure of specific clusters.

19.3 Bivariate Local Join Count Statistic

In a bivariate binary context, the two variables under consideration, say x and z, again only take on a value of 0 and 1.

Two different situations are considered, referred to as *co-location* and *no co-location*. In the first, it is possible for both variables to take the same value at a location. In other words, at location i, it is possible to have $x_i = z_i = 1$. For example, this would be the case when the variables pertain to two different types of events that can occur in all locations, such as the presence of several non-exclusive characteristics (e.g., a city block that is both low-density and commercial, or a housing unit that is classified as single family/multifamily and owned/rented).

In the no co-location case, whenever $x_i = 1$, then $z_i = 0$ and vice versa. In other words, the two *events* cannot happen in the same location. This would be case where the classification of a location is exclusive, i.e., if a location is of one type, it cannot be of another type (e.g., majority ethnicity, a zoning classification, or a case-control design). As before, it is important to make sure that the *second* variable, z, is a rare occurrence, appearing in less than half of the sample.

In `GeoDa`, the no-colocation case is referred to as **Bivariate Local Join Count**, since this is the only setting in which it is operational. As conceived, the no co-location setup does not work for more than two variables.

In its general form, the expression for a Bivariate Local Join Count statistic given in Anselin and Li (2019) is:

$$BJC_i = x_i(1 - z_i) \sum_j w_{ij} z_j (1 - x_j),$$

with w_{ij} as unstandardized (binary) spatial weights.

The roles of x and z may be reversed, but the statistic is not symmetric, so that the results will tend to differ whether x or z is the focus. In addition, it is important to keep in mind that the statistic will only be meaningful when the proportion of the second variable (for the neighbors) in the population is small.

Since the condition of no co-location ensures that $1 - z_i = 0$ whenever $x_i = 1$, and vice versa, the statistic simplifies to:

$$BJC_i = x_i \sum_j w_{ij} z_j,$$

whenever $x_i = 1$, which are the only locations of interest.

Inference can be based on a hypergeometric distribution, considering P observations with $x_i = 1$ and Q observations with $z_j = 1$. Again, a more robust alternative is based on a permutation approach.

A pseudo p-value can be obtained from a one-sided conditional permutation test. This is implemented by carrying out a series of k_i draws (with k_i as the number of neighbors for i) for each location i where $x_i = 1$ (and thus $z_i = 0$). The draws are without replacement from $n - 1$ data tuples (x_j, z_j) of which Q observations have $z = 1$ (since $z_i = 0$) and $P - 1$ observations have $x = 1$. In practice, we only need to draw the z_j, since the matching x_j in the tuple are zero by construction. The number of times the resulting local join count statistic equals or exceeds the observed value yields a pseudo p-value, in the usual way.

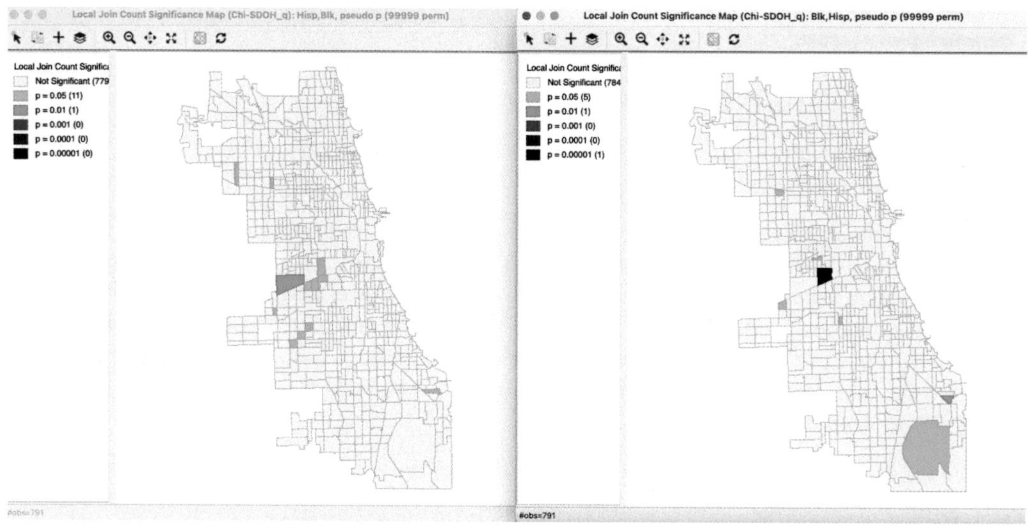

Figure 19.3: Bivariate local join count significance map

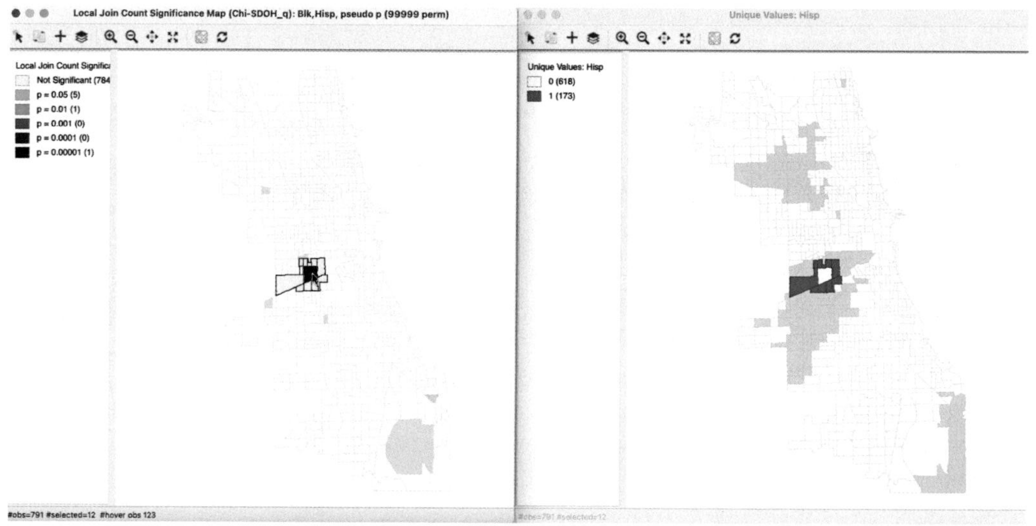

Figure 19.4: Local join count spatial outlier

19.3.1 Implementation

The Bivariate Local Join Count statistic is invoked from the third group in the drop-down list associated with the **Cluster Maps** toolbar icon, or, from the menu, as **Space > Bivariate Local Join Count**. Next is a **Variable Settings** dialog from which the two binary variables are selected. The **First Variable (X)** (x) is selected from the left-hand column in the dialog, the **Second Variable (Y)** (z) is taken from the right-hand column.

Two cases are illustrated in Figure 19.3. In the left-hand panel, the first variable is **Hisp** with the second variable as **Blk**. In the right-hand panel, it is the other way around. Both significance maps are based on queen contiguity (**Chi_SDOH_q**) and use 99,999 permutations with a 0.05 significance cut-off. Clearly, the statistic is not symmetric.

The statistic picks up the rare occasions where a tract with an ethnic majority of one type is surrounded by tracts with a majority of the other type. In the example, this is only the case in very few instances. **Hisp** is surrounded by **Blk** in 12 locations, but 11 of those are only a p = 0.05. **Blk** is surrounded by **Hisp** in only 7 instances, but only one of these is highly significant at p < 0.00001. As shown in Figure 19.4, where the neighbors are selected in the unique values map on the left, this is a clear example of a spatial outlier, where a Black majority tract is surrounded by all Hispanic majority tracts.

All the same options as before are available, including saving the results.

19.4 Co-Location Local Join Count Statistic

The second extension of the Local Join Count statistic to multiple variables considers co-location. This allows two or more variables to take a value of 1 at the same location, i.e., $x_i = z_i = 1$.

For two variables, a co-location cluster requires that an observation for which $x_i = z_i = 1$ coincides with neighbors for which $x_j = z_j = 1$ as well. For two variables, the corresponding Local Join Count statistic takes the form (Anselin and Li, 2019):

$$CLC_i = x_i z_i \sum_j w_{ij} x_j z_j,$$

with w_{ij} as unstandardized (binary) spatial weights. As before, there are P observations with $x_i = 1$ and Q observations with $z_i = 1$ out of a total of n.

A conditional permutation approach can be constructed for those locations with $x_i = z_i = 1$. The permutation consists of draws of k_i pairs of observations (x_j, z_j) from the remaining set of $n - 1$ tuples, which contain $P - 1$ observations with $x_j = 1$ and $Q - 1$ observations with $z_j = 1$. In a one-sided test, the number of times are counted where the statistic equals or exceeds the observed join count value at i.

The extension to more than two variables is mathematically straightforward. At each location i, k variables are considered, i.e., x_{hi}, for $h = 1, \ldots, k$, with $\Pi_{h=1}^{k} x_{hi} = 1$, which enforces the co-location requirement.

The corresponding statistic is then:

$$CLC_i = \Pi_{h=1}^{k} x_{hi} \sum_j w_{ij} \Pi_{h=1}^{k} x_{hj}.$$

The implementation of a conditional permutation strategy follows as a direct generalization of the two-variable co-location case. However, for a large number of variables, such co-locations become less and less likely, and a different conceptual framework may be more appropriate.

19.4.1 Implementation

The Co-Location Local Join Count statistic is invoked from the third group in the drop-down list associated with the **Cluster Maps** toolbar icon, or, from the menu, as **Space > Co-location Local Join Count**. This is followed by a **Multi-Variable Settings** dialog, where the variables to be considered can be specified. At the bottom of the dialog is the customary drop-down list with the spatial weights.

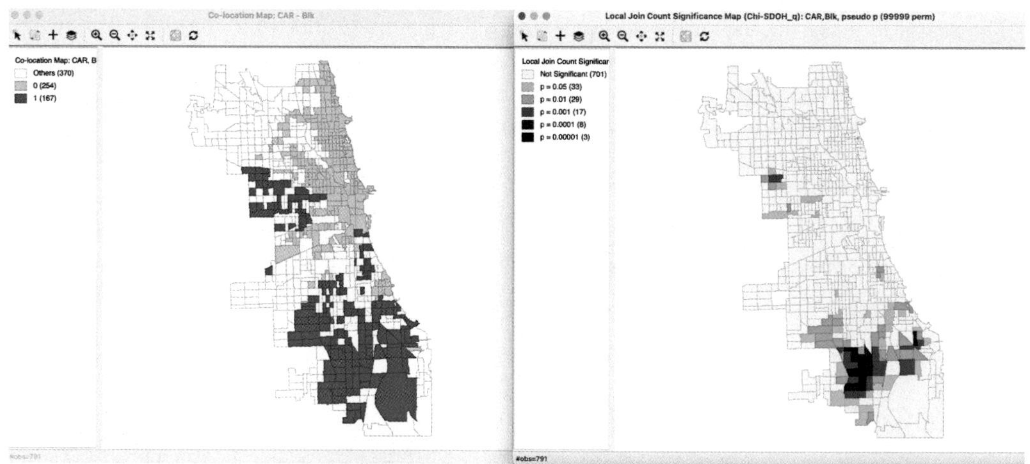

Figure 19.5: Co-location local join count

To illustrate this statistic, the two variables are **Blk**, tracts with majority Black population, and **CAR**, tracts where more than 50% of the commutes happen by car. The spatial weights are again queen contiguity, **Chi_SDOH_q**.

A co-location map of the two binary variables is shown in the left-hand panel of Figure 19.5. Of the 791 tracts, 167 are both majority Black and majority commute by car. Of the remainder, 254 are neither and 370 show a mismatch of the majorities. The corresponding Co-Location Local Join Count significance map is shown in the right-hand panel, using 99,999 permutations and a cut-off of 0.05. At 0.05, there are 90 cores of clusters that overlap, out of the 167, but at 0.01, only 57 of those remain. They confirm the impression of high dependence on commuting by car in majority Black neighborhoods.

All the usual options are available.

19.5 Quantile LISA

The Local Join Count statistics are designed to deal with binary variables. The principles behind these statistics can be readily extended to the situation where a subset of observations on a continuous variable is selected based on a specific value interval. The most straightforward example is when all the observations in a given quantile are assigned a value of 1, and the remaining observations become 0. As a result, assessing local spatial autocorrelation in this case can be approached as an extension of the Local Join Count statistics.

In Anselin (2019b), this case is referred to as *Quantile LISA* or quantile local spatial autocorrelation. In a bivariate or multivariate setting, the quantile LISA often serves as a viable alternative to a Bivariate Local Moran or a Multivariate Local Geary, especially when the focus is on *extremes* in the distribution.

As discussed in Section 18.3, the continuous (linear) association between two variables that is measured by the Bivariate Local Moran suffers from the problem of in-situ correlation. The quantile local spatial autocorrelation sidesteps this problem by converting the continuous variable to a binary variable that takes the value of 1 for a specific quantile. By considering either co-location (positive in-situ correlation) or no co-location (negative in-situ correlation), the correlation between the two variables can be controlled for.

For example, the spatial association between two continuous variables can be simplified by focusing on the local autocorrelation between binary variables for the upper quintile. Since the bivariate (or multivariate) Local Join Count statistic enforces co-location, the problem of in-situ correlation is controlled for. This provides an alternative to other bivariate and multivariate local spatial autocorrelation statistics, albeit with a loss of information, associated with going from the continuous variable to a binary category.

In addition, the quantile LISA approach can also be employed to assess negative spatial autocorrelation, such as between the upper and lower quantile of the same variable. This becomes a special case of a No-Colocation Bivariate Local Join Count statistic.

The quantile approach focuses on a subset of a distribution and thus constitutes a loss of information. However, in practice, this loss of information is often compensated for by superior insight and a focus on the most important *interesting* locations.

Formally, a continuous variable y with a cumulative density function $F(y)$ yields a sequence of ranked observations as $y_1, y_2, \ldots, y_n$. A new binary variable x is created that takes on the value $x_i = 1$ for all y_i for which

$$y_{ql} \leq y_i < y_{qu},$$

with y_{ql} and y_{qu} as lower and upper bounds for a given quantile.[3] For all other observations, $x_i = 0$.

The new x variable (or set of such variables in a multivariate case) then forms the basis for analysis by means of a Local Join Count statistic, or one of its multivariate extensions.

19.5.1 Implementation

The Univariate and Multivariate Quantile LISA are invoked from the next to last block in the drop-down list associated with **Cluster Maps** icon on the toolbar, or, from the menu, as **Space > Univariate Quantile LISA** and **Space > Multivariate Quantile LISA** respectively.

The **Quantile LISA Dialog** contains several options. In the univariate case, the *continuous* variable needs to be specified, together with the spatial weights. Next follow the criteria to create the binary form of the variable: the **Number of Quantiles**, **Select a Quantile for LISA** and **Save Quantile Selection in Field**. The default number of quantiles is **5**, for quintiles. The quantile selection consists of a drop-down list with appropriate values, i.e., in this example **1** to **5**, with **1** for the lowest quantile (in this case, quintile), and **5** for the highest. The resulting binary variables is added to the data table with a default variable name of **QT**.

In the multivariate case, a similar interface allows for the creation of one binary variable at a time, with for each the number of categories and the order of the selected category, saved under default variables **QTk**, with **k** as the sequence number. Each newly created variable must be moved to the right-hand panel of the dialog by means of the > button (and, conversely, can be removed by means of the < button).

[3]In general, this may be applied to any interval, not just a given quantile.

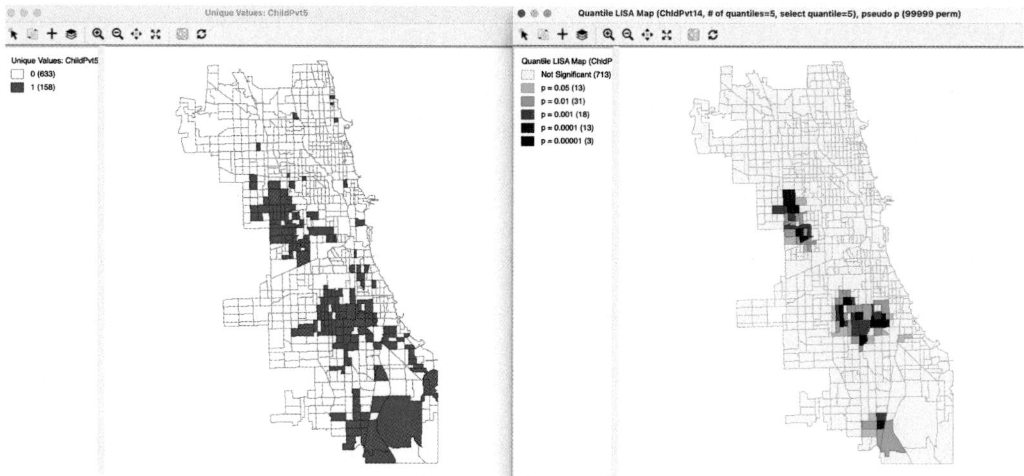

Figure 19.6: Univariate Quantile LISA

A check box determines whether **No co-location** must be enforced. The default is to allow co-location. A warning message results in case of no overlap when there should be overlap, and vice versa.

With the spatial weights selected, the analysis is run and a significance map is created. This is illustrated with three specific cases: a univariate Quantile LISA, and a bivariate and multivariate example. All the standard options are available.

19.5.2 Univariate Quantile LISA

To illustrate the univariate Quantile LISA case, consider the upper quintile of the variable **ChldPvt14**, the children poverty rate. A quartile map (but not a quintile map) was shown earlier in Figure 18.1. The upper quintile is converted into a binary variable **ChildPvt5**, shown in the unique values map in the left-hand panel of Figure 19.6.

The right-hand panel shows the significance map using 99,999 permutations with a p-value cut-off of 0.05, for queen contiguity (**Chi_SDOH_q**). The highlighted locations form a subset of the cores identified by means of the Local Moran map in the left-hand panel of Figure 18.3. Of the 101 High-High cluster cores at p=0.01 in the Local Moran cluster map, 61 are retained in the Quantile LISA significance map. Three of these are highly significant at p=0.00001.

Obviously, Low-Low clusters and spatial outliers are not considered. Nevertheless, the Quantile LISA approach identifies essentially the same *interesting* High-High cluster locations as the analysis that uses the full continuous distribution.

19.5.3 Bivariate and Multivariate Quantile LISA

The bivariate case allows for both no co-location (box checked) and co-location, as specified through the **Multivariate Quantile LISA Dialog**. The **No co-location** option only works in the bivariate case and provides a way to address negative spatial autocorrelation, i.e., spatial outliers. For example, this can be used to assess whether observations in a top quantile are surrounded by locations in a bottom quantile, or vice versa (as pointed out, the statistic is not symmetric).

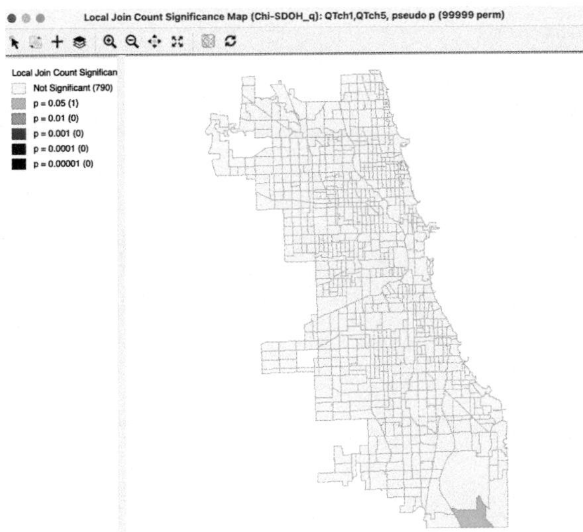

Figure 19.7: Spatial outliers with Quantile LISA

The co-location option is the default and it allows to identify *clusters* of observations that belong to specified quantiles for different variables. Typically, this will be the top or bottom quantiles, but the tool is sufficiently flexible to allow any combination (provided it makes sense). A critical constraint is that there need to be co-location of the quantiles for *all* the variables considered. For example, if three variables are taken into account, then there must be locations (at least one), where the respective quantiles coincide. Just as for the multivariate local join count, this becomes harder to satisfy as more variables are being considered.

19.5.3.1 Spatial outliers

The No co-location option is illustrated in Figure 19.7 for the lower and upper quintiles of the variable **ChldPvt14**, using 99,999 permutations with $p = 0.05$ and queen contiguity spatial weights. Only one census tract can be identified that belongs to a bottom quintile and is surrounded by more neighbors from the top quintile than likely under spatial randomness. Moreover, the evidence is very weak and only holds for $p = 0.05$. Note that the same location was also identified as a Low-High spatial outlier in the Local Moran cluster map in Figure 18.3.

The reverse assessment, for top quintiles surrounded by bottom quintiles, does not yield a single significant location.

19.5.3.2 Multivariate co-location

The multivariate co-location case is illustrated with the same three variables as used for the Multivariate Local Geary in Section 18.4.1, but now for queen contiguity spatial weights. The three variables, child poverty (**ChldPvt14**), crowded housing (**EP_CROWD**) and lack of health insurance (**EP_UNINSUR**) are converted into indicator variables with a value of 1 for those observations belonging to the top quintile. The corresponding Multivariate Quantile LISA significance map is shown in the left-hand panel of Figure 19.8, for 99,999 permutations and $p = 0.05$.

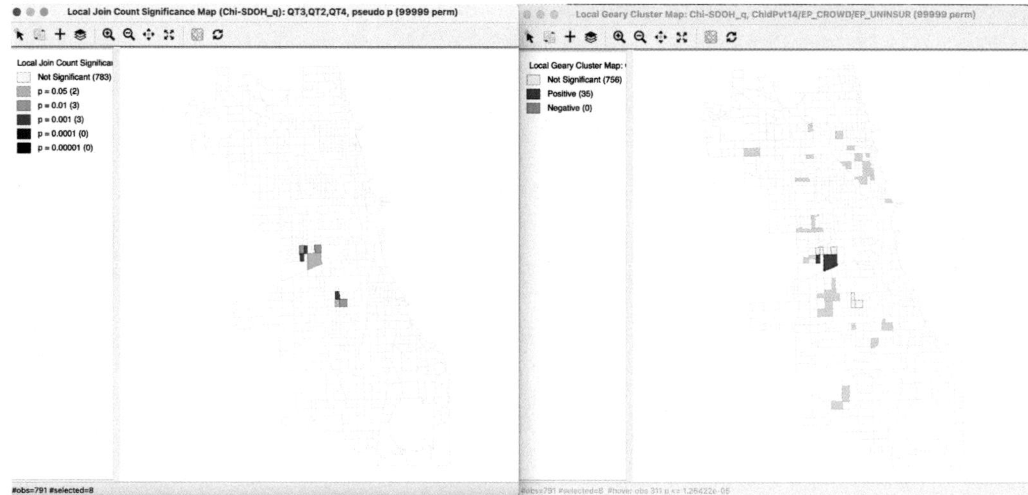

Figure 19.8: Multivariate Quantile LISA and Multivariate Local Geary

Eight cluster cores are identified, of which three at p = 0.001. The cluster map on the right shows the significant locations for the corresponding Multivariate Local Geary (using the Bonferroni bound for 0.01 and 99,999 permutations). The Multivariate Local Geary cluster map includes 35 cluster cores. Of those, only two are also identified by the Multivariate Quantile LISA. The six others, are not found to be significant in the Multivariate Local Geary cluster map.

In contrast to the large number of *significant* locations obtained with the Multivariate Local Geary, the Quantile version focuses on a much smaller (sub)set of significant observations and may thus provide clearer insight into the *interesting* locations.

20

Density-Based Clustering Methods

In this last chapter dealing with local patterns, density-based clustering methods are considered. These approaches search the data for high density subregions of arbitrary shape, separated by low-density regions. Alternatively, the *elevated* regions can be interpreted as *modes* in the spatial distribution over the support of the observations.

The methods covered form a transition between the local spatial autocorrelation statistics and the regionalization methods considered in Volume 2. They pertain primarily to point patterns, but can also be extended to a full multivariate setting. At first sight, density-based clustering methods may seem similar to spatially constrained clustering, but they are not quite the same. In contrast to the regionalization methods considered in Volume 2, the result of density based clustering does not necessarily yield a complete partitioning of the data. In a sense, the density-based cluster methods are thus similar in spirit to the identification of clusters by means of local spatial autocorrelation statistics, although they are not formulated as hypothesis tests. Therefore, these methods are included in the discussion of local spatial autocorrelation, rather than with the regionalization methods considered in Volume 2.

Attempts to discover high density regions in the data distribution go back to the classic paper on mode analysis by Wishart (1969), and its refinement in Hartigan (1975). In the literature, these methods are also referred to as *bump hunting*, i.e., looking for bumps (high regions) in the data distribution.

In this chapter, the focus is on the application of density-based clustering methods to the geographic location of points, but the methods can be generalized to locations in high-dimensional attribute space as well.

Four approaches are considered. First is a simple *heat map* as a uniform density kernel centered on each location. The logic behind this graph is similar to that of Openshaw's *Geographical Analysis Machine* (Openshaw et al., 1987) and the approach taken in spatial scan statistics (Kulldorff, 1997), i.e., a simple count of the points within the given radius. This is also the main idea behind the Getis-Ord local statistics considered in Chapter 17.

The remaining methods are all related to DBSCAN (Ester et al., 1996), i.e., *Density-Based Spatial Clustering of Applications with Noise*. Both the original DBSCAN is outlined, as well as its *improved* version, referred to as DBSCAN*, and its *Hierarchical* version, referred to as HDBSCAN, or, sometimes, HDBSCAN* (Campello et al., 2013, 2015).

The methods are illustrated with the *Italy Community Banks* sample data set that contains the locations of 261 community banks.

DOI: 10.1201/9781003274919-20

20.1 Topics Covered

- Interpret the results of a uniform kernel heat map
- Understand the principles behind DBSCAN
- Set the parameters for a DBSCAN approach
- Interpret the clustering results yielded by DBSCAN
- Understand the difference between DBSCAN and DBSCAN*
- Analyze the dendrogram generated by DBSCAN*
- Understand the logic underlying HDBSCAN
- Interpret the condensed tree and clustering results yielded by HDBSCAN
- Understand soft clustering and outlier identification in HDBSCAN

GeoDa Functions

- Map option > Heat Map
- Clusters > DBscan
- Clusters > HDBscan

Toolbar Icons

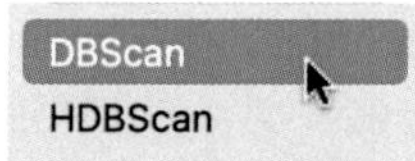

Figure 20.1: Clusters > DBSCAN | HDBSCAN

20.2 Heat Map

The Heat Map as implemented in `GeoDa` is a simple uniform kernel for a given bandwidth. A circle with radius equal to the bandwidth is centered on each point in turn, and the number of points within the radius is counted. There is no correction for distance decay. The density of the points is reflected in the color shading.

The Heat Map follows the same principle as the idea underlying the *Geographical Analysis Machine* or GAM of Openshaw et al. (1987), where the uniform kernel is computed for a range of bandwidths to identify high intensity *clusters*. However, in the implementation considered here, no significance is assigned, and the heat map is used primarily as a visual device to highlight differences in the density of the points.

20.2.1 Implementation

Unlike the other methods considered in this Chapter, the Heat Map is not invoked from the cluster menu, but as the **Heat Map** option on any point map. This brings up a dialog with various options. The top item is to **Display Heat Map**, which, when checked, brings up the default, with the **Heat Map on Top** item active.

The default Heat Map is constructed for the max-min bandwidth that ensures that each location has at least one other location within its circle. This is the same default as used to

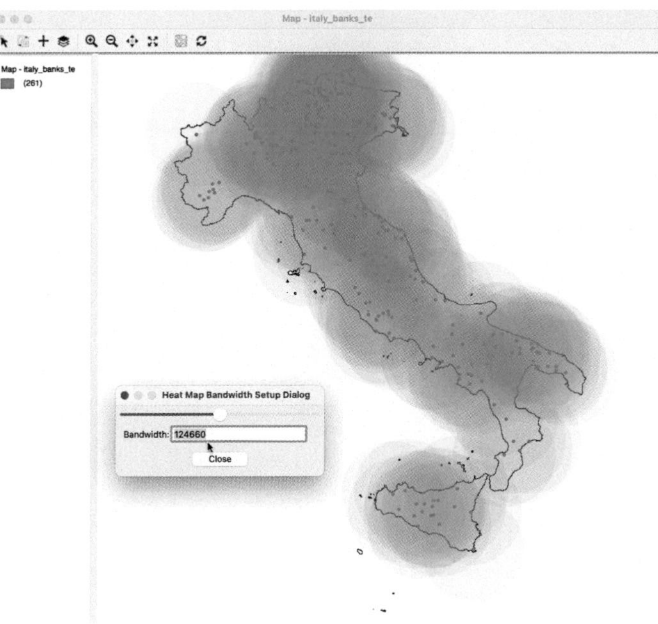

Figure 20.2: Default heat map

specify the distance-band spatial weights (see Section 11.3.1). In most applications, this is not very informative, as shown in Figure 20.2, which depicts the locations contained in the *Italy Community Banks* sample data set.[1] The **Heat Map Bandwidth Setup Dialog** in the example indicates a value of 124,660 (meters), the same as for the distance-band spatial weights used as an illustration in Chapter 11. The map can be customized using the various available options, of which the most important is to **Specify Bandwidth**.

20.2.1.1 Options

In Figure 20.3, a customized version is shown, where advantage is taken of several of the options. First, the bandwidth is set to 73,000 (meters), the same as in Section 11.4.2 (Figure 11.8). In addition, the **Change Fill Color** is used to set the color to blue, with the **Change Transparency** to 0.95.

These settings provide a much clearer indication of the varying density over the map, with different gradations of the color corresponding to high and low density regions. In particular, the area in the North of the country is highlighted as a high density location.

A final option, **Specify Core Distance**, only applies to the output of an HDBSCAN clustering routine (see Section 20.5.5.3).

[1] The map shown in the figure has the default point colors changed to black, with the point size as 3. In addition a second layer consisting of the outlines of the Italian regions is added, of which the **Only Map Boundary** option is selected.

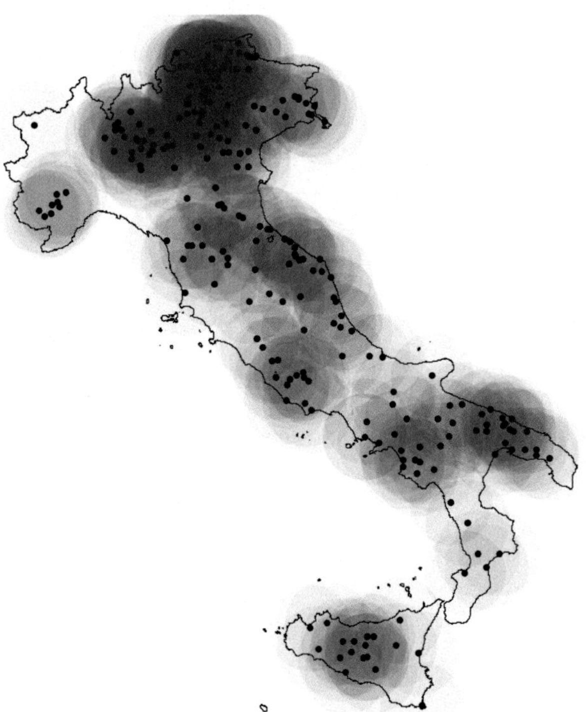

Figure 20.3: Heat map with bandwidth 73 km

20.3 DBSCAN

The DBSCAN algorithm was originally outlined in Ester et al. (1996) and Sander et al. (1998) and was more recently elaborated upon in Gan and Tao (2017) and Schubert et al. (2017). Its logic is similar to that just outlined for the uniform kernel. In essence, the method again consists of placing circles of a given radius on each point in turn, and identifying those groupings of points where a lot of locations are within each others range.

20.3.1 Important concepts

The DBSCAN algorithm introduces some specialized terminology to characterize the connections between points as well as the ensuing network structure.

20.3.1.1 Core, Border and Noise points

Points are classified as *Core*, *Border* or *Noise* depending on how many other points are within a critical distance band, the so-called *Eps neighborhood.* To visualize this, Figure 20.4 contains nine points, each with a circle centered on it with radius equal to *Eps.*[2] Note that in DBSCAN, any distance metric can be used, not just Euclidean distance as in this illustration.

[2]The figure is loosely based on Figure 1 in Schubert et al. (2017).

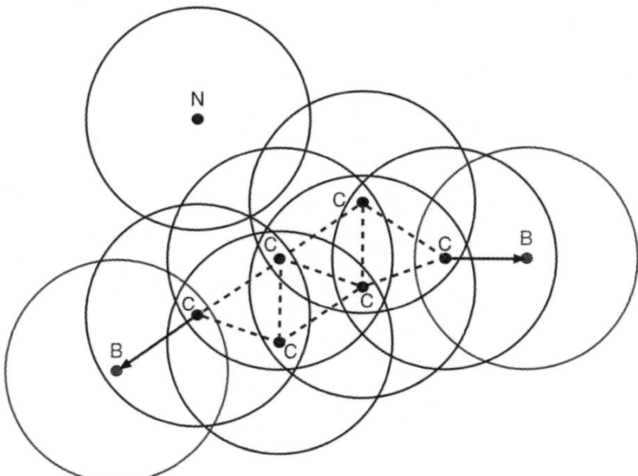

Figure 20.4: DBSCAN Core, Border and Noise points

A second critical concept is the number of points that need to be included in the distance band in order for the spatial distribution of points to be considered as dense. This is the so-called minimum number of points, or *MinPts* criterion. In the example, this is set to four. However, in contrast to the convention used before in defining k-nearest neighbors (Section 11.3.2), the *MinPts* criterion includes the point itself. So a *MinPts* = 4 corresponds to a point having 3 neighbors within the *Eps* radius .[3]

In Figure 20.4, all red points with associated red circles have at least four points within the critical range (including the central point). They are labeled as *Core* points (the letter C in the graph) and are included in the cluster (connected by the dashed blue lines).

The points in magenta (labeled B), with associated magenta circles have some points within the distance range (one, to be precise), but not sufficient to meet the *MinPts* criterion. They are potential *Border* points and may or may not be included in the cluster.

Finally, the blue point (labeled N) with associated blue circle does not have any points within the critical range and is labeled *Noise*. Such a point cannot become part of any cluster.

20.3.1.2 Reachability

A point is *directly density reachable* from another point if it belongs to the *Eps* neighborhood of that point and is one of *MinPts* neighbors of that point. This is not a symmetric relationship.

In the example, any red point C is directly density reachable from at least two other red points. Any such point pairs are within each others critical range. However, for border points B, the relationship only holds in one direction, as shown by the arrow. They are directly density reachable from a neighbor C, but since the B points only have one neighbor, their range does not meet the minimum criterion. Therefore, the neighbor C is *not* directly density reachable from B.

[3]This definition of *MinPts* is from the original paper. In some software implementations, the minimum points pertain to the number of nearest neighbors, i.e., *MinPts* − 1. `GeoDa` follows the definition from the original papers.

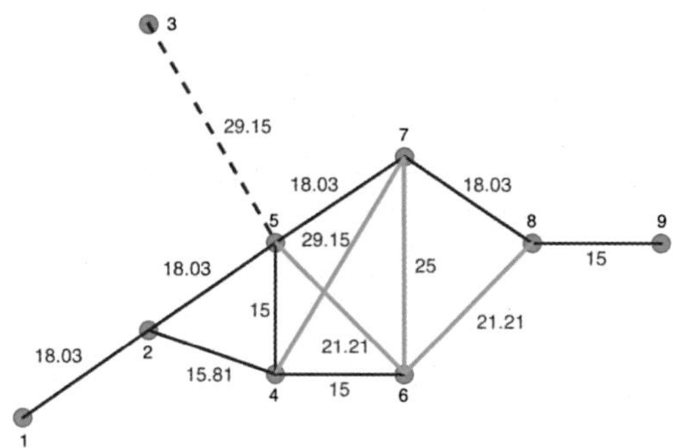

Figure 20.5: Connectivity graph for Eps = 20

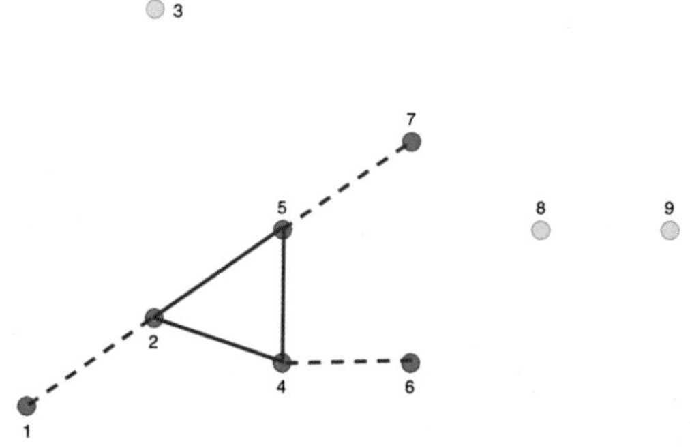

Figure 20.6: DBSCAN clusters for Eps = 20 and MinPts = 4

A chain of points in which each point is directly density reachable from the previous one is called *density reachable.* In order to be included, each point in the chain has to have at least *MinPts* neighbors and could serve as the core of a cluster. All the points labeled C in the figure are density reachable, highlighted by the dashed connections between them.

In order to decide whether a border point should be included in a cluster, the concept of *density connected* is introduced. A point becomes density connected if it is connected to a density reachable point. For example, the points B are each within range of a point C that is itself density reachable. As a result, the *Border* points B become included in the cluster.

20.3.2 DBSCAN algorithm

In DBSCAN, a cluster is defined as a collection of points that are density connected and maximize the density reachability. The algorithm starts by randomly selecting a point and determining whether it can be classified as *Core* – with at least *MinPts* in its *Eps* neighborhood – or instead as *Border* or *Noise.* A *Border* point can later be included in a cluster if it is *density connected* to another *Core* point. It is then assigned the cluster label of the core point and no longer further considered. One implication of this aspect of the algorithm is that once a *Border* point is assigned a cluster label, it cannot be assigned to a different cluster, even though it might actually be closer to the corresponding core point. In a later version, labeled DBSCAN* (considered in Section 20.4), the notion of border points is dropped, and only *Core* points are considered to form clusters.

The algorithm systematically moves through all the points in the data set and assesses their range. The search for neighbors is facilitated by using an efficient spatial data structure, such as an R* tree. When two clusters are density connected, they are merged. The process continues until all points have been evaluated.

Two critical parameters in the DBSCAN algorithm are the distance range and the minimum number of neighbors. In addition, sometimes a tolerance for noise points is specified as well. The latter constitute zones of low density that are not deemed to be interesting.

In order to avoid any noise points, the critical distance must be large enough so that every point is at least density connected. An analogy is the specification of a max-min distance band in the creation of spatial weights, which ensures that each point has at least one neighbor. In practice, this is typically not desirable, but in some implementations a maximum percentage of noise points can be set to avoid too many low density areas.[4]

Ester et al. (1996) recommend that *MinPts* be set at 4 and the critical distance adjusted accordingly to make sure that sufficient observations can be classified as *Core*. As in other cluster methods, some trial and error is typically necessary. Having to find a proper value for *Eps* is often considered a major drawback of the DBSCAN algorithm.

20.3.2.1 Illustration

To illustrate how the DBSCAN proceeds, a toy example of nine points is depicted in Figure 20.5. The inter-point distances are included on the graph. The typical point of departure is to consider a connectivity structure that ensures that each point has at least one neighbor, i.e., the max-min critical distance. In the example, this distance is 29.1548, with the corresponding number of neighbors in the connectivity graph ranging from 1 to 5.

The concept of *Noise* points (i.e., unconnected points for a given critical distance) becomes clear when the *Eps* distance is set to 20. In Figure 20.5, this results in one unconnected point, labeled 3, shown by the dashed blue line between 3 and 5. The green lines correspond with inter-point distances that do not meet the *Eps* criterion either, but they do not result in isolates. The effective connectivity graph for *Eps* = 20 is shown in solid blue.

With *MinPts* as 4 (i.e., 3 neighbors for a core point), the algorithm proceeds through each point, one at a time. For example, it could start with point 1, which only has one neighbor (point 2) and thus does not meet the *MinPts* criterion. Therefore, it is initially labeled as *Noise*. Next, point 2 is considered. It has 3 neighbors, meaning that it meets the *Core* criterion, and therefore it is labeled cluster 1. All its neighbors, i.e., 1, 4 and 5 are also labeled cluster 1 and are no longer considered. Note that this changes the status of point 1 from *Noise* to *Border*. Point 3 has no neighbors and is therefore labeled *Noise*.

Next is point 6, which is a neighbor of point 4 that belongs to cluster 1. Since 6 is therefore *density connected*, it is added to cluster 1 as a *Border* point. The same holds for point 7, which is similarly added to cluster 1.

Point 8 has two neighbors, which is insufficient to reach the *MinPts* criterion. Even though it is connected to point 7, which belongs to the cluster, 7 is not a *Core* point, so therefore 8

[4]`GeoDa` currently does not support this option.

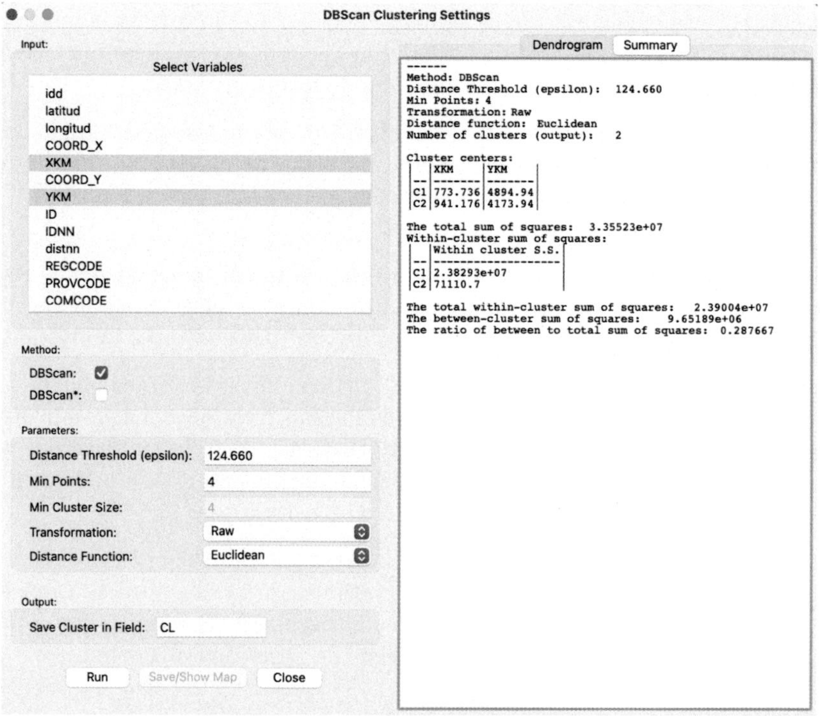

Figure 20.7: DBSCAN clustering settings

is not density connected and is labeled *Noise*. Finally, point 9 has only point 8 as neighbor and is therefore labeled *Noise* as well.

This results in one cluster consisting of 6 points, shown as the dark blue points in Figure 20.6. The solid blue lines show the connections between the *Core* points, the dashed lines show the *Border* points. The light blue points for 3, 8 and 9 indicate *Noise* points.

20.3.3 Implementation

DBSCAN is invoked from the cluster toolbar icon (Figure 20.1) as the first item in the density cluster group, or from the menu as **Clusters > DBScan** . This brings up the **DBScan Clustering Settings** dialog, which combines two functions. One is the specification of variables and various cluster options, shown in the left-hand panel of Figure 20.7. The other is the presentation of **Summary** characteristics of the resulting clusters, and, for DBSCAN*, the **Dendrogram** (see Section 20.4.2.1).

The **Select Variables** panel includes the variables that must be chosen as X and Y coordinates. For the *Italy Community Banks* sample data set, the variables **XKM** and **YKM** are specified, which correspond with the projected coordinates expressed in kilometers (the original is in meters). The selected **Method** is **DBScan**.

The selection of the coordinates immediately populates some of the fields in the **Parameters** panel. The default **Transformation** is **Standardize (Z)**, with the **Distance Threshold (epsilon)** as 0.443177. However, in this instance, the original coordinates should be used, hence the **Raw** option is specified in Figure 20.7. This yields a critical distance of 124.659981.

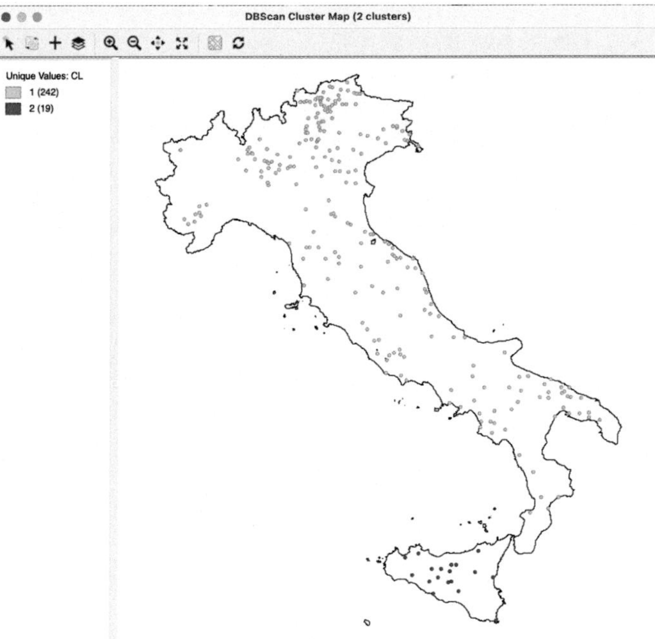

Figure 20.8: Default DBSCAN cluster map

As it turns out, a slight correction is needed, and the value is rounded to 124.660 in the dialog.

The other options consist of the **Min Points**, which is initially left to the default of **4**, and the **Distance Function**, set to **Euclidean** (the other option is **Manhattan** distance). The **Min Cluster Size** option is not available for DBSCAN.

A final item to be specified is the variable that will contain the cluster identifier (**Save Cluster in Field**), with **CL** as the default. In `GeoDa` the identifiers are assigned in decreasing order of the size of the cluster, i.e., with 1 for the largest cluster. Points that remain as *Noise* points are assigned a value of 0.

20.3.3.1 Cluster map and summary

The resulting cluster map for the default settings is shown in Figure 20.8. Only two clusters are identified, one very large one, consisting of 242 bank locations on the mainland, the other, a very small one of 19 banks on the island of Sicily. Given the choice of the max-min bandwidth, there are no *Noise* points.

The **Summary** in the right-hand panel of Figure 20.7 lists the parameters used, the cluster centers and within-cluster sum of squares, as well as an overall summary of the fit, the ratio of between to total sum of squares. The value of 0.288 is very low, which is not surprising, given the large number of points contained in the main cluster, which results in a high within-cluster sum of squares. The specifics of these summary measures are discussed in more detail in the treatment of spatial clustering in Volume 2.

20.3.3.2 Sensitivity analysis

In most applications, the default parameters are just a point of departure and considerable sensitivity analysis is needed to assess the effect of changes in the settings. Specifically, the

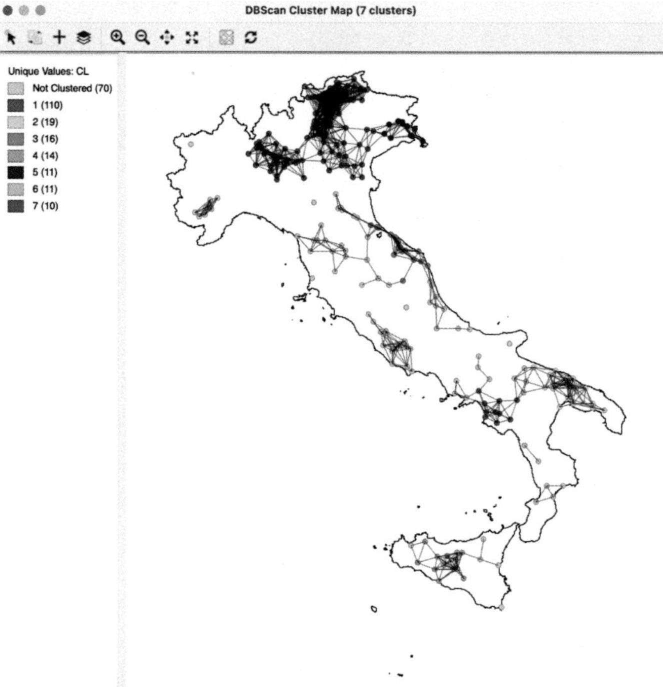

Figure 20.9: DBSCAN cluster map, Eps=50, Min Pts=10

two main parameters of **Distance Threshold** and **Min Points** play a crucial role. As an illustration, the results of a cluster analysis with a bandwidth of 50 km and the minimum points set to 10 are given in Figure 20.9. In addition to the cluster points, identified by different colors, the connectivity graph for a 50 km distance band is shown as well. Seven clusters are identified, ranging in size from 110 to 10 observations. A total of 70 locations are labeled *Noise*, either because they become isolates (not connected to the graph), or they have insufficient neighbors. The overall fit of the clusters is much better, with a between to total ratio of 0.957 (not shown), due to a grouping of locations that are relatively close by.

20.3.3.3 Border points and minimum points

In some instances, the outcome of a DBSCAN algorithm can lead to seemingly counter intuitive results. For example, using the Italian bank data with a 50 km bandwidth and 8 minimum points yields 10 clusters and a good fit of 0.960. However, cluster 10 consists of only 6 observations, less than the minimum points requirement (results not shown). This is a result of the peculiar way in which *Border* points are treated. Specifically, as mentioned above, they are assigned to the first set of density reachable points to which it is density connected, even though it may be closer to a later cluster. The specific implementation of the algorithm does not change the allocation, but does recognize the later clusters as such, even with less than minimum points.

20.4 DBSCAN*

One of the potentially confusing aspects of DBSCAN is the inclusion of so-called *Border* points in a cluster. The DBSCAN* algorithm, outlined in Campello et al. (2013) and further elaborated upon in Campello et al. (2015), does away with the notion of border points, and only considers *Core* and *Noise* points.

Similar to the approach in DBSCAN, a *Core* object is defined with respect to a distance threshold (*Eps*) as the center of a neighborhood that contains at least *Min Points* other observations. All non-core objects are classified as *Noise* (i.e., they do not have any other point within the distance range *Eps*). Two observations are *Epsilon reachable* if each is part of the Epsilon-neighborhood of the other. *Core* points are *density-connected* when they are part of a chain of Epsilon-reachable points. A *cluster* is then any largest (i.e., all eligible points are included) subset of density connected pairs. In other words, a cluster consists of a chain of pairs of points that belong to each others Epsilon-neighborhoods.

20.4.1 Mutual reachability distance

An alternative perspective on the concept of Epsilon-neighborhoods is to consider the duality between density of points within a given radius and the distance needed to reach a given number of neighbor points. More specifically, the requirement of having k neighbors (*Minimum Points* − 1) within a given Epsilon distance range can be replaced by considering the k-nearest neighbor distance for a point. The smaller the distance required to reach the k-nearest neighbor, the denser the point distribution will be, and vice versa. In DBSCAN* and HDBSCAN (see Section 20.5), this is called the *core distance* (d_{core}) for a given k. The larger the core distance, the less dense is the point distribution around the reference location.

This consideration leads to a new concept of distance between two points A and B, defined as the *mutual reachability distance*. This is a compromise between the actual inter-point distance, d_{AB} and the density of the points, reflected in the respective *core distances*, $d_{core}(A)$ and $d_{core}(B)$. Specifically, even though A and B may be close, if they are in low density regions, their inter-point distance will be replaced by the larger of $d_{core}(A)$ and $d_{core}(B)$, effectively *pushing* the points further away from each other. Formally, the *mutual reachability distance* between A and B is defined as:

$$d_{mr}(A, B) = max[d_{core}(A), d_{core}(B), d_{AB}].$$

This mutual reachability distance replaces the original distance d_{AB} in the connectivity graph that forms the basis for the determination of clusters (e.g., a graph like Figure 20.5). Unless A and B are mutual k-nearest neighbors, this effectively replaces the inter-point distances by a *core* distance.[5]

In DBSCAN*, the adjusted connectivity graph is reduced to a *minimum spanning tree* or MST (see Section 3.4) to facilitate the derivation of clusters. Rather than finding a single solution for a given distance range as in DBSCAN, the longest edges in the MST are cut *sequentially* to provide a solution for each value of inter-point *mutual reachability distance*. This boils down to finding a cut for smaller and smaller values of the core distance. It yields a hierarchy of clusters for decreasing values of the core distance.

[5]The largest k-nearest neighbor distance will always be larger than the inter-point distance, unless A and B are mutual k-nearest neighbors, in which case $d_{mr}(A, B) = d_{core}(A) = d_{core}(B) = d_{AB}$.

Figure 20.10: DBSCAN-Star dendrogram, Eps=50, Min Pts=10

In practice, one starts by applying a cut between two edges that are the furthest apart. This either results in a single point splitting off (a *Noise* point), or in the single cluster to split into two (two subtrees of the MST). Subsequent cuts are applied for smaller and smaller distances. Decreasing the critical distance is the same as increasing the density of points, referred to as λ in this context (see Section 20.5).

As Campello et al. (2015) show, applying cuts to the MST with decreasing distance (or increasing density) is equivalent to applying cuts to a dendrogram obtained from single linkage hierarchical clustering using the mutual reachability distance as the dissimilarity matrix (hierarchical clustering methods are considered in Volume 2).

DBSCAN* derives a set of *flat* clusters by applying a *cut* to the dendrogram associated with a given distance threshold. The main difference with the original DBSCAN is that *border* points are no longer considered. In addition, the clustering process is visualized by means of a dendrogram. However, in all other respects it is similar in spirit, and it still suffers from the need to select a distance threshold.

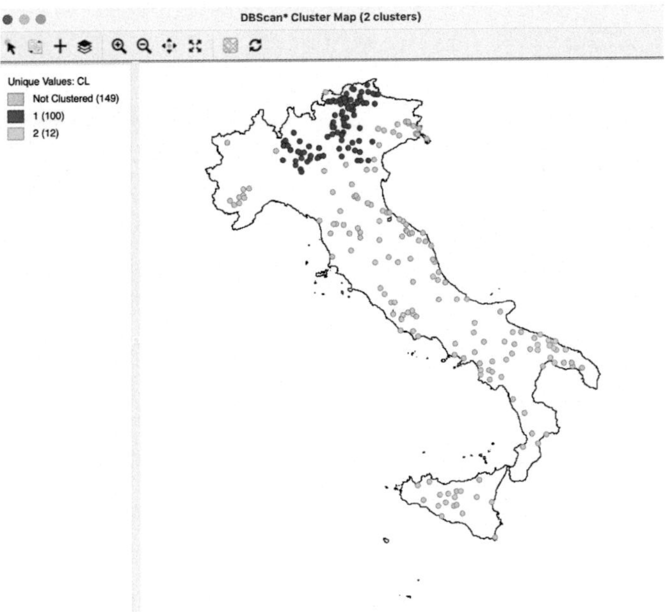

Figure 20.11: DBSCAN-Star cluster map, Eps=50, Min Pts=10

20.4.2 Implementation

DBSCAN* is invoked in the same way as DBSCAN, but with **DBScan*** checked as the **Method** in the **DBScan Clustering Settings** dialog (Figure 20.7). All other settings are the same, except that now the **Min. Cluster Size** option becomes available. However, in practice, this is typically set equal to the **Min Points** parameter and it is seldom used.

The functionality operates largely the same as for DBSCAN, except that the first result for a given **Min Points** is a **Dendrogram**, shown in the right panel of the dialog. Figure 20.10 depicts the dendrogram for the same settings as in Figure 20.9, i.e., with a threshold distance of 50 km and 10 minimum points. However, the results are very different. Only two clusters are identified, compared to 7 clusters before. They are visualized by selecting the **Save/Show Map** button at the bottom of the dialog.

The resulting cluster map is shown in Figure 20.11. One cluster is very large, containing 100 observations, the other is small, with only 12 observations. A total of 149 observations are classified as *Noise*. Relative to the DBSCAN solution in Figure 20.9, the fit declined to a ratio of 0.894.

20.4.2.1 Exploring the dendrogram

The main advantage of DBSCAN* over DBSCAN is a much more straightforward treatment of *Border* points (they are all classified as *Noise*), and an easy visualization of the sensitivity to the threshold distance by means of a dendrogram. Clusters can be explored for different core distances. As long as **Min Points** (and the minimum cluster size) remains the same, the dendrogram can be used to assess the cluster structure for different threshold distance cut values. This is implemented by moving the cut-off line in the dendrogram. The distance on the horizontal axis is the *mutual reachability distance* or *core distance*, and it replaces the original inter-point distances.

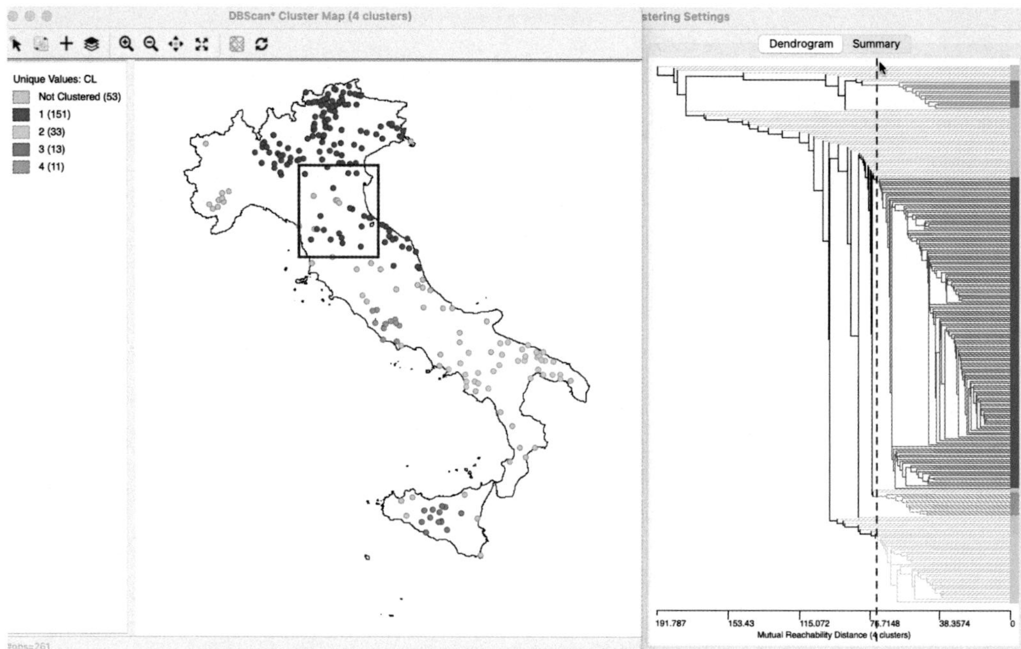

Figure 20.12: Exploring the dendrogram

For example, in the right-hand panel of Figure 20.12, the dashed red line corresponding to the cut-off distance is moved to the left, to a distance of 73 km (also used in Section 11.4.2). The corresponding cluster map is shown in the left-hand panel. The number of clusters has doubled to four, ranging in size from 151 to 11 observations, with 53 unclustered *Noise* points. The fit ratio decreases to 0.855.

The map may give the impression that seemingly disconnected points are contained in the same cluster. In the example, the points within the red rectangle contain both observations that belong to cluster 1 (dark blue) as well as unclustered points (light blue). This seeming contradiction is due to the fact that the *core distance* is used in the dendrogram, and not the actual inter-point distance.

The dendrogram needs to be reconstructed for each different value of **Min Points**, since this critically affects the mutual reachability and core distances.

The main disadvantage of DBSCAN*, just as for DBSCAN, remains that the same fixed threshold distance must be applied to all clusters, yielding a so-called *flat* cluster. This is remedied by HDBSCAN.

20.5 HDBSCAN

HDBSCAN was originally proposed by Campello et al. (2013), and more recently elaborated upon by Campello et al. (2015) and McInnes and Healy (2017).[6] As in DBSCAN*, the

[6]This method is variously referred to as HDBSCAN or HDBSCAN*. Here, for simplicity, the former will be used.

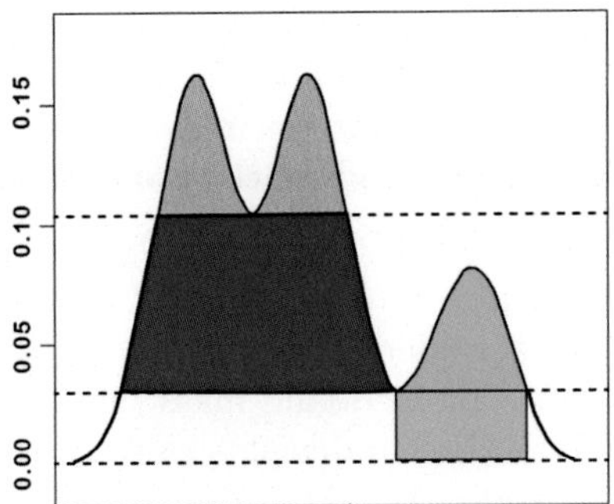

Figure 20.13: Level set

algorithm implements the notion of *mutual reachability distance*. However, rather than applying a fixed value of the cut-off distance to produce a cluster solution, a hierarchical process is implemented that finds an optimal cluster combination, using a different critical cut-off distance for each cluster. In order to accomplish this, the concept of cluster *stability* or *persistence* is introduced. Optimal clusters are selected based on their *relative excess of mass* value, which is a measure of their persistence.

HDBSCAN keeps the notion of *Min Points* from DBSCAN and also uses the concept of *core distance* of an object (d_{core}) from DBSCAN*. The core distance is the distance between an object and its k-nearest neighbor, where k = Min Points − 1 (in other words, as for DBSCAN, the object itself is included in *Min Points*).

Intuitively, for the same k, the core distance will be much smaller for densely distributed points than for sparse distributions. Associated with the core distance for each object x is the concept of *density* or λ_p, as the inverse of the core distance. As the core distance becomes smaller, λ increases, indicating a higher density of points.

As in DBSCAN*, the mutual reachability distance is employed to construct a minimum spanning tree, or, rather, the associated dendrogram from single linkage clustering. From this, HDBSCAN derives an optimal solution, based on the notion of *persistence* or *stability* of the cluster.

20.5.1 Important concepts

Several new concepts are introduced in the treatment of HDBSCAN.

20.5.1.1 Level set

A first general concept is that of a *level set* associated with a given density level λ. This is introduced in the context of *bump hunting*, or the search for high density regions in a general distribution f. Formally, the level set associated with density λ in the associated distribution is:

$$L(\lambda; f) = \{x | Prob(x) > \lambda\},$$

i.e., the collection of data points x for which the probability density exceeds the given threshold. A subset of such points that contains the maximum number of points that are *connected* (so-called connected components) is called a *density-contour cluster* of the density function. A *density-contour tree* is then a tree formed of nested clusters that are obtained by varying the level λ (Hartigan, 1975; Müller and Sawitzki, 1991; Stuetzle and Nugent, 2010).

To visualize this concept, the data distribution can be represented as a two-dimensional profile, as in Figure 20.13 (loosely based on the example in Stuetzle and Nugent, 2010). A level set consists of those data points (i.e., areas under the curve) that are above a horizontal line at p. This can be thought of as islands sticking out above the ocean, where the ocean level corresponds with p.

For the lowest p-values in the figure, up to around 0.03, all the land mass is connected and it appears as a single entity. As the p-value increases, part of the area gets submerged and two separate islands appear, with a combined land mass (level set) that is smaller than before. Each of these land masses constitutes a density-contour cluster. As the p-value becomes larger than around 0.08, the island to the right disappears, and only a single island remains, corresponding to a yet smaller level set. Finally, for $p > 0.10$, this splits into two separate islands.

The level set establishes the connection between the *density* and the cluster size.

20.5.1.2 Persistence and excess of mass

As the value of the density λ increases, or, equivalently, as the core distance becomes smaller, larger clusters break into sub-clusters (the analogy with rising ocean levels), or shed *Noise* points. In the current context and in contrast to the previous discussion, a *Noise* point dropping out is not considered a split of the cluster. The cluster remains, but only becomes smaller.

The objective is to identify the most prominent clusters as those that continue to exist longest. In the Figure, it corresponds with the area under the curve while a particular cluster configuration exists. This is formalized through the concept of *excess of mass* (Müller and Sawitzki, 1991), or the total density contained in a cluster after it is formed. The point of formation is associated with a minimum density, or λ_{min}.

In the island analogy, λ_{min} would correspond to the ocean level where the land bridge is first covered by water, such that the islands appear for the first time as separate entities. The excess of mass would be the volume of the island above that ocean level, as illustrated in Figure 20.13.

The *stability* or *persistence* of a cluster is formalized through a notion of *relative excess of mass* of the cluster. For each cluster C_i, there is a minimum density, $\lambda_{min}(C_i)$, where it comes into being. In addition, for each point x_j that belongs to the cluster, there is a maximum density $\lambda_{max}(x_j, C_i)$ after which it no longer belongs to the cluster. For example, in Figure 20.13, the cluster on the left side comes into being at $p = 0.03$. At $p = 0.10$, it splits into two new clusters. So, for the large cluster, λ_{min}, or the point where the cluster comes into being is 0.03. The cluster stops existing (it splits) for $\lambda_{max} = 0.10$. In turn, $p = 0.10$ represents λ_{min} for the two smaller clusters. In this example, all the members of the cluster exit at the same time, but in general, this will not be the case. Individual points that fail to meet the critical distance as it decreases (λ increases) will fall out. Such points will therefore have a different $\lambda_{max}(x_j, C_i)$ from those that remain in the cluster. They are treated as *Noise* points, while the cluster itself continues to exist.

The *persistence* of a given cluster C_i is then the sum of the difference between the λ_{max} and λ_{min} for each point x_j that belongs to the cluster at its point of creation:

$$S(C_i) = \sum_{x_j \in C_i} [\lambda_{max}(x_j, C_i) - \lambda_{min}(C_i)],$$

ID	Core_d	Lambda_p
1	18.03	0.055463
2	15.81	0.063251
3	29.15	0.034305
4	15.00	0.066667
5	15.00	0.066667
6	15.00	0.066667
7	18.03	0.055463
8	18.03	0.055463
9	15.00	0.066667

Figure 20.14: Core distance and lambda for each point

where $\lambda_{min}(C_i)$ is the minimum density level at which C_i exists, and $\lambda_{max}(x_j, C_i)$ is the density level after which point x_j no longer belongs to the cluster.

The *persistence* of a cluster becomes the basis to decide whether to accept a cluster split or to keep the larger entity as the cluster.

20.5.1.3 Condensed dendrogram

A *condensed dendrogram* is obtained by evaluating the stability of each cluster relative to the clusters into which it splits. If the sum of the stability of the two descendants (or children) is larger than the stability of the parent node, then the two children are kept as clusters and the parent is discarded. Alternatively, if the stability of the parent is larger than that of the children, the children and all their descendants are discarded. In other words, the large cluster is kept and all the smaller clusters that descend from it are eliminated. Here again, a split that sheds a single point is not considered a cluster split, but only a shrinking of the original cluster (which retains its label). Only *true* splits that result in sub-clusters that contain at least two observations are considered in the simplification of the tree.

Sometimes a further constraint is imposed on the minimum size of a cluster, which would pertain to the splits as well. As suggested in Campello et al. (2015), a simple solution is to set the minimum cluster size equal to *Min Points.*

For example, in the illustration in Figure 20.13, the area in the left cluster between 0.03 and 0.10 (in blue) is larger than the sum of the areas in the smaller clusters obtained at 0.10 (in purple). As a result, the latter would not be considered as separate clusters. The condensed dendrogram yields a much simpler tree representation and an identification of *clusters* that maximize the sum of the individual cluster stabilities.

20.5.2 HDBSCAN algorithm

The HDBSCAN algorithm has the same point of departure as DBSCAN*, i.e., a minimum spanning tree or dendrogram derived from the *mutual reachability distance.* To illustrate the logic, the same toy example is employed as in Figure 20.5. To keep matters simple, the **Min Cluster Size** and **Min Points** are set to **2**. As a result, the *mutual reachability distance* is the same as the nearest neighbor distance. Since the core distance for each point is its smallest distance to have one neighbor, the maximum of the two core distances and the actual distance will always be the actual distance. While this avoids one of the innovations of the algorithm, it still allows to highlight its distinctive features.

The nearest neighbor distances can be taken from the connectivity graph in Figure 20.5, which yields the core distance for each point. The λ_p value associated with this core distance is its inverse. The results for each point in the example are summarized in Figure 20.14.

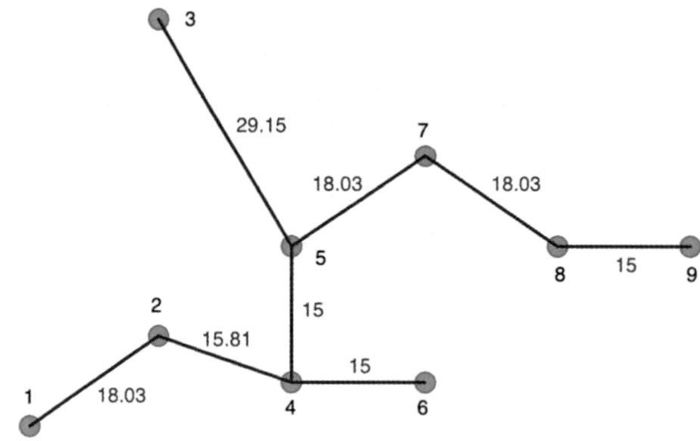

Figure 20.15: Minimum spanning tree connectivity graph

The inter-point distances form the basis for a minimum spanning tree (MST), which is essential for the *hierarchical* nature of the algorithm. The corresponding connectivity graph is shown in Figure 20.15, with both the node identifiers and the edge weights listed (the distance between the two nodes the edge connects). The HDBSCAN algorithm essentially consists of a succession of *cuts* in the connectivity graph. These cuts start with the highest edge weight and move through all the edges in decreasing order of the weight. The process of selecting cuts continues until all the points are singletons and constitute their own cluster. Formally, this is equivalent to carrying out single linkage hierarchical clustering based on the mutual reachability distance matrix.

20.5.2.1 Minimum spanning tree

An alternative visualization of the pruning process is in the form of a tree, shown in Figure 20.16. In the *root* node, all the points form a single cluster. The corresponding λ level is set to zero. As λ increases, or, equivalently, the *core distance* decreases, points fall out of the cluster, or the cluster splits into two sub-clusters. This process continues until each point is a *leaf* in the tree (i.e., it is by itself).

In Figure 20.15, the longest edge is between 3 and 5, with a core distance of 29.15 for point 3. The corresponding λ is 0.034, for which point 3 is removed from the tree and becomes a leaf. In the `GeoDa` implementation of HDBSCAN, singletons are not considered to be valid clusters, so node C1 becomes the new root, with λ_{min} reset to zero.

The next meaningful cut is for $\lambda = 0.055$, or a distance of 18.03. This creates two *clusters*, one consisting of the six points 1–7, but without 3, labeled C2, and the other consisting of 8 and 9, labeled C3. The process continues for increasing values of λ, although in each case the split involves a singleton and no further valid clusters are obtained. This leads to the consideration of the *stability* of the clusters and the condensed tree.

20.5.2.2 Cluster stability and condensed tree

A closer investigation of the cuts in the tree reveals that the λ_p values associated with each point (Figure 20.14) correspond to the λ value where the corresponding point *falls out* of a clusters in the tree. For example, for point 7, this happens for $\lambda = 0.055$, or a core distance of 18.03.

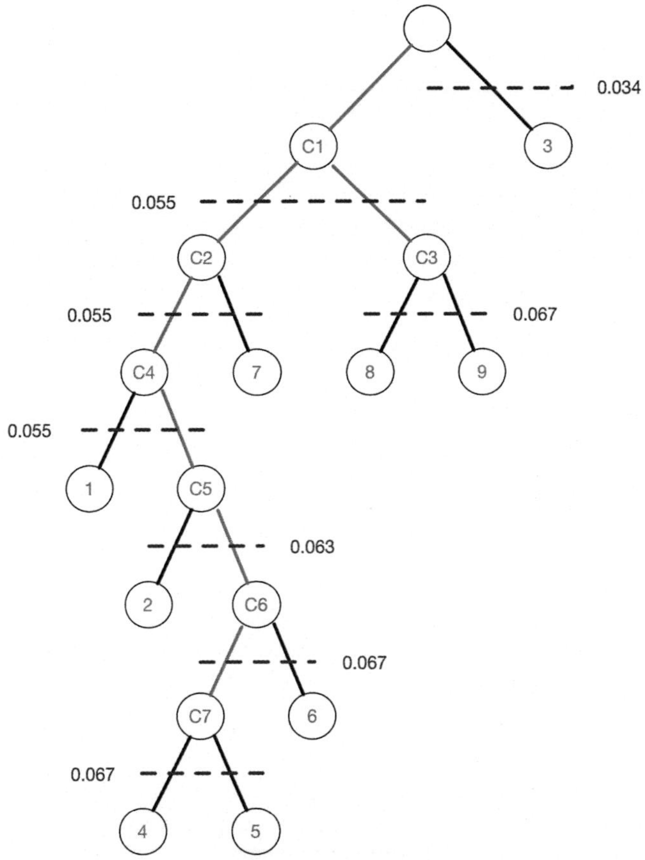

Figure 20.16: Pruning the MST

ID	C1	C2	C3
1	0.055463	0.000000	
2	0.055463	0.007788	
3	0.034305		
4	0.055463	0.011204	
5	0.055463	0.011204	
6	0.055463	0.011204	
7	0.055463	0.000000	
8	0.055463		0.011204
9	0.055463		0.011204
	0.47801025	0.041400	0.022408

Figure 20.17: Cluster stability

The stability or persistence of each cluster is defined as the sum over all the points of the difference between λ_p and λ_{min}, where λ_{min} is the λ level where the cluster forms.

For example, cluster C3 is formed (i.e., splits off from C1) for $\lambda = 0.055463$ and consists of points 8 and 9. Therefore, the value of λ_{min} for C3 is 0.055463. The value of λ_p for points 8 and 9 is 0.066667, corresponding with a distance of 15.0 at which they become singletons (or leaves in the tree). The contribution of each point to the persistence of the cluster C3 is thus 0.066667 − 0.055463 = 0.011204. The total value of the persistence of cluster C3 is therefore the sum of this contribution for both points, or 0.022408. Similar calculations yield the persistence of cluster C2 as 0.041400. This is summarized in Figure 20.17. For the sake of completeness, the results for cluster C1 are included as well, although those are ignored by the algorithm, since C1 is reset as *root* of the tree.

The *condensed* tree only includes C1 (as root), C2 and C3. For example, the split of C2 into C4 and 7 is based on the following rationale. Since the persistence of a singleton is zero, and C4 includes one less point than C2, but with the same value for λ_{min}, the sum of the persistence of C4 and 7 is less than the value for C2, hence the condensed tree stops there.

The condensed tree allows for different values of λ to play a role in the clustering mechanism, potentially corresponding with a separate critical cut-off distance for each cluster.

20.5.3 Cluster membership

As outlined in the previous section, a point becomes part of a cluster for λ_{min}, which corresponds of a cut-off distance d_{max}. Once a point is part of a cluster, it remains there until the threshold distance is small enough such that it becomes *Noise*, which happens for $d_{p,core}$. This may be the same or be different from the smallest distance where the cluster still exists, or d_{min}. The *strength* of the cluster membership of point p can now be expressed as the ratio of the final cut-off distance to the core distance for point p, $d_{max}/d_{p,core}$, or, equivalently, as the ratio of the density that corresponds to the point's core distance to the maximum density, or, λ_p/λ_{max}.

The ratio gives a measure of the degree to which a point *belongs* to a cluster. For example, for points that meet the d_{min} thresholds, i.e., observations in the highest density region, the ratio will be 1. *Noise* points are assigned 0. More interestingly, for points that are part of the cluster when it forms, but drop out during the pruning process, the ratio will have a value smaller than 1. Therefore, the higher the ratio, the stronger the membership of the point in the corresponding cluster.

This ratio can be exploited as the basis of a *soft* or *fuzzy* clustering approach, in which the degree of membership in a cluster is considered. This is beyond the current scope. Nevertheless, the ratio, or *probability* of cluster membership, remains a useful indicator of the extent to which the cluster is concentrated in high density points.

20.5.4 Outliers

Related to the identification of density clusters is a concern to find points that do *not* belong to a cluster and are therefore classified as *outliers*. Parametric approaches are typically based on some function of the spread of the underlying distribution, such as the well-known 3-sigma rule for a normal density.[7]

An alternative approach is based on non-parametric principles. Specifically, the distance of an observation to the nearest cluster can be considered as a criterion to classify outliers. Campello et al. (2015) proposed a post-processing of the HDBSCAN results to characterize the degree to which an observation can be considered an outlier. The method is referred to as GLOSH, which stands for *Global-Local Outlier Scores from Hierarchies*.

[7]These approaches are sensitive to the influence of the outliers on the estimates of central tendency and spread. This works in two ways. On the one hand, outliers may influence the parameter estimates such that their presence could be *masked*, e.g., when the estimated variance is larger than the true variance. The reverse effect is called *swamping*, where observations that are legitimately part of the distribution are made to look like outliers.

The logic underlying the outlier detection is closely related to the notion of cluster membership just discussed. In fact, the probability of being an outlier is the complement of the cluster membership.

The rationale behind the index is to compare the density threshold for which the point is still attached to a cluster (λ_p in the previous discussion) and the highest density threshold λ_{max} for that cluster. The GLOSH index for a point p is then:

$$\text{GLOSH}_p = \frac{\lambda_{max} - \lambda_p}{\lambda_{max}},$$

or, equivalently, $1 - \lambda_p/\lambda_{max}$, the complement of the cluster membership for all but *Noise* points. For the latter, since the cluster membership was set to zero, the actual λ_{min} needs to be used.

Given the inverse relationship between density and distance threshold, an alternative formulation of the outlier index is as:

$$\text{GLOSH}_p = 1 - \frac{d_{min}}{d_{p,core}}.$$

20.5.5 Implementation

HDBSCAN is invoked from the Menu or the cluster toolbar icon (Figure 20.7) as the second item in the density clusters subgroup, **Clusters > HDBSCAN**. The overall setup is very similar to that for DBSCAN, except that there is no threshold distance. Again, the coordinates are taken as **XKM** and **YKM**.

The two main parameters are **Min Cluster Size** and **Min Points**. These are typically set to the same value, but there may be instances where a larger value for **Min Cluster Size** is desired. In the illustration, these values are set to **10**. The **Min Points** parameter drives the computation of the core distance for each point, which forms the basis for the construction of the minimum spanning tree. As before, **Transformation** is set to **Raw** and **Distance Function** to **Euclidean**.

The **Run** button brings up a cluster map and the associated dendrogram. The cluster map, shown in Figure 20.18, reveals 7 clusters, ranging in size from 67 to 10, with 69 *Noise* points. The **Summary** button reveals an overall fit of 0.970, which is the best value achieved so far.

In and of itself, the raw dendrogram is not of that much interest. One interesting feature of the dendrogram in `GeoDa` is that it supports linking and brushing, which makes it possible to explore what branches in the tree subsets of points or clusters belong to (selected in one of the maps). The reverse is supported as well, so that branches in the tree can be selected and their counterparts identified in other maps and graphs.

In Figure 20.19, the 19 observations are selected that form cluster 5, consisting of bank locations on the island of Sicily. The corresponding branches in the dendrogram are highlighted. This includes all the splits that are possible *after* the formation of the final cluster, in contrast to the *condensed tree*, considered next.

20.5.5.1 Exploring the condensed tree

An important and distinct feature of the HDBSCAN method is the visualization of cluster persistence (or stability) in the *condensed tree* or condensed dendrogram. This is invoked by means of the middle button in the results panel. The tree visualizes the relative excess of mass that results in the identification of clusters for different values of λ. Starting at the

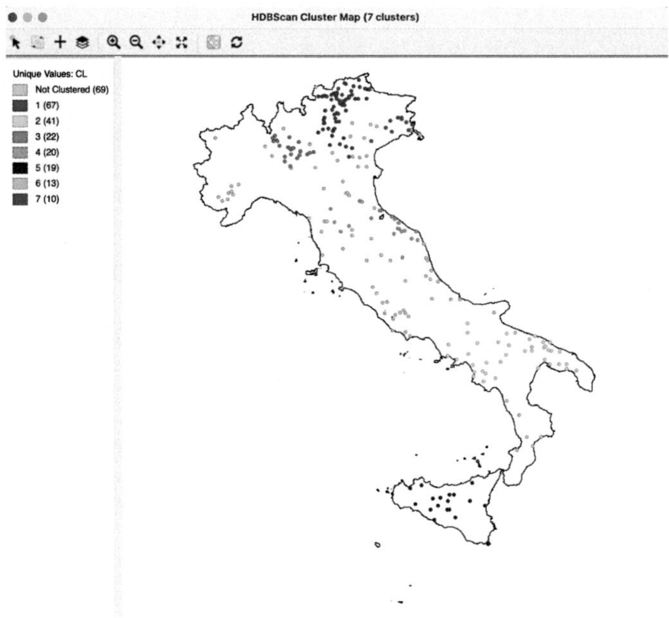

Figure 20.18: HDBSCAN cluster map

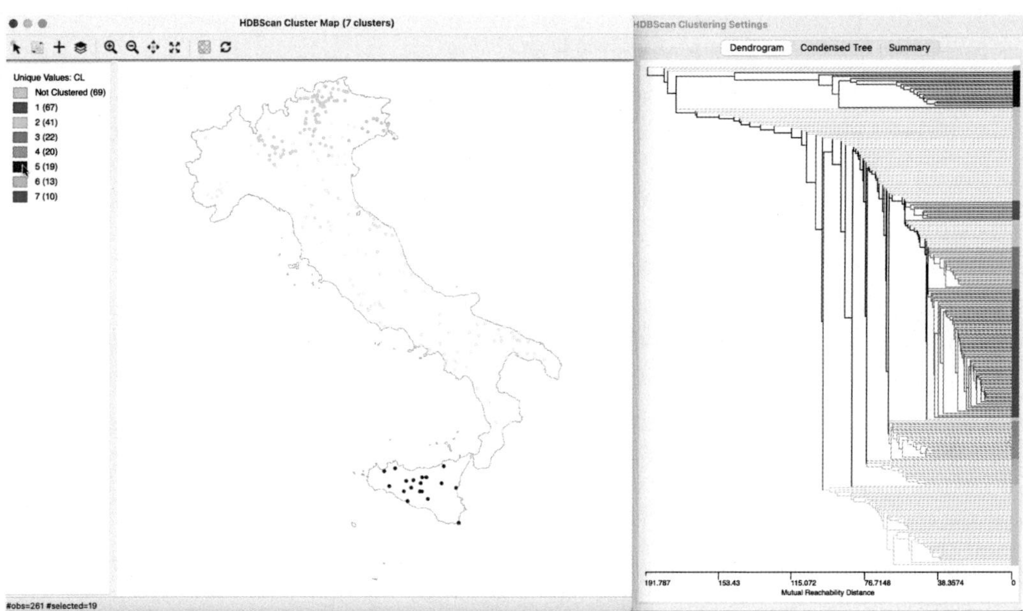

Figure 20.19: HDBSCAN dendrogram

root (top), with $\lambda = 0$, the left hand axis shows how increasing values of λ (going down) results in branching of the tree. The shaded areas are not rectangles, but they gradually taper off, as single points are shredded from the main cluster.

The optimal clusters are identified by an oval with the same color as the cluster label in the cluster map. The shading of the tree branches corresponds to the number of points contained

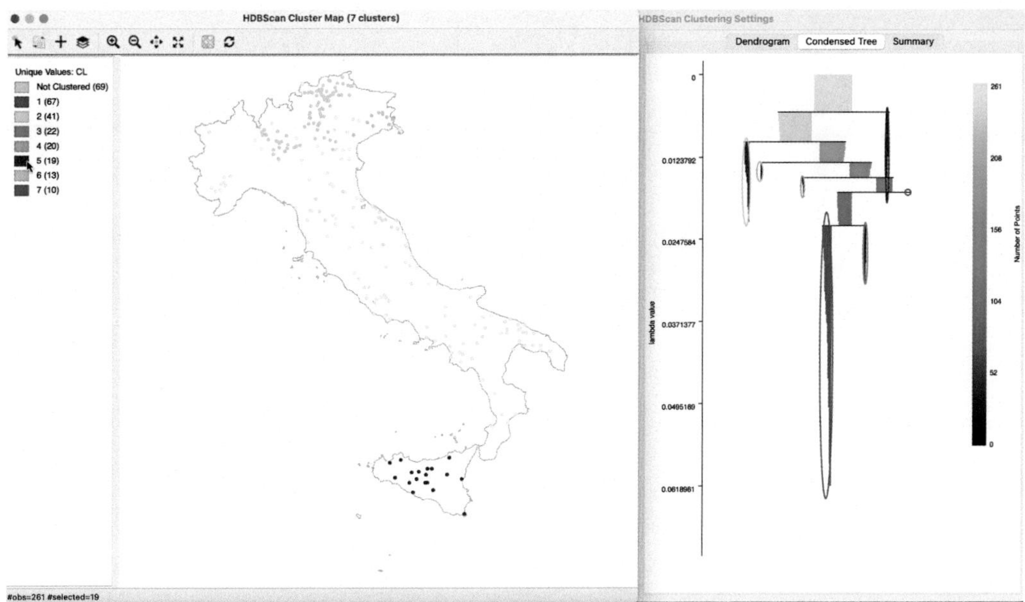

Figure 20.20: HDBSCAN condensed tree

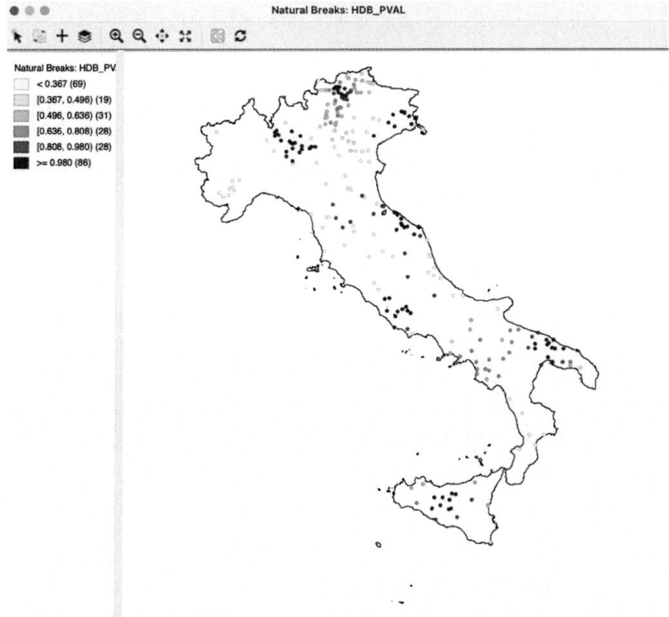

Figure 20.21: HDBSCAN cluster membership

in the branch, with lighter suggesting more points. In Figure 20.20, the same 19 members of cluster 5 are selected as before. In contrast to the dendrogram, no splits occur below the formation of the cluster, highlighted in red.

In `GeoDa`, the condensed tree supports linking and brushing as well as a limited form of zoom operations. The latter may be necessary for larger data sets, where the detail is difficult

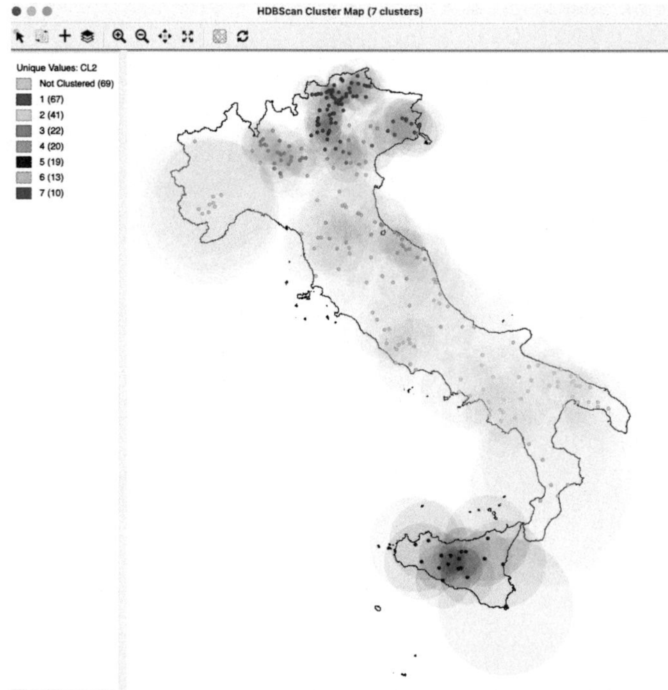

Figure 20.22: HDBSCAN core distance heat map

to distinguish in the overall tree. Zooming is implemented as a specialized mouse scrolling operation, slightly different in each system. For example, on some track pads, this works by moving two fingers up or down. There is no panning, but the focus of the zoom can be moved to specific subsets of the graph.

20.5.5.2 Cluster membership

The **Save** button provides the option to add the core distance, cluster membership probability and outlier index (GLOSH) to the data table. The default variable names are **HDB_CORE** for the core distance associated with each point, **HDB_PVAL** for the degree of cluster membership and **HDB_OUT** for the outliers. These variables can be included in any type of map. For example, Figure 20.21 shows the degree of cluster membership as a natural breaks map with 6 intervals. Clearly, the observations near the center of the cluster (smaller core distances) carry a darker color and locations further away from the core, are lightly colored.

A similar map can be created for the outlier probability. This will show the reverse shading, with more remote points colored darker.

20.5.5.3 Core distance map

With the the core distance for each point saved to the data table (**HDB_CORE**), in `GeoDa` it is possible to create a **Heat Map** of the core distances on top of the Cluster Map.[8] This

[8]Note that for this to work properly, the coordinates used for the point map and the coordinates employed in the HDBSCAN algorithm must be the same. In the current example, this means that **COORD_X** and **COORD_Y** should be used instead of their kilometer equivalents. The cluster solutions are the same, since it is a simple rescaling of the coordinates.

is one of the **Heat Map** options, invoked by right clicking on the cluster point map. The core distance variable must be specified in the **Specify Core Distance** option. This then yields a heat map as in Figure 20.22, where the core distances associated with each cluster are shown in the corresponding cluster color.

The overall pattern in the figure corresponds to gradations in the cluster membership map.

Part VI

Epilogue

21

Postscript – The Limits of Exploration

In this first volume, devoted to *exploring* spatial data, I have outlined a progression of techniques to aid in finding *interesting patterns* of dependence and heterogeneity in geospatial data, supported by the `GeoDa` software. The highly interactive process of discovery moves from simple maps and graphs to a more structured visualization of clusters, outliers and indications of potential structural breaks.

The main topic in this volume is the investigation of spatial correlation, i.e., the match or tension between attribute similarity and locational similarity. In that quest, the primary focus is on the identification of univariate locations of clusters and spatial outliers by means of a range of local indicators of spatial association (LISA).

For a single variable, Tobler's first law of geography suggests the omnipresence of positive spatial autocorrelation, with a distance decay effect for the dependence. This is indeed observed for many variables in empirical applications. However, Tobler's law does not necessarily generalize to a bivariate or multivariate setting, where the situation is considerably more complex (Anselin and Li, 2020). As discussed in more detail in Chapter 18, in a multivariate setting, not only is there the tension between attribute and locational similarity, but there is an additional dimension of inter-attribute similarity, which is typically not uniform across space.

The `GeoDa` software is designed to make this process of discovery easy and intuitive. However, this may also unintentionally facilitate potentially pointless clicking through the range of interfaces until one finds an outcome one likes, without fully understanding the limitations of the techniques involved. The extensive discussion of the methods in the book is aimed at remedying the second aspect. Nevertheless, some caution is needed.

First and foremost, exploration is not the same as explanation. Interesting patterns may be discovered, but that does not necessarily identify the process(es) that yielded the patterns. As stated strongly by Heckman and Singer (2017): "data never speak for themselves." As argued in the discussion of global spatial autocorrelation in Section 13.5.3, spatial data are characterized by the *inverse problem*, in the sense that the same pattern may be generated by very different spatial processes. More specifically, there is an important distinction between *true contagion* and *apparent contagion*. In cross-sectional data, both types of processes yield a clustered pattern. However, the pattern that follows from apparent contagion is generated by a spatially heterogeneous process and not a dependent process, as is typically assumed. Without further information, such as a time dimension, it is impossible to distinguish between the two generating processes.

The interpretation and validation of outcomes of an exploratory (spatial) data analysis has been the subject of growing discussion in the literature, as reviewed in Section 4.2.1. The exploratory process is neither inductive nor deductive, but rather *abductive*, involving a move back and forth between data analysis, hypothesis generation and reformulation, as

DOI: 10.1201/9781003274919-21

well as the addition of new information, an approach sometimes referred to as the "Sherlock Holmes method" (Gahegan, 2009; Heckman and Singer, 2017).

How such a discovery process is carried out has become increasingly important with the advent of *big data*, and the associated argument for data-driven discovery as the *fourth paradigm* in scientific reasoning (see, e.g., Hey et al., 2009; Gahegan, 2020).

In this final chapter, I want to briefly consider three broader issues that run the risk of getting lost in the excitement of the discovery process driven by interactive software such as `GeoDa`: (1) the potential pitfalls of data science and their implications for scientific reasoning; (2) the limitations intrinsic to *spatial* analysis; and (3) reproducible research.

A critical aspect of the abductive approach is how to deal with surprising results, or with results that run counter to pre-conceived ideas. As outlined in detail in Nuzo (2015), among others, it is easy to fool oneself by finding patterns where there are none, by focusing on explanations that fit one's prior convictions, or by failing to consider potential alternative hypotheses. This can result in confirmation bias (i.e., finding what one sets out to find), as well as disconfirmation bias (tendency to reject results that are counter to one's priors). In `GeoDa`, the extensive use of the permutation approach to represent spatial randomness in the data is one way to partially address this concern, but by itself, it is insufficient.

Even without nefarious motivations (e.g., driven by the pressure to publish), problems such as p-hacking (searching until significant results are found), *HARKing* (hypotheses after the results are known), *JARKing* (justifying after the results are known), and the like can unwittingly infiltrate an exploratory spatial data analysis. This has led to extensive discussions in the literature as to how these issues affect the process of scientific discovery (among others, Kerr, 1998; Simmons et al., 2011; Gelman and Loken, 2014; Gelman and Hennig, 2017; Rubin, 2017). Sound scientific reasoning is a way of reasoning that looks for being wrong, a systematic and enduring search for what might be wrong and an iterative process that goes back and forth between potential explanations and evidence.[1] Ideally, the process of data exploration should be guided by these principles.

A critical notion in this regard is the *researcher degrees of freedom*, a reference to the many decisions one makes with respect to the data that are included, the hypotheses considered (and not considered), and methods selected, in the so-called *garden of forking paths* (Gelman and Loken, 2014). In the context of the methods considered here and implemented in `GeoDa`, aspects such as the selection of spatial scale, how to deal with outliers, whether to apply imputation for missing values, the choice of spatial weights, the treatment of unconnected observations (islands or isolates), and various tuning parameters are prime examples of decisions one has to make that may affect the outcome of the data exploration. Careful attention to these decisions, ideally accompanied by a sensitivity analysis can partially remedy the problem.

Finally, as in any data science, spatial data science similarly may suffer from well-known generic *pitfalls*, for example as outlined in the book by Smith and Cordes (2019). Their book specifically lists "using bad data, putting data before theory, worshipping math, worshipping computers, torturing data, fooling yourself, confusing correlation with causation, regression toward the mean and doing harm" as examples of such pitfalls. Any serious student of exploratory spatial data analysis should become familiar with these potential traps and learn to recognize and avoid them.

[1]For examples of how this applies to data science and spatial data science, see, e.g., https://puttingscienceintodatascience.org.

In addition to these aspects associated with data science in general, *spatial* data science also faces its own special challenges. These include the ecological fallacy (Robinson, 1950; Goodman, 1953; 1959), the modifiable areal unit problem, or MAUP (Openshaw and Taylor, 1979), the change of support problem (Gotway and Young, 2002), as well as the more general issue of the importance of spatial scale (Goodchild, 2011; Oshan et al., 2022). Many of these are just special cases of well-known challenges associated with any type of statistical analysis. For example, the ecological fallacy was first raised in sociology and it cautions against interpreting the results of aggregate (spatial) analysis to infer individual behavior. The change of support problem concerns the combination of data at various spatial scales, such as point data (e.g., measurements by environmental sensors) and areal data (e.g., health outcomes at the census tract level), and associated issues of data aggregation and imputation, shared with the attention to scale in geographical analysis.

The MAUP stands out as particular to spatial analysis. It includes both aspects of data aggregation as well as spatial arrangement, or, *zonation.* In essence, MAUP suggests that different spatial scales and different areal boundaries will yield different and sometimes conflicting statistical results. In many instances in the social sciences, the boundaries are pre-set (e.g., administrative districts) and there is little one can do about it, other than being careful in phrasing the findings of an analysis. This is particularly important when areal boundaries do not align with the spatial scale of the processes investigated. For example, in the U.S., census tracts are typically assumed to correspond with neighborhoods, even though this is seldom the case. Similarly, counties are identified with labor markets, but this is clearly invalid in metro areas (consisting of multiple counties) or in the sparsely populated large counties in the west.

In terms of the spatial exploration, this means that the discovery of clusters needs to be interpreted with caution. A cluster may be nothing more than an indication of poor spatial scale (e.g., spatial units of observation much smaller than the process of interest), especially when the size of the areal units is heterogeneous across the data set. On the other hand, when individual spatial observations are available, the regionalization methods from Volume 2 may be applied to group them into meaningful spatial aggregates.

MAUP and its associated challenges does not invalidate spatial analysis, but it may limit the relevance of some of the findings. Where possible, sensitivity analysis should be implemented.

A second type of limit to spatial analysis follows from the enormous advances made in machine learning and artificial intelligence. For example, deep learning techniques (e.g., Goodfellow et al., 2016) are able to identify patterns without much prior (spatial) structure, as long as they are *trained* on very large data sets. GeoAI pertains to the application of these new methods to the analysis of geospatial data (Janowicz et al., 2020). Most current applications have been in the physical domain, such as land cover classification by means of remote sensing, landscape feature extraction and way finding. The focus in GeoAI has been on so-called feature engineering, i.e., identifying those spatial characteristics that should be included in the training data sets to yield good predictive performance. To date, much of GeoAI can be viewed as applying data science to spatial data, rather than spatial data science as conceived of in this book. Nevertheless, this raises an important question about the relevance of spatial constructs and spatial thinking in AI. Does this imply that spatial analysis as considered here will become irrelevant?

While it may be tempting to reach this conclusion, the powerful new deep learning methods are not a panacea for all empirical situations. Specifically, in order to perform well, these techniques require very large training data sets, consisting of millions (and even billions) of data points. Such data sets are still rather rare in the contexts in empirical practice, especially in the social sciences. In addition, the objective in exploratory spatial data analysis is to discover patterns and generate hypotheses, not prediction, the typical focus in AI. So, the insights gained from the methods covered in this book will remain relevant for some time. Nevertheless, the extent to which spatial structure and explicit spatial features contribute to the performance of deep learning methods, or, instead becomes discovered by these methods (and thus irrelevant) still remains to be answered.

A final concern pertaining to the exploration of spatial data is the extent to which it can be reproducible and replicable. Reproducibility refers to obtaining the same results in a repetition of the analysis on the same data (by the researchers themselves or by others), replicability to obtaining the same type of results using different data.

The growing demands for transparency and openness in the larger scientific community (so-called TOP guidelines, for transparency and openness promotion) have resulted in requirements of open data, open science and open software code. These concerns have also been echoed in the context of spatial data science (among others, by Rey, 2009; 2023; Singleton et al., 2016; Brunsdon and Comber, 2021; Kedron et al., 2021). Common approaches to accomplish this in spatial data analysis is the codification of workflows by means of visual modeling languages, e.g., as implemented in various GIS software (see also Kruiger et al., 2021). More formally, data, text and code are combined in *notebooks*, such as the well-known Jupyter (originally for Python) and RStudio (originally for R) implementations (see Rowe et al., 2020).

To what extent does the highly dynamic and interactive visualization in `GeoDa` meet these requirements? At one level, it does not and cannot. The thrill of discovery can easily result in a rapid succession of linked graphs and maps, in combination with various cluster analyses, in a fairly unstructured manner that does not lend itself to codification in workflows. However, this intrinsic lack of reproducibility can be remedied to some extent.

For example, it may be possible to embed elementary analyses (e.g., a Local Moran) into a specific workflow (e.g., select variable, specify weights, identify clusters), although this will necessarily result in a loss of spontaneity. Alternatively, as outlined in Appendix C, the `geodalib` architecture allows specific analyses to be embedded in notebook workflows through the new `RGeoDa` and `PyGeoda` wrappers.

In addition, there are some simple ways to obtain a degree of reproducibility for major aspects of the analysis, such as the specification of spatial weights, custom classifications and various variable transformations through the *project file* (see, e.g., Sections 4.6.3, 9.2.1.1 and 10.3.5). Nevertheless, obtaining full reproducibility for dynamic and interactive visualization remains a challenge.

In spite of these challenges, I believe that the exploratory perspective developed in the book and implemented in the `GeoDa` software can be very efficient at generating useful insights. As long as the approach is applied with caution, such as by including ample sensitivity analyses, pitfalls can be avoided. Time will tell.

The second volume of this *introduction to spatial data science* is devoted to what is referred to in the machine learning literature as *unsupervised learning*. The objective is to reduce the complexity in multivariate data by means of dimension reduction, both in the attribute dimension (e.g., principal components and multidimensional scaling), as well as in the observational dimension (clustering and regionalization). The distinctive perspective offered is to emphasize the relevance of spatial aspects, by spatializing the results of classic methods, either through mapping in combination with linking and brushing, or by imposing spatial constraints in the clustering algorithms themselves. This builds upon the foundations developed in the current volume.

A

Appendix A – GeoDa Preference Settings

The **Preferences** item on the main GeoDa menu includes several options that can be set to fine tune the appearance and performance of the software. They appear under two different tabs, one for **System** and one for **Data**. The contents are shown in Figures A.1 and A.2. In a MAC OSX operating system, the preferences are accessed under the **GeoDa** item on the menu bar. In Windows and Linux, the item is under **File**.

The **System** preferences pertain to detailed settings for the map and plot properties, support for multiple languages, general appearance and computational customization.

The map and plot appearance settings control the transparency of selected items, font size, and an option to revert to the legacy method of highlighting selected observations using cross-hatching (this was the method used in *legacy GeoDa* version 0.9.5-i).

In addition to the default English, GeoDa currently has language customization in all menus and dialogs for Simplified Chinese, Russian, Spanish, Portuguese and French, with more languages slated to be added in the future.

Several minor options determine the appearance of a few dialogs dealing with data entry, and other options such as automatic crash detection and software updates, as well as Retina display support (on Mac only). Finally, some features are the default settings for the random number seed, and the number of cores and/or GPU used in computation. Most of these are best left to their default specification.

Under the **Data** tab, the most useful setting is the supported formats for dates and time (see also Section 2.4.1.2 in Chapter 2). The other items pertain to the interaction with remote data base servers, as well as the numerical stopping criterion used in the auto-weighting method for spatially constrained clustering (covered in Volume 2). Here as well, the default should be fine for most situations in practice.

DOI: 10.1201/9781003274919-A

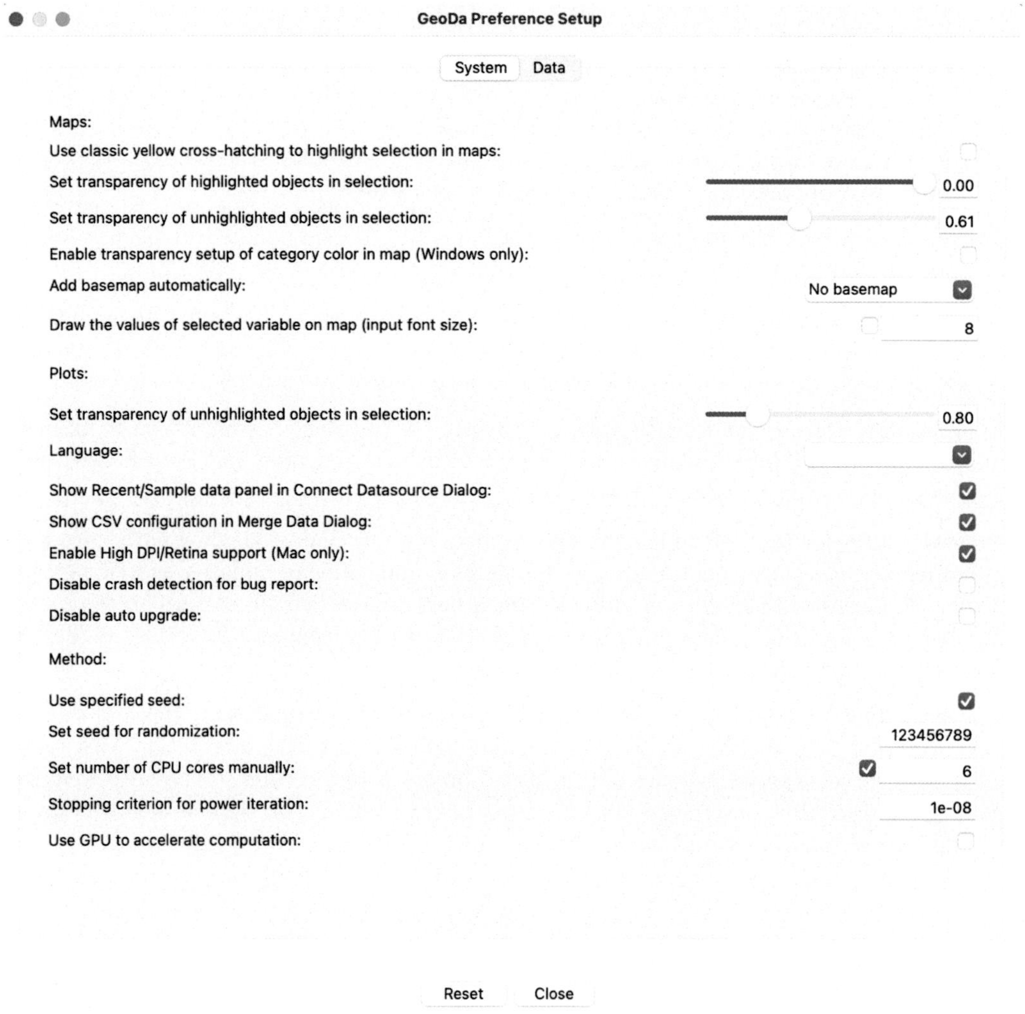

GeoDa Preference Setup

System Data

Maps:

Use classic yellow cross-hatching to highlight selection in maps:

Set transparency of highlighted objects in selection: 0.00

Set transparency of unhighlighted objects in selection: 0.61

Enable transparency setup of category color in map (Windows only):

Add basemap automatically: No basemap

Draw the values of selected variable on map (input font size): 8

Plots:

Set transparency of unhighlighted objects in selection: 0.80

Language:

Show Recent/Sample data panel in Connect Datasource Dialog:

Show CSV configuration in Merge Data Dialog:

Enable High DPI/Retina support (Mac only):

Disable crash detection for bug report:

Disable auto upgrade:

Method:

Use specified seed:

Set seed for randomization: 123456789

Set number of CPU cores manually: 6

Stopping criterion for power iteration: 1e-08

Use GPU to accelerate computation:

Reset Close

Figure A.1: Preferences settings – System

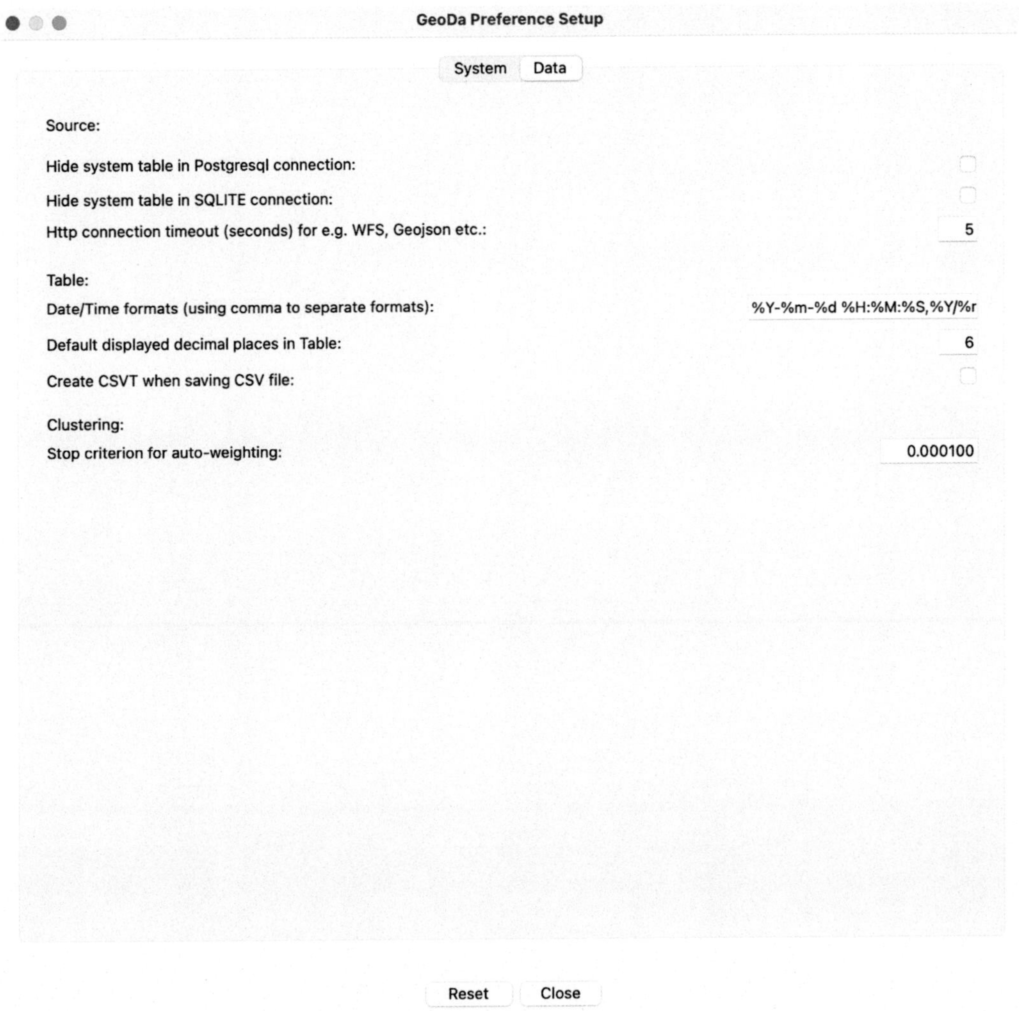

GeoDa Preference Setup

System Data

Source:

Hide system table in Postgresql connection:

Hide system table in SQLITE connection:

Http connection timeout (seconds) for e.g. WFS, Geojson etc.: 5

Table:

Date/Time formats (using comma to separate formats): %Y-%m-%d %H:%M:%S,%Y/%r

Default displayed decimal places in Table: 6

Create CSVT when saving CSV file:

Clustering:

Stop criterion for auto-weighting: 0.000100

Reset Close

Figure A.2: Preferences settings – Data

B

Appendix B – Menu Structure

This Appendix lists the complete menu structure of `GeoDa`, with the exception of items pertaining to spatial regression (covered in Anselin and Rey, 2014), but including the clustering methods covered in Volume 2. The listing below shows the organization for a MacOSX operating system. In Windows and Linux, **Preferences** appear under the **File** menu. Otherwise, the structure is identical across operating systems.

- GeoDa
  - About GeoDa
  - Preferences
  - Quit GeoDa
- File
  - New from Recent
  - New
  - Save
  - Save As
  - Save Selected As
  - Open Project
  - Save Project
  - Project Information
  - Close
- Edit
  - Time
  - Category
- Tools
  - Weights Manager
  - Shape
    - Points from Table
    - Create Grid
  - Spatial Join
  - Dissolve
- Table
  - Aggregate
  - Merge
  - Selection Tool
  - Invert Selection
  - Clear Selection
  - Save Selection
  - Move Selected to Top
  - Calculator
  - Add Variable
  - Delete Variable(s)

DOI: 10.1201/9781003274919-B

  - Edit Variable Properties
  - Encode
    * (encoding options)
  - Setup Number Formatting
- Map
  - Themeless Map
  - Quantile Map
    * (number of quantiles)
  - Percentile Map
  - Box Map (Hinge=1.5)
  - Box Map (Hinge=3.0)
  - Standard Deviation Map
  - Unique Values Map
  - Co-location Map
  - Natural Breaks Map
    * (number of categories)
  - Equal Intervals Map
    * (number of categories)
  - Custom Breaks
    * Create New Custom
  - Rates-Calculated Map
    * Raw Rate
    * Excess Risk
    * Empirical Bayes
    * Spatial Rate
    * Spatial Empirical Bayes
  - Conditional Map
  - Cartogram
  - Map Movie
- Explore
  - Histogram
  - Box Plot
  - Scatter Plot
  - Scatter Plot Matrix
  - Bubble Chart
  - 3D Scatter Plot
  - Parallel Coordinate Plot
  - Averages Chart
  - Conditional Plot
    * Map
    * Histogram
    * Scatter Plot
    * Box Plot
- Clusters
  - PCA
  - MDS
  - t-SNE
  - K Means
  - K Medians
  - K Medoids
  - Spectral

  - Hierarchical
  - DBScan
  - HDBScan
  - SC K Means
  - SCHC
  - skater
  - redcap
  - AZP
  - max-p
  - Cluster Match Map
  - Make Spatial
  - Validation
- Space
  - Univariate Moran's I
  - Bivariate Moran's I
  - Differential Moran's I
  - Moran's I with EB Rate
  - Univariate Local Moran's I
  - Bivariate Local Moran's I
  - Differential Local Moran's I
  - Local Moran's I with EB Rate
  - Local G
  - Local G*
  - Univariate Local Join Count
  - Bivariate Local Join Count
  - Co-location Join Count
  - Univariate Local Geary
  - Multivariate Local Geary
  - Univariate Quantile LISA
  - Multivariate Quantile LISA
  - Local Neighbor Match Test
  - Spatial Correlogram
  - Distance Scatter Plot
- Time
  - Time Player
  - Time Editor

C

Appendix C – Scripting with GeoDa via the geodalib Library

The architecture of the GeoDa software is based on a very tight integration of the graphical user interface and the actual computations. While effective in a traditional desktop environment, this approach becomes less efficient when moving the functionality to a different computing platform, such as a browser in a web-GIS, or a cyberGIS environment (Anselin et al., 2004; Wang et al., 2013).

An alternative to the GUI-driven approach toward spatial data science taken in desktop GeoDa is to leverage open-source software development environments, such as R (Pebesma and Bivand, 2023) or Python (Rey et al., 2023). Such environments facilitate scripting and stress reproducibility, which has become of increased relevance and importance in spatial data science (see the discussion in Chapter 21).

The GeoDa desktop environment does not lend itself to scripting. It is also ill-suited for repeated execution of the same application, such as in a simulation experiment. Finally, apart from the limited record in the *project file*, there is no explicit way to ensure reproducibility.

In light of these limitations of the original design, a major refactoring effort was embarked upon to separate the user interaction in the software from the core computational functionality, and to collect the latter in a library, named libgeoda. The library contains the same C++ code as in the computations underlying desktop GeoDa, but has a more limited range of functionality. The focus has been on methods that are (still) unique to GeoDa, such as some of the recent LISA statistics. In addition, applications are included where the reliance on C++ yields large performance improvements in terms of speed and scalability. Examples include weights creation, and permutation tests from this Volume, as well as regionalization methods covered in Volume 2.

The libgeoda library has a clearly defined *Application Programming Interface* (API), which allows other C++ code to access its functionality directly. In fact, this is what currently happens under the hood for part of desktop GeoDa, and in the experimental web-GeoDa (jsgeoda, implemented through javascript). In addition to achieving a more flexible interaction with different graphical user interface implementations, the API also allows other software, such as R or Python programs to access the functionality through well-defined *wrapper* code. The overall architecture is illustrated in Figure C.1.

The primary focus in this effort so far has been to create an R package, rgeoda, and a Python module, pygeoda. These provide easy access to the functionality in libgeoda through a native interface and designated middleware. The interaction between R and Python and the C++ library is implemented under the hood, so that from a user's perspective, everything works natively as in any other R package or Python module.

As shown in Figure C.1, the core of the libgeoda library consists of three broad categories of functionality: spatial weights, LISA, and spatial clustering (regionalization). In addition,

DOI: 10.1201/9781003274919-C

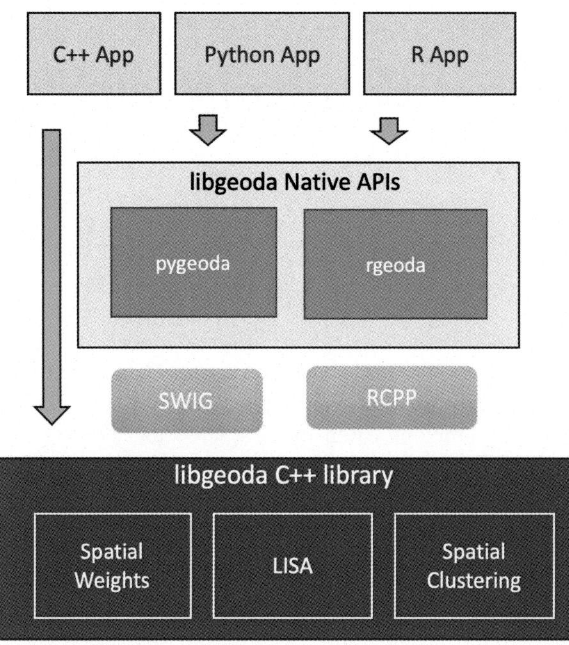

Figure C.1: libgeoda architecture

there are a number of helper functions, such as support for different map classifications (to facilitate visualization) and variable standardization (for use in the cluster routines). The functionality in `pygeoda` and `rgeoda` is the same, with only minor differences to reflect the particular characteristics of each software environment. For example, in `Python`, methods are attributes of a class (e.g., a spatial weights class) and invoked as such. In contrast, in `R`, the typical approach is to apply a function to an object to extract the relevant information (e.g., spatial weights characteristics).

Extensive details and specific examples can be found in Anselin et al. (2022) and in the documentation on the GitHub site. Since the functionality of `libgeoda` mimics what has been covered in the book, it is not further considered here. The software development effort is ongoing.

Bibliography

Akbari, K., Winter, S., and Tomko, M. (2023). Spatial causality: A systematic review on spatial causal inference. *Geographical Analysis*, 55: 56–89.

Algeri, C., Anselin, L., Forgione, A. F., and Migliardo, C. (2022). Spatial dependence in the technical efficiency of local banks. *Papers in Regional Science*, 101:385–416.

Amaral, P., de Carvalho, L. R., Rocha, T. A. H., da Silva, N. C., and Vissoci, J. R. N. (2019). Geospatial modeling of microcephaly and zika virus spread patterns in Brazil. *PLoS ONE*, 14.

Andrienko, G., Andrienko, N., Keim, D., MacEachren, A., and Wrobel, S. (2011). Challenging problems of geospatial visual analytics. *Journal of Visual Languages and Computing*, 22:251–256.

Andrienko, N., Lammarsch, T., Andrienko, G., Fuchs, G., Keim, D., Miksch, S., and Rind, A. (2018). Viewing visual analytics as model building. *Computer Graphics Forum*, 37:275–299.

Angrist, J. and Pischke, J.-S. (2015). *Mastering 'Metrics, The Path from Cause to Effect.* Princeton University Press, Princeton, New Jersey.

Anselin, L. (1988). *Spatial Econometrics: Methods and Models.* Kluwer Academic Publishers, Dordrecht, The Netherlands.

Anselin, L. (1990). What is special about spatial data? Alternative perspectives on spatial data analysis. In Griffith, D. A., editor, *Spatial Statistics, Past, Present and Future*, pages 66–77. Institute of Mathematical Geography, (IMAGE), Ann Arbor, MI.

Anselin, L. (1992). *SpaceStat, a Software Program for Analysis of Spatial Data.* National Center for Geographic Information and Analysis (NCGIA), University of California, Santa Barbara, CA.

Anselin, L. (1994). Exploratory spatial data analysis and geographic information systems. In Painho, M., editor, *New Tools for Spatial Analysis*, pages 45–54. Eurostat, Luxembourg.

Anselin, L. (1995). Local indicators of spatial association — LISA. *Geographical Analysis*, 27:93–115.

Anselin, L. (1996). The Moran scatterplot as an ESDA tool to assess local instability in spatial association. In Fischer, M., Scholten, H., and Unwin, D., editors, *Spatial Analytical Perspectives on GIS in Environmental and Socio-Economic Sciences*, pages 111–125. Taylor and Francis, London.

Anselin, L. (1998). Exploratory spatial data analysis in a geocomputational environment. In Longley, P. A., Brooks, S., Macmillan, B., and McDonnell, R., editors, *Geocomputation: A Primer*, pages 77–94. John Wiley, New York, NY.

Anselin, L. (1999). Interactive techniques and exploratory spatial data analysis. In Longley, P. A., Goodchild, M. F., Maguire, D. J., and Rhind, D. W., editors, *Geographical*

Information Systems: Principles, Techniques, Management and Applications, pages 251–264. John Wiley, New York, NY.

Anselin, L. (2005a). *Exploring Spatial Data with GeoDa: A Workbook.* Spatial Analysis Laboratory (SAL). Department of Geography, University of Illinois, Urbana-Champaign, IL.

Anselin, L. (2005b). Spatial statistical modeling in a GIS environment. In Maguire, D. J., Batty, M., and Goodchild, M. F., editors, *GIS, Spatial Analysis and Modeling*, pages 93–111. ESRI Press, Redlands, CA.

Anselin, L. (2012). From spacestat to cybergis, twenty years of spatial data analysis software. *International Regional Science Review*, 35:131–157.

Anselin, L. (2019a). A local indicator of multivariate spatial association, extending Geary's c. *Geographical Analysis*, 51:133–150. doi:10.1111/gean.12164.

Anselin, L. (2019b). Quantile local spatial autocorrelation. *Letters in Spatial and Resource Sciences*, 12:155–166.

Anselin, L. (2020). Spatial data science. In Richardson, D., editor, *The International Encyclopedia of Geography: People, the Earth, Environment, and Technology.* Wiley, New York, NY.

Anselin, L. (2021). Spatial models in econometric research. In Hamilton, J., editor, *Oxford Research Encyclopedia of Economics and Finance.* Oxford University Press.

Anselin, L. and Arribas-Bel, D. (2013). Spatial fixed effects and spatial dependence in a single cross-section. *Papers in Regional Science*, 92:3–17.

Anselin, L., Kim, Y.-W., and Syabri, I. (2004). Web-based analytical tools for the exploration of spatial data. *Journal of Geographical Systems*, 6:197–218.

Anselin, L. and Li, X. (2019). Operational local join count statistics for cluster detection. *Journal of Geographical Systems*, 21:189–210.

Anselin, L. and Li, X. (2020). Tobler's law in a multivariate world. *Geographical Analysis*, 52:494–510.

Anselin, L., Li, X., and Koschinsky, J. (2022). GeoDa, from the desktop to an ecosystem for exploring spatial data. *Geographical Analysis*, 54:439–466.

Anselin, L., Lozano-Gracia, N., and Koschinky, J. (2006a). Rate transformations and smoothing. Technical report, Spatial Analysis Laboratory, Department of Geography, University of Illinois, Urbana, IL.

Anselin, L. and Rey, S. J. (2014). *Modern Spatial Econometrics in Practice, A Guide to GeoDa, GeoDaSpace and PySAL.* GeoDa Press, Chicago, IL.

Anselin, L. and Smirnov, O. (1996). Efficient algorithms for constructing proper higher order spatial lag operators. *Journal of Regional Science*, 36:67–89.

Anselin, L., Syabri, I., and Kho, Y. (2006b). GeoDa, an introduction to spatial data analysis. *Geographical Analysis*, 38:5–22.

Anselin, L., Syabri, I., and Smirnov, O. (2002). Visualizing multivariate spatial correlation with dynamically linked windows. In Anselin, L. and Rey, S., editors, *New Tools for Spatial Data Analysis: Proceedings of the Specialist Meeting.* Center for Spatially Integrated Social Science (CSISS), University of California, Santa Barbara. CD-ROM.

Assuncao, R. M. and Reis, E. A. (1999). A new proposal to adjust Moran's I for population density. *Statistics in Medicine*, 18:2147–2161.

Banerjee, S., Carlin, B. P., and Gelfand, A. E. (2015). *Hierarchical Modeling and Analysis for Spatial Data, 2nd Edition.* Chapman & Hall/CRC, Boca Raton.

Bavaud, F. (1998). Models for spatial weights: A systematic look. *Geographical Analysis*, 30:153–171.

Becker, R. A. and Cleveland, W. (1987). Brushing scatterplots. *Technometrics*, 29:127–142.

Becker, R. A., Cleveland, W., and Shyu, M.-J. (1996). The visual design and control of Trellis displays. *Journal of Computational and Graphical Statistics*, 5:123–155.

Becker, R. A., Cleveland, W., and Wilks, A. (1987). Dynamic graphics for data analysis. *Statistical Science*, 2:355–395.

Bellman, R. E. (1961). *Adaptive Control Processes.* Princeton University Press, Princeton, N.J.

Benjamin, D. and 72 others (2018). Redefine statistical significance. *Nature Human Behavior*, 2:6–10.

Benjamini, Y. and Hochberg, Y. (1995). Controlling the false discovery rate: A practical and powerful approach to multiple testing. *Journal of the Royal Statistical Society B*, 57:289–300.

Bjornstad, O. N. and Falck, W. (2001). Nonparametric spatial covariance functions: Estimation and testing. *Environmental and Ecological Statistics*, 8:53–70.

Blommestein, H. J. (1985). Elimination of circular routes in spatial dynamic regression equations. *Regional Science and Urban Economics*, 15:121–130.

Brewer, C. A. (1997). Spectral schemes: controversial color use on maps. *Cartography and Geographic Information Systems*, 49:280–294.

Brewer, C. A. (2016). *Designing Better Maps. A Guide for GIS Users, 2nd Edition.* ESRI Press, Redlands, CA.

Brewer, C. A., Hatchard, G., and Harrower, M. A. (2003). ColorBrewer in print: a catalog of color schemes for maps. *Cartography and Geographic Information Science*, 30:5–32.

Brock, W. A. and Durlauf, S. N. (2001). Discrete choice with social interactions. *Review of Economic Studies*, 59:235–260.

Brunsdon, C. and Comber, L. (2015). *Geocomputation, a Practical Primer.* Sage, Thousand Oaks, CA.

Brunsdon, C. and Comber, L. (2019). *An Introduction to R for Spatial Analysis and Mapping, 2nd Edition.* Sage, Thousand Oaks, CA.

Brunsdon, C. and Comber, L. (2021). Opening practice: supporting reproducibility and critical spatial data science. *Journal of Geographical Systems*, 23:477–496.

Buja, A., Cook, D., Hofmann, H., Lawrence, M., Lee, E.-K., Swayne, D. F., and Wickham, H. (2009). Statistical inference for exploratory data analysis and model diagnostics. *Philosophical Transactions of the Royal Society A*, 367:4361–4383.

Buja, A., Cook, D., and Swayne, D. (1996). Interactive high dimensional data visualization. *Journal of Computational and Graphical Statistics*, 5:78–99.

Campello, R. J., Moulavi, D., and Sander, J. (2013). Density-based clustering based on hierarchical density estimates. In Pei, J., Tseng, V. S., Cao, L., Motoda, H., and Xu, G., editors, *Advances in Knowledge Discovery and Data Mining. PAKDD 2013. Lecture Notes in Computer Science, Vol. 7819*, pages 160–172.

Campello, R. J., Moulavi, D., Zimek, A., and Sandler, J. (2015). Hierarchical density estimates for data clustering, visualization, and outlier detection. *ACM Transactions on Knowledge Discovery from Data*, 10,1.

Carr, D. B. and Pickle, L. W. (2010). *Visualizing Data Patterns with Micromaps*. Chapman & Hall/CRC, Boca Raton, FL.

Case, A. (1991). Spatial patterns in household demand. *Econometrica*, 59:953–965.

Case, A., Rosen, H. S., and Hines, J. R. (1993). Budget spillovers and fiscal policy interdependence: Evidence from the states. *Journal of Public Economics*, 52:285–307.

Chen, C., Härdle, W., and Unwin, A. (2008). *Handbook of Data Visualization*. Springer, Berlin, Germany.

Chilès, J.-P. and Delfiner, P. (1999). *Geostatistics, Modeling Spatial Uncertainty*. John Wiley & Sons, New York, NY.

Chow, G. (1960). Tests of equality between sets of coefficients in two linear regressions. *Econometrica*, 28:591–605.

Clayton, D. and Kaldor, J. (1987). Empirical Bayes estimates of age-standardized relative risks for use in disease mapping. *Biometrics*, 43:671–681.

Cleveland, W. S. (1979). Robust locally weighted regression and smoothing scatterplots. *Journal of the American Statistical Association*, 74:829–836.

Cleveland, W. S. (1993). *Visualizing Data*. Hobart Press, Summit, NJ.

Cleveland, W. S., Grosse, E., and Shyu, W. M. (1992). Local regression models. In Chambers, J. M. and Hastie, T. J., editors, *Statistical Models in S*, pages 309–376. Wadsworth and Brooks/Cole, Pacific Grove, CA.

Cleveland, W. S. and McGill, M. (1988). *Dynamic Graphics for Statistics*. Wadsworth, Pacific Grove, CA.

Cliff, A. and Ord, J. K. (1973). *Spatial Autocorrelation*. Pion, London.

Cliff, A. and Ord, J. K. (1981). *Spatial Processes: Models and Applications*. Pion, London.

Comber, L. and Brunsdon, C. (2021). *Geographical Data Science and Spatial Analysis. An Introduction in R*. Sage, Thousand Oaks, CA.

Cressie, N. (1993). *Statistics for Spatial Data*. Wiley, New York.

Dasu, T. and Johnson, T. (2003). *Exploratory Data Mining and Data Cleaning*. John Wiley, Hoboken, NJ.

de Berg, M., Cheong, O., van Kreveld, M., and Overmars, M. (2008). *Computational Geometry, Algorithms and Applications, 3rd Edition*. Springer Verlag, Berlin.

de Castro, M. C. and Singer, B. H. (2006). Controlling the false discovery rate: An application to account for multiple and dependent tests in local statistics of spatial association. *Geographical Analysis*, 38:180–208.

Deutsch, C. and Journel, A. (1998). *GSLIB: Geostatistical Software Library and User's Guide.* Oxford University Press, New York, NY.

Dorling, D. (1996). *Area Cartograms: Their Use and Creation.* CATMOG 59, Institute of British Geographers.

Dray, S., Saïd, S., and Débias, F. (2008). Spatial ordination of vegetation data using a generalization of Wartenberg's multivariate spatial correlation. *Journal of Vegetation Science*, 19:45–56.

Dykes, J. (1997). Exploring spatial data representation with dynamic graphics. *Computers and Geosciences*, 23:345–370.

Dykes, J., MacEachren, A. M., and Kraak, M.-J. (2005). *Exploring Geovisualization.* Elsevier, Oxford, United Kingdom.

Efron, B. and Hastie, T. (2016). *Computer Age Statistical Inference. Algorithms, Evidence, and Data Science.* Cambridge University Press, Cambridge, UK.

Ester, M., Kriegel, H.-P., Sander, J., and Xu, X. (1996). A density-based algorithm for discovering clusters in large spatial databases with noise. In *KDD-96 Proceedings*, pages 226–231.

Farah Rivadeneyra, I. (2017). A geospatial analysis of Mexico's distal effects on food insecurity. Master's thesis, University of Chicago, Chicago, IL.

Fisher, W. D. (1958). On grouping for maximum homogeneity. *Journal of the American Statistical Association*, 53:789–798.

Fotheringham, A. S., Brunsdon, C., and Charlton, M. (2002). *Geographically Weighted Regression.* John Wiley, Chichester.

Friendly, M. (2008). A brief history of data visualization. In Chun-houh Chen, C., Härdle, W., and Unwin, A., editors, *Handbook of Data Visualization*, pages 15–56. Springer, Berlin, Germany.

Gahegan, M. (2009). Visual exploration and explanation in geography: analysis with lights. In Miller, H. and Han, J., editors, *Geographic Data Mining and Knowledge Discovery*, pages 291–315. Taylor and Francis, Boca Raton, FL.

Gahegan, M. (2020). Fourth paradigm GIScience? prospects for automated discovery and explanation from data. *International Journal of Geographical Information Science*, 34:1–21.

Gan, J. and Tao, Y. (2017). On the hardness and approximation of Euclidean DBSCAN. *ACM Transactions on Database Systems (TODS)*, 42:14.

Gao, S. (2021). *Geospatial artificial intelligence (GeoAI).* Oxford Bibliographies, Oxford University Press.

Geary, R. (1954). The contiguity ratio and statistical mapping. *The Incorporated Statistician*, 5:115–145.

Gelman, A., Carlin, J. B., Stern, H. S., Dunson, D. B., Vehtari, A., and Rubin, D. B. (2014). *Bayesian Data Analysis, 3rd Edition.* Chapman & Hall, Boca Raton, FL.

Gelman, A. and Hennig, C. (2017). Beyond subjective and objective in statistics. *Journal of the Royal Statistical Society A*, 180:967–1033.

Gelman, A. and Loken, E. (2014). The statistical crisis in science. *American Scientist*, 102:460–465.

Getis, A. (1984). Interaction modeling using second-order analysis. *Environment and Planning A*, 16:173–183.

Getis, A. (2009). Spatial weights matrices. *Geographical Analysis*, 41:404–410.

Getis, A. and Franklin, J. (1987). Second-order neighborhood analysis of mapped point patterns. *Ecology*, 68:473–477.

Getis, A. and Ord, J. K. (1992). The analysis of spatial association by use of distance statistics. *Geographical Analysis*, 24:189–206.

Good, I. (1983). The philosophy of exploratory data analysis. *Philosophy of Science*, 50:283–295.

Goodchild, M. F. (2011). Scale in GIS: an overview. *Geomorphology*, 130:5–9.

Goodfellow, I., Bengio, Y., and Courville, A. (2016). *Deep Learning*. MIT Press, Cambridge, MA.

Goodman, L. A. (1953). Ecological regression and behavior of individuals. *American Sociological Review*, 18:663–664.

Goodman, L. A. (1959). Some alternatives to ecological correlation. *American Journal of Sociology*, 64:610–625.

Gotway, C. A. and Young, L. J. (2002). Combining incompatible spatial data. *Journal of the American Statistical Association*, 97:632–648.

Greenbaum, R. T. (2002). A spatial study of teacher's salaries in Pennsylvania school districts. *Journal of Labor Research*, 23:69–86.

Hall, P. and Patil, P. (1994). Properties of nonparametric estimators of autocovariance for stationary random fields. *Probability Theory and Related Fields*, 99:399–424.

Harris, C. (1954). The market as a factor in the localization of industry in the United States. *Annals of the Association of American Geographers*, 64:315–348.

Harris, R., Moffat, J., and Kravtsova, V. (2011). In search of 'W'. *Spatial Economic Analysis*, 6:249–270.

Harrower, M. A. and Brewer, C. A. (2003). ColorBrewer.org: an online tool for selecting color schemes for maps. *The Cartographic Journal*, 40:27–37.

Hartigan, J. A. (1975). *Clustering Algorithms*. John Wiley, New York, NY.

Haslett, J., Bradley, R., Craig, P., Unwin, A., and Wills, G. (1991). Dynamic graphics for exploring spatial data with applications to locating global and local anomalies. *The American Statistician*, 45:234–242.

Hastie, T., Tibshirani, R., and Friedman, J. (2009). *The Elements of Statistical Learning, 2nd Edition*. Springer, New York, NY.

Heckman, J. and Singer, B. (2017). Abducting economics. *American Economic Review, Papers and Proceedings*, 107:298–302.

Hey, T., Tansley, S., and Tolle, K. (2009). *The Fourth Paradigm: Data-Intensive Scientific Discovery*. Microsoft Research, Redmond, WA.

Hubert, L. J., Golledge, R., and Costanzo, C. M. (1981). Generalized procedures for evaluating spatial autocorrelation. *Geographical Analysis*, 13:224–233.

Hubert, L. J., Golledge, R., Costanzo, C. M., and Gale, N. (1985). Measuring association between spatially defined variables: An alternative procedure. *Geographical Analysis*, 17:36–46.

Hullman, J. and Gelman, A. (2021). Designing for interactive exploratory data analysis requires theories of graphical inference. *Harvard Data Science Review*, 3.

Inselberg, A. (1985). The plane with parallel coordinates. *Visual Computer*, 1:69–91.

Inselberg, A. and Dimsdale, B. (1990). Parallel coordinates: A tool for visualizing multidimensional geometry. *Proceedings of the IEEE Visualization 90*, pages 361–378.

Isaaks, E. H. and Srivastava, R. M. (1989). *An Introduction to Applied Geostatistics.* Oxford University Press, New York, NY.

Isard, W. (1960). *Methods of Regional Analysis.* MIT Press, Cambridge, MA.

Isard, W. (1969). *General Theory.* MIT Press, Cambridge, MA.

James, W. and Stein, C. (1961). Estimation with quadratic loss. *Proceedings of the Fourth Berkeley Symposium on Mathematical Statistics and Probability*, 1:361–379.

Janowicz, K., Gao, S., McKenzie, G., Hu, Y., and Bhaduri, B. (2020). GeoAI: spatially explicit artificial intelligence techniques for geographic knowledge discovery and beyond. *International Journal of Geographical Information Science*, 34:625–636.

Jenks, G. (1977). Optimal data classification for choropleth maps. Occasional. paper no. 2, Department of Geography, University of Kansas, Lawrence, KS.

Journel, A. and Huijbregts, C. (1978). *Mining Geostatistics.* Academic Press, London.

Kafadar, K. (1996). Smoothing geographical data, particularly rates of disease. *Statistics in Medicine*, 15:2539–2560.

Kafadar, K. (1997). Geographic trends in prostate cancer mortality: An application of spatial smoothers and the need for adjustment. *Annals of Epidemiology*, 7:35–45.

Kaufman, L. and Rousseeuw, P. (2005). *Finding Groups in Data: An Introduction to Cluster Analysis.* John Wiley, New York, NY.

Kedron, P., Frazier, A. A., Trgovac, A. B., Nelson, T., and Fotheringham, A. S. (2021). Reproducibility and replicability in geographical analysis. *Geographical Analysis*, 53:135–147.

Kelejian, H. H. and Prucha, I. R. (2007). HAC estimation in a spatial framework. *Journal of Econometrics*, 140:131–154.

Kerr, N. (1998). HARKing: hypothesizing after the results are known. *Personality and Social Psychology Review*, 2:196–217.

Kessler, F. C. and Battersby, S. E. (2019). *Working with Map Projections: A Guide to Their Selection.* CRC Press, Boca Raton, FL.

Kielman, J., Thomas, J., and May, R. (2009). Foundations and frontiers in visual analytics. *Information Visualization*, 8:239–246.

Kolak, M. and Anselin, L. (2020). A spatial perspective on the econometrics of program evaluation. *International Regional Science Review*, 43:128–153.

Kolak, M., Bhatt, J., Park, Y. H., Padrón, N. A., and Molefe, A. (2020). Quantification of neighborhood-level social determinants of health in the continental united states. *JAMA Network Open*, 3(1):e1919928–e1919928.

Kraak, M. and MacEachren, A. (2005). Geovisualization and GIScience. *Cartography and Geographic Information Science*, 32:67–68.

Kraak, M. and Ormeling, F. (2020). *Cartography. Visualization of GeoSpatial Data, 4th Edition.* CRC Press, Boca Raton, FL.

Kruiger, J., Kasalica, V., Meerlo, R., Lamprecht, A.-L., Nyamsuren, E., and Schneider, S. (2021). Loose programming of GIS workflows with geo-analytical concepts. *Transactions in GIS*, 25:424–449.

Kulldorff, M. (1997). A spatial scan statistic. *Communications in Statistics – Theory and Methods*, 26:1481–1496.

Lancaster, T. (2000). The incidental parameter problem since 1948. *Journal of Econometrics*, 95:391–413.

Law, M. and Collins, A. (2018). *Getting to Know ArcGIS Desktop, 5th Edition.* Environmental Systems Research Institute, Redlands, CA.

Lawson, A. B., Browne, W. J., and Rodeiro, C. L. V. (2003). *Disease Mapping with WinBUGS and MLwiN.* John Wiley, Chichester.

Lee, J. A. and Verleysen, M. (2007). *Nonlinear Dimensionality Reduction.* Springer-Verlag, New York, NY.

Lee, S.-I. (2001). Developing a bivariate spatial association measure: An integration of Pearson's r and Moran's I. *Journal of Geographical Systems*, 3:369–385.

Lin, J. (2020). A local model for multivariate analysis: extending Wartenberg's multivariate spatial correlation. *Geographical Analysis*, 52:190–210.

Loader, C. (1999). *Local Regression and Likelihood.* Springer-Verlag, Heidelberg.

Loader, C. (2004). Smoothing: Local regression techniques. In Gentle, J. E., Härdle, W., and Mori, Y., editors, *Handbook of Computational Statistics: Concepts and Methods*, pages 539–563. Springer-Verlag, Berlin.

Lovelace, R., Nowosad, J., and Muenchow, J. (2019). *Geocomputation with R.* CRC Press, Boca Raton, FL.

MacEachren, A. M. and Kraak, M.-J. (1997). Exploratory cartographic visualization: Advancing the agenda. *Computers and Geosciences*, 23:335–343.

MacEachren, A. M., Wachowicz, M., Edsall, R., Haug, D., and Masters, R. (1999). Constructing knowledge from multivariate spatiotemporal data: Integrating geographical visualization with knowledge discovery in database methods. *International Journal of Geographical Information Science*, 13(4):311–334.

Madry, S. (2021). *Introduction to QGIS: Open Source Geographic Information System.* Locate Press, Chugiak, AK.

Mantel, N. (1967). The detection of disease clustering and a generalized regression approach. *Cancer Research*, 27:209–220.

Marshall, R. J. (1991). Mapping disease and mortality rates using Empirical Bayes estimators. *Applied Statistics*, 40:283–294.

McCann, P. (2001). *Urban and Regional Economics.* Oxford University Press, New York, NY.

McInnes, L. and Healy, J. (2017). Accelerated hierarchical density clustering. In *2017 IEEE International Conference on Data Mining Workshops (ICDMW)*, New Orleans, LA.

Monmonier, M. (1989). Geographic brushing: Enhancing exploratory analysis of the scatterplot matrix. *Geographical Analysis*, 21:81–4.

Monmonier, M. (1993). *Mapping it Out: Expository Cartography for the Humanities and Social Sciences.* University of Chicago Press, Chicago, IL.

Monmonier, M. (2018). *How to Lie with Maps, 3rd Edition.* University of Chicago Press, Chicago, IL.

Moran, P. A. (1948). The interpretation of statistical maps. *Journal of the Royal Statistical Society, B*, 10:243–251.

Müller, D. and Sawitzki, G. (1991). Excess mass estimates and tests for multimodality. *Journal of the American Statistical Association*, 86:738–746.

Munasinghe, R. L. and Morris, R. D. (1996). Localization of disease clusters using regional measures of spatial autocorrelation. *Statistics in Medicine*, 15:893–905.

Newman, M. (2018). *Networks.* Oxford University Press, Oxford, United Kingdom.

Nuzo, R. (2015). Fooling ourselves. *Nature*, 526:182–185.

Oden, N. L. and Sokal, R. R. (1986). Directional autocorrelation: an extension of spatial correlograms to two dimensions. *Systematic Zoology*, 35:608–617.

Okabe, A., Boots, B., Sugihara, K., and Chiu, S. N. (2000). *Spatial Tessellations: Concepts and Applications of Voronoi Diagrams, 2nd Edition.* Wiley, New York, NY.

Openshaw, S., Charlton, M. E., Wymer, C., and Craft, A. (1987). A mark I geographical analysis machine for the automated analysis of point data sets. *International Journal of Geographical Information Systems*, 1:359–377.

Openshaw, S. and Taylor, P. (1979). A million or so correlation coefficients. In Wrigley, N., editor, *Statistical Methods in the Spatial Sciences*, pages 127–144. Pion, London.

Ord, J. K. and Getis, A. (1995). Local spatial autocorrelation statistics: Distributional issues and an application. *Geographical Analysis*, 27:286–306.

Ord, J. K. and Getis, A. (2001). Testing for local spatial autocorrelation in the presence of global autocorrelation. *Journal of Regional Science*, 41:411–432.

Oshan, T. M., Wolf, L., Sachdeva, M., Bardin, S., and Fotheringham, A. S. (2022). A scoping review of the multiplicity of scale in spatial analysis. *Journal of Geographical Systems*, 24:293–324.

Pebesma, E. and Bivand, R. (2023). *Spatial Data Science: With Applications in R.* Chapman & Hall/CRC, Boca Raton, FL.

Preston, S. H., Heuveline, P., and Guillot, M. (2001). *Demography, Measuring and Modeling Population Processes.* Blackwell Publishers Ltd., Oxford, UK.

Prim, R. (1957). Shortest connection networks and some generalizations. *Bell System Technical Journal*, 36:1389–1401.

Racine, J. (2019). *An Introduction to the Advanced Theory of Nonparametric Econometrics. A Replicable Approach Using R.* Cambridge University Press, Cambridge, United Kingdom.

Rattenbury, T., Hellerstein, J. M., Heer, J., Kandel, S., and Carreras, C. (2017). *Principles of Data Wrangling. Practical Techniques for Data Preparation.* O'Reilly, Sebastopol, CA.

Reich, B. J., Yang, S., Guan, Y., Giffin, A. B., Miller, M. J., and Rappold, A. (2021). A review of spatial causal inference methods for environmental and epidemiological applications. *International Statistical Review*, 89:605–634.

Rey, S. J. (2009). Show me the code: Spatial analysis and open source. *Journal of Geographical Systems*, 11:191–207.

Rey, S. J. (2023). Big code. *Geographical Analysis*, 55:211–224.

Rey, S. J., Anselin, L., Pahle, R., Kang, X., and Stephens, P. (2013). Parallel optimal choropleth map classification in PySAL. *International Journal of Geographical Information Science*, 27:1023–1039.

Rey, S. J., Arribas-Bel, D., and Wolf, L. J. (2023). *Geographic Data Science with Python.* CRC Press, Boca Raton, FL.

Rey, S. J., Stephens, P., and Laura, J. (2017). An evaluation of sampling and full enumeration strategies for Fischer Jenks classification in big data settings. *Transactions in GIS*, 21:796–810.

Rhyne, T., MacEachren, A., and Dykes, J. (2006). Exploring geovisualization. *IEEE Computer Graphics and Applications*, 26:20–21.

Robinson, W. (1950). Ecological correlations and the behavior of individuals. *American Sociological Review*, 15:351–357.

Rowe, F., Maier, G., Arribas-Bel, D., and Rey, S. (2020). The potential of notebooks for scientific publication, reproducibility and dissemination. *REGION*, 7:E1–E5.

Rubin, M. (2017). When does HARKing hurt? identifying when different types of undisclosed post hoc hypothesizing harm scientific progress. *Review of General Psychology*, 21:308–320.

Sander, J., Ester, M., Kriegel, H.-P., and Xu, X. (1998). Density-based clustering in spatial databases: The algorithm GDBSCAN and its applications. *Data Mining and Knowledge Discovery*, 2:169–194.

Schubert, E., Sander, J., Ester, M., Kriegel, P., and Xu, X. (2017). DBSCAN revisited, revisited: why and how you should (still) use DBSCAN. *ACM Transactions on Database Systems (TODS)*, 42:19.

Shellito, B. A. (2020). *Introduction to Geospatial Technologies, 5th Edition.* Macmillan Learning, New York, NY.

Silverman, B. (1986). *Density Estimation for Statistics and Data Analysis.* Chapman & Hall, New York, NY.

Simmons, J., Nelson, L., and Simonsohn, U. (2011). False-positive psychology: undisclosed flexibility in data collection and analysis allows presenting anything as significant. *Psychological Science*, 22:1359–1366.

Singleton, A. and Arribas-Bel, D. (2021). Geographic data science. *Geographical Analysis*, 53:61–75.

Singleton, A., Spielman, S., and Brunsdon, C. (2016). Establishing a framework for open geographic information science. *International Journal of Geographical Information Science*, 30:1507–1521.

Slocum, T. A., McMaster, R. B., Kessler, F. C., and Howard, H. H. (2023). *Thematic Cartography and Geovisualization, 4th Edition.* CRC Press, Boca Raton, FL.

Smith, G. and Cordes, J. (2019). *The 9 Pitfalls of Data Science.* Oxford University Press, Oxford, UK.

Snyder, J. P. (1993). *Flattening the Earth: Two Thousand Years of Map Projections.* University of Chicago Press, Chicago, IL.

Sokal, R. R. (1979). Testing statistical significance of geographic variation patterns. *Systematic Zoology*, 28:227–232.

Sokal, R. R., Oden, N. L., and Thompson, B. A. (1998a). Local spatial autocorrelation in a biological model. *Geographical Analysis*, 30:331–354.

Sokal, R. R., Oden, N. L., and Thompson, B. A. (1998b). Local spatial autocorrelation in biological variables. *Biological Journal of the Linnean Society*, 65:41–62.

Stein, M. L. (1999). *Interpolation of Spatal Data, Some Theory for Kriging.* Springer-Verlag, New York.

Stewart, J. Q. (1947). Empirical mathematical rules concerning the distribution and equilibrium of population. *Geographical Review*, 37:461–485.

Stuetzle, W. (1987). Plot windows. *Journal of the American Statistical Association*, 82:466–475.

Stuetzle, W. and Nugent, R. (2010). A generalized single linkage method for estimating the cluster tree of a density. *Journal of Computational and Graphical Statistics*, 19:397–418.

Thomas, J. and Cook, K. (2005). *Illuminating the Path: The Research and Development Agenda for Visual Analytics.* IEEE Computer Society Press, Los Alamitos, CA.

Tiefelsdorf, M. (2002). The saddlepoint approximation of Moran's I and local Moran's I_i's reference distribution and their numerical evaluation. *Geographical Analysis*, 34:187–206.

Tiefelsdorf, M. and Boots, B. (1995). The exact distribution of Moran's I. *Environment and Planning A*, 27:985–999.

Tobler, W. (1970). A computer movie simulating urban growth in the Detroit region. *Economic Geography*, 46:234–240.

Tobler, W. (1973). Choropleth maps without class intervals? *Geographical Analysis*, 5:262–265.

Tobler, W. (2004). Thirty five years of computer cartograms. *Annals of the Association of American Geographers*, 94:58–73.

Tomlin, C. D. (1990). *Geographic Information Systems and Cartographic Modeling.* Prentice-Hall, Englewood Cliffs, NJ.

Tufte, E. R. (1983). *The Visual Display of Quantitative Information.* Graphics Press, Cheshire, CT.

Tufte, E. R. (1997). *Visual Explanations: Images and Quantities, Evidence and Narrative.* Graphics Press, Cheshire, CT.

Tukey, J. (1962). The future of data analysis. *Annals of Mathematical Statistics*, 33:1–67.

Tukey, J. (1977). *Exploratory Data Analysis.* Addison Wesley, Reading, MA.

Tukey, J. and Wilk, M. B. (1966). Data analysis and statistics: an expository overview. In *Proceedings of the November 7-10, 1966, Fall Joint Computer Conference*, pages 695–709. ACM.

Wall, P. and Devine, O. (2000). Interactive analysis of the spatial distribution of disease using a geographic information system. *Journal of Geographical Systems*, 2(3):243–256.

Waller, L. A. and Gotway, C. A. (2004). *Applied Spatial Statistics for Public Health Data.* John Wiley, Hoboken, NJ.

Wang, S. (2010). A cyberGIS framework for the synthesis of cyberinfrastructure, GIS, and spatial analysis. *Annals of the Association of American Geographers*, 100:535–557.

Wang, S., Anselin, L., Badhuri, B., Crosby, C., Goodchild, M., Liu, Y., and Nyerges, T. (2013). CyberGIS software: a synthetic review and integration roadmap. *International Journal of Geographical Information Science*, 27:2122–2145.

Wartenberg, D. (1985). Multivariate spatial correlation: A method for exploratory geographical analysis. *Geographical Analysis*, 17:263–283.

Wegman, E. J. (1990). Hyperdimensional data analysis using parallel coordinates. *Journal of the American Statistical Association*, 85:664–675.

Wegman, E. J. and Dorfman, A. (2003). Visualizing cereal world. *Computational Statistics and Data Analysis*, 43(4):633–649.

Wickham, H. (2016). *ggplot2 Elegant Graphics for Data Analysis.* Springer Nature, London, UK.

Wickham, H., Cook, D., Hofmann, H., and Buja, A. (2010). Graphical inference for Infovis. *IEEE Transactions on Visualization and Computer Graphics*, 16:973–979.

Wishart, D. (1969). Mode analysis: a generalization of nearest neighbor which reduces chaining effects. In Cole, A., editor, *Numerical Taxonomy*, pages 282–311, New York, NY. Academic Press.

Worboys, M. and Duckham, M. (2004). *GIS, A Computing Perspective.* CRC Press, Boca Raton, FL.

Index

Page numbers in *italic* refer to figures.

Printed in the United States
by Baker & Taylor Publisher Services